Biologic Markers
in
Reproductive Toxicology

Subcommittee on Reproductive
and Neurodevelopmental Toxicology
Committee on Biologic Markers

Board on Environmental Studies and Toxicology

Commission on Life Sciences

National Research Council

NATIONAL ACADEMY PRESS
Washington, D.C. 1989

NATIONAL ACADEMY PRESS 2101 Constitution Avenue, NW Washington, DC 20418

NOTICE: The project that is the subject of this report was approved by the Governing Board of the National Research Council, whose members are drawn from the councils of the National Academy of Sciences, the National Academy of Engineering, and the Institute of Medicine. The members of the committee responsible for the report were chosen for their special competences and with regard for appropriate balance.

This report has been reviewed by a group other than the authors according to procedures approved by a Report Review Committee consisting of members of the National Academy of Sciences, the National Academy of Engineering, and the Institute of Medicine.

The National Academy of Sciences is a private, nonprofit, self-perpetuating society of distinguished scholars engaged in scientific and engineering research, dedicated to the furtherance of science and technology and to their use for the general welfare. Upon the authority of the charter granted to it by the Congress in 1863, the Academy has a mandate that requires it to advise the federal government of scientific and technical matters. Dr. Frank Press is president of the National Academy of Sciences.

The National Academy of Engineering was established in 1964, under the charter of the National Academy of Sciences, as a parallel organization of outstanding engineers. It is autonomous in its administration and in the selection of its members, sharing with the National Academy of Sciences the responsibility for advising the federal government. The National Academy of Engineering also sponsors engineering programs aimed at meeting national needs, encourages education and research, and recognizes the superior achievements of engineers. Dr. Robert M. White is president of the National Academy of Engineering.

The Institute of Medicine was established in 1970 by the National Academy of Sciences to secure the services of eminent members of appropriate professions in the examination of policy matters pertaining to the health of the public. The Institute acts under the responsibility given to the National Academy of Sciences by its congressional charter to be an adviser to the federal government and, upon its own initiative, to identify issues of medical care, research, and education. Dr. Samuel O. Thier is president of the Institute of Medicine.

The National Research Council was organized by the National Academy of Sciences in 1916 to associate the broad community of science and technology with the Academy's purposes of furthering knowledge and advising the federal government. Functioning in accordance with general policies determined by the Academy, the Council has become the principal operating agency of both the National Academy of Sciences and the National Academy of Engineering in providing services to the government, the public, and the scientific and engineering communities. The Council is administered jointly by both Academies and the Institute of Medicine. Dr. Frank Press and Dr. Robert M. White are chairman and vice chairman, respectively, of the National Research Council.

The project was supported by the Environmental Protection Agency; the National Institute of Environmental Health Sciences; the Air Force Office of Scientific Research; and the Comprehensive Environmental Response, Compensation, and Liability Act Trust Fund through cooperative agreement with the Agency for Toxic Substances and Disease Registry, U.S. Public Health Service, Department of Health and Human Services.

Library of Congress Cataloging-in-Publication Data

Biologic markers in reproductive toxicology / [Board on Environmental
  Studies and Toxicology].
       p.   cm.
    Bibliography: p.
    Includes index.
    ISBN 0-309-03930-4 (cloth); ISBN 0-309-03937-7 (paper)
    1. Reproductive toxicology. 2. Biochemical markers. I. National
Research Council (U.S.). Board on Environmental Studies and
Toxicology.
RA1224.2B56 1989
616.6′507--dc19   89-3071
                                                            CIP
                                                          (Rev.)

Printed in the United States of America

Subcommittee on Reproductive and Neurodevelopmental Toxicology

Donald R. Mattison, *Chairman*, University of Arkansas for Medical Sciences, Little Rock, Arkansas, and National Center for Toxicological Research

Panel on Male Reproductive Toxicology

Larry L. Ewing, *Chairman*, Johns Hopkins School of Hygiene and Public Health, Baltimore, Maryland
Walderico M. Generoso, Oak Ridge National Laboratory, Oak Ridge, Tennessee
C. Alvin Paulsen, Pacific Medical Center, Seattle, Washington
Bernard Robaire, McGill University School of Medicine, Montreal, Quebec
Richard Sherins, National Institute for Child Health and Human Development, Bethesda, Maryland
Andrew J. Wyrobek, Lawrence Livermore National Laboratory, University of California, Livermore, California

Panel on Female Reproductive Toxicology

Maureen C. Hatch, *Chairman*, Columbia University, New York, New York
Robert E. Canfield, Columbia University, New York, New York
Caleb Finch, University of Southern California, Los Angeles, California
Arthur F. Haney, Duke University Medical Center, Durham, North Carolina
Neena Schwartz, Northwestern University, Evanston, Illinois

Panel on Pregnancy

Richard K. Miller, *Chairman*, University of Rochester, Rochester, New York
J. David Erickson, Centers for Disease Control, Atlanta, Georgia
W. Page Faulk, Methodist Hospital of Indiana Medical Research, Indianapolis, Indiana
Stanley R. Glasser, Baylor College of Medicine, Houston, Texas
Lawrence D. Longo, Loma Linda University, Loma Linda, California

Committee on Biologic Markers

Bernard Goldstein, *Chairman*, UMDNJ-Robert Wood Johnson Medical School, Piscataway, New Jersey
James Gibson, Chemical Industry Institute of Toxicology, Research Triangle Park, North Carolina
Rogene F. Henderson, Lovelace Biomedical and Environmental Research Institute, Albuquerque, New Mexico
John E. Hobbie, Marine Biological Laboratory, Woods Hole, Massachusetts
Philip J. Landrigan, Mount Sinai Medical Center, New York, New York
Donald R. Mattison, University of Arkansas for Medical Sciences, Little Rock, Arkansas, and National Center for Toxicological Research
Frederica Perera, Columbia University, New York, New York
Emil A. Pfitzer, Hoffmann-La Roche, Inc., Nutley, New Jersey
Ellen K. Silbergeld, Environmental Defense Fund, Washington, D.C.

Project Staff

Senior Staff:

Devra Lee Davis
Alvin G. Lazen
Lee R. Paulson
Andrew M. Pope
Richard D. Thomas
Diane K. Wagener

Research Staff:

Robin Bowers
Victor Miller
Linda Miller Poore
Anne M. Sprague
Leslye B. Wakefield
Bernidean Williams

Editors:

Norman Grossblatt
Lee R. Paulson

Support Staff:

Beulah S. Bresler
Mireille Mesias
Erin Schneider
Susan Tawfik
Julie Walker

v

Board on Environmental Studies and Toxicology

Commission on Life Sciences

Sponsors

National Institute of Environmental Health Sciences
U.S. Air Force Office of Scientific Research
U.S. Environmental Protection Agency
U.S. Public Health Service, Agency for Toxic Substances and Disease Registry

Government Liaison Group

John R. Fowle III, *Chairman,* U.S. Environmental Protection Agency, Washington, D.C.
Henry Falk, Centers for Disease Control, Atlanta, Georgia
W. Harry Hannon, Centers for Disease Control, Atlanta, Georgia
Suzanne Hurd, National Heart, Lung and Blood Institute, Bethesda, Maryland
Dennis Jones, Agency for Toxic Substances and Disease Registry, Atlanta, Georgia
James Lamb, U.S. Environmental Protection Agency, Washington, D.C.
George Lucier, National Institute of Environmental Health Sciences, Research Triangle Park, North Carolina
Carol Mapes, Food and Drug Administration, Washington, D.C.
Michael D. Waters, U.S. Environmental Protection Agency, Washington, D.C.

Preface

Biologic markers are powerful tools that can be useful in many ways to environmental health scientists. Markers that indicate the occurrence of an internal dose or a biologically effective dose or the presence of an incipient disease can be useful in hazard identification, for example, as the qualitative step that causally associates an environmental agent with an adverse effect. Markers can also be used to determine dose-response relationships and to estimate risk, especially at the low doses relevant to most environmental chemicals. Thus, the development of biologic markers could enable scientists to make better use of laboratory animal data (usually obtained at high-dose exposures) in estimating the effects of low-dose exposures in humans. Another major role of markers is clarification of the extent of exposure in human populations. Methods of direct or indirect measurement of total exposure through analysis of body fluids are far more likely to be of value in epidemiologic studies than are most of the modeling and ambient monitoring approaches now in use. Biologic markers of exposure also hold the promise of demonstrating which individuals in a potentially affected population (e.g., residents in the neighborhood of a hazardous waste dump) have been exposed to a potentially harmful extent. Developments in the field of biologic markers are also likely to lead to a more accurate determination of the proportion of highly susceptible people within the population and of the results of human exposure.

In 1986, the National Academy of Sciences/National Research Council (NAS/NRC) was asked by the Environmental Protection Agency (EPA), the National Institute of Environmental Health Sciences, and the Agency for Toxic Substances and Disease Registry to conduct a study of the scientific basis, current state of development, validation, and use of biologic markers in environmental health research. The project was designed to be conducted by four subcommittees within NRC's Board on Environmental Studies and Toxicology. These subcommittees would evaluate the status of biologic markers for specific biologic systems: markers of reproductive and neurodevelopmental effects; pulmonary system markers of exposure, effect, and susceptibility; markers of immunotoxicity; and markers of ecologic toxicity, including markers of ecosystem exposure and altered processes.

As part of the project, the first subcommittee, the Subcommittee on Reproductive and Developmental Toxicology, convened a symposium on January 12-13, 1987, in Washington, DC, as an information-gathering activity. Invited speakers described their research and its possible application to the development and use of biologic markers in reproductive and developmental toxicology.   The proceedings of that symposium were published in *Environmental Health Perspectives* in October 1987 and served as a starting point for this, the subcommittee's report.

In this report, the oversight committee (the Committee on Biologic Markers) sets forth in general terms the broad concepts and definitions of biologic markers and, in the introduction, discusses the use of markers in environmental health research. Those presentations are followed by the subcommittee's report, which applies the concepts and definitions to reproductive and neurodevelopmental toxicology.

To say that biologic markers have generated interest and controversy in recent years would be an understatement. Biologic markers represent the newest and most promising substrate for future developments in environmental health research, and we hope that this report helps catalyze those developments.

Finally, the committee expresses its appreciation for the vigilant and effective assistance of the NRC staff with whom it worked in producing this report.

Bernard Goldstein
Chairman, Committee on Biologic Markers

Donald Mattison
Chairman, Subcommittee on Reproductive
and Neurodevelopmental Toxicology

Dedication

Robert L. Dixon was educated at the University of California, Davis, Idaho State University, and the University of Iowa, where he received his PhD in pharmacology/toxicology in 1963. His postdoctoral training was in the Laboratory of Chemical Pharmacology of the National Cancer Institute, where he served as a senior investigator. After four years as assistant and then associate professor in the Department of Pharmacology of the University of Washington School of Medicine, Bob returned to the NCI, where he was appointed chief of the Laboratory of Environmental Toxicology in the chemotherapy program. From 1972-1984 he served the National Institute of Environmental Health Sciences, first as chief of the Laboratory of Environmental Toxicology and later as chief of the Laboratory of Reproductive and Developmental Toxicology. Along with his duties as laboratory chief, he was assistant to the director of the Institute's international programs. During this period, Bob also served on detail as a senior policy analyst in the Office of Science and Technology Policy of the Carter administration.

Moving to the Environmental Protection Agency in 1984, Bob was named director of the Office of Health Research, in which position he was responsible for research activities at laboratories in Research Triangle Park, North Carolina, and Cincinnati, Ohio. In 1985 he was recruited by Sterling-Winthrop Research Institute to serve as senior director of toxicology; and in 1987, he became vice-president for drug safety in the Sterling Research Group (SRG), the research and development wing of Sterling Drug, Inc., a subsidiary of Eastman Kodak Company. At the time of his death, he had responsibility worldwide for drug safety related to SRG drug discovery and development efforts.

Bob received many awards for the excellence and commitment he brought to toxicology, including the Society of Toxicology Achievement Award (1972), the NIEHS Director's Award (1977), and the EPA Distinguished Career Award (1987). He was also president of the Society of Toxicology during 1982-1983. Bob belonged to 18 scientific societies and was a member of many scientific committees and editorial boards. He was an active writer and during his career published 67 journal articles and 49 conference proceedings or chapters. He was also editor-in-chief of the Target Organ Toxicity Monograph Series of Raven Press.

Bob was known for bringing a critical scientific mind to his work and exercising leadership in the discharge of his responsibilities. Perhaps as important, his warmth and humanity had a uniquely positive impact on many scientists. Interactions with him were cherished, from both the scientific and the human perspective. The loss of Bob Dixon and his friendship will be deeply felt.

The Committee on Biologic Markers and its Subcommittee on Reproductive and Neurodevelopmental Toxicology are proud to dedicate this report to the memory of Robert L. Dixon, who died on August 28, 1988.

Contents

Tables and Figures

TABLES

FIGURES

Biologic Markers
in
Reproductive Toxicology

Executive Summary

Reproduction and neurodevelopment are processes on which the continuation of any species depends. For humans, reproductive processes carry substantial emotional weight. We want healthy children born without impairments that hinder their structural or functional development, and we want reproduction to be successful at the appropriate time in life.

In the United States, approximately 250,000 babies are born with birth defects each year. Twenty percent of these birth defects are attributed to multiple causes, 15% to intrauterine infections, and 5% to a mutant gene. Environmental factors are identified as a cause with relative certainty in only 2-3% of the total number of cases. This leaves nearly 60% of birth defects for which the etiology is unknown but in which environmental exposure might play a role. For every 3,000,000 U.S. births annually, at least 600,000 embryos or fetuses are aborted spontaneously before the 20th week, and some 24,000 fetuses die before birth. Of live births, nearly 8% are premature, and approximately 7% have low birthweight. Another 3-7% possess some type of malformation.

Although the overall incidence of infertility remained stable between 1965 and 1982, infertility among married couples in which wives were ages 20 to 24 increased from 4% to 10%, and more than 2 million American couples who want to have a baby are unable to do so. This increase appears to be linked to several factors, including changes in the incidence of sexually transmitted diseases, but other factors, such as xenobiotic exposures, have not been well studied, and may contribute to reproductive impairment. The adverse effects on human reproduction of high doses of ionizing radiation, polychlorinated biphenyls, dibromochloropropane, cancer chemotherapy, and alcohol are well established, but the consequences of lower doses of these and other materials to which humans might be exposed environmentally have not been well studied.

Despite the substantial expenditures that have been made on the study of basic reproductive and neurodevelopmental processes, knowledge of these processes and their relationship to environmental exposures remains disappointing. However, it is clear that the situation can be improved if science can identify biologic markers of the various steps in the processes.

In light of these longstanding and continuing concerns, the Board on Environmental Studies and Toxicology (BEST) of the National Research Council's Commission on Life Sciences undertook a major investigation of the use of biologic markers in environmental health research. At the

1

request of the Office of Health Research of the U.S. Environmental Protection Agency, the National Institute of Environmental Health Sciences, and the Agency for Toxic Substances and Disease Registry, the Markers Oversight Committee was formed to clarify the concepts and definitions of biologic markers that could be applied by two subcommittees: the Subcommittee on Markers of Pulmonary Toxicity to review and assess promising markers of the pulmonary system in a separate volume; the Subcommittee on Reproductive and Neurodevelopmental Toxicology developed this report with individual panels on male reproduction, female reproduction, pregnancy, and neurodevelopment.

ORGANIZATION OF THIS REPORT

This report has four major sections that correspond to the work of the panels on male, female, pregnancy, and neurodevelopmental toxicology. Each section ends with a summary on conclusions and recommendations. This executive summary encapsulates the major points of the four sections and is followed by an introductory chapter that presents concepts, definitions, and selected applications of biologic markers, which reflects the efforts of the oversight committee to create a framework for the overall project.

CONCEPTS AND DEFINITIONS

The oversight committee's task was to clarify and define these concepts such that subsequent subcommittees could apply them to reviews of the status and potential use of biologic markers in specific areas of environmental health research.

Biologic markers are indicators signaling events in biologic systems or samples. It is useful to classify biologic markers into three types—exposure, effect, and susceptibility—and to describe the events particular to each type. A biologic marker of *exposure* is an exogenous substance or its metabolite(s) or the product of an interaction between a xenobiotic agent and some target molecule or cell that is measured in a compartment within an organism. A biologic marker of *effect* is a measur-

able biochemical, physiologic, or other alteration within an organism that, depending on magnitude, can be recognized as an established or potential health impairment or disease. A biologic marker of *susceptibility* is an indicator of an inherent or acquired limitation of an organism's ability to respond to the challenge of exposure to a specific xenobiotic substance. Biologic markers of susceptibility are discussed in the report only insofar as they also can serve as markers of exposure or effect.

Once exposure has occurred, a continuum of biologic events may be detected. These events may serve as markers of the initial exposure, dose (half-life, circulating peak, or cumulative dose), biologically effective dose (dose at the site of toxic action, dose at the receptor site, or dose-to-target macromolecules), altered structure/function with no subsequent pathology, or potential or actual health impairment. Even before exposure occurs, biologic differences among humans might cause some individuals to be more susceptible to environmentally induced disease. Biologic markers, therefore, are tools that can be used to clarify the relationship, if any, between exposure to a xenobiotic compound and health impairment.

Markers of Exposure

Exposure is the sum of xenobiotic material presented to an organism, whereas dose is the amount of the xenobiotic compound that is actually absorbed into the organism.

Blood flow, capillary permeability, transport into an organ or tissue, the number of receptor sites, and route of administration (which determines the path of the parent compound or its metabolites in the body) all can influence absorbed or biologically effective dose. An inhaled carcinogen might produce tumors in the lung, but if the same material were ingested and eliminated via the kidney, renal tumors might be produced. If the parent compound is responsible for the observed toxicity, the amount of metabolite reaching the target may be of no consequence. If metabolites are responsible, however, metabolism in the liver, the target organ, or elsewhere as a result of metabolic cooperation between several tissues is an important determinant of absorbed and biologically effective dose.

Markers of Effect

For present purposes, the effects on or responses of an organism to an exposure are considered in the context of the relationship of exposure to health impairment or the probability of health impairment. An effect is defined as an actual health impairment or recognized disease, an early precursor of a disease process that indicates a potential for impairment of health, or an event peripheral to any disease process but correlated with it and thus predictive of development of impaired health.

A biologic marker of an effect or response, then, can be any change that is qualitatively or quantitatively predictive of health impairment or potential impairment resulting from exposure. Biologic markers are also useful to identify an endogenous component or a system function that is considered to signify normal health, e.g., blood glucose. It is important to realize, however, that the concentration or presence of these markers represent points on a continuum. Therefore, the boundaries between health and disease may change as knowledge increases.

Markers of Susceptibility

Some biologic markers indicate individual or population differences that affect the biologically effective dose of or the response to environmental agents independent of the exposure under the study. An intrinsic genetic or other characteristic or a pre-existing disease that results in an increase in the absorbed dose, the biologically effective dose, or the target tissue response can be markers of increased susceptibility. Such markers may include inborn differences in metabolism, variations in immunoglobulin levels, low organ-reserve capacity, or other identifiable genetically determined or environmentally induced variations in absorption, metabolism, and response to environmental agents. Other factors that may affect individual susceptibilities include nutritional status of the organism, the role of the target site in overall body function, condition of the target tissue (present or prior disease), and compensation by homeostatic mechanisms during and after exposure. The reserve capacity of an organ to recover from an insult at the time of exposure may also play an important role in determining the extent of an impairment.

EXTRAPOLATION FROM ANIMALS TO HUMANS

Extrapolations to humans are to be based on the most sensitive animal species tested, barring clear evidence that the species is toxicologically distinct from humans. Within the past 2 years, EPA issued guidelines for evaluating reproductive studies. These provide a means to estimate data quality and stipulate segment II developmental toxicity requirements of two species and three treatment groups with 20 rodents or 10 nonrodent mammals. As this report went to press, EPA issued additional guidelines on evaluating reproductive studies.

Laboratory animals and humans can differ in toxicokinetics. Thus, use of data from animals to determine health risks in humans must be assessed carefully, especially when the data are derived from monitoring of external exposure.

The toxicity of some chemicals is mediated either by activation or by detoxification biotransformation reactions. Inasmuch as biotransformation differs among species, it is important to establish whether the routes and rates of human and animal metabolic pathways are similar.

Health risks often are associated with combinations of effects in humans. For example, cardiovascular disease in humans can encompass atherosclerosis and hypertension. Although swine provide the most suitable animal model for studying spontaneous atherosclerosis, young rats might be most appropriate for studying hypertension. For humans, estimating the diseases necessarily would entail some appropriate combination of the relevant animal test systems.

A common source of uncertainty in risk assessment is the dose-response relationship at low doses or for rare effects. It is often impractical to conduct studies of effects at low doses, because large numbers of animals are required to detect a low incidence of effects. Demonstrable health effects in humans, given the limits of epidemiology, often are associated with high doses and hence high risk. Sensitive molecular markers being developed will permit study of the relationship between exposure to chemicals at low ambient concentrations and the formation of a molecular marker predictive of human risk. The development of biologic markers might enable scientists to make better use of laboratory animal data in estimating the effects of chemicals in humans.

As a 1986 NRC study on drinking water and health observed, the timing of exposure and the patterns of dose response obtained in animal studies have important implications for extrapolating data to humans. Most often neglected in such extrapolations in the reproductive arena is fetal growth retardation, which has obvious relevance to low birth weights in humans. In the absence of a significant reduction in maternal weight gain, fetal growth retardation can be an important event.

The NRC also reported that good evidence of dose-response relationships exist for a number of well-studied developmental toxicants. An FDA review of the literature showed that of 38 compounds having demonstrated or suspected teratogenic activity in humans, all except one tested positive in at least one animal species. More than 80% were positive in more than one species. Despite this concordance of findings, qualifications must be placed on their direct application to risk assessment for humans.

QUALITY AND QUANTITY OF DATA

Quantity of data required can be determined by statistical power considerations and resource considerations. Statistical power is related to the number of subjects in a group, rarity of the end point studied, and variability in the frequency of the end point occurrence. The greater the expected relative risk, the smaller the population that will need to be studied.

The study of reproduction and development poses major resource and logistic problems for those working with laboratory animals. For instance, manageable sample populations do not reveal increases in toxic events of less than 5 to 10%. For some health effects associated with reproduction and neurodevelopment, such as mutagenesis and teratogenesis, incidences in a human population of 3 per 10,000 are significant. Obviously, these effects cannot be well defined in whole-body studies of thousands of experimental animals at a time. Classic toxicology studies of rodents involve the exposure and pathological analyses of 200 animals for 2 years. In such assays, each animal is a surrogate for 1,000,000 people. Birth defects undetectable in this rodent population could be epidemic within ten generations, if they occurred in humans.

BIOLOGIC MARKERS ASSOCIATED WITH REPRODUCTIVE AND NEURODEVELOPMENTAL TOXICOLOGY

From the outset, the subcommittee grappled with the knowledge that the processes under its consideration are complex and normal, and therefore, markers of functioning might well provide the only markers for study. (In contrast, a study of markers of disease would seek to identify signals of adverse health or pathophysiology.)

Reproduction and neurodevelopment are complex, stepwise processes that begin with gametogenesis; continue through gamete interaction, implantation, embryonic development, growth, parturition, and postnatal adaptation; and are completed with sexual and developmental maturation of the newly formed individual. The study of these areas subsumes the disciplines of reproductive and developmental biology, toxicology, teratology, and pharmacology, as well as epidemiology, occupational and environmental health, and medicine.

Interest is growing in the use of biologic markers to study the human health effects of exposure to environmental toxicants in clinical medicine, epidemiology, toxicology, and related biomedical fields. Clinical medicine uses markers to allow early detection and treatment of disease; epidemiology uses markers as indicators of absorbed dose or of health effects; toxicology uses markers to help determine underlying mechanisms of diseases, develop better estimates of dose-response relationships, and improve the technical bases for assessment of risks at low levels of exposure.

This report focuses on the identification of indicators of differences between individuals or between cells that might be related to the reproductive potential of adults or the development of children. Few such biologic markers have been demonstrated to identify early stages of health impairment or toxicologically relevant absorbed doses. The detection of increased alpha-fetoprotein in a pregnant woman's serum and in amniotic fluid has been used to identify fetuses at risk of neural tube defects. Concentrations of lead in serum have been correlated with neurologic changes. Those two examples of biologic markers predict adverse health effects if measured concentrations are extreme. But for only a few other biologic markers have particular values or ranges of values been demonstrated to be predictive of adverse health effects of specific toxic exposures. Therefore, this report discusses a broad range of biologic markers and their use in studies of reproductive and developmental toxicology. The subcommittee does not discuss their utility for predicting adverse health effects of toxic exposures, because research results relevant to that interpretation are not available.

The use of biologic markers in reproductive and developmental toxicology is changing quickly, which reflects the rapid incorporation of sophisticated measurement techniques at the subcellular levels. Markers now used in clinical studies are sometimes also used in population studies merely because there is some knowledge of their utility and their shortcomings. Their use today does not imply that they will be used tomorrow in their present form or even at all. The subcommittee determined that few reproductive and developmental markers have been used in environmental health research. Therefore, in applying the oversight committee's definitions of biologic markers, the subcommittee described biologic markers of reproductive function and neurodevelopment in general, and classified them according to their immediate and potential utility in environmental health research:

• Biologic markers that could be recommended today for assessment of an exposure and reproductive or developmental response to that exposure.
• Biologic markers found promising in laboratory studies that could be recommended for further research aimed at use in clinical and epidemiologic studies.
• Biologic markers that could be studied only in animal models or only in some specific situations in humans (e.g., at autopsy or in surgical specimens).

For some assessments, the reproductive or neurodevelopmental health effects associated with an abnormal test result might not be defined, the range of normal values of a biologic marker might be poorly characterized, or information might be lacking. For example, if xenobiotic concentrations in the exposed population are low, the population's risk of adverse reproductive or developmental health effects might be concluded to be small. However, such a conclusion might be based on the absence of studies of low-dose exposures, rather than on the presence of detailed epidemiologic research that demonstrates safety. These important distinctions must be explicitly taken into account in any assessment of exposure.

Ideally, tests selected to detect the occurrence of xenobiotic exposure and resulting effects are sensitive, specific, inexpensive, minimally invasive, attended by low risk, and easily used in large populations. It is sometimes possible to use a multistage testing process to detect exposures: the earlier stages would use sensitive but easily applied tests to identify a group that is at high risk of exposure, and more definitive tests—which might be more expensive, complicated, and invasive—would be used with the group identified by earlier tests as likely to have been exposed. The subcommittee encourages the development of improved tests.

When protocols are being designed, biologic markers that are still being developed should be included. New markers can then be validated against those in use. In addition, long-term storage of biologic and environmental samples always should be considered a resource for future studies; they can be analyzed retrospectively for markers that will be developed later.

Not all biologic markers are relevant to therapeutic intervention. Some might identify an exposure or the presence of disease but not provide information relevant to a therapeutic decision. Thus, in some circumstances, diseased or otherwise affected persons would not benefit from the information yielded but would function as biologic markers for the community, allowing public health measures to prevent further exposure or disease.

BIOLOGIC MARKERS ASSOCIATED WITH MALE REPRODUCTION

Information on the frequency of naturally occurring reproductive disorders in the human male and reproductive abnormalities induced by toxic chemicals is sparse. Reproductive health outcomes of two major classes are of concern when human males are exposed to toxic chemicals: pathophysiologic changes that might be associated with alterations in fertility and genetic damage that will be conveyed to future generations.

Several markers are used to assess whether pathophysiologic changes in human males have occurred in response to toxic chemicals. Such markers are measures of potency, measures of fertility, testicular size, serum concentrations of gonadal and pituitary steroids, and semen characteristics, such as spermatozoal concentration, motility, and structure. Those markers are of varied utility for identifying toxic effects. For instance, fertility is unlikely to be a sensitive indicator of moderate effects of toxic insult in humans, because average daily sperm production exceeds what is required. In addition, many elements of the male reproductive system (e.g., epididymal function) can be altered in response to toxic chemicals without affecting fertility.

Markers of genetic damage in the human male (or female) genome include pregnancy outcome, presence of sentinel phenotypes (e.g., achondroplasia), and changes in macromolecules in offspring, such as electrophoretic variants of red-cell enzymes. Changes in rates of heritable mutation have not been demonstrated in human populations exposed to agents that are known to induce heritable mutations in animal models.

Clearly, a battery of biologic markers is needed that reflects a wide array of pathophysiologic changes and genetic damage. Research on or development of the following areas are important.

Markers of Physiologic Damage

• In vitro sperm assays that measure the extent of the capacity of animal and human sperm to fertilize oocytes.

• Sperm function assays that use monoclonal antibodies of specific domains on spermatozoal plasma membranes.

• Computer-based and automated techniques for measuring and sorting spermatozoa on the basis of such characteristics as counts, motility, structure, domains, and enzyme function.

• Specific markers and their cDNA probes for the testes and individual accessory sex organs.

• Markers that reflect epididymal function, particularly sperm motility and the acquisition of fertilizing ability.

• Probes for testicular RNA, DNA, and other macromolecules, to use in developing assays for quantifying specific steps in the development of germ cells.

• Assays for inhibin, androgen-binding protein, and Müllerian-inhibiting factor and other polypeptides that constitute markers of Sertoli cell function.

• Assays that allow enumeration of spermatogenic stem cells.

• Comparative analyses of human and laboratory-animal germ cell responses to toxicants and identification of useful laboratory animal models for studying effects on human beings.

• Mechanisms of toxicity in laboratory animals and in vitro culture systems.

Markers of Genetic Damage and Heritable Mutations

• Human semen markers of genetic toxicity and induced mutations, including methods for detecting gene mutations, aneuploidy, chromosomal aberrations, and DNA adducts in mature sperm and in immature germ cells.

• Techniques to use DNA markers to monitor human heritable mutations such as Lerner gels, restriction-fragment-length polymorphisms, subtractive hybridization, and RNase digestion.

• Improved genetic and cytogenetic methods for detecting induced heritable aneuploidy and molecular methods for detecting heritable DNA changes in laboratory mammals.

• Basic mechanisms of induction and molecular nature of heritable mutations in laboratory mammals.

Identifying biologic markers to assess toxic agents or effects of toxic agents on specific male reproductive functions and heritable genetic damage is only the first step. Biologic markers to assess exposure must be distinguished from markers to assess effect. Variations in markers must be correlated with doses of specific toxicants and with health outcomes, mechanisms of toxicity must be elucidated, and strategies for risk assessment must be developed.

BIOLOGIC MARKERS ASSOCIATED WITH FEMALE REPRODUCTION

Information on female reproductive toxicology is even more sparse than that on the male, because of differences in gametogenesis and accessibility of germinal cells and because of the cyclic nature of female reproductive function. The differences are found in experimental animals, as well as humans.

Among possible biologic markers of female reproduction, some clinical measures are available for assessing human female sexual development and maturation and cyclic ovarian function. Those measures, based on serum concentrations of ovarian steroids and pituitary hormones, have not yet been used in population-based studies of reproductive toxicology. Development and improvement of hormonal markers in easily obtained biologic specimens, such as urine and saliva, are needed for application in field studies. In light of diurnal and cyclic fluctuations in most reproductive hormones, sampling schemes that are valid and practical have to be worked out. Most applications to date have involved markers of ovulation (such as luteinizing hormone—LH) in the context of fertility research. The results have suggested that it is feasible to incorporate hormonal assessments of cyclic function into population studies. Conventional self-reported epidemiologic measures of female reproduction are also applicable, including pubertal development, cycle length and characteristics, fertility, and age at menopause. Results of work with conventional epidemiologic measures of reproductive performance suggest that the processes of sexual maturation and ovarian function are vulnerable to toxic insult. The use of objective and more precise biologic markers should provide additional data on sites of action and mechanisms of toxicity.

Appropriately sensitive and specific urinary human chorionic gonadotropin (hCG) assays of pregnancy are available for assaying fertility and have been field tested. However, the best of the current assays are labor-intensive, so further refinements will be needed to allow large-scale applications. hCG is a valid marker of pregnancy following implantation, but a marker of conception in the preimplantation period also is needed, so that the frequency and fate of human conception can be estimated accurately and the determinants of embryonic loss in humans fully elucidated.

As to genetic damage, the relative inaccessibility of ovarian material makes it necessary to take advantage of special clinical opportunities to obtain tissues or make observations. Materials from spontaneous abortions, in vitro fertilization/embryo transfer centers and related technologies could be made available to researchers and would lend themselves to systematic study. Such specimens have been used to generate valuable data on germinal exposure to pollutants and to estimate the frequency of chromosomal anomalies in the oocyte and conceptus. The issue of female germ cell damage is of the utmost importance, because most aneuploidy in humans originates in the oocyte. Measurement of oocyte genetic damage now depends on human samples of convenience, such as material from surgical or infertility patients, and on research with laboratory animals. Heritable damage can be assessed in offspring, but research has not yet identified an environmental influence on heritable damage, at least among the common exposures that have been studied in humans.

Laboratory investigation of female reproductive toxicology, as well as female reproductive biology in general has suffered from the lack of a practical animal model appropriate for all aspects of reproduction. For example, the mouse is a good model of primate oogenesis, but lacks a functional corpus luteum, except during pregnancy, and thus is not useful for evaluating the effects of toxicants on luteal function. A systematic assessment might identify relevant animal models for different aspects of the female reproductive process that could be simultaneously studied to evaluate environmental health effects.

Germ Cell Damage

• Use of new reproductive technologies could provide human materials for examining relationships among oocyte cytogenetics and follicular fluid, serum, and tissue concentrations of pollutants.
• Estimates of oocyte chromosomal anomalies need to be validated based on comparison of infertility patients with normal females.
• Mechanisms of aneuploidy and assays that identify agents that cause it need study and development.

• Markers of genotoxicity (e.g., DNA-adduct formation) should be applied to ovarian materials (oocytes and granulosa cells).
• Molecular techniques to measure alterations in ovarian function, such as DNA probes for follicular regulatory factors, should be developed.

Development and Aging

• Critical periods in sexual differentiation should be identified further.
• Methods of applying markers of pubertal onset in epidemiologic studies should be evaluated.
• Menstrual cycle-length changes as indicators of maturation and aging should be assessed.
• Oocyte depletion should be assessed (perhaps through the use of gonadotropins, imaging analysis, and inhibin assays) as a potential indicator of exposure to xenobiotics.
• Neurotransmitters in cerebrospinal fluid could be measured in relation to reproductive changes.
• Hypothalamic dysfunction should be assessed in field studies (perhaps through the use of gonadotropic hormones).

Cyclic Ovarian Function

• The use of urinary progesterone metabolite assays as markers of ovulation in exposed populations should be encouraged.
• Assays of salivary progesterone and other steroid hormones need validation.
• Gonadotropin assays with increased sensitivity and the ability to measure bioactivity should be developed.
• Data on cycle regularity should be correlated with data on markers of ovulatory function.
• Endometrial development in the preimplantation period should be assessed (perhaps through the use of uterine washings).

Fertilization, Implantation, and Early Loss

• The role of tubal motility in gamete and conceptus transport needs to be evaluated.

• hCG assay methods (including, but not limited to, methods based on radio-activity) should be further refined to reduce laboratory time, and quantities of urine and antibodies required and thus maximize utility in field studies.

• Markers of pregnancy in the preimplantation period need to be developed.

• New reproductive technologies to evaluate relationships among follicular fluid, blood, and tissue concentrations of pollutants, and fertilization, early embryo transplant, implantation (as measured by hCG production), clinically apparent pregnancy, and pregnancy outcome need to be studied.

• In vivo and in vitro fertilization assessments should be extended to oocyte function (both oocyte fertilizability and control of post-fusion events).

BIOLOGIC MARKERS ASSOCIATED WITH PREGNANCY

Identifying biologic markers of toxic exposures that take place during pregnancy is a multidisciplinary challenge that requires integration of basic and clinical sciences. We need to learn how an adverse state (such as premature labor and delivery or spina bifida) is produced and how a xenobiotic agent can influence biologic processes to produce such a state. When persons in their reproductive years are exposed to a xenobiotic agent that is a teratogen, an important concern should be whether they intend to conceive a child in the near future or whether the woman is already pregnant.

Various periods of fetal development are differentially sensitive to insult. They include the particularly vulnerable times before and around implantation and the times of development of various organ systems throughout gestation, during which exposure to xenobiotic agents can cause specific functional and structural impairment. The critical postimplantation windows of developmental sensitivity have been associated with specific defects, such as phocomelia from the maternal ingestion of thalidomide and reproductive tract anomalies from the maternal ingestion of diethylstilbestrol. As our knowl-

edge base increases, it is important to reevaluate our understanding of critical periods for functional and structural development.

Not only the conceptus is at risk. The mother's health can be compromised by pregnancy itself and by exposure to a xenobiotic. A detailed clinical history (genetic, reproductive, and menstrual) is essential, not only to obtain an exact chronology of the pregnancy, but also to determine potential risks for both mother and conceptus. The timing of known exposures must be established, including the time of the most recent exposure. All the above information is used to interpret markers of exposure, which could be reflected in the concentrations of xenobiotic agents and their metabolites in maternal blood, urine, hair, or fat at various times during gestation.

Each gestational stage is associated with its own problems of assessment and physiologic evaluation. Inaccessibility of fetal and placental tissues and fluids is one of the most important difficulties, especially in the embryonic stage of a continuing pregnancy. Because of that inaccessibility, much effort has been expended on the evaluation of more readily available fluids and tissues from the mother, especially before 8 weeks of gestation. In the fetal period, more invasive techniques have been used to demonstrate (usually after the fact) damage to the conceptus. A goal for the study of markers is to discover a point at which therapy might reverse or prevent damage to the various fetal organ systems.

Markers of exposure and effect at three periods during pregnancy have been evaluated: around implantation, during organogenesis, and during the fetal and neonatal periods.

The current application of biologic markers during the peri-implantation period includes the use of sensitive hCG assays to document implantation. Other markers should be developed to document the differentiation of the inner cell mass to the embryo and of the trophectoderm to the fetal placenta. Early pregnancy factor assays and immunologic assays also are potentially useful in limited studies to identify women at risk of spontaneous abortion. It is hoped that results of those tests will be integrated into the set of peri-implantation markers to document some of the mechanisms of spontaneous abortion. hCG concentrations—widely used to document normal, ectopic, and failing pregnancies—are not particularly useful until after implantation. The specificity of several early pregnancy factors needs to be evaluated.

Concentration of hCG continues to be a useful marker during organogenesis to document pregnancy. Sonography has provided noninvasive measures of growth, location, size, and movement of the conceptus; increasing resolution will provide additional information on embryonic organization. Although these markers are nonspecific and insensitive to physiologic changes, they are the only biologic markers that now can be used to assess reproductive toxicity to the conceptus during organogenesis.

Many more biologic markers have been used routinely during the fetal period after 8 weeks of gestation. However, they do not assess variation specific for particular xenobiotic exposures. Sonography can be used to detect dysmorphisms (such as limb defects, anencephaly, and renal agenesis), central nervous system (CNS) function, and fetal chest-wall movement. Other kinds of biophysical monitoring can be used to assess fetal cardiovascular function, e.g., Doppler monitoring for uterine blood flow velocity, umbilical blood flow velocity, and fetal cardiac function. Plasma markers, such as concentrations of human placental lactogen and alpha-fetoprotein may be used to assess placental growth and the potential for neural tube defects, respectively. Invasive tests, such as amniocentesis and chorionic villus sampling, yield additional substantive information concerning genetics (through karyotyping), exposure (through measurement of xenobiotic agents with biologic assays or direct analysis), and markers of effect (e.g., through measurement of alpha-fetoprotein or cholinesterase in the amniotic fluid to document a neural tube defect). The continued use of specific assays of selected organs is encouraged, although they are not agent-specific.

Thus, important biologic markers related to pregnancy have not been linked with exposures to specific xenobiotic agents. The only biologic marker that has diagnostic utility after exposure to a xenobiotic agent is the concentration of alpha-fetoprotein in maternal serum and amniotic fluid after exposure to the drug valproate, for which increased concentration of alpha-fetoprotein is evidence of spina bifida in the fetus resulting from the exposure.

Research is needed on the following important subjects:

• Knowledge on endometrial-trophoblastic attachment in animals should be extended to humans, to identify measurable markers of implantation and placentation.
• The use of products of conception obtained after spontaneous or induced abortion to document exposure should be developed.
• Embryonic or fetal tissues should be analyzed for genetic or environmental factors that alter development.
• Fluorescence-activated flow cytometry to isolate embryonic (fetal) cells in maternal circulation at various critical periods of gestation should be investigated, to determine the early impact of environmental exposure without invasive techniques.
• The utility and risk of new noninvasive means of detecting defects—such as magnetic resonance imaging—should be determined.
• The application of current molecular biologic techniques, such as use of DNA probes or detection of DNA adducts, to assess xenobiotic interactions with the conceptus should be studied. (Some DNA adducts, for example, appear to provide specific fingerprints of exposure in selected fetal tissues in animals and in human placenta; other adducts are not specific, but reflect exposure to general classes of toxicants. Molecular genetic technology combined with earlier chorionic villus sampling might provide powerful assessments not only of exposure, but also of effect, by examining enzyme induction,

genetic abnormalities, and general metabolic function of embryonic tissue, especially the trophoblast.)

• The application of new testing procedures during the first 8 weeks of gestation should be studied. (In spite of current limitations, assessment of pregnancies beyond 8 weeks has been substantially improved; however, many new techniques might be especially applicable during the first 8 weeks of gestation, a period that lacks thorough evaluation.)

• Detailed epidemiologic data bases with exacting quality-assurance standards need to be developed.

• Markers related to pregnancy that can be assessed using or with in vitro studies of human material or in animals should be investigated.

• The use of chorionic villus sampling material for cytogenetic research should be investigated.

BIOLOGIC MARKERS ASSOCIATED WITH NEURODEVELOPMENT

Neurodevelopmental toxicity includes any detrimental neurological effect produced by exposures during embryonic or fetal stages of development. Complex interactions of lifestyles, pre- and postnatal environmental factors, education, peers, and social class all profoundly effect the ultimate development and well being of children. To decrease the personal and social burdens of reproductive losses, developmental impairment, and mental retardation, scientific progress must be made in unraveling the complexities of development, and articulating the factors that affect it. At present, even careful neuropathologic studies cannot account for most causes of neurologic dysfunction. Although a few pathologic correlates of mental retardation have been identified, they seldom have been studied systematically.

The interactions among neurodevelopmental, ecologic, medical, and sociopolitical factors are complex and not well understood. Little information has been published on the effect of particular environmental chemicals on reproduction or development in experimental animals,

and less is available on effects in humans. Where data are available on experimental animals, questions abound concerning extrapolation of results across species or from high to low doses.

The process of development affects the absorption, distribution, metabolism, excretion, and effects of toxicants. Age-dependent changes in body composition and xenobiotic clearance have profound effects on interpretation of markers of absorbed dose and of biologic effect. Unique physiologic variables in the newborn and during puberty also can alter markers quantitatively and qualitatively.

For compounds with very long half-lives (e.g., polychlorinated biphenyls), effects of in utero exposure noted later in life might be due either to persistence of the xenobiotic or to effects secondary to physiologic changes that occur during earlier critical periods of development. Such considerations have important implications for potential therapeutic interventions designed to alter xenobiotic clearance or otherwise modify toxic outcome.

Establishing causal relationships in neurotoxicology requires both experimental and epidemiologic studies; neither is sufficient in itself. Studies of animal responsiveness are needed to validate epidemiologic data, to determine mechanisms, and to establish dose-response relationships. Three approaches to the application of animal data to human observations are used: investigation of the underlying mechanisms of functional changes observed in animals, examination of comparable end points in humans and animals and a search for homologous mechanisms.

In this report, the CNS is discussed as a model of developing biologic systems. The CNS develops through multiple morphogenetic processes, including cell death and cell migration. Two toxic agents—ionizing radiation and lead—have been shown to influence cell death and cell migration and show dose and stage specificity with respect to particular biologic end points. If granule cells of the cerebellum or hippocampus are killed before their final migration, the deficit in cell number produces highly specific behavioral aberrations, especially hyperactivity. Therefore, these behaviors serve as biologic markers that are correlated with lead and irradiation exposures.

Minor physical abnormalities (MPAs), which are primarily of ectodermal origin, have been associated with various behavioral aberrations. Increased MPAs are found among schizophrenic patients, autistic children, and boys. Boys with multiple MPAs tend to be hyperactive; girls tend to be behaviorally inhibited and intractable. A dose-dependent relationship between lead concentration in the umbilical cord blood and MPAs has been reported. And an inverse association between number of MPAs and verbal IQ score has been reported. These findings suggest that the presence of MPA indicates impaired CNS development.

Neuronal communication is determined by neurochemical factors. Most of the neurotransmitters released in the CNS are unknown or poorly characterized. In addition to their role as chemical messengers, many neurotransmitters have a role in CNS development. Access to CNS neurotransmitters is constrained; only a small portion of neurotransmitters measured peripherally in the blood are related to CNS activity. Cerebrospinal fluid analysis offers a closer view of CNS metabolism, but this material is not readily available.

Few data are available on neurochemical markers of toxicant exposure. In children with lead intoxication, high 24-hour excretion of homovanillic acid has been reported. Experimental administration of N-methyl-4-phenyltetrahydropyridine (MPTP), an agent known to produce a syndrome similar to parkinsonism, causes decreased excretion of dopamine metabolites.

Investigators of behavioral neurotoxicity can use a broad repertoire of behavioral assessments in evaluating subjects exposed to toxicants—from measures of psychophysical functions (such as reaction time, hearing thresholds, and nerve conduction time) to measures of more complex functions (such as psychometric intelligence, visual motor integration, and social behavior).

Research is needed on the following important subjects:

- Neurotoxic exposures should be correlated with behavioral measures and with physiologic assessments that use such techniques as magnetic resonance imaging, positron emission tomography (PET), and brainstem evoked potentials.
- The specificity and sensitivity of markers of exposure and effect need to be estimated.
- The use of such tissues as cerebrospinal fluid and CNS cells to correlate exposure and effect markers with tissue concentrations of neurochemicals needs to be studied.
- Interactions between neuroendocrine and neuroimmune functions and their relation to pollutant exposure and effect should be examined.
- Integrated assessments of exposure and outcome are needed for estimating toxic effects on development.
- The use of early developmental markers of exposure and effect needs to be adapted to the study of brain plasticity and the prediction of later effects.
- Relationships among minor physical abnormalities, toxicant exposure, and behavioral effects should be established.

GENERAL RECOMMENDATIONS AND CONCLUSIONS

Assessment of reproductive and developmental health is a task that cuts across several scientific disciplines and international boundaries. In the United States, this complex field comes under the purview of several federal agencies with responsibilities for regulation, research, or monitoring. It is not within the scope or expertise of this committee to evaluate the administration of scientific research, but broad international collaboration in addressing the scientific issues related to environmental hazards to reproduction probably would be advantageous and appropriate. International collaborations could increase the statistical power of particular studies by combining data and diversifying the assessments performed, as well as providing a broader base of funding and permitting international health-registry data to be used. Several structures for international scientific cooperation have been successful (e.g., the World Health Organization and the International Agency for Research on Cancer), and they might be considered as a means for coordinating and promoting research in reproductive and developmental toxicology through multidisciplinary symposia, research, and public education.

Another substantial need in reproductive and developmental toxicology is for normative data and better methods to follow the developmental and reproductive health of specific persons throughout their lives. Scandinavian countries have assigned a personal identification number (similar to a Social Security number) at birth and use it on hospital, employment, and other records. This has allowed the development of data bases that can be used to explore relationships among environmental or occupational exposures and many facets of reproductive and developmental health. This approach might not be acceptable in the United States, but alternative approaches, such as exposure registries and selective pre-employment sampling, should be explored.

Collection of data on individual and population exposure to xenobiotics and health outcomes is vital. Major issues of ethics, confidentiality, and impact on the legal system and insurance industry must be addressed, and interdisciplinary discussion of such issues is urgent.

Consideration should be given to performance of more detailed cross-sectional and longitudinal studies of reproduction and development. Models for such studies include the National Health and Nutrition Examination Survey and the Collaborative Perinatal Project in the United States and the Birthday Trust in the United Kingdom. Lifetime studies of a selected cohort should be considered for the collection of data on the development, expression of mature function, and performance of the reproductive system.

It is important to remember that noninvasive techniques might themselves have reproductive or developmental effects, that physiologic changes in pregnancy might alter maternal toxic responses, and that existing invasive diagnostic or therapeutic techniques (e.g., amniocentesis, follicular aspiration, chorionic villus biopsy, and therapeutic abortion) might provide unique materials for the development and validation of biologic markers.

The use of batteries of tests for population investigations is encouraged. Given the complex multiorgan nature of reproduction and development, the use of tests to assess only a single facet of reproductive or developmental health is likely to fail more often than succeed in the identification of reproductive or developmental toxicants.

A series of general guidelines warrant attention when a previously identified, potentially useful biologic marker is to be validated:

- Establish normal baseline values and distribution for the marker in laboratory animals and humans.
- Evaluate the sensitivity and specificity of the markers in predicting a health outcome (e.g., infertility) or genetic damage.
- Understand in detail the time course of response of the marker to a toxic chemical, with special attention to the recovery process.
- Develop a strategy for and a consensus on the use of multiple species in toxicologic studies.
- Develop human assays that use semen, saliva, or urine, rather than tissue or blood, whenever possible.
- Use noninvasive techniques, such as ultrasound or magnetic resonance imaging, whenever possible.
- Consider a battery of markers that reflects a wide array of physiologic functions and genetic damage and then relate the marker in question to others in the battery.
- Identify populations at high risk for reproductive or developmental health impairment (perhaps populations exposed to drugs with reproductive or developmental toxicity, aging populations, or offspring of women treated with specific drugs during pregnancy), to serve as test subjects for the initial assessment and validation of biologic markers.
- Include among high-exposure populations those with special or unique occupational exposures (e.g., agricultural groups).

• Encourage and support institutions in the development of sample banks, to speed the identification and validation of markers.

• Establish a task force to develop and coordinate strategy.

Few areas have changed in medicine as rapidly as those that are the subject of this report. The problems of reproductive and developmental toxicology remain complex, because advances in our understanding of toxic chemicals have not kept pace with the introduction of new materials into the environment. A significant proportion of human reproductive wastage remains of unknown etiology. Important research opportunities identified in this volume likely will provide significant improvements in promoting reproductive and developmental health. The subcommittee recommends specific research be pursued as outlined above and offers one general observation: those concerned with the complex ethical questions of protecting life at its origin need to appreciate the importance of scientific research in furthering their aims.

1

Report of the Oversight Committee

Reproduction and neurodevelopment are processes on which the continuation of any species depends. For humans, these processes carry a substantial emotional aura. We want our children to be born healthy with no impairment that would hinder their structural or functional development, and we want reproduction to be successful at the appropriate time or times in life.

In the United States, approximately 250,000 babies are born with birth defects each year. Twenty percent of these birth defects are attributed to multiple causes, 15% intrauterine infections, and 5% to a mutant gene. Environmental factors are identified as a cause with relative certainty in only 2-3% of the total number of cases. This leaves nearly 60% of birth defects that might involve unknown environmental factors. For every 3,000,000 U.S. births annually, at least 600,000 embryos or fetuses are aborted spontaneously before the 20th week, and some 24,000 fetuses die before birth. Of live births, nearly 8% are premature and approximately 7% have low birthweight. Another 3-7% possess some type of malformation.

Although the overall incidence of infertility remained stable between 1965 and 1982, infertility among married couples in which wives were ages 20 to 24 increased from 4% to 10%, and more than 2 million American couples who want to have a baby are unable to do so. This increase appears to be linked to several factors, including changes in the incidence of sexually transmitted diseases, but other factors, such as xenobiotic exposures, have not been well studied, and may contribute to reproductive impairment. The adverse effects on human reproduction of high doses of polychlorinated biphenyls, dibromochloropropane, and alcohol are well established, but the consequences of lower doses of these and other materials have not been well studied.

The economic commitment to healthy children is difficult to measure, but Americans spent about $1 billion on medical care in 1987 to overcome infertility. Furthermore, the amount spent on remediating developmental problems is large. In 1985, neurologic and communicative disorders alone were estimated to have affected 42 million Americans and cost $114 billion (Freeman, 1985). Despite these expenditures, knowledge of basic reproductive and developmental processes and the environmental causes of adverse reproductive outcomes remains staggeringly disappointing.

Against this background desire for perfect families with perfect children, concern is developing that environmental exposures might impair reproductive or developmental processes.   Congress expressed this concern in the recent Superfund Amendments and Reauthorization Acts (SARA) by establishing the Agency for Toxic Substances and Disease Registry and recommending that epidemiologic studies be conducted in populations having high risks of exposure to toxic substances from dumps. Also, beginning in the summer of 1988, the SARA requires federal, state, and local governments and industry to make publicly available an inventory of toxic chemical emissions from certain facilities. Through the Geographic Environmental Monitoring system (GEMS), the public will have access to name, location, type of business, and quantity of chemicals entering each environmental medium annually.

This type of information may cause the public to seek additional information on health effects and to request more local studies be performed.   Such searches or studies undoubtedly will reveal that, in many cases, the impact of an environmental exposure on reproduction or development can be very difficult to determine. There is often little published information on the effect of an environmental chemical on reproduction or development in experimental animals and less on effects in humans. By way of example, the committee reviewed evidence on an outbreak of premature sexual maturity in children in Puerto Rico in the early 1980s, where no definitive cause could be identified. Where data are available on experimental animals, questions about experimental methods or about extrapolation of results across species or from high to low dose might hinder understanding of the potential adverse human reproductive or developmental effects.

The adverse effects of high doses of polychlorinated biphenyls, dibromochloropropane, and alcohol for human reproduction are well established, but the consequences of lower doses of these and other materials have not been well studied.

Biologic markers, broadly defined, are indicators of variation in cellular or biochemical components or processes, structure, or function that are measurable in biologic systems or samples. For most purposes in environmental health research, the reason for interest in biologic markers is a desire to identify the early stages of health impairment and to understand basic mechanisms of exposure and response in research and medical practice.

The growth of molecular biology and biochemical approaches to medicine has resulted in the rapid development of markers for understanding disease, predicting outcome, and directing treatment. Many diseases are now defined, not by overt signs and symptoms, but by the detection of biologic markers at the subcellular or molecular level.   For example, liver and kidney diseases are often diagnosed by measuring enzymes in blood or proteins in urine; lead poisoning can be diagnosed on the basis of blood lead concentrations and such biologic changes as increases in heme biosynthesis components in red cells and urine; and many inborn errors of metabolism, such as phenylketonuria, are diagnosed on the basis of cell biochemistry, rather than expressed dysfunction. The identification, validation, and use of markers in medicine and biology depend fundamentally on increased understanding of mechanisms of action and the role of molecular and biochemical processes in cell biology.

It is important to recognize that markers represent signals on a continuum between health and disease and that their definitions might shift as our knowledge of the fundamental processes of disease progression increases.   That is, today's markers of exposure may become tomorrow's markers of early biologic effect.   What are perceived at first to be early signals of risk could come to be considered health impairments themselves because the predictive relationship is so strong; i.e., the early signal could represent an effect at a stage in the progression at which, it is difficult to prevent a health impairment from occurring.   Thus, biologic markers can be valuable in the prevention, early detection, and early treatment of disease. Figure 1-1 depicts the continuum involved.

Recent advances in laboratory techniques in molecular biology have been accompanied by increasing emphasis on the

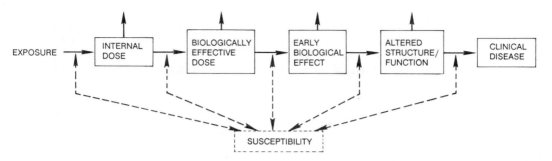

FIGURE 1-1 Simplified flow chart of classes of biological markers (indicated by boxes). Solid lines indicate progression, if it occurs, to the next class of marker. Dashed lines indicate that individual susceptibility influences the rates of progression, as do other variables described in the text. Biological markers represent a continuum of changes, and the classification of change may not always be distinct. Source: Committee on Biological Markers of the National Research Council, 1987.

use of markers in epidemiology. In 1977, Higginson described molecular epidemiology and its use of markers as the application of sophisticated techniques to the epidemiologic study of biologic material. Perera and Weinstein (1982) later defined molecular cancer epidemiology as an approach combining analytic epidemiology and biochemical or molecular techniques to identify the role of exogenous agents or host factors in the causation of human cancer; these techniques included those for identifying carcinogens in human tissues, cells, or fluids and measuring early morphologic, biochemical, or functional responses to carcinogens. At the same time, Lower and Kanarek (1982) published an extensive discussion of the mechanisms of neoplastic disease and described molecular epidemiology as the measurement of molecular parameters related to neoplastic disease. In June 1984, the National Institute of Environmental Health Sciences convened a major task force on research needs in environmental health. The report from that task force included recommendations for study of biochemical and cellular markers of chemical exposure and preclinical indicators of disease (NIEHS, 1985). Other publications that examined the use of biologic markers in environmental health research include those of Reff and Schneider (1982), Fowle (1984), IARC (1984), Silbergeld (1985), and Wogan and Gorelick (1985).

There is growing interest in the use

of biologic markers to study the human health effects of exposure to environmental toxicants in clinical medicine, epidemiology, toxicology, and related biomedical fields. Clinical medicine uses markers to allow earlier detection and treatment of disease; epidemiology uses markers as indicators of internal dose or of health effects; toxicology uses markers to help determine underlying mechanisms of diseases, develop better estimates of dose-response relationships, and improve the technical bases for assessing risks at lower levels of exposure.

This report focuses on the identification of indicators of differences between individuals or between cells that might be related to the reproductive potential of adults or the development of children. Few such biologic markers have been demonstrated to identify early stages of health impairment or toxicologically relevant internal doses. The detection of increased alpha-fetoprotein in a pregnant woman's serum and in amniotic fluid has been used to identify fetuses at risk of neural tube defects. Concentrations of lead in serum have been correlated with neurologic changes. Those two examples of biologic markers predict adverse health effects if measured concentrations are extreme. But for only few other biologic markers have particular values or ranges of values been demonstrated to be predictive of adverse health effects of specific toxic exposures. Therefore, this report discusses a broad range of biologic markers and their use in studies of reproductive and developmental toxicology. The subcommittee does not discuss their utility for predicting adverse health effects of toxic exposures, because research results relevant to that interpretation are not available.

This introductory chapter presents concepts, definitions, and selected applications of biologic markers, which reflects the efforts of the committee to create a framework for the overall project. The following four sections discuss biologic markers associated with male reproductive, female reproductive, pregnancy, and neurodevelopmental toxicology.

Given the current status of research and development with respect to biologic markers in these areas, the approach to describing them varies. In each section, biologic markers are discussed in terms of their immediate and potential utility in environmental health research. The final chapter of each section is a summary of conclusions and recommendations.

CONCEPTS AND DEFINITIONS

As we use the term here, markers can be signals or indicators of normal physiology or forerunners of health impairment. A specific biologic marker can serve several purposes and is best defined by the use to which it is put in a particular context. Markers can indicate susceptibility, exposure to an exogenous agent, internal dose, biologically effective dose (dose at receptor site), early biologic effect, structural or functional alteration, physiologic status, or disease. Figure 1-1 shows the relationship among these and indicates that a biologically effective dose can itself alter susceptibility. The choice of a marker and its interpretation depend on the purpose of its use, and its intended use depends on characteristics specific to an exogenous agent in question, to the individual organism, and sometimes to a target organ or tissue (see Table 1-1). When the goal is prevention, the major emphasis would be on markers that identify biologic changes that are predictive of health impairment or overt disease.

TABLE 1-1 Examples of Characteristics of Exogenous Agents, Organisms, or Targets That Influence Choice of Biologic Marker

Agent-specific characteristics
  Physicochemical properties
  Interactions
  Routes of exposure
  Exposure
  Exposure concentration
  Pattern of exposure
  Metabolism
  Activation
  Detoxification

Organism-specific characteristics
  Species
  Age
  Sex
  Physiologic state
  Pharmacokinetic characteristics
  Genetic factors
  Life-style factors

Organ- or tissue-specific characteristics
  Location
  Blood flow
  Membrane permeability
  Transport
  Receptors
  Function
  Homeostasis
  Structure
  Physiologic state

The committee has found it useful to define three general categories of biologic markers: those of exposure to chemical or physical agents, those of effects of exposure, and those of susceptibility to the effects of exposure. A biologic marker of exposure is an exogenous substance or its metabolite(s) or the product of an interaction between a xenobiotic agent and some target molecule or cell that is measured in a compartment within an organism. A biologic marker of effect is a measurable alteration of an endogenous component within an organism that, depending on magnitude, can be recognized as a potential or established health impairment or disease. A biologic marker of susceptibility is an indicator of an inherent or acquired limitation of an organism to respond to the challenge of exposure to a specific xenobiotic substance.

Biologic Markers of Exposure

External exposure is the amount or concentration of xenobiotic material in the environment of an organism; internal dose is the amount of a xenobiotic material that is transferred or absorbed into the organism. Biologically effective dose, in general terms, is the internal dose that is quantitatively correlated with an identifiable biologic effect; however, it is more precisely considered to be the amount of xenobiotic material that has interacted with a critical cellular or tissue receptor or target where the biologic effect is initiated. Because such receptor sites are often not known, or are not accessible for sampling, it is frequently necessary to use a surrogate site for which the dose has been correlated with the biologically effective dose or the identifiable biologic effect at the target.

There is a continuing need for the development of more accurate markers of internal dose that reflect the biologically effective dose. The amount of a xenobiotic that is actually absorbed is usually not known. Biologically effective dose might depend on individual characteristics, which account for a large part of observed differences in effect. Markers of exposure can be based on steady-state or pharmacokinetic measures, such as circulating peak concentration, cumulative dose, or plasma half-life. Individual variations in physiologic characteristics—such as sex, age, blood flow, membrane permeability, and respiratory rate—can significantly affect the absorption and distribution of a chemical and its metabolites (Table 1-2) (Doull, 1980). For example, physiologic alterations in blood flow during pregnancy significantly alter distribution of drugs to the target tissue (Mattison, 1986). Also, age and health status, such as disease, can alter respiratory rates and thus the pulmonary dose of a toxicant..

Exposure concentration (inhalation), size of delivered dose (ingestion), and dose rate also affect internal dose. When absorption capacities are exceeded, alternate pathways of clearance come into

TABLE 1-2 A Classification of Toxicity-Influencing Factors

Factors related to the toxic agent
Chemical composition (pH, choice of anion, etc.)
Physical characteristics (particle size, method of formulation, etc.)
Presence of impurities or contaminants
Stability and storage characteristics of the toxic agent
Solubility of the toxic agent in biologic fluids
Choice of the vehicle
Presence of excipients: adjuvants, emulsifiers, surfactants, binding agents, coating agents, coloring agents, flavoring agents, preservatives, antioxidants, and other intentional and nonintentional additives

Factors related to the exposure situation
Dose, concentration, and volume of administration
Route, rate, and site of administration
Duration and frequency of exposure
Time of administration (time of day, season of the year, etc.)

Inherent factors related to the subject
Species and strain (taxonomic classification)
Genetic status (littermate, siblings, multigenerational effects, etc.)
Immunologic status
Nutritional status (diet factors, state of hydration, etc.)
Hormonal status (pregnancy, etc.)
Age, sex, body weight, and maturity
Central nervous sytem status (activity, crowding, handling, presence of other species, etc.)
Presence of disease or specific organ pathology

Environmental factors related to the subject
Temperature and humidity
Barometric pressure (hyper- and hypobaric effects)
Ambient atmospheric composition
Light and other forms of radiation
Housing and caging effects
Noise and other geographic influences
Social factors
Chemical factors

play. High vapor concentrations may be "blown off," not absorbed. Species differences in metabolism can drastically alter internal doses of reactive metabolites.

The internal dose of an xenobiotic can vary with route of exposure, chemical species, and physical form. To make qualitative or quantitative estimates of exposures with biologic markers, the concentration, duration and pattern of exposure, and physicochemical nature of a toxicant must be considered in the selection of an appropriate marker of exposure (Gibaldi and Perrier, 1982). Other environmental factors, such as temperature, can affect exposure by changing amounts of water consumption and thus waterborne pollutants ingested. Diet alters intestinal motility and gastric emptying time, as well as the transport of specific substances, for example, diets low in iron appear to facilitate intestinal uptake of lead (Silbergeld et al., 1988).

The presence of active mechanisms of transport into an organ or tissue and the density of receptor sites can all influence internal dose and biologically effective dose. Dependence on metabolic activation is critical; the tissue distribution of metabolizing enzymes is also an important determinant of biologically effective dose. Interpretation of dosimetric data also involves understanding the role of the receptor in overall cell-organ-organism function, of coexisting or pre-existing stresses on the organism, and of the existence and availability of compensation during and after exposure (Doull, 1980). The determination of the significance of a biologically effective dose depends on an understanding of how a predicted effect is induced. For example, prepubertal males and females appear to be less sensitive than sexually mature persons to the effects of alkylating agents on gonadal function. In the female, that is probably related to the greater number of oocytes in the ovary before the onset of ovulation (Mattison, 1985); in the male, it appears to be due to the lower rates of cell proliferation and blood flow and lower capillary permeability that are characteristic of the sexually immature testis (Blatt et al., 1981).

Another important biologic marker of exposure is body burden, which is the total internal dose or that which has accumulated over time in the organism. Depending on the toxicokinetics of the agent, body burden might be a biologic marker of exposures that occurred recently or in the distant past. Body burden includes the dose at the target receptor sites, but it can also include amounts of xenobiotic material stored in other, nontarget compartments. Although the body burden in remote compartments might or might not be relatively inert biologically, it has the potential to be released under conditions of metabolic stress. Such release could result in a highly detrimental biologically effective dose long after the original external exposure.

Biologic Markers of Effect

For purposes of environmental health research, biologic markers of effects in an organism after exposure to a toxicant are considered in the context of their relationship to health status—from normal health, through health impairment, to overt disease. In that context, an effect is defined as any of the following:

- An alteration in a tissue or organ.
- An early event in a biologic process that is predictive of development of a health impairment.
- A health impairment or clinically recognized disease.
- A response peripheral or parallel to a disease process, but correlated with it and thus usable in predicting development of a health impairment.

Thus, a biologic marker of an effect can be any qualitative or quantitative change that is predictive of health impairment resulting from exposure to an exogenous agent. The same biologic marker might also be useful as an indicator of normal physiology, e.g., a particular range of blood glucose concentration.

Markers of early biologic effects include alterations in the functions of target tissues after exposure. As early-warning signals, such markers can be useful dosimeters to guide intervention aimed at reducing or preventing further exposure. Such early-warning signals might also be observed in organs or tissues other than the sites that are critical for toxic action.

A tissue affected by a toxicant might exhibit altered function even if the exposed person has no overt manifestations. Such altered function can in some cases be determined by testing, particularly with biochemical methods. Biologic markers of such altered functions are most useful if related to a specific organ or function—e.g., β-microglobulin for kidney function and luteinizing hormone for ovarian function.

If exposure to a toxicant and internal dose is great enough, disease will develop, because the biologically effective dose will be sufficient to affect some function irreversibly or for a substantial period. Disease that occurs soon after exposure might be directly linked to the toxicant. Disease that occurs long after exposure might be difficult to relate to a toxicant (e.g., ovarian or testicular failure or cancer), unless the findings are pathognomonic, i.e., are relatively specific to a particular type of exposure (such as mesothelioma) or are rare in unexposed persons (such as angiosarcoma or vaginal adenocarcinoma).

The transition to overt disease can depend on properties of the toxicant, the nature of exposure, the disease process itself, or individual susceptibilities. Because people respond differently to toxicants, it is not surprising that only some members of a population similarly exposed to a given environmental agent will develop a given disease.

Although scientists tend to divide biologic markers into groups, it seems evident that there is a continuum between health and disease, and advances in toxicology have demonstrated a continuum between exposure and effect. Accordingly, what once appeared to be more or less discrete groups of biologic markers are now more difficult to discern. Biologic markers are best divided operationally, depending on how they are assessed and how they will be used, but the divisions should not be interpreted to imply mechanistic distinctions.

If a biologically effective dose is correlated with an effect or concentration at a peripheral site, this can also function usefully as a surrogate for the dose or effect that is occurring in the target tissue. Such surrogates can be used as markers of exposure and effect at the site of action. They include indicators of the dose of indirectly acting toxicants–such as signals of altered hepatic metabolism of sex hormones, which can affect fertility (Mattison, 1985)—and signals from surrogate compartments, such as measurements of red blood cell δ-aminolevulinic acid dehydrase (ALAD), an enzymatic marker of a biologic effect of lead (Hernberg, 1980; Singhal and Thomas, 1980). An example of a marker that is closely related to external dose, biologically effective dose, and health status is the use of lymphocyte DNA adducts as markers of absorbed dose, dose at the molecular site of action, and likelihood of cancer (Perera and Weinstein, 1982). Carboxyhemoglobin (COHb) concentration after exposure to carbon monoxide has also bepen used to indicate internal dose and to predict effects. A major goal of biologic marker research is to develop surrogates that link exposure and effect.

It can be difficult to establish an association between biologic marker of exposure and a marker of effect. For example, blood lead content might appear to be a more direct indicator of internal lead dose than is a lead-induced increase in free erythrocyte protoporphyrin (FEP), which is clearly an effect. In some clinical situations, however, FEP is a more valid marker of total lead body burden than is blood lead content, and it might also provide more accurate information on the biologically effective dose of lead to target organs, such as the brain (Lauwerys, 1983). Although the presence of hemoglobin and lymphocyte DNA adducts of mutagenic alkylating xenobiotic compounds reflects a biochemical effect, they might also be considered markers of biologically effective doses of carcinogens that are uniformly distributed (Perera and Weinstein, 1982; Osterman-Golkar and Ehrenberg, 1983; Shamsuddin et al., 1985; Wogan and Gorelick, 1985). Unless the lymphocyte itself is the precursor of a tumor, the white-cell DNA adducts (and their surrogates, the hemoglobin adducts) are appropriately considered to be indirect markers of biologically effective dose in the target organ (NIEHS, 1985; NRC, in press).

Biologic Markers of Susceptibility

Some biologic markers indicate individual or population factors that can affect response to environmental agents. These factors are independent of whether exposure has occurred, although exposure sometimes increases susceptibility to the effects of later exposures (e.g., sensitization to formaldehyde). An intrinsic characteristic or pre-existing disease state that increases the internal dose or the biologically effective dose or that amplifies the effect at the target tissue can be a biologic marker of increased susceptibility (NIEHS, 1985; Omenn, 1986). Such markers can include inborn differences in metabolism, variations in immunoglobulin concentrations, low organ reserve capacity, or other identifiable genetically or environmentally induced factors that influence absorption, metabolism, detoxification, and effect of environmental agents. We do not discuss these types of biologic markers fully here, but cover them only to the extent that markers of susceptibility also serve as markers of exposure or effect.

PRINCIPLES OF SELECTION OF MARKERS

The selection of biologic markers of exposure or effect is based on a wide array of background data from in vitro and in vivo experimental studies, from epidemiologic studies, and from measurements of exposure and on physicochemical properties of the toxicant in question, as shown in Table 1-1.

Biologic markers of exposure can be obtained by measuring the concentration of a particular toxicant or its metabolites alone or bound to DNA, RNA, proteins or receptors in body tissues or fluids, and in excretory products. The use of markers can be complemented by the use of questionnaires that call for estimates of duration and magnitude of exposure, such as work-history questionnaires or activity time-budget questionnaires.

Markers of effect can be obtained by such procedures as biochemical analyses for organ-specific events.

Biologic Considerations

A mechanistic approach to the basic events that result in an adverse health effect must be taken in the selection of an appropriate biologic marker. The mechanistic approach should yield biologic markers that identify the initial stages of disease. These markers are valuable tools for developing strategies to prevent progression of disease.

Practical Limitations

Ideally, the use of biologic markers to screen human populations involves minimally invasive techniques. Organ analysis, high-dose x irradiation, autoradiography, or covalent binding assays can be used to identify sites of toxic action in laboratory animals, but cannot be readily applied to human populations. Less invasive methods, such as nuclear magnetic resonance imaging, might eventually make it possible to estimate the concentrations of specific chemicals and specific types of effects (e.g., changes in cellular energetics and phosphorylation (Cohen et al., 1983) in remote target tissues in humans). In the meantime, detection must be done in surrogate compartments.

The use of biologic markers also should involve test procedures that are readily acceptable by subjects. Unless a test is readily available, uncomplicated, and acceptable to the general public, participation will be low. For example, fetal monitoring by ultrasonography, transabdominal amniocentesis, and the karyotyping of amniotic or chorionic villus cells for chromosomal abnormalities is relatively safe, but those procedures are not acceptable to the general population for routine biologic monitoring. Field studies—whether performed in the home, in the workplace, or elsewhere—often have the lowest refusal rate for human studies.

In assessing the predictive value of biologic markers, it is necessary to account for the heterogeneity of the human population, which is composed of persons who differ in age, genetic constitution, nutritional status, and general health. It is also necessary to identify persons who are likely to exhibit the earliest and most severe effects of exposure to environmental agents. Animal models can sometimes be used to study the biologic mechanisms of hypersusceptibility and to examine in detail the effects of environmental toxicants on hypersusceptible groups.

Batteries of Biologic Markers

Limitations are associated with most biologic markers of exposure and of effect; therefore, it might be reasonable to use a battery of markers. And, it might be useful to develop markers of both exposure and effect for chemicals of concern. For example, to determine whether there is a potential or actual adverse effect of dimethylformamide on male reproductive function, a battery of functional tests of the male reproductive system might be tried, including concentration of formamide in urine or in seminal fluid (Kennedy, 1986); sperm density in the ejaculate; sperm motility; sperm morphology (seminal cytology), including sperm-head shapes and Y-body test (Kapp and Jacobson, 1980); interspecies sperm penetration assay (e.g., human sperm and Syrian golden hamster ovum) as an indicator of sperm chromosomal aberrations; and determination of plasma concentrations of hormones (e.g., testosterone). Although each of these tests has inherent limitations of specificity and sensitivity and varying predictive value as a biologic marker of male reproductive function (e.g., sperm density fluctuates daily, and sperm motility is difficult to measure), a battery of tests can be a powerful tool for indicating dysfunction and its association with exposure to a chemical.

Ethical Issues

The use of biologic markers has raised a number of important ethical issues (Ashford et al., 1984; Ashford, 1986; Samuels, 1986; Yodaiken, 1986). Although the principal purpose of this project is to address scientific aspects of biologic markers, it is appropriate to draw attention to the broader questions and issues that need to be addressed soon by society.

These are related particularly to markers of susceptibility.

Does society have an obligation to protect people beyond informing them of risks? Can an employee be forced to leave his or her job once a susceptibility marker has been detected or a biologically effective dose has been received? There is a concern that focusing on the detection of susceptible persons could replace efforts to remove toxic chemicals from the workplace. Other ethical considerations arise from the degree to which susceptibility markers are accurate predictors. For instance, it is important to distinguish between markers that are totally predictive of an adverse effect, reasonably predictive, or only minimally predictive.

Ethical issues are also pertinent in the consideration of using biologic markers as a basis for making decisions about consumer products. For example, should an item of value or convenience to the general public be withdrawn from commerce because a few persons are susceptible to adverse effects of the item, or should those susceptible be responsible for avoiding contact with the item, given adequate labeling?

Developments in science and technology have posed many ethical questions. As we move rapidly into an era of greater understanding of the interactions between genetic material and exogenous chemicals and other biologic interactions, we must anticipate and be prepared to address the ethical issues that will certainly arise.

VALIDATION OF BIOLOGIC MARKERS

Sensitivity and Specificity

To validate the use of a biologic measurement as a marker, it is necessary to understand the relationship between the marker and the event or condition of interest, e.g., potential for actual health impairment, health impairment, or susceptibility. Sensitivity and specificity are critical components in the process of validation (MacMahon and Pugh, 1970). Sensitivity is the quality of an epidemiologic test method that relates to the ability to identify correctly those who have the disease or condition of interest. Specificity relates to identifying correctly those who do not have the disease or condition of interest. Thus, markers of exposure or effect must be validated in terms of their ability to assess the true exposure or disease (sensitivity) and their ability to assess the lack of exposure or disease (specificity).

Particularly critical for markers is the strength of biologic plausibility that allows an association between a change in a specified signal (designated as a marker) and the occurrence of a specific exposure or a change in probability of a specific outcome. A major purpose of markers in environmental health research is to identify exposed persons, so that risk can be predicted and disease prevented; therefore, validation involves both forward and backward processes of association—i.e., from marker backward to exposure and from marker forward to effect.

A complete understanding of the limitations of a given biologic marker is crucial for its appropriate application and interpretation.

Uses

Validation of a specific marker also depends on its expected use. Biologic markers observed well before the onset of disease might have low predictive value for the disease itself, nevertheless function acceptably as criteria for defining exposed populations and thus be useful for long-term followup. For example, measurement of concentrations of pesticides, such as polychlorobiphenyls (PCBs) in human breast milk is clearly useful for exposure assessment and epidemiologic research, although its relationship to disease outcome might be difficult to determine. Conversely, an effect marker that is expressed long after exposure could be of relatively little use in exposure assessment, but be very important in predicting progression of disease or calculating risk. For example, a prenatal exposure that results in altered structure or function in the child or adult might be

difficult to identify on the basis of markers associated with the altered structure or function.

Animal Models

In validating biologic markers, animal models are useful for understanding mechanistic bases of the expression of markers and relationships among exposure, early effects, and disease. If a disease can be satisfactorily induced in experimental animals, then potential biologic markers for predicting eventual disease can be explored, and early indicators of the disease might be identified for use in epidemiologic studies. Also, markers of exposure can be explored for utility as markers of effect by relating the concentrations in an accessible compartment (or surrogate) to concentrations at the actual receptor site. The goal is to develop markers that reliably indicate an early stage in the development of a disease in humans when effective intervention is still possible.

A useful approach to the validation of marker data is to conduct experimental studies in animals and clinical studies in humans to develop information that permits interspecies comparisons. Markers of acute effects of short-term, low-dose, or high-dose exposures to a pollutant can be investigated in both animals and humans. Comparison of the results with information on markers of chronic effects of long-term low-dose exposure of animals to the same pollutant could lead to the development of markers that are predictive of health effects in chronically exposed humans.

Quality Assurance

Quality assurance and quality control are fundamental to the objective development and application of accurate and verifiable biologic markers. The objective of laboratory quality-assurance practices is to ensure that findings reported by one laboratory are in fact verifiable and within acceptable limits of measurement error, that they accurately indicate the concentrations or presence of materials reported to have been found, and that they are objective and free from sources of bias introduced through the analytic process.

General issues of quality assurance and quality control have been addressed by documents produced by the Food and Drug Administration (FDA), the U.S. Environmental Protection Agency, the Organization for Economic Co-operation and Development, and other regulatory organizations. FDA developed a set of guidelines known as Good Laboratory Practices—GLPs (U.S. FDA, 1988), which are now incorporated into the standard procedures of most testing and analytic laboratories. GLPs are intended to reduce the chance of contamination (particularly important in the measurement of biologic markers of exposure) or of changes in biologic variables introduced by sample storage, processing, or measurement (Zeisler et al., 1983). The application of GLPs to analysis of biologic samples, especially human tissue, has been reviewed by operational units of the Centers for Disease Control, the National Bureau of Standards, and various clinical laboratories (ACS, 1980; NCCLS, 1981, 1985). Issues of quality assurance related to screening for mutagens and reproductive toxins are discussed in Bloom (1981).

In establishing guidelines for quality assurance, the usual sequence is to develop methods of increasing authority based on accumulation of experience. Experience permits a method to be standardized, once it has been shown to be feasible, reproducible, and accurate when used in various laboratories. In some cases, cost-effectiveness is also a factor, particularly for clinical measurements intended to be used for screening, rather than research. This approach to standardized guidelines has been followed for some biologic markers.

Standardized reference methods, which have been well tested in the field, are available for measuring blood lead concentration as a marker of exposure. In fact, a standardized procedure for interlaboratory comparison of blood lead content has been developed and used by the Centers for Disease Control (Annest et al., 1983). Such a procedure, sometimes known as round-robin testing, can be used to verify the performance of various testing laboratories.

The preparation of biologic standards for measuring markers of effect is more complex. Cell-culture systems might provide particular types of standards; in some cases, the chemicals that constitute the marker (metabolic product, intermediate, or other material, such as protoporphyrin) can be synthesized or derived from other biologic materials and incorporated into an appropriate biologic matrix to meet criteria of quality assurance.

Sensitivity of Measurements

Other general quality-assurance issues are related to sensitivity and specificity. Estimations of sensitivity must include considerations of the so-called background rate of events or concentrations likely to be found in persons without particular exposures, as well as considerations of the magnitude of external exposure or internal dose likely to be received by the population being sampled. In the presence of relative uncertainty as to the nature and extent of exposures in the reference or control population, decisions concerning sensitivity can be difficult.

Markers of Specific Exposures

Ideally, we seek to correlate a biologic marker with a specific exposure. In order to measure concentrations of chemicals in body tissues and fluids, analytic techniques must be validated in the biologic media measured. Identification with atomic-absorption spectrophotometry, spectrophotofluorometry, or electrochemistry can sometimes be confirmed by gas or liquid chromatography in conjunction with mass spectrometry or nuclear magnetic resonance imaging.

Biologic markers might be specific with respect to the system being investigated, but of unknown specificity with respect to the exposure or end point being studied. That is the case with markers of reproductive function, such as plasma concentration of human chorionic gonadotropin (a marker of early stages of gestation) or amniotic fluid concentration of alpha-fetoprotein (a marker of integrity of fetal development). These are highly specific markers of reproductive status, but of unknown and probably variable specificity with respect to exposure to xenobiotic agents. Other biologic markers are even less specific, such as some serum enzyme concentrations, which can reflect a wide variety of organ-level phenomena, including hepatic metabolism, renal clearance, turnover, release, and cytotoxicity.

Determination of specificity involves consideration of variation, including age, sex, time of day, etc. When genetic markers are being measured, information on the possible impacts of heterozygous variation must be considered relevant. Age-matched and sex-matched cohorts should be established to control for age- and sex-related differences. Diurnal variation might be unknown for some markers, but is clearly important for some reproductive, nervous-system, and immunologic markers.

The handling of many specificity issues will improve as information is gained on aspects of human biology pertinent to understanding the impact of xenobiotics. In the absence of such knowledge, it is critical to gather as much information on potentially confounding variables as possible through comprehensive history-taking and other methods. Whenever possible, the collection of material for long-term storage should be encouraged, so that nested case-control studies can be conducted later, as new measures are developed, and hypotheses related to risk and outcome can be validated.

ECOLOGIC MARKERS

The biologic markers approach has great potential relevance to assessing and predicting not only effects of exposure to xenobiotics, but effects of environmental modifications on ecologic systems and nonhuman target organisms. Biologic markers of the status and function of an ecologic system include morphologic and biochemical observations on individual members of the system, observations that are not

different from those in humans (although some tissues obtained from nonhuman organisms might be unobtainable from humans because of ethical constraints). Markers might also be derived from functional groups, or communities, within ecosystems. Such markers would reflect the biologic consequences of exposure, such as shifts in the fixation of nitrogen or photosynthesis by trees. They might involve control of rates of processes, such as nitrification by a single species in the forest floor, or of the total photosynthesis of algal populations in a body of water. Physiologic effects in individuals in ecosystems can combine to affect a larger process, such as reproduction of field mice in a grassland system. As a consequence of such effects, the relative abundance of species, the types of species, or the individuals within a species can change over time.

The loss of a species is an ecologic marker of ecosystem damage, but it is unlikely to provide early warning of potential effects before substantial damage has been sustained. Earlier, more subtle biologic markers in the sequence of events that leads to such a relatively drastic outcome should be sought.

USE OF BIOLOGIC MARKERS IN RISK ASSESSMENT

Cellular and molecular markers can be powerful new tools for the assessment of risks associated with exposure to environmental toxicants. Markers that indicate the receipt of an internal, biologically effective dose or the induction of a disease process can be useful in hazard identification—i.e., as the qualitative step by which an environmental agent is causally associated with an adverse effect. Biologic markers can also be used to determine dose-response relationships, especially at the low doses relevant to exposure to most environmental chemicals. As Ehrenberg has demonstrated (1988), it is possible to measure concentrations of ethylene oxide-derived DNA adducts that correspond to an increased cancer risk of approximately one in a million, an increase often used to justify environmental regulation. The use of biologic markers indicating exposure and dose in molecular epidemiology studies provides the opportunity to determine the shape of the lower end of the dose-response curve in humans, an opportunity not available with standard epidemiologic or animal carcinogenicity testing approaches. Another major role of biologic markers pertinent to risk assessment is in clarification of the extent and distribution of exposure and effect in human populations, as well as of the variability and susceptibility among individuals in a population (Fowle, 1984; Perera et al., 1986).

EXTRAPOLATION FROM ANIMALS TO HUMANS

The validity of a specific biologic marker for the identification of an adverse health effect depends on the reliability of studies that provide the background data, particularly on mechanisms. In addition, the direct relevance of studies in animals to humans needs to be assessed carefully.

Laboratory animals and humans can differ in toxicokinetics. Thus, use of data from animals to determine health risks in humans might be inappropriate if the data are derived from monitoring of external exposure.

The toxicity of some chemicals is mediated either by activation or by detoxification biotransformation reactions. Inasmuch as biotransformation differs among species, it is important to establish whether the routes and rates of human and animal metabolic pathways are similar.

Health risks are often associated with combinations of effects in humans. For example, cardiovascular disease in humans can encompass both atherosclerosis and hypertension. Although swine are the most suitable animal model for studying spontaneous atherosclerosis, young rats might be most appropriate for studying hypertension. For humans, estimating these diseases will necessarily entail some appropriate combination of the relevant animal test systems.

A common source of uncertainty in risk assessment is the dose-response curve relationship at low doses or for rare effects (NRC, 1983, 1986). It is often impractical to conduct studies of effects at low doses, because large numbers of animals are required to detect a low incidence of effects (Wilkinson, 1987). Demonstrable health effects in humans, given the limits of epidemiology, are often associated with high doses and hence high risk. Sensitive molecular markers being developed will permit study of the relationships between exposure to chemicals at low ambient concentrations and the formation of a molecular marker predictive of human risk. The development of biologic markers might enable scientists to make better use of laboratory animal data in estimating the effects of chemicals in humans.

QUALITY AND QUANTITY OF DATA

Extrapolations to humans are to be based on the most sensitive animal species tested, barring clear evidence that the species is pharmacokinetically distinct from humans.

Within the past 2 years, EPA has issued guidelines for evaluating reproductive studies (U.S. EPA, 1987). These provide a means to estimate data quality and stipulate segment II developmental toxicity requirements of two species, three treatment groups, with 20 rodents or 10 nonrodent mammals. As this report went to press, EPA issued additional guidelines on animal reproductive studies.

Quantity of data required can be determined by statistical power considerations, and resource considerations. Statistical power is related to the number of subjects in a group, to the rarity of the end point studied, and to the variability in the frequency of the end point's occurrence. The greater the expected relative risk, the smaller the population that will need to be studied.

The study of reproduction and development poses major resource and logistic problems for those working with laboratory animals. For instance, manageable sample populations do not reveal increases in toxic events of less than 5 to 10 percent. For some health effects, such as mutagenesis and teratogenesis, incidences in a human population of 3 per 10,000 are significant. Obviously, these effects cannot be well studied in whole-body studies of thousands of experimental animals at a time. Classic toxicology studies of rodents involve the exposure and pathological analyses of 200 animals for 2 years. In such assays, each animal is a surrogate for 1,000,000 people. Birth defects that would be undetectable in this rodent population, if occurring in humans, could provide an epidemic within 10 generations.

Approaches currently used for measuring human exposure can be divided into three major categories: modeling, ambient measurement, and biologic monitoring. The development of models for grossly estimating human exposure is a major accomplishment of environmental scientists in the last few decades. Models, however, do not accurately take into account major sources of variation that affect dose received, e.g., variations in terrain or in human life styles. Ambient measurements, particularly those related closely to actual human exposures (e.g., personal monitoring for air pollutants), can be useful techniques to estimate exposure. However, simplifying assumptions about respiration rates and other variables that affect internal dose are necessary in deriving estimates of human exposure from ambient measurements of air, water, food, or contaminants in soil.

Biologic monitoring is the measurement of exposures or effects of exposure directly in receptor organisms, such as humans. The application of this approach has been limited principally by its expense and by the problems associated with sampling of humans. Furthermore, because the kinetics and metabolism of most environmental chemicals are not known, it has been difficult to develop strategies for measurement. Even when metabolic information is available, there might be no analytic methods to measure metabolites. Advances in analytic chemistry and molecular biology leading to such techniques as the use of monoclonal and polyclonal antibodies and the chemical analysis of DNA adducts have made it possible to detect biologic markers of exposure to a number of compounds in humans. Those techniques should prove valuable for determining the extent of exposure to toxic chemicals and for establishing priorities for effort and resources; e.g., they might be directed at persons who have received the greatest exposure or who are at highest risk in connection with some environmental hazard.

IMPLEMENTATION OF BIOLOGIC MARKERS IN POPULATION STUDIES

The identification of biologic markers that indicate exposure, effect, or susceptibility is a complicated process involving studies in animals, refinements in laboratory assays, and studies in special human populations. Moreover, even when a marker has been validated in such studies, its use in larger populations is not straightforward.

The framework for implementing an identified, potentially informative biologic marker in large population studies would include the following steps:

- Establish normal baseline values and distribution for the marker in laboratory animals and humans.
- Evaluate the sensitivity and specificity of the marker in predicting a health outcome (e.g., infertility or genetic damage).
- Understand in detail the time course of response of the marker to a toxic chemical, with special attention to the recovery process.
- Develop a strategy for and a consensus on the use of multiple species in toxicologic studies.
- Develop human assays that use semen, saliva, or urine, rather than tissue or blood, whenever possible.
- Use noninvasive techniques, such as ultrasound or magnetic resonance imaging, whenever possible.
- Consider a battery of markers that reflect a wide array of physiologic functions and genetic damage and relate the marker in question to others in the battery.
- Identify populations at high risk for reproductive or developmental health

impairment (perhaps populations exposed to drugs with reproductive or developmental toxicity, aging populations, or offspring of women exposed to diethylstilbestrol), to serve as test subjects for the initial assessment and validation of biologic markers.

• Include among high-exposure populations those with special or unique occupational exposures (e.g., agricultural groups).

• Encourage and support institutions in the development of sample banks, to speed the identification and validation of markers.

• Establish a task force to develop and coordinate strategy.

LONG-TERM TISSUE AND CELL STORAGE FOR RETROSPECTIVE ANALYSIS

Long-term tissue and cell storage for retrospective analysis of exposures to previously unknown environmental toxicants has been used successfully by several organizations, e.g., the American Type Culture Collection and the Environmental Specimen Bank Program managed by the National Institutes of Standards and Technology.

The oversight committee reviewed available information on long-term tissue storage but found that national programs in this area are limited and inconsistent. The subcommittee on reproductive and developmental markers also attempted to identify national storage programs for tissues from the reproductive tract; except for a few recently established sperm and ovary storage banks, no consistent programs were found. Some hospitals and medical research centers have attempted to store placenta; however, the protocols for storage are not developed sufficiently to make the tissue samples generally useful in reproductive markers research.

The oversight committee and the subcommittee recognize that many difficulties are associated with long-term tissue archiving. For example, establishing a standard protocol and training personnel to use it for collection and storage of samples is not easy to accomplish, nor can adherence to a protocol be ensured easily. Collecting samples under sterile conditions; storing tissues in sterile, metal-free containers; and proper freezing (usually at the temperature of liquid nitrogen) upon collection and during storage are vital for preserving the cell and tissue integrity necessary for reliable future chemical analyses. In addition, space considerations and stable institutional commitment are necessary for the success of any banking program. The National Research Council's Committee on National Monitoring of Human Tissues, which is reviewing the EPA's National Human Monitoring Program (an adipose tissue bank), is considering how many of these difficulties affect the collection, and ultimate analysis of archived tissues.

Fixed and imbedded tissues generally are retained after toxicological studies are completed and might have some application in future biological markers research. Their use has been very limited because the processes of fixing and imbedding produce changes in the molecular constituents. However, recent developments in molecular biological techniques, including the polymerase chain reaction, show promise in detecting alterations in DNA and RNA in archival formaldehyde-fixed paraffin imbedded tissues (Burmer and Loeb, in press). In addition, antibodies to certain constituents also might work on aldehyde-fixed specimens.

USE OF BIOLOGIC MARKERS IN REPRODUCTIVE AND DEVELOPMENTAL TOXICOLOGY

With the above general description of the concepts and definitions of biologic markers as background, the remainder of this report will focus on the use of biologic markers in reproductive and neurodevelopmental toxicology. Hypothetical examples of the utility of biologic markers follow. These are meant to be instructive and do not represent the limits of utility of biologic markers in reproductive or developmental toxicology.

Studies of Male Reproductive Toxicants

Consider as a male reproductive toxicant a metal that is used in an industrial process and emitted into the air (Fig. 1-2). Assuming that uptake occurs by inhalation and ingestion (i.e., via the mucociliary elevator), biologic markers of internal dose can include sputum, blood, urine, sperm, and semen concentrations of the metal or its metabolites. For some metals, it might be possible, with noninvasive monitoring techniques, to measure the amount of metal in tissues, such as bone or testis. The biologic effects of the metal on male reproductive function might be measured on the basis of testicular volume (e.g., testicular swelling or shrinkage), circulating concentrations of protein or steroid hormones, alterations in sperm morphology or function, alterations in male reproductive behavior, changes in fertility and increased heritable mutations.

Studies of Female Reproductive Toxicants

Consider that a small-molecule organic compound is suspected as a female reproductive toxicant (Fig. 1-3), and exposure is through ingestion (drinking water), skin (bathing and swimming), and inhalation (water droplets). Biologic markers of internal dose can include blood and urinary concentrations of the parent compound or metabolites. Measurements of exhaled parent compound or metabolites might also be of value. More invasive procedures will be required to determine the concentrations of the xenobiotic or its metabolites in adipose tissue or follicular fluid. The biologic effects of the exposure could include alterations in the mean concentration or pulse frequency or amplitude of circulating steroid or protein hormones. Functional alterations in the female reproductive system might be reflected in menstrual irregularity, a change in ovulatory frequency, and a decrease in fecundity.

Studies of Pregnancy Toxicants

Assume that an inorganic metal that is found in air, water, and some foods is a putative pregnancy toxicant (Fig. 1-4) External exposure occurs through air, water ingestion and contact, and dietary practices or preferences. The dose to a pregnant woman is the sum of inhaled, ingested, and transdermal uptake. Individual variables that alter the dose include respiratory rate and volume, water and food ingestion, and contact with contaminated water. Some of these factors can vary considerably in pregnancy. For example, minute volume increases by approximately 40% and blood flow to the skin increases by as much as a factor of 6 during pregnancy. Biologic markers of internal dose might include maternal and fetal blood concentrations, amounts excreted by the mother in urine and feces, and concentrations in maternal or fetal tissues, including placenta. It might be possible to monitor for some metals in the target tissue with noninvasive techniques such as neutron activation or magnetic resonance imaging. If the metal is toxic to placental function, the concentration of placental steroid or protein hormones in maternal serum or the rate of their excretion in maternal urine could change in response to placental concentration of the metal. The biologic effects of exposure either before or during pregnancy might include alterations in placental function, such as transport of gases, carbohydrates, amino acids, or other essential nutrients from maternal to fetal circulation. Alteration in placental transport might impair fetal growth and later functional development or functional capacity of the child or adult. Exposure early in gestation might lead to fetal malformation, fetal death, or spontaneous abortion.

Studies of Neurodevelopmental Toxicants

Assume that an organometallic compound found as a contaminant in food is a putative developmental toxicant (Fig. 1-5). Exposure of an infant or child is determined by dietary practices, amounts of contami-

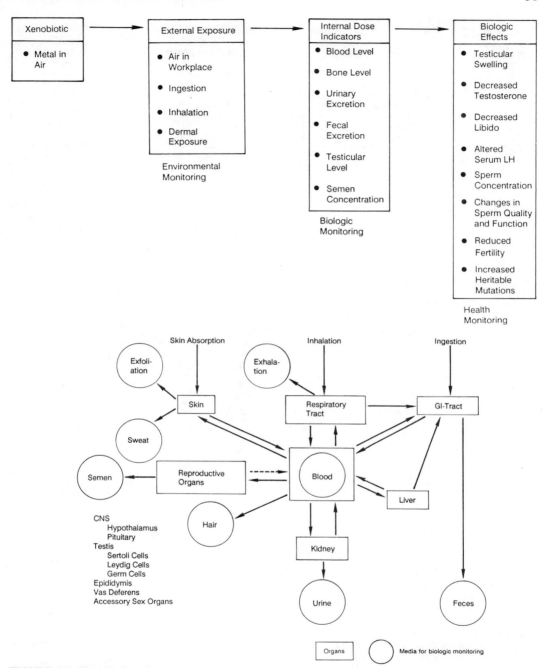

FIGURE 1-2 Hypothetical male reproductive toxicant used in an industrial process and emitted into air. Top, potential monitoring media and markers. Bottom, metabolic model of hypothetical metal toxicant. Arrows indicate transfer of toxicant or metabolite. Source: Adapted from Committee on Biological Markers of the National Research Council, 1987.

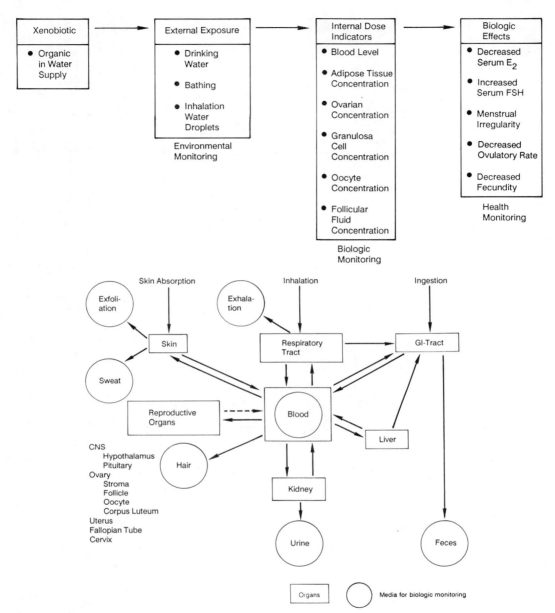

FIGURE 1-3 Hypothetical female reproductive toxicant with exposure through ingestion, skin, and inhalation. Top, potential environmental monitoring media and markers. Bottom, metabolic model of hypothetical small-molecule organic toxicant. Arrows indicate transfer of toxicant or metabolite. Source: Adapted from Committee on Biological Markers of the National Research Council, 1987.

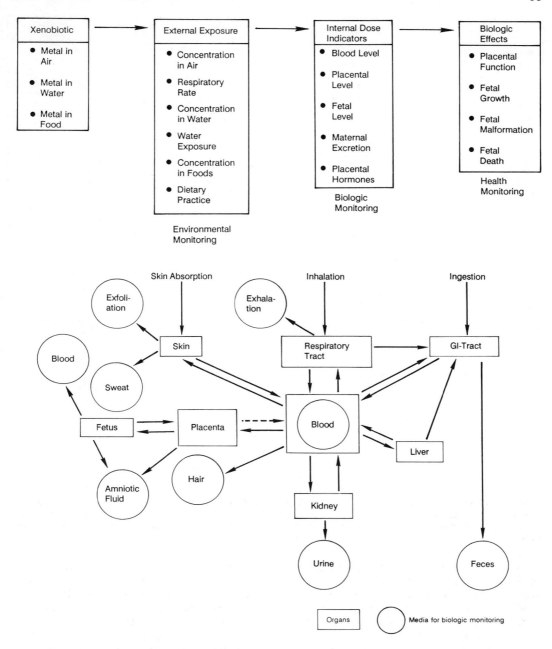

FIGURE 1-4 Hypothetical pregnancy toxicant found in air, water, and some foods. Top, potential environmental monitoring media and markers. Bottom, metabolic model of hypothetical inorganic toxicant. Arrows indicate of toxicant or metabolite. Source: Adapted from Committee on Biological Markers of the National Research Council, 1987.

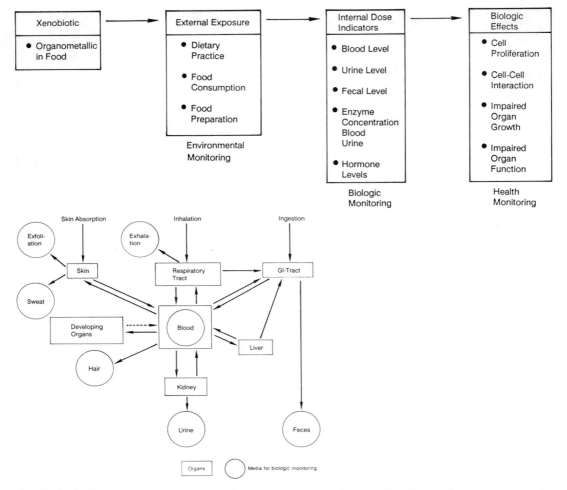

FIGURE 1-5 Hypothetical developmental toxicant found as a contaminant in food. Top, potential environmental monitoring media and markers. Bottom, metabolic model of hypothetical organometallic toxicant. Arrows indicate toxi of toxicant or metabolite. Source: Adapted from Committee on Biological Markers of the National Research Council, 1987.

nated foods ingested, and effects of preparation on the concentration of parent organometallic compound in the ingested foods. Markers of internal dose can include blood, urinary, or fecal concentrations of the parent compound or metabolites. If the organometallic xenobiotic alters endocrine function, it might be possible to determine the biologically effective dose indirectly on the basis of alterations in the concentration or pulse frequency of hormones. The biologic effects of the organometallic compound could include altered rates of cell proliferation or cell-cell interactions. Cellular effects might not be directly measurable in some organs or systems. However, in the central nervous system, the effects may be discernable with sensitive diagnostic measures, including electroencephalography, computerized axial tomography, and magnetic resonance imaging. Later tests of organ or system function might also reflect disordered development, but latency could make identification of the etiologic xenobiotic difficult.

SUMMARY

It is the job of the laboratory worker to develop tests that are as sensitive and specific as possible. It is the job of the clinician or public health worker to ensure that the benefits of using (invariably) imperfect tests outweigh the difficulties that arise from low predictive values when the tests are used in individuals or communities with low a priori probability of exposure or disease.

Careful consideration must be given to how a test for a biologic marker of exposure will perform in the field. It is not generally appreciated that a key factor that will affect performance is the frequency of exposure in the population in which the marker test is used. The use of even a good test in a population in which exposure is rare will result in a low predictive value of a positive test result, that is, many false-positives results. Widespread application of such tests must be carefully considered to ensure that the benefits outweigh the risks.

Few areas have changed in medicine as rapidly as those that are the subject of this report. The problems of reproductive and developmental toxicology remain complex because advances in our understanding of toxic chemicals have not kept pace with the introduction of new materials into the environment. A significant proportion of human reproductive wastage remains of unknown etiology.

I

Biologic Markers in Male Reproductive and Gametic Genetic Toxicology

2

Introduction

In the past, some believed that most problems with fertility, fetal damage, and congenital malformations were due to reproductive dysfunctions in the female. Recent years, however, have seen the accumulation of a considerable body of knowledge regarding male-mediated effects on development and the effects of environmental agents on male reproductive function (see Strobino et al., 1978; Soyka and Joffe, 1980; Wyrobek et al., 1983a; and Schrag and Dixon, 1985 for reviews).

The reproductive functions of the male mammal are to produce sperm, to attract receptive and fertile females, and to deposit adequate numbers of genetically normal sperm in a manner and at a time suitable for fertilization. Those functions involve numerous organ systems and complex brain functions. In human beings, social and psychologic factors are important for reproductive success. In this context, we attempt to develop markers for quantifying key aspects of human male reproductive processes and for detecting dysfunction associated with subfertility or abnormalities resulting from exposure to xenobiotic agents.

The male's role in reproduction can be divided broadly into physiological and genetic functions. Any change in either function can reduce the ability of sperm to fertilize an egg. Genetic defects in male germ cells occurring during spermatogenesis or during their passage through the efferent ducts may persist and lead to infertility, early or late pregnancy loss, congenital malformations, perinatal problems, and heritable mutations (chromosomal or genic) that may cause disease later in life and be passed on to future generations.

BIOLOGIC MARKERS OF MALE PHYSIOLOGIC DAMAGE

Biologic markers of male reproductive physiology have at least four major applications, as follows:

• Development and evaluation of safe and acceptable male contraception methods would be greatly facilitated by the existence of reliable biologic markers of normal male reproduction.

• About 15% of couples are infertile; in about 40% of infertile couples, the infertility is in the male (Mosher, 1980). Reliable biologic markers would help determine biochemical mechanisms for infertility and might help monitor treatment.

• Global industrialization has led to increased use of and dependence on chemicals. There are no human or animal data on the reproductive effects of most chemicals. Reliable human biologic markers

39

would permit direct measurements of reproductive effects in people exposed to xenobiotic agents. Direct human studies would circumvent problems associated with extrapolation of results of toxicity studies from animals to humans.

• Comparable markers in animals and human beings would provide a quantitative means for extrapolating animal data to man and would allow investigations of physiologic and toxicologic mechanisms.

The development and validation of markers for human male reproductive health typically require a multidisciplinary approach that includes basic research in animal and human reproductive biology, engineering and statistical development of automated and quantitative procedures, clinical studies of human factors that affect variation and of the predictive value of individual markers, and epidemiologic studies of populations exposed to xenobiotic agents.

Fertility potential, which is a combined function of the male and female, is difficult to assess in humans. Hence, the predictive value of abnormal ranges from biologic marker assessments requires knowledge of mechanisms. In the sections that follow, numerous ways of assessing normal and abnormal physiologic function and genetic variation in male germ cells and reproductive organs are discussed. In most instances, their utility as markers of exposure or markers of effect have not been assessed fully.

The prevalence of humans exposed to environmental, occupational, and therapeutic agents that are potential reproductive toxins argues strongly for the development of validated methods for measuring germinal and reproductive damage directly in people. Methods used in the evaluation of the reproductive health of human males are in three broad classes: personal history, physical examination, and laboratory analysis. The clinical application of personal history and physical examination in fertility assessment is important in screening populations and evaluating laboratory analyses. Laboratory analyses include testicular biopsy, hormonal analyses, and semen analyses.

Markers differ in the numbers and kinds of assays available to measure them, in the degree of quantitation attainable so far, in the extent to which their underlying mechanisms are understood, and in their feasibility for human studies. They also differ in sensitivity, specificity, and predictive value from assay to assay and from use to use. For example, as many as three applications of some of the markers (e.g., sperm concentration, motility, and structure) have been proposed: as markers of sperm production, as indicators of fertility status, and as indicators of exposure to a reproductive toxin. The validity of each marker depends on its specific application (e.g., see Chapter 7 for a discussion of sperm number).

A multistep process is required to validate all new markers of male reproductive health. Marker validation requires the description of measurement statistics of well-characterized groups and the understanding of the biologic and technical factors that affect measurement variability. Validation also requires a critical and quantitative assessment of a marker's ability to discriminate, e.g., between men with normal sperm production and men with abnormal sperm production, fertile men and infertile men, and exposed men and unexposed men. However, the use of semen markers to discriminate the effects of environmental, therapeutic, and occupational exposures does not necessarily require that a marker be associated with fertility status. In the latter applications, distributional characteristics of semen values in exposed and unexposed cohorts can be compared with each other and with historical controls to identify exposed populations and to evaluate the effect of exposure.

BIOLOGIC MARKERS OF GENETIC DAMAGE AND HERITABLE MUTATIONS IN HUMAN GERM CELLS

Tests that measure the potential for mutagenicity in humans are important, because some populations are being exposed to drugs, as well as environmental and occupational chemicals, and the mutagenicity of those chemicals in laboratory animals, such as mice, is well established. Although induction of germinal mutations by mutagens is well documented in animals, there is no firm evidence that any agent has induced germinal mutations in people, and monitoring for increased mutations in humans has proved unsuccessful.

From the animal literature, at least two broad types of induced genetic damage in exposed people—gene mutations and chromosomal alterations (in either chromosome structure or number)—can be expected. As discussed in Chapter 9, current human methods might be inadequate to detect induced mutations among offspring using the sizes of exposed cohorts evaluated to date. Also, increasing evidence points to induced genetic damage in human male germ cells, especially for ionizing radiation.

New approaches are under development to improve the detection of induced germinal mutations in people. These employ recent recombinant DNA and molecular techniques and use two sources of tissue for analyses: sperm of exposed men and somatic tissue from offspring of exposed individuals. These innovations promise the increased sensitivity needed to detect genetic defects in the germ cells of small cohorts of mutagenized people.

Germinal mutations are rare events and, as described in Chapter 9, the development of human assays for measuring genetic damage in germ cells presents special validation challenges, including a precise understanding of the spectrum of mutational damage detected, as well as an understanding of underlying mechanisms and assay responsiveness. Once developed, these detection assays would provide a means to identify human germinal mutagens and to manage human exposure so that the associated risk of inherited genetic defects and diseases could be reduced. Also, investigations of germinal mutations in laboratory animals, including mice (Chapter 8), is continuing to increase understanding of the relative sensitivity of germ cell stages, mutational mechanisms in germ cells, and the spectrum of genetic lesions induced by mutagen exposure.

Germinal exposures to mutagens clearly are not a male issue solely. In animals, male and female germ cells are known to be sensitive to germinal mutagens. Chapter 9, which discusses human germinal mutagens is included in this report because part of the progress in new technologies involves sperm-based assays. However, any new mutational assay that uses offspring tissue clearly would be applicable for studies of either or both human male and female exposures.

IMPORTANCE OF ANIMAL STUDIES IN MARKER DEVELOPMENT

Humans would be the species of choice for all investigations of human reproductive health and of factors leading to infertility. However, human studies are constrained by patients' needs, human subjects' rights, and the difficulties of controlling genetic, environmental, and exposure factors.

Most discoveries in human reproductive biology and the markers in use today were based on earlier investigations in animals. Animal studies of basic biochemical mechanisms, cellular processes, and the effects of genetic and environmental factors require continued support. Multigenerational studies in animals are the cornerstone of reproductive toxicity testing of chemicals (Zenick and Cleeg, 1989). Animal end points include markers of gonadal, extragonadal, seminal, and hormonal pathophysiology and markers of offspring quantity and quality. Animal experiments permit control of such variables as age and genotype (i.e., pharmacokinetics and metabolism), as well as exposure routes, dosages, and durations of exposure. Animal studies are not limited to noninvasive markers, as are human studies. The effects of many chemicals have been evaluated in animals with varied study designs and exposure conditions, and animal data have been used to provide presumptive evidence of human reproductive toxicity. However, animal data on a given chemical are typically incomplete, and it is difficult to come to a definite conclusion regarding reproductive effects. In addition, there are uncertainties in quantitative interspecies comparisons, and large safety factors are involved in extrapolation of risk to humans.

Animals and humans can differ markedly in their responses to chemical exposure. That is well illustrated by comparing the germinal effects of 1,2-dibromo-3-chloropropane (DBCP) in animals and humans. Human exposure to DBCP, a highly effective nematocide, resulted in male infertility and germ-cell aplasia at doses that showed no other signs of organ or system toxicity (Whorton et al., 1977). Some data suggest increased frequencies of spontaneous abortions among the wives of exposed workers, and there is indirect evidence that DBCP is a human germinal mutagen (Wyrobek et al., in press). However, the response among animals is highly species-dependent. At one extreme, mice are resistant to DBCP; essentially no induced germ-cell killing or germinal mutagenesis has been observed after varied exposures of different strains. Clearly, the negative germinal-mutagenicity data from the mouse are not relevant for human mutagenic risk assessment. In contrast, DBCP exposure of rats at similar and lower doses induced extensive germ-cell killing, subfertility, and dominant lethality. Those results suggest that further efforts are needed to develop the rat and other mammals as models for assessing human germinal toxicity and mutagenicity.

Species differences underscore the need for improved strategies for extrapolating reproductive effects from animals to humans. Ideally, animals with metabolism and biologic effects most similar to humans' would provide the most reliable data for extrapolation to humans. Detailed molecular comparisons of metabolites, adduct formation, and molecular damage (e.g., DNA strand breakage) might provide a means for comparing responsiveness quantitatively among mammals. For example, certain types, quantities, and kinetics of the formation and removal of DBCP metabolites, adducts, and other molecular damage in mouse, rat, and human, may be associated with induced germinal cell killing, infertility, and mutagenicity. As part of this research, improved techniques to detect adducts, metabolites, and molecular damage are required (the use of monoclonal antibodies, high-performance liquid chromatography, etc.). Sensitive detection methods would benefit studies of both animals and humans and ultimately the assessment of human germinal risk.

ORGANIZATION OF MALE REPRODUCTION SECTION

This section evaluates markers that could be used to assess reproductive effects of pathophysiologic changes and heritable genetic damage in males. The markers discussed are at varying stages of validation, ranging from markers that already are used to assess the effects of human exposure to reproductive toxins (e.g., sperm number) to markers that are only promising concepts and very early in their development. The section begins with a review of the clinical procedures for evaluating male infertility, including medical history, physical examination, and semen analyses (Chapter 3). This is followed by detailed evaluations of available methods and promising research related to markers of the structure and function of the testis, epididymis, accessory sex organs, and semen and sperm (Chapters 4-7). Semen analysis is discussed in several chapters because it is a noninvasive means of obtaining information regarding testicular, epididymal, and accessory organ function. Chapter 8 discusses the concept and status of genetic risk assessment. It is followed by a discussion of methods for detecting germinal and heritable mutations in human beings (Chapter 9).

Relevant research questions and promising concepts that may lead to future improved markers of male reproductive and genetic toxicity are identified throughout the section and are summarized in Chapter 10. Detailed consideration of sexual behavior, sexual differentiation, and puberty is beyond the scope of this report.

3

Clinical Evaluation of Male Infertility

This chapter briefly discusses assessments that should be included in a detailed medical evaluation. Such evaluations might be used to describe a population or used in conjunction with biologic markers described in later chapters. Some of the assessments noted here are discussed in detail in later chapters, notably, organ size and semen analysis.

MEDICAL HISTORY

Markers used to assess pubertal changes include growth spurt, development of facial hair, time when shaving began and shaving frequency, penile growth, nocturnal emission, and desire to masturbate. Although delay in puberty is usually the result of exposure to noxious agents either in utero (e.g., in the fetal alcohol syndrome) or in the neonatal period, chemical exposure somewhat later might delay the onset of puberty in connection with central nervous system damage or (less probably) direct testicular damage. The more likely causes of delay or impairment in puberty are genetic disorders, such as hypogonadotropic eunuchoidism (an autosomal dominant disorder) and Klinefelter's syndrome (typically with XXY sex chromosomal configuration) (Matsumoto, 1988). Appropriate laboratory tests can clarify whether a sex chromosomal abnormality exists; and

the presence or absence of anosmia or hyposmia, with a detailed family history, should assist in assessing the presence of hypogonadotropic eunuchoidism (Parks, 1988).

Traditionally, detection by medical history of changes in testicular function that occur after puberty have relied on such markers as decrease in libido, decrease in sexual potency, decrease in facial hair growth, diminution in muscular strength, and postpubertal onset of infertility. Large decreases in testosterone production are required before the first four of these markers are manifest (Clark, 1988). Moreover, alterations in libido and sexual potency are usually related to psychologic or other systemic disorders, e.g., coronary insufficiency and organic brain syndrome (Walsh and Wilson, 1987). But exceptions do exist, such as the reported link between decrease in libido and central nervous system damage resulting from lead poisoning (Lancranjan et al., 1975; Zenz, 1988) and the association of alcohol and marijuana abuse with changes in sexuality (Sherins and Howards, 1986; Reich, 1987).

To assess infertility as a marker, careful attention should be paid to the spouse's general health and reproductive function. Prior fertility should be noted, and, if there are no young children, the voluntary or involuntary aspect of that status should be documented. The man's work history should be recorded, including at least current and longest jobs (e.g., he might have worked in a chemical or insecticide plant or in a laboratory). Finally, systemic disorders associated with chronic negative nitrogen balance can adversely affect spermatogenesis; these disorders include ulcerative colitis, Crohn's disease, and poorly controlled diabetes mellitus (Jequier, 1986). A history of viral orchitis, mumps (epidemic parotitis), or infectious mononucleosis should not be ignored (Sherins and Howards, 1986).

The problem of erectile dysfunction is considered separately, because it has some unique features. As with other markers of male reproductive function, erectile dysfunction is nonspecific in origin; in fact, many men who have it also have psychologic problems. As to specific etiologic factors, most of the men who seek medical assistance are on antihypertensive medications, usually in conjunction with diuretic agents. Guanethidine severely disrupts the sympathetic nervous system and so-called dry ejaculation (actually, retrograde ejaculation, in which not all the sperm and seminal fluid are ejaculated, but some is taken up into the bladder) is a common complaint (Walsh and Wilson, 1987). Propranolol and other beta-blocking agents affect the sympathetic nervous system to a smaller degree, but still can cause erectile dysfunction (Walsh and Wilson, 1987). Some patients with diabetes mellitus and neuropathy also report erectile dysfunction (Walsh and Wilson, 1987). Any chemical that causes nerve damage can affect the erectile and orgasmic responses, such as lead poisoning and dioxin.

PHYSICAL EXAMINATION

Clinical signs that might be used as markers include eunuchoidal skeletal measurements; pattern of facial, body, and pubic hair; penile growth; size of testes; and size of prostate. When damage to the reproductive system begins before or around puberty, eunuchoidal features might be present, including an abnormal ratio of arm span to height (arm span at least 5 cm greater than height and distance from symphysis pubis to floor at least 5 cm greater than distance from symphysis pubis to top of skull). The presence of such features requires a marked impairment in testosterone production, because that permits a delay in epiphyseal closure (Griffin and Wilson, 1987).

Effects on pattern of hair growth are obvious if testicular function failed before puberty. But if puberty occurred normally and Leydig cell function became impaired, it could take 2 years or longer for changes in hair growth, including frequency of shaving, to appear (Griffin and Wilson, 1987).

Penile size is normally variable, but is usually greater than 2 cm in diameter and 4 cm in length. The scrotal skin is normally pigmented, and rugal folds are present. The eunuchoidal state is easily detected; but, if testicular function fails after puberty, no discernible change in genitalia occurs.

Testicular size is dictated primarily by the volume of seminiferous tubules. The testis normally is 4.6 cm long (range, 3.5-5.5 cm) and 2.6 cm wide (range, 2.1-3.2 cm) (Griffin and Wilson, 1987). Even in the adult, there is a noticeable decrease in testicular size if damage occurs, particularly if it is accompanied by fibrosis. One exception is temporary damage, such as in single-dose ionizing-radiation exposure (Matsumoto, 1988), which damages specific cell types but does not cause a reduction in testis size.

Prostatic size generally increases gradually with age. If damage to the reproductive system occurs with a sharp decrease in testosterone concentration, the prostate might shrink to almost nonpalpable dimensions; the lateral lobes can no longer be felt, and there is no median sulcus. Even if testosterone concentration is normal, the prostate might fail to develop normally or might atrophy if 5α-reductase is not functioning; the reason is the lack of conversion of testosterone to dihydrotestosterone, which is the principal androgen for normal prostatic growth.

SEMEN CHARACTERISTICS

It is logical to examine semen for assessment of testicular function, because it is easily obtained and contains, in addition to spermatozoa, many biochemical components that can be measured. Moreover, if a person has ingested or otherwise been exposed to therapeutic agents, such as tetracycline and various chemicals (e.g., TRIS, a flame retardant that was formerly applied to children's pajamas), they can be concentrated and appear in the semen (Hudec et al., 1981).

Sperm concentration or total sperm count is variable and depends in part on frequency of ejaculation, so the patient should be instructed to abstain from ejaculating at least 2 or 3 days before semen collection. The patient might collect his semen by masturbation (without lubricants) at home and then submit it to a laboratory within 4 hours. The sample should be kept at ordinary room temperature and protected from extremes in temperature (but not refrigerated) before and during transport. Because the ordinary daily variation is so large, the patient should submit at least three semen samples at separate times (at least a week apart) before conclusions about his fertility are made. The physician should note whether the patient has been ill and has had a body temperature of over 100°F during the preceding 3 months. If so, the sperm count might be temporarily decreased by the illness.

Ejaculate volume of a healthy man varies between 2.0 and 6.0 ml. If ejaculate volume is less than 1.0 ml, possible loss of some of the sample or retrograde ejaculation should be suspected.

The motility of sperm can be assessed with a light microscope, but dark-field illumination or phase-contrast illumination provides better discrimination (Sherins and Howards, 1986). This procedure should be carried out at 37°C. A drop of semen is used with a coverslip. A 400x microscope objective should be used. There are various ways to classify sperm motility, but it is most important to determine the proportion of sperm that are moving forward in a rapid, apparently purposeful

fashion. At least 10 fields should be viewed for the assessment. If more sophisticated hardware and software are not available, automated technologies can be obtained commercially, as well as special counting chambers, to facilitate the determination of motility (Jequier, 1986). Normally, at least 45% of the sperm should be moving forward rapidly. When the percentage is lower, supravital staining should be used to document the ratio of living to dead sperm (Jequier, 1986).

The concentration and especially the total numbers of sperm in the semen are important markers of the integrity of testicular function, but there are others. If numbers of sperm are small by usual fertility standards, but their motility and morphology are excellent, then fertility might be normal, especially if the female partner is of high fertility; if numbers of sperm are large, but their motility and morphology are poor, the person might be infertile (WHO, 1980a). The probability of pregnancy is the product of the innate fertility of the male and the female, not the property of either partner. This issue might seem elementary, but many researchers lose sight of it and focus almost exclusively on the concentration or total number of sperm in the ejaculate (WHO, 1987).

In general, sperm concentration in normal men ranges from about 20×10^6/ml to 200×10^6/ml. There is large variation among and within people. Moreover, the variation does not have a normal distribution; log transformation is commonly used for statistical analyses (WHO, 1987), but some investigators believe that square-root or cube-root transformation is more useful.

It is common to obtain one semen sample from each person in a group and then use group data to compare with the frequency distribution in a control population to determine whether there has been a hazardous exposure (WHO, 1980a; 1987). However, that procedure will reveal an adverse effect only if a large population of the men at risk shows azoospermia or oligospermia or if longitudinal studies show a consistent pattern or a return to normal values.

Priority should be given to developing a battery of valid in vitro tests to determine the fertilizing capacity of human sperm. When the sperm penetration test was developed by Yanagimachi and colleagues (1976), there were high hopes for its use in this manner. However, prospective studies have pointed out its deficiencies when it is used as a model. Its accuracy in identifying sperm from fertile men is only about 50%. A search for more suitable biochemical markers is needed to overcome the problems in this kind of assessment.

4

Biologic Markers of Testicular Function

This chapter focuses on physical and chemical markers of the testes; stereologic and biochemical assessments of Leydig cells, Sertoli cells, and germ cells; and molecular biologic analyses of DNA and RNA in germ cells. Some markers other than semen analysis (discussed in Chapter 7) are noninvasive or minimally invasive and can be used to assess testicular function in human males exposed to toxicants. In addition, several promising noninvasive or minimally invasive markers using new imaging techniques, molecular biologic assays, and biochemical assays of saliva, serum, and urine are identified.

The testis has two compartments: the interstitium and the seminiferous tubules. The interstitium contains Leydig cells that produce the male hormone testosterone. Testosterone causes the differentiation and development of the fetal reproductive tract, the neonatal organization of what will become androgen-dependent target tissues in puberty and adulthood, the masculinization of the male at puberty, and the maintenance of growth and function of androgen-dependent organs in the adult.

The seminiferous tubules contain germinal epithelium and supporting cells. The supporting cells include Sertoli cells; in adults, Sertoli cells are static nonproliferating cells that are inti-

mately associated with and support germinal cells involved in spermatogenesis, the production of spermatozoa. The germinal epithelium is populated by cells that give rise to spermatozoa. Spermatogenesis encompasses a phase during which primitive spermatogonia divide either to replace their number (stem cell renewal) or produce new spermatozoa that are committed after additional mitotic divisions to become spermatocytes; a meiotic phase during which spermatocytes undergo the first and second meiotic divisions that result in haploid spermatids; and a spermiogenic phase during which spermatids undergo a dramatic metamorphosis in size and shape to form spermatozoa.

Figure 4-1 shows the human germ cell types, the life span of each, and the time required for each to reach the ejaculate. The figure illustrates three important points. First, the human testis has 14 recognizable types of germ cells. Second, toxic effects on a specific germ cell might not be manifested in the semen for days or weeks, because more mature unaffected cells will continue to develop and appear in the ejaculate. Third, the time of appearance of defective, immotile, or reduced numbers of spermatozoa in the ejaculate provides important information about the germ cell type affected by a toxicant. Figure 4-2 presents an example to

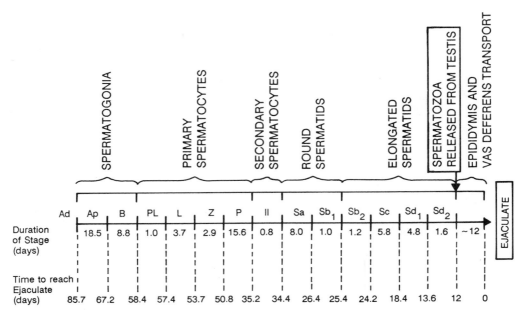

FIGURE 4-1 Spermatogenesis in man, showing life span of each cell and time necessary to reach ejaculate. Redrawn from da Cunha et al., 1982.

illustrate those points. Spermatozoal concentration and motility were monitored at various times after 12 courses of treatment of a melanoma patient with AMSA [4'-(9-acridinylamino)methane-sulfon-*m*-anisidide]. The slow decline in the number of motile spermatozoa in the ejaculate was interpreted by the investigators as suggesting that there was no immediate damage to spermatozoa transport or the epididymis. They believed that primarily postspermatogonial germ cells were killed by the AMSA therapy. (Initially, the cells that were killed were primary spermatocytes; later, type B spermatogonia were also killed.) They interpreted the rapid recovery of sperm concentration and motility in the semen 13 weeks after the first nine courses of AMSA treatment as showing that type A (stem) spermatogonia were not irreversibly affected.

PHYSICAL AND CHEMICAL MARKERS OF TESTICULAR FUNCTION

Markers in Use

Testicular consistency and size have been considered important in the clinical evaluation of human male fertility. With tonometers, one can measure testicular consistency through the stretched scrotal skin. Tonometer readings have been correlated with clinical impressions of testicular consistency in human males (Lewis et al., 1985) and with sperm morphology in bulls (Hahn et al., 1969). The advantages of testicular consistency as a biologic marker are that it is noninvasive and simple to measure. The disadvantages are that it is unrelated to a specific defect in testicular function and that large changes are required to detect effects of treatment.

Testicular weight is directly correlated with testicular volume, which can be estimated from testicular size in many vertebrates (Bailey, 1950; Kenagy, 1979; Handelsman and Staraj, 1985). In mammals, the bulk of testicular volume and therefore

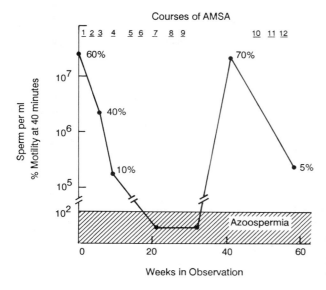

Courses of AMSA

FIGURE 4-2 Sperm concentration and motility during chemotherapy with AMSA [(4'-9-acridinyl-amino) methane-sulfon-*m*-anisidide]. Source: da Cunha et al., 1982.

weight is accounted for by germ cells (Amann, 1970a). Estimated testicular volume based on measured size is a simple and noninvasive marker of sperm production (Foote, 1969; Chubb and Nolan, 1985). However, it is imprecise, because errors are inherent in the measurement process itself and in the calculation of volume of nonspherical objects. The use of weight measurements, although precise, has the disadvantage of being invasive, in that the testes must be removed.

Markers Requiring Research and Development

Organ chemiluminescence can give readily detectable, continuously monitorable, noninvasive signals of oxidative metabolism (Boveris et al., 1980). It is possible that chemiluminescence can be used to monitor the effect of toxicants on the radical reactions of lipid peroxidation in testes in situ or in testes perfused in vitro.

Kopp et al. (1986) used phosphorus-31 magnetic resonance imaging (MRI) to determine functional metabolic correlates, temporal relationships, and intracellular actions of cardiotoxic chemicals nondestructively in isolated intact perfused rat hearts. They obtained useful information on biochemical mechanisms respon-

sible for the cardiotoxic actions of xenobiotics. This approach can probably be applied profitably to an in vitro perfused rat testis model immediately and perhaps to the human testis in situ eventually. One potential problem is that MRI might alter testicular temperature. Other indirect measures of testicular size and function are promising but have received little attention, including ultrasound, positron-emission tomography, and computed axial tomography.

LEYDIG CELLS

Leydig cells are the principal source of testosterone in the mammalian male (Ewing and Zirkin, 1983). Leydig cell growth and differentiated function depend on anterior pituitary production of luteinizing hormone (LH) (Ewing and Zirkin, 1983). Toxicants can interfere with testosterone production indirectly by interfering with gonadotropin-releasing hormone (GnRH) stimulation of pituitary gonadotropes, by interfering with LH production by pituitary gonadotropes, or by interfering with receptor-mediated LH stimulation of testosterone secretion by Leydig cells. Toxicants might also inhibit the Leydig cell steroidogenic apparatus directly.

Markers in Use

Stereologic techniques allow determination of Leydig cell numbers per testis at the light microscopic level and of Leydig cell cytoplasmic organelle volume and membrane surface area (e.g., area of inner mitochondrial membrane and smooth endoplasmic reticulum) at the electron microscopic level in both humans (Mori et al., 1982) and experimental animals (Mori and Christensen, 1980). Although they are invasive and tedious to carry out, these biologic markers of Leydig cell structure have been shown to be highly correlated with Leydig cell steroidogenic function under a variety of experimental and physiologic conditions (Christensen and Peacock, 1980; Zirkin et al., 1980; Ewing et al., 1981). For example, prolonged exposure to lead causes a diminution in testosterone production and in the surface area of smooth endoplasmic reticulum in rat Leydig cells (Zirkin et al., 1985a). In addition, Leydig cell numbers diminish with advancing age in humans (Kaler and Neaves, 1978). These morphologic markers are particularly helpful in studies aimed at understanding mechanisms of toxicity.

Testosterone is commonly secreted episodically by Leydig cells in the mammalian testis (Ewing et al., 1980). There are annual and diurnal rhythms in testosterone production, and a diurnal rhythm and frequent sporadic bursts of testosterone in some species. The episodic nature of testosterone secretion in man is blunted, compared with that in many species. The variation complicates the assessment of Leydig cell steroidogenic activity by measurement of peripheral blood testosterone concentration, because a regimen of frequent sampling must be followed, especially in experimental animals (Ismail et al., 1986). Measurement of testosterone concentration in blood serum at 10-minute intervals for 8 hours will usually indicate whether a toxicant has altered the entire hypothalamo-adenohypophysial-testicular axis in experimental animals. But toxic effects in humans might be diagnosable with less frequent or even a single serum testosterone measurement, because testosterone production is not so episodic. The function of the hypothalamo-adenohypophysial link can also be assessed by measuring serum LH concentrations every 3-10 minutes for 24 hours. Measurement of LH and testosterone concentrations in peripheral blood for less than 24 hours but at the same frequency will impart the same information regarding the hypothalamo-hypophysial function. Measurement of serum LH concentration might be more useful than testosterone pulses for diagnosing defects in the hypothalamo-adenohypophysial-testicular axis in humans, because of the attenuation of testosterone pulses.

In both men and experimental animals, the capacity of pituitary gonadotropes to respond to GnRH can be assessed by injecting a bolus of GnRH either intravenously or subcutaneously and then measuring LH in peripheral blood. Similarly, the capacity of Leydig cells to respond to LH can be assessed by injecting a bolus of human chorionic gonadotropin (hCG) intravenously and measuring testosterone in peripheral blood. Again, measurement of LH and testosterone concentrations in peripheral blood for a shorter period but at the same frequency will impart the same information; with less frequent sampling, there is a loss of sensitivity and precision in detecting an effect of exposure to a toxicant. The advantages of measuring the concentration of LH and testosterone in peripheral blood serum of humans or experimental animals are that it directly monitors the gonadotrope and Leydig cell function, respectively, and that repetitive measurements can be obtained for assessing temporal effects of a treatment. Disadvantages of measuring LH and testosterone in peripheral blood are that only small amounts of blood can be collected from small rodents, unless red blood cells are replaced and that, because LH and testosterone production are episodic, several measurements are required.

Most testosterone (over 90%) in peripheral blood is bound to albumin and testosterone estradiol-binding globulin and therefore is biologically inert. If it is important, free biologically active testosterone can be measured by separating free and protein-bound testosterone

(Vermeulen et al., 1971). This issue becomes important when increased sex steroid-binding globulin results in a decrease in free testosterone and causes hypogonadism. An example is the effect of chronic alcoholism in human males (Van Thiel et al., 1974).

Inhibition of Leydig cell steroidogenesis in vivo causes a diminution in accessory sex organ weight, discussed in Chapter 6 as a bioassay for peripheral blood testosterone concentration in experimental animals (Dorfman and Shipley, 1956; Chubb and Nolan, 1985; Coffey, 1986).

Finally, Leydig cell steroidogenesis can be monitored by removing testes from experimental animals and measuring their capacity to produce testosterone in vitro. For example, testes of some species (e.g., mice and rats) can be dispersed with collagenase and the production of testosterone by testicular cells measured (Bordy et al., 1984). Alternatively, testes of numerous species have been perfused in vitro (Ewing et al., 1981). These techniques are invasive and limited to a few hours duration in vitro. But they can be particularly helpful in studies aimed at understanding mechanisms of toxicity. Clearly, this approach is impractical in humans.

Markers Requiring Research and Development

It has been suggested that measurement of testosterone in saliva is an excellent biologic marker for Leydig cell testosterone production, because saliva can be collected from humans repetitively in a nonstressful manner and because salivary testosterone concentration is correlated closely with the biologically active free testosterone in blood (Riad-Fahmy et al., 1982). Studies have borne this idea out: human salivary testosterone concentration has been shown to exhibit a circadian rhythm (Magrini et al., 1986), to increase after hCG stimulation (Nahoul et al., 1986), and to be highly correlated with pathophysiologic conditions that result from modifications in serum testosterone concentrations (Riad-Fahmy et al., 1982). To our knowledge, this technique has not

been used to monitor the effect of xenobiotics on Leydig cell function in humans. It should be added to the armamentarium of reproductive toxicologists and epidemiologists, because it provides a noninvasive, specific, accurate, and sensitive marker to monitor exposure to or effects of a toxic agent on Leydig cell steroidogenic function in human males.

Measurement of serum or salivary testosterone concentration does not represent the integrated 24-hour rate of testosterone production. That can still be achieved only by sophisticated study of metabolic clearance rate, which is tedious, time-consuming, expensive, and impractical for application to humans. However, considerable evidence shows that it is possible to monitor ovarian function and pregnancy status by measuring gonadotropin and gonadal steroids and/or steroid metabolites in morning or random urine specimens from females of numerous species (Lasley et al., 1980; Czekala et al., 1981). The advantage of this approach is that the marker of interest accumulates in urine, thus obscuring the episodic nature of hormone production and making it possible to estimate integrated hormone production over time with fewer samples. There are shortcomings (Edwards et al., 1969; Curtis and Fogel, 1970), and a careful evaluation will determine whether specific hormones can be measured in urine samples as biologic markers of the functional status of the hypothalamo-adenohypophysial-Leydig cell axis in experimental animals and humans.

Leydig cells probably have functions other than testosterone production. Therefore, considerable research is required to uncover potentially new and useful Leydig cell markers. The testis contains peptides that are also formed elsewhere in the body. There is evidence of a GnRH-like factor, thyrotropin-releasing hormone, arginine vasopressin (AVP), somatomedins, oxytocin, mitogens, epidermal growth factor, and several pro-opiomelanocortin (POMC)-derived peptides (Hsueh and Schaeffer, 1985; Boitani et al., 1986; Kasson et al., 1986) in mammalian testes. The most experimental detail is available on the GnRH-like factor and AVP, which have been shown to alter Leydig cell steroidogenesis, and on POMC-derived peptides, which apparently are produced in Leydig cells. It is beyond the scope of this report to discuss each of these testicular peptides in detail. We focus our attention here on the POMC-derived peptides, for three reasons: concepts established for the use of one peptide as a biologic marker might have application to others; the GnRH-like peptide seems to be restricted only to the rat testis, whereas POMC-derived peptides seem to be more widely distributed; and more information is available on the physicochemical characteristics of POMC-derived peptides than on those of the other testicular peptides. The reader is referred to comprehensive reviews for more information on testicular GnRH-like factor (Hsueh and Schaeffer, 1985) and AVP (Cooke and Sullivan, 1985; Kasson et al., 1986).

POMC-derived peptides have been localized by immunocytochemical procedures in Leydig cells, but not in myoid and Sertoli cells of rat, mouse, hamster, guinea pig, and rabbit testes (Tsong et al., 1982a,b). mRNA for POMC-like proteins was localized in Leydig cells of mouse testes by in situ hybridization (Gizang-Ginsberg and Wolgemuth, 1985). It was later shown that genes for POMC-like proteins and the concentration of Leydig cell POMC-derived peptides are regulated by LH (Shaha et al., 1984; Boitani et al., 1986; Valenca and Negro-Vilar, 1986). Together, those results suggest that Leydig cells might produce POMC-like proteins. The function of such molecules is unknown. Nevertheless, it is possible that a Leydig cell-specific POMC-derived peptide can be secreted by testes and that its measurement might constitute a biologic marker for some as yet poorly understood Leydig cell function.

SEMINIFEROUS TUBULES

The seminiferous tubules in the adult mammalian testis contain germ cells in various developmental phases and nonproliferating Sertoli cells. The reader is referred to several comprehensive reviews

(Roosen-Runge, 1969; Clermont, 1972; Ewing et al., 1980; Griswold, 1988) for detailed descriptions of spermatogenesis and the structure and function of Sertoli cells. Briefly, the primary function of the seminiferous tubules is the production of spermatozoa.

A major difficulty in elucidating the site or mechanism of action of a toxicant on spermatogenesis in mammals is that, as germ cells differentiate, they physically interact with and are affected by each other, somatic cells of the seminiferous tubules (e.g., Sertoli cells), and indirectly through chemical signals (e.g., from follicle-stimulating hormone (FSH), LH, testosterone). It is extremely difficult to ascertain whether a toxicant acts directly on a specific cell type in the germinal epithelium (e.g., a spermatogonium or spermatid) or indirectly via the Sertoli cells, Leydig cells, or (even more indirectly) cells in the hypothalamus or adenohypophysis.

Toxicologic elucidation is further complicated by the presence of Sertoli-Sertoli junctional complexes that subdivide the seminiferous epithelium into a basal compartment and an adluminal compartment in many species, including humans. It is believed that these specialized junctional complexes constitute the principal site of the blood-testis barrier that restricts the free movement of specific chemicals between the blood and seminiferous tubular fluid. Apparently, spermatogonia and young spermatocytes are outside the permeability barrier, in the basal compartment next to the basement membrane of the seminiferous tubule, and presumably exposed to xenobiotics in blood and lymph. In contrast, mature spermatocytes and spermatids are sequestered within the permeability barrier, in the adluminal compartment. A practical consideration is the possibility of differential drug access to the cells sequestered behind the barrier. However, studies with labeled alkylating agents in mice have indicated that spermatocytes and spermatids can be exposed and that exposure can result in cell-killing and induced mutations. Despite these complex interactions, we have divided the following discussion of biologic markers of the seminiferous tubules into a section on Sertoli cells and a section on the germinal epithelium to simplify the presentation.

Sertoli Cells

The structure and function of Sertoli cells have been the subject of several comprehensive reviews (Fawcett, 1975; Dym et al., 1977, Ewing et al., 1980; Griswold, 1988). Briefly, Sertoli cells in the adult mammal are nondividing or slowly dividing cells that rest against the basement membrane with projections into the lumen of the seminiferous tubule. Their shape is complex and constantly changing, depending on the stage of the cycle of the seminiferous epithelium. Generally, however, Sertoli cell shape is characterized by an irregular nucleus, prominent nucleolus, and filamentous cytoplasm.

Although the structure of the Sertoli cell has been described for numerous species, its functions remain enigmatic, because of its intimate association with a population of germ cells that change over time and space (Ewing et al., 1980). Sertoli cells must be involved at least in germ cell division and differentiation, in view of their direct and specialized membrane contact with germ cells, their formation and presumed control of the milieu of the adluminal compartment of the seminiferous tubule via the tight junctions between adjacent Sertoli cells, and their apparent transduction of hormonal signals (e.g., from FSH and testosterone) that are known to regulate spermatogenesis. Although Sertoli cells can be counted, few biologic markers of Sertoli cell function have been developed, and their used is complicated by differential and variable secretion from Sertoli cells into the blood/lymph or seminiferous tubule fluid draining into the rete testes and epididymal and seminal fluids.

Markers in Use

Numerous stereologic procedures have been used to learn the number of Sertoli cells per testis (Wing and Christensen, 1982; Johnson et al., 1984a; Johnson, 1986) and the numbers of several specific germ cell types associated with an average Sertoli cell (Wing and Christensen, 1982; Johnson et al., 1984a). Knowing the former allows an investigator to test the effect of a toxicant on the viability of Sertoli cells; knowing the latter allows one to test the effect on the functional capacity of Sertoli cells to support each germ cell type. To our knowledge, however, these markers have not been used to test the effect of a xenobiotic chemical on Sertoli cell number or structure, probably because they require biopsy or autopsy specimens, are tedious to use, and are subject to artifacts of tissue preparation. These morphologic markers should be particularly helpful in studies of mechanism of toxicity in experimental animals, because they can provide information about germ-Sertoli cell interaction.

Sertoli cells cultured in vitro secrete at least 60 proteins, as measured by the incorporation of radioactive amino acids into spots on two-dimensional gels (Wright et al., 1981). Sertoli cells secrete both serum proteins and testis-specific proteins. Serum proteins identified include transferrin, ceruloplasmin, somatomedin C, and sulfated glycoproteins 1 and 2; testis-specific proteins include androgen-binding protein (ABP), inhibin, Müllerian-inhibiting substance (MüIS), Sertoli-derived growth factors, and cyclic proteins-2 (Griswold, 1988). With the exception of MüIS, and perhaps transferrin and ABP, these proteins have poorly understood functions. However, each is a potentially useful marker of Sertoli cell function in vitro. It is important to note that these data are derived largely from the culture of Sertoli cells from immature, rather than mature, rats. Therefore, extrapolation of the results from immature rat Sertoli cells in vitro to the human in vivo situation must be made cautiously.

Markers Requiring Research and Development

Transferrin, ABP, MüIS, and inhibin are the most extensively characterized Sertoli cell products and therefore the best candidates for in vivo markers of the pathophysiologic state of Sertoli cells. Transferrin's usefulness is limited, because it also is synthesized in the liver (Skinner et al., 1984). The other three hold considerable promise as biologic markers, because they are specific products of the Sertoli cell, because they probably are secreted into the peripheral blood, and because considerable progress has been made in cloning genes for them.

The discovery of ABP (Ritzen et al., 1971; Hansson and Djoseland, 1972) and its identification in peripheral blood of rats (Gunsalus et al., 1980) suggest that peripheral blood concentrations of ABP might serve as a marker of the pathophysiology of Sertoli cells in vivo. It was recently shown (Orth and Gunsalus, 1987) that ABP concentration in the peripheral blood of rats is highly correlated with Sertoli cell number. That finding was possible because the rat has no testosterone-estradiol binding globulin (TEBG), which is probably identical with ABP in other species. It will require considerable research and development, however, to validate a radioimmunoassay that differentiates between ABP and TEBG in species other than the rat, to elucidate the relationship between serum ABP derived directly from Sertoli cells and ABP derived indirectly from the epididymis, and finally to describe the correlation between Sertoli cell function and serum or seminal ABP concentrations.

A number of studies (McCullagh, 1932; Rich and De Kretser, 1977; de Jong, 1979) have shown selective increases in serum FSH after destruction of the germinal epithelium. Sertoli cells secrete inhibin, a factor that diminishes the release of FSH from cultured pituitary cells (Steinberger and Steinberger, 1976). Further research and development are required, however, to purify inhibin from testes, to raise a specific antibody against subunits of inhibin, to validate a radioimmunoassay for inhibin, and finally to elucidate the relationship between Sertoli cell pathophysiology and serum inhibin concentration. These studies are under way as this report is written (see, e.g., de Jong, 1987).

MüIS is a glycoprotein that causes regression of the Müllerian duct (Picard et al., 1986). The bovine and human genes for MüIS were recently isolated, and the human gene can be expressed in animal cells (Cate et al., 1986). Its C-terminal domain shows a marked homology with human transforming growth factor β and the β chain of porcine inhibin (Mason et al., 1985). Considerable research must be completed to determine whether MüIS is secreted by adult testes, whether it has any function in adults, and whether it can serve as a biologic marker of Sertoli cell function.

Germ Cells

Spermatogonia are the most undifferentiated germ cells in the seminiferous epithelium. Spermatogenesis is the process by which undifferentiated spermatogonia divide and differentiate into spermatozoa. The spermatogonia undergo a number of mitotic divisions, enter the meiotic phase of spermatogenesis as diploid spermatocytes, and undergo the first and second meiotic divisions to produce haploid spermatids, which differentiate into spermatozoa. Spermatogonia are of two major types: one is a stem cell that divides occasionally to replenish itself, or two, produces a committed spermatogonium that undergoes several mitotic divisions and then forms spermatocytes, and eventually gives rise to spermatozoa. As spermatogonia divide and differentiate, they are also replenished by a process termed stem cell renewal.

Spermatogenesis consists of a series of events that takes a different amount of time in different species (Clermont, 1972). Table 4-1 compares several characteristics of spermatogenesis and sperm production in mice, hamsters, rats, rabbits, beagles, rhesus monkeys, and humans. Species differ substantially in characteristics of spermatogenesis and sperm production; they differ less in epididymal transit time. Amann (1986) concluded that no species is identical with the human in these respects, so care must be taken when extrapolating from animals to humans. The potential for a toxic effect on spermatogenesis might be greater in humans because sperm production per gram of testis

TABLE 4-1 Species Differences in Spermatogenesis, Daily Sperm Production, and Epididymal Transit Time[a]

	Mouse	Hamster	Rat	Rabbit	Dog (Beagle)	Monkey (Rhesus)	Man
Duration of spermatogenesis, days	34-35	35-36	48	48-51	62	70	72-74
Duration of cycle of seminiferous epithelium, days	8.9	8.7	12.9	10.7	13.6	9.5	16.0
Life span, days							
B spermatogonia	1.5	1.6	2.0	1.3	4.0	2.9	6.3
Leptotene	2.0	0.8	1.7	2.2	3.8	2.1	3.8
Pachytene spermatocytes	8.0	8.1	11.9	10.7	12.4	9.5	12.6
Golgi spermatids	1.7	2.3	2.9	2.1	6.9	1.8	7.9
Cap spermatids	3.6	3.5	5.0	5.2	3.0	3.7	1.6
Testicular weight (total)	0.2	3.0	3.7	6.4	12.0	49.0	34.0
Daily sperm production, millions							
Per gram of testis	28	24	24	25	20	23	4.4
Per male	5.6	72	89	160	240	1,127	150
Sperm reserves in caudae epididymes (at sexual rest), millions	49	1,020	440	1,600	2,100	5,700	420
Epididymal transit time (at sexual rest), days		14.8	8.1	12.7	11.3	10.5	5.5-12

[a]Data derived largely from a table constructed by Amann (1986).

in humans is approximately one-fourth to one-sixth that in the other species. Daily sperm production in all species probably is in excess of that required for fertility. For example, a 90% reduction in fertile sperm available for ejaculation did not suppress fertility in rats (Amann, 1986). Thus, especially in animal species, but even in humans, fertility is unlikely to be a sensitive indicator of toxic insult of spermatogenesis.

Therefore, our task is to elucidate biologic markers that reflect the exposure to or effect of toxicants on the number and function of germ cells, from primitive spermatogonia to fully formed spermatozoa. The complexity and species variation of spermatogenesis suggest that a battery of markers will be required to detect toxic effects on spermatogenesis. For example, one toxic chemical might act on cells undergoing rapid mitosis, and another primarily on meiotic cells in which complex genetic rearrangements occur. Slowly dividing stem spermatogonia might be resistant to cytotoxic chemicals, but vulnerable to DNA alterations. The nondividing condensed spermatids are resistant to direct effects of toxic chemicals on development and function, but sensitive to point mutations and chromosomal breakage. Detection of each alteration might require a different biologic marker.

Markers in Use

In any given region of a seminiferous tubule, germ cells are differentiating. That combination of phenomena creates a complex histologic appearance at the light microscopic level, where adjacent cross sections through seminiferous tubules generally appear quite different. Clermont and coworkers (Leblond and Clermont, 1952; Heller and Clermont, 1964) showed that distinct cellular associations or stages exist and that their number depends on the species—e.g., 6 in humans and 14 in rats. The complex histologic cell association pattern allows trained observers to evaluate subjectively whether external factors have specific effects on germ cells in different steps of differentiation. This constitutes a relatively simple marker, which is particularly applicable to studies with experimental animals, whose testicular cytoarchitecture can be well preserved with large tissue samples and in which enough spermatogenesis remains to allow staging of species-specific cellular associations. Studies with human biopsy specimens are complicated by the small amount of tissue available, by artifacts induced by the biopsy procedure, and by the fact that a cross section of a human seminiferous tubule usually contains more than one stage of the cycle of the germinal epithelium (Heller and Clermont, 1964).

A simple quantitative approach is to determine the percentage of seminiferous tubules with mature spermatids lining the lumen versus the percentage of tubules without spermatids lining the lumen. That approach assumes that a chemical simply does or does not have a toxic effect on a tubule and, therefore, a clear judgment can be made as to which category a particular tubule belongs to. Alternatively, it is possible to measure the minor diameter of 15 seminiferous tubules, which diminish with the loss of germ cells (Courot, 1964); the advantage of this quantification is that it is sensitive, easy to measure, and provides a spectrum of values from a maximal to a minimal diameter of seminiferous tubules. The disadvantages of both approaches are that they are nonspecific and relatively insensitive and require autopsy or biopsy specimens. Again, great care must be taken to prevent artifacts associated with fixing, embedding, and sectioning the testicular tissue (Amann, 1981).

It is possible to count stem cells in histologic sections of rodent testes (Oakberg, 1978). But the technique is subjective and laborious, because stem cells are difficult to identify and are few. Alternatively, a functional test for stem cell renewal can be used.

Methods to measure the effect of toxicants on germ cells in the testis have been described and reviewed (Amann, 1970a; Berndtson, 1977; Amann, 1981). These methods are applicable to humans and experimental animals, provided that the testicular specimens can be fixed appropriately and biopsy artifacts can be prevented. One widely used method involves counting germ cells in a fixed number of tubule cross sections of one cellular association (Amann, 1970b; Berndtson, 1977). The germ cell numbers are generally but not always expressed per Sertoli cell nucleolus, to correct for seminiferous tubule shrinkage caused by histologic processing and experimental treatment. The method has proved sensitive in assessing the effect of antimitotic agents on spermatogenesis (Amann, 1981). However, the numbers generated are relative, rather than absolute; and the results might be nonspecific, because numerous agents can cause a morphologically identical pattern of response (Russell et al., 1981). In a simplified version of the procedure, the mean number of spermatids per tubule from a testicular biopsy has been used to predict sperm count in humans (Silber and Rodriguez-Rigau, 1981).

Another method applies stereologic principles (Van Dop et al., 1980a,b; Wing and Christensen, 1982; Jones and Berndtson, 1986) to obtain quantitative information on the diameter and volume of the seminiferous epithelium and lumen and to learn the absolute, rather than relative, numbers of Sertoli and germ cells (from preleptotene primary spermatocytes to step 10 spermatids). The method allows an experimenter to determine precisely the effect of a treatment on the number of cell types in the germinal epithelium. It has disadvantages: it is invasive, labor-intensive, and subject to numerous artifacts of tissue preparation. To our knowledge, the latter method has not been used to test the effect of xenobiotic chemicals on the seminiferous tubular epithelium in animals or humans.

An alternative to direct counting of stem cells is to count the stem-dependent cells after a toxic insult. One way to do that is to count the number of repopulating and nonrepopulating seminiferous tubular cross sections in a series of experimental animals killed at different times after treatment with a toxic chemical (Meistrich, 1986). It is assumed that each repopulating cross section of seminiferous tubules results from the presence of at least one surviving stem cell and that nonrepopulating cross sections result from the absence of stem cells. In a second assay used in mice, sperm heads in the testis 50 days after exposure to a toxic insult are counted (Meistrich, 1986). Both methods are simple and rapid. But both are invasive and therefore not readily applicable to humans; and repopulating tubules cannot be counted in low-dose situations, because extensive killing of stem cells is required for empty tubular cross sections to be produced.

One method of counting spermatid nuclei in homogenates of testicular parenchyma is based on the fact that late spermatid nucleoprotein becomes highly condensed and therefore resistant to homogenization (Amann and Lambiase, 1969). It is simple to use and its results are highly correlated with daily sperm output in rabbits (Amann, 1970b; Amann, 1981; Berndtson, 1977). It is also applicable to humans (Amann and Howards, 1980; Johnson et al., 1980a, 1984b). The disadvantage of the technique is that it is invasive (or requires autopsy specimens) and therefore has only limited human application as a biologic marker.

The amount of testicular $LDHC_4$, a germ cell-specific isozyme of lactate dehydrogenase, is proportional to the numbers of meiotic and postmeiotic germ cells in testes of mice and can be used to estimate the survival of spermatogonial stem cells after treatment with toxicants (Meistrich, 1982). The method is indirect, more difficult than measuring spermatid numbers in testicular homogenates, and invasive. But, it holds promise, because the $LDHC_4$:sperm ratio in seminal plasma of human males might serve as an indicator of the function of the seminiferous epithelium (Eliasson and Virji, 1985; Virji, 1985. The approach should be investigated thoroughly, not only with $LDHC_4$, but with other proteins specific for meiotic and postmeiotic germ cells.

Severe damage to the germinal epithelium in the testes of many species results in increased serum FSH concentrations (McCullagh, 1932; Rich and De Kretser, 1977;

de Jong, 1979), in part because a Sertoli cell product (inhibin), which regulates FSH secretion, is produced in low amounts in azoospermic animals and in high amounts in normospermic animals. Major insults to spermatogenesis can be monitored indirectly by measuring FSH concentration in peripheral blood. The principal advantage of this biologic marker is that it can be measured in peripheral blood samples, which are easy to collect repetitively. The disadvantages are that the inverse relationship between FSH concentration in serum and germinal epithelium damage are *not* tightly coupled; FSH production is episodic, so the results are variable. That renders the marker relatively insensitive to changes in spermatogenesis, despite the sensitivity of the radioimmunoassay. In addition, very low sperm counts are required for FSH increase to become evident. Finally, gonadal steroids (testosterone and estradiol) can account for selective FSH increase when testosterone production is low (Sherins et al., 1982). Consequently, FSH concentration might not always reflect *only* inhibin production. The marker is nonspecific, in that any toxic insult that depletes germ cells causes an increase in FSH production and testosterone and estradiol also partially control FSH production.

Markers Requiring Research and Development

The application of molecular biology to the study of mammalian testicular differentiation is providing investigators with insights into many of the molecular mechanisms that regulate male germ cell formation. Gene expression during spermatogenesis is temporally and spatially regulated with precision; many macromolecules and organelles are synthesized in specific cell types during the continuum of testicular cell differentiation (Hecht, 1987). Although variants of ubiquitous enzymes and structural proteins are expressed in many organs, the testis appears to be an especially rich source of isozymes (Goldberg, 1977). Presumably because of the specialized requirements for producing a spermatozoon, unique testicular isozymes that code for proteins such as lactate dehydrogenase, phosphoglycerate kinase, and cytochrome c—have evolved (Fig. 4-3). In addition to the testicular isozymes, many structural proteins of the maturing spermatid and the spermatozoon have been identified.

Results of biochemical analyses of the differentiating haploid germ cells that transform into the highly polarized spermatozoon suggest that the definition of all the possible sperm-specific molecu-

lar markers has only begun. As a result of such studies and the well-characterized sequence of events leading to the formation of spermatozoa, efforts to monitor the effects of toxicants on male germ cells can be based on substantial knowledge and use the numerous DNA probes already available for the mammalian testis to study the mechanism of toxic action of individual chemicals on spermatogenesis.

Most DNA cloning efforts during spermatogenesis have been directed toward two

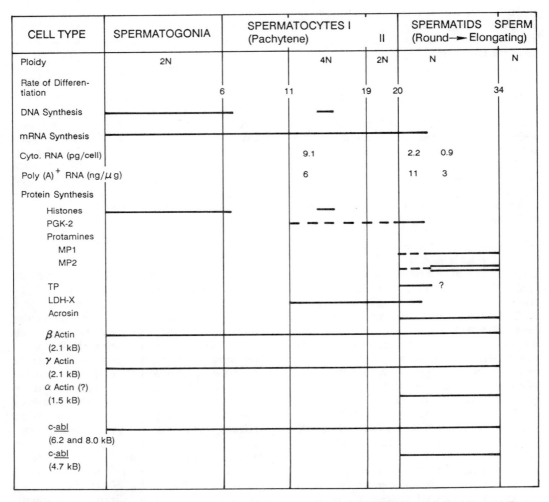

FIGURE 4-3 Periods of active synthesis of DNA, RNA, and proteins diagrammed for various cell types. Source: Hecht, 1987b.

intervals of spermatogenesis: meiosis and spermiogenesis (Kleene et al., 1983; Dudley et al., 1984; Fujimoto et al., 1984). That is because the meiotic pachytene spermatocytes and the haploid round spermatids represent two critical periods for gene expression during male germ cell development. Moreover, highly enriched populations of pachytene spermatocytes and round spermatids can be obtained readily with cell separation techniques, because of their marked size differences and abundance in the sexually mature testis (Romrell et al., 1976; Meistrich, 1977; Hecht, 1987b).

Almost all DNA probes for genes ex pressed in testis cells have been isolated from animals, such as mice, rats, and bulls. The origin of these probes will not pose a problem for application to human DNA. Because DNA probes for homologous genes in humans and other vertebrates share substantial sequence homology, the animal probes can be used to isolate equivalent human sequences from available human cDNA or genomic DNA libraries, and investigators committed to human toxicology studies should seriously consider the direct isolation and characterization of human DNA probes. For cases in which both rodent and human probes are available, the animal studies can be used to validate the DNA probe marker for human use and, more important, to reveal the mechanisms by which specific chemicals interact with the genome.

Meiotic DNA Probes. Lactate dehydrogenase C_4 ($LDHC_4$) is one of the best-characterized proteins in the testis (Goldberg, 1977). It appears to be testis-specific and is synthesized initially during meiosis and in decreasing amounts during early spermiogenesis (Meistrich, 1977). $LDHC_4$ mRNA makes up as much as 0.18% of total functional mRNA in mouse pachytene spermatocytes (Wieben, 1981). $LDHC_4$ is also present on the surface of mature spermatozoa, so it has been used extensively in immunocontraceptive studies (Wheat et al., 1985). The protein sequence of rodent LDHC4 has been known for some time (Pan et al., 1983), but only recently has a cDNA probe for human $LDHC_4$ been isolated (Millan et al., 1987).

Two distinct forms of phosphoglycerate kinase (PGK) have been characterized in mammals. PGK-1 is an X-linked gene that is expressed in somatic cells, whereas the autosomally derived isozyme, PGK-2, is specific to the testis (Kramer and Erickson, 1981). Although PGK-2 is synthesized during spermiogenesis, the gene appears to be initially transcribed during meiosis, with increased synthesis of PGK-2 mRNA in spermatids (Gold et al., 1983). Recent elegant studies of the human PGK multigene family have produced detailed sequence knowledge and DNA probes for these important enzymes (Michelson et al., 1985).

Cytochrome c, the electron-transport protein from the mitochondrial respiratory chain, exists in two forms in the testis (Goldberg et al., 1977). One variant, cytochrome c_t, is restricted to the testis; the other variant, cytochrome c_s, is presumably found in all tissues. Indirect immunofluorescence with monospecific antibodies first detects cytochrome c_t in the mitochondria of pachytene spermatocytes and in later stages of spermatogenesis, whereas cytochrome c_s is found in the mitochondria of interstitial cells, Leydig cells, and spermatogonia. Sequence analysis of the two mouse cytochrome c molecules has revealed that cytochrome c_t differs from cytochrome c_s in 13 amino acid residues (Hennig, 1975). Although DNA probes exist only for the c_s variant of cytochrome c, knowledge of the sequence of the testicular form of cytochrome c would allow appropriate oligonucleotide DNA probes to be prepared. Such oligonucleotides would facilitate the isolation of cDNA or genomic DNA probes for human cytochrome c_t.

During spermatogenesis, a dramatic reorganization of the germ cell nucleus occurs. The transformation ultimately produces a sperm nucleus with highly compacted DNA and accompanied by the replacement of histones with a group of transiently associated nuclear proteins and finally with protamines (Hecht, in press). In addition to the standard complement of histone molecules found in mammalian cells, the meiotic pachytene spermatocyte contains several additional histone variants believed to be peculiar to the testis. DNA sequence analysis of one testis-specific histone variant, Hlt, has revealed it to be a unique gene product, and not a posttranslational modification of an existing histone (Cole et al., 1986). The availability of a specific DNA probe for one of these meiotic histones of the rat provides a means to obtain the equivalent probe for the human male meiotic histone gene. Probes for other testis-specific histones can also be isolated.

Postmeiotic DNA Probes. The predominant proteins in mammalian spermatozoa are the protamines, a group of small arginine-rich DNA-binding proteins that aid in nuclear DNA compaction during spermiogenesis (Hecht, 1987c). With the exception of a few species, most mammalian spermatozoa have been reported to contain one type of protamine. In the mouse, two protamine variants, MP1 and MP2, have been identified by DNA sequence analysis of isolated cDNA clones (Kleene et al., 1985; Yelick et al., 1987). Protein sequence studies have identified similar P1 and P2 human protamine variants (McKay et al., 1985; Ammer et al., 1986; McKay et al., 1986). Although the P1 and P2 protamines differ substantially in size and sequence, in the mouse they are closely linked on chromosome 16 and are temporally and translationally regulated during spermiogenesis (Hecht et al., 1986a). The human protamine genes probably are also chromosomally linked. The P1 human protamine was recently shown to be on human chromosome 16 (R.H. Reeves, Johns Hopkins University, and N.B. Hecht, Tufts University, unpublished observation, 1987).

Northern blots of RNA from prepubertal testes and from isolated meiotic and postmeiotic testicular cell types have revealed that the protamines are expressed solely during the haploid interval of spermatogenesis (Kleene et al., 1983, 1984; Hecht et al., 1986a,b). Moreover, changes in length of protamine mRNAs during spermatogenesis allow the protamine probes to be used as molecular markers to evaluate the extent of spermiogenesis in wild-type, mutant, or chemically induced sterile animals (Kleene et al., 1984). For instance, if the MP1-cDNA probes are used, no MP1-mRNA is detected in testicular ex-

tracts of prepubertal mice up to 20 days old (a time when spermatogenesis has advanced to meiosis), whereas a 580-nucleotide form of MP1-mRNA is present in the testes of 22-day-old mice (early spermatids are present by day 22) and a heterogeneous population of 580- and 450-nucleotide MP1-mRNAs is present in the testes of sexually mature animals (Hecht et al., 1986b). Results of cell-separation studies confirm that no MP1-mRNA is present in pachytene spermatocytes, a 580-nucleotide MP1-mRNA is found in round spermatids, and elongating spermatids contain an additional 450-nucleotide mRNA (Kleene et al., 1984). These mRNA length changes result from a partial deadenylation of the protamine mRNAs that takes place when they move from the ribonucleoprotein particle fraction of the cytoplasm (in round spermatids) to polysomes (in elongating spermatids). Similar size changes occur in the MP2 mouse protamine and in P1 and P2 rat and hamster protamine mRNAs (Bower et al., in press).

The protamine genes appear to be excellent candidates to serve as probes to monitor genomic defects induced during spermiogenesis. They are two single-copy genes that express abundant postmeiotic testicular mRNAs essential for sperm function. Moreover, the cDNA probes show much homology to the DNA and RNA of many other vertebrates, including humans. Several laboratories are seeking to isolate human probes for these male-specific DNA-binding proteins.

In mammals, histones are not directly replaced by protamines, but by a presumably heterogeneous group of basic proteins called testis-specific proteins (TP) (Hecht, 1987c). TPs are associated with the spermatid nucleus during its transition from its nucleosome-like structure to the smooth branching fibril of the spermatozoan nucleus. TPs are replaced by protamine during spermiogenesis. Recently, cDNA probes for the mouse and rat TPs have been identified (Hecht et al., 1986b; M.A. Heidaran and W.S. Kistler, University of South Carolina, personal communication). Phylogenetic studies have indicated a strong sequence conservation of TPs in rodents and humans. The identical pattern of expression of mouse TP and MP1 and MP2 suggests that these three genes are coordinately and temporally regulated in the postmeiotic testicular cells and could be used together to monitor postmeiotic gene expression.

In mammals, actins are encoded by a multigene family that expresses at least six distinct but closely related forms of actin. In addition to the general role that actin plays in cell motility and division, secretion, organelle movement, and maintenance of cellular cytoarchitecture, testicular actins are likely to be involved in chromosomal movement during meiosis, in shaping specific nuclear structures during spermiogenesis, and in spermatozoal function (Hecht et al., 1984). Although the mRNAs coding for the cytoplasmic β and γ actin isotypes have been detected in all testicular cell types throughout spermatogenesis, mRNA that encodes an additional actin variant is first detected during spermiogenesis (Waters et al., 1985). It should be possible to obtain several distinct DNA probes for actin isotypes that are expressed constitutively during spermatogenesis and of other actin isotypes that are expressed temporally in specific stages or cell types.

Microtubules consist of heterodimers of α and β tubulin. The α and β tubulin subunits are distinct sequences, each encoded by multigene families. In the testis, the tubulins are involved in mitotic and meiotic divisions, in changes in cell shape and structure, in the species-specific shaping of the sperm nucleus, and in the synthesis of the axoneme of sperm tails. Results of protein gel electrophoresis and DNA cloning studies have suggested that multiple isoforms of α and β tubulin are expressed during spermatogenesis (Hecht et al., 1984). The availability of cDNAs for a number of mouse testicular α tubulins and a detailed study of the expression of the β tubulin multigene family will provide a set of useful probes to monitor the differential expression of several cytoplasmic structural genes during testicular germ cell development.

Results of in vivo and in vitro postmeiotic protein synthesis studies and the many morphologic changes in cell shape and structure that occur during spermiogenesis have indicated that many additional unique macromolecules are synthesized during spermiogenesis. Continuing studies in many laboratories suggest that such proteins as acrosin, hyaluronidase, a sperm-specific enolase, and sperm tail proteins (e.g., the dyneins and outer dense-fiber proteins) will provide additional sources of stage-specific DNA probes for this critical interval of spermatogenesis. DNA probes are also available for some proto-oncogenes, such as c-*abl* and c-*myc*, that are differentially expressed during spermatogenesis. Although c-*abl* mRNA is present in premeiotic, meiotic, and postmeiotic cell types, a novel c-*abl* mRNA of distinct size is first detected in postmeiotic cells (Ponzetto and Wolgemuth, 1985). Because of its unique size, a probe specific to this shortened c-*abl* transcript could be prepared. In contrast, the proto-oncogene c-*myc* appears not to be expressed in testicular germ cells (Stewart et al., 1984).

Stem Cell DNA Probes. Spermatogonia make up only a few percent of the cells found in the sexually mature mammalian testis. It has therefore been difficult to work biochemically with this cell type, and no DNA probes peculiar to spermatogonia have yet been isolated for this critical stage of spermatogenesis. Because genetic alterations in stem cell DNA will produce persistent heritable defects, a major effort needs to be commenced to obtain an armamentarium of DNA probes specific to animal and human testicular stem cells. Recent improvements in testicular cell separation methods make this possible, in that highly enriched populations of several types of spermatogonia can be obtained from the prepubertal testis. With poly(A)$^+$ RNA isolated from enriched populations of spermatogonia, radiolabeled cDNAs can be prepared and a differential-hybridization approach similar to that used previously to obtain postmeiotic cDNAs can be conducted to isolate stem cell-specific cDNAs (Kleene et al., 1983). In brief, a total testicular cDNA library or a cDNA library enriched with spermatogonial cDNAs would be differentially hybridized with radiolabeled cDNAs prepared from spermatogonial, meiotic, or postmeiotic cell types. The cDNAs that appear to be preferentially expressed in spermatogonia would be isolated and their temporal appearance confirmed by the early appearance of RNA in the testes of prepubertal-staged mice (Kleene et al., 1983). DNA sequence analysis could be used to help to identify the proteins coded for by the stem cell cDNAs. One possible candidate DNA probe for a protein expressed in spermatogonia would be the DNA that codes for the H2A histone stem cell variant found in mouse embryonic spermatogenic cells (Gizang-Ginsberg and Wolgemuth, 1985).

Clearly, these macromolecular probes at several cell stages represent an excellent collection of biologic markers for the dynamic process of spermatogenesis. They can be used to assess the effect of a toxicant on the genome that directs specific biochemical events in spermatogenesis. The approach would be particularly useful in studies of the mechanism of action of toxicants in spermatogenesis. Disadvantages include the invasiveness of the present techniques and their requirement for autopsy or biopsy specimens.

Development of these probes holds out the possibility of analyzing potential toxic effects at the DNA level in ejaculated spermatozoa. However, that is unlikely to occur soon. The primary limitations on the application of DNA probe technology to evaluate genomic DNA alterations in spermatozoa are three: DNA probes are available for only a very small portion of the genome; current procedures severely limit the number of probes that can be assayed at one time, thereby restricting the percentage of the genome examined in each analysis; and the DNA from single cells, such as spermatozoa, cannot be analyzed. Theoretically, the limitations can be overcome. Procedures can be developed to monitor large fragments of genomic DNA with a battery of DNA probes that cover vast regions of the human genome. Advances in fluorescence detection of DNA combined with computer imaging of samples will aid in the analysis of DNA from single cells.

5

Biologic Markers of Epididymal Structure and Function

This chapter focuses on markers of epididymal function. Very few direct markers of human epididymal function are available. However, there is a useful array of indirect markers. The chapter discusses markers of tissue structure; markers of structural, biochemical, membrane compositional, and functional changes of the maturing spermatozoa; biochemical markers of the luminal fluid; and histologic and biochemical markers of the epithelium. Many of the assessments must be performed on biopsied or autopsied tissue or in animals. These studies are essential to the understanding of basic mechanisms of action of toxicants.

The epididymis is a single highly convoluted duct whose length varies from 3-4 meters in man to 80 meters in horses (Maneely, 1959). The duct begins where the efferent ducts—the ductuli efferentes, numbering from 4 to 20, depending on species (Nistal Martín de Serrano and Paniagua Gómez-Alvarez, 1984; Hemeida et al., 1978)—come together. The epididymal duct continues as a straight tube, the vas deferens, that is surrounded by a thick muscular layer. The vas deferens connects with the urethra, which empties outside the body. The epididymis is usually divided into three gross anatomic segments: head (caput), body (corpus), and tail (cauda). An initial segment lies between the efferent ducts and the remainder of

the caput epididymis, has a characteristic histologic appearance and function and is also often identified (Benoit, 1926; Robaire and Hermo, 1987).

Interest in the epididymis has grown over the past 2 decades, since the demonstration that it was in this tissue that spermatozoa mature enough to become able to fertilize eggs (Bedford, 1967; Orgebin-Crist, 1967a). On leaving the testis, spermatozoa have a light microscopic appearance similar to that of spermatozoa in semen, but are incapable of fertilizing eggs. Several reviews on various facets of epididymal structure and function have appeared in the past 15 years (Hamilton, 1972; Bedford, 1975; Hamilton, 1975; Neaves, 1975; Orgebin-Crist et al., 1975; Turner, 1979; Hinton, 1980; Courot, 1981; Orgebin-Crist, 1981; Brooks, 1982; Glover, 1982; Howards, 1983; Orgebin-Crist, 1984; Cooper, 1986; Amann, 1987; Robaire and Hermo, 1987).

To monitor whether a toxicant has compromised male fertility, it is essential to determine not only whether the normal physiologic functions of the epididymis are still being performed, but also whether the toxicant has so altered the spermatozoa while they were in the epididymis that they cannot fertilize eggs or can fertilize eggs but produce only nonviable or abnormal offspring.

A discussion of potentially useful markers of epididymal function that are available, are under development, or ought to be developed is presented below. We begin by discussing markers of the tissue as a whole and then focus on markers that reflect changes in maturing spermatozoa, the implications of the makeup of the fluid in the luminal compartment that bathes maturing spermatozoa, and the activities of the epithelial compartment of the epididymis.

MARKERS OF EPIDIDYMAL TISSUE

Few noninvasive direct determinations are used to assess whether a substance has deleteriously affected the epididymis. The tissue can be palpated and the size, consistency, and shape assessed. Such simple markers are sometimes of value in diagnosing tubal obstruction or epididymitis, but they are highly subjective and do not lend themselves to assessment of variability, limits of detection, and so forth. The more recently developed imaging techniques are only now being tested as tools in assessing the epididymis; these probably will yield a reliable index of such characteristics as size and shape.

If the tissues are available, the weight of the epididymis is, in fact, a simple and useful biologic marker. The epididymis is a tissue with two compartments: epithelium and lumen. The lumen is filled with fluid and spermatozoa and, under normal conditions in many mammals, makes up approximately 50% of the weight of the tissue (e.g., Robaire et al., 1977); small decreases in epididymal weight are usually associated with proportionately larger decreases in epididymal sperm reserves. Accessory sex tissues become markedly hypertrophic in the presence of large excesses of androgen, even in a castrated animal. However, the epididymis does not become hypertrophic. Thus, changes in epididymal weight can be a valuable index of the status of the tissue.

As with other hormone-dependent tissues, blood is the major source of regulatory substances that reach the epididymal epithelium. Measurement of serum concentrations of hormones, especially testosterone, is useful and routine in monitoring the biologic regulation of epididymal function. The epididymis contains receptors for a variety of hormones, e.g., testosterone, estradiol, prolactin, and vitamin D; but the value, with respect to epididymal function, of measuring the serum concentrations of the hormones as markers has not been established. The topography of the blood vessels entering and leaving the testis is complex and species-specific (Chubb and Desjardins, 1982). It is evident from numerous castration and hormone-replacement studies that blood vessels are a major route by which hormonal signals are received by the epididymis. Our knowledge of capillary flow and its impact on tissue function is sparse, but it is likely that we will soon be able to monitor blood flow in epididymal capillaries. Whether such measurement will become an important marker of epididymal function will depend on the ability of the capillary system to adapt to deleterious actions of toxicants.

Very little is known about epididymal lymphatic input and drainage. As the nature of the blood/epididymis barrier becomes better understood and the use of immunosuppressive agents (which alter the immune system and lymph content) grows, we will increase our understanding of this system, although no lymphatic markers regarding epididymal function are likely to be developed soon.

Neuronal input is responsible for the basal epididymal contractility, which is at least partially responsible for the transport of sperm and fluid down the duct system from the testis. Assessment of whether drugs that affect the autonomic nervous system can modulate epididymal contractility and hence the rate of sperm transit and the acquisition of fertilizing ability in vivo has not been attempted. It would be surprising if they could not; but such drugs affect an array of other systems, so their effects on the epididymis might be masked. Although the exact epididymal effects of modulating neuronal activity would be of interest, measurements of neuronally mediated epididymal contractility and of the factors that regulate it will probably not become selective markers.

Some epididymal functions apparently are regulated in a paracrine manner, i.e., by factors from the testis that enter the epididymal lumen directly (Robaire and Hermo, 1987). Factors secreted by the testis into the efferent ducts that directly modulate epididymal function are probably of Sertoli cell origin (Scheer and Robaire, 1980; Robaire and Zirkin, 1981). Their identities are not resolved, but androgen-binding protein has been proposed as a likely candidate. Determining the concentrations of various factors in rete testis fluid is difficult for many species, because it is technically difficult to obtain the fluid, only small volumes can be obtained, and the procedure cannot be repeated in most small animals and cannot be done in humans. However, it is apparent that some key regulators of epididymal function, which are also probably useful epididymal markers, are to be found in this fluid. The presence and concentrations of such compounds in semen might reflect not only testicular but also epididymal function, and research to identify them and to develop means of monitoring them should be strongly encouraged, although it will certainly take several years for such markers to become readily available.

CHANGES IN MATURING SPERMATOZOA

A number of morphologic and biochemical changes reported to occur in spermatozoa during transit through the epididymis and vas deferens have been reviewed extensively (Bedford, 1975; Bedford, 1979; Olson and Orgebin-Crist, 1982; Eddy et al., 1985; Cooper, 1986; Robaire and Hermo, 1987).

Structural Changes

The most consistent morphologic change that takes place in spermatozoa during ductal transit is the migration of the cytoplasmic droplet from the neck region of the flagellum to the end of the midpiece of the sperm (mitochondrial sheath). The change has been noted in snakes and birds (Bedford, 1979), in many mammals (Branton

and Salisbury, 1947; Phillips, 1975; Kaplan et al., 1984), and in humans (Hafez and Prasad, 1976). In rats, lysosomal and other degradative enzymes are present in the droplet (Dott and Dingle, 1968; M.L. Roberts et al., 1976); however, the functional importance of the enzymes and of the droplet remains to be resolved.

The percentage of spermatozoa that retain cytoplasmic droplets or the number of cytoplasmic droplets per spermatozoon might well be a useful indication of the state of maturation of spermatozoa and might also reflect the activity of clear cells in some species. It might be a sensitive measure, but there will probably be large variability in it. This marker has not yet been validated as a good correlate of epididymal function, but it should be fairly straightforward to do so. Such a measure could be particularly useful in selected cases in which clear cell function or droplet shedding is impaired.

Spermatozoa undergo a change in acrosomal size, shape, and internal structure as they pass through the duct system. This has been demonstrated in many species (Fawcett and Hollenberg, 1963; Fawcett and Phillips, 1969; Bedford and Nicander, 1971; Jones et al., 1974; Bedford and Millar, 1978). In animals in which acrosomal shape clearly changes during epididymal transit, observation of spermatozoa should reveal whether exposure to a given chemical has altered this marker of spermatozoal maturation. Further studies are needed with primates to resolve whether such changes are correlated with spermatozoal quality, i.e., fertilizing ability.

Biochemical Changes

An array of biochemical alterations in spermatozoa as they traverse the ducts has been reported (Bedford, 1975; Bedford and Cooper, 1978; Olson and Orgebin-Crist, 1982). Only the characteristics with the greatest potential for use as markers are discussed here. In spite of numerous detailed studies on how sperm change during epididymal transit, it is still not possible to dissociate factors that cause the maturation of spermatozoa from factors that result from maturation or are incidental to it. Hence, any of the markers suggested above might correlate well with changes in spermatozoa that alter their fertilizing potential or their ability to produce normal offspring under a given set of conditions without having the desired predictive value.

An increase in the relative number of disulfide bonds in the nuclei of spermatozoa as they reach the cauda epididymis has been observed in many species (Calvin and Bedford, 1971; Bedford et al., 1973; Bedford, 1975; Johnson et al., 1980b). The increase is associated with a more condensed (cross-linked) spermatozoal nucleus—spermatozoa taken from the cauda epididymis are harder to decondense than those taken from the initial segment (Zirkin et al., 1985b). The state of chromatin condensation can be assessed by determining how long it takes for detergent-treated nuclei to dissolve in the presence of a given concentration of a disulfide reducing agent. If chemicals can affect spermatozoa by preventing the proper chromatin condensation that takes place while spermatozoa are in the epididymis, then measuring such rates might provide a simple and useful biologic marker.

The anionic charge on spermatozoa increases as they reach the cauda epididymis (Bedford, 1975; Toowicharanont and Chulavatnatol, 1983); the increase is probably acquired during passage through the corpus epididymis (Fain-Maurel et al., 1983). The exact cause of the change in charge has not been resolved, but it is most likely due to changes in glycoprotein composition of the sperm plasma membrane that take place during epididymal transit.

The membrane proteins of spermatozoa obtained from different segments of the epididymis present a markedly different pattern after separation on polyacrylamide gel electrophoresis (Jones et al., 1981; Chulavatnatol et al., 1982; Brown et al., 1983; Dacheux and Voglmayr, 1983; Jones et al., 1983; Brooks and Tiver, 1984). During epididymal transit, the number of concanavalin A binding sites on the sperm membrane decreases (Lewin et al., 1979; Nicolson and Yanagimachi, 1979; Olson and Orgebin-Crist, 1982), the ability of spermatozoa to activate comple-

ment decreases (Witkin et al., 1983), and the activity of protein methyltransferase in spermatozoa decreases dramatically (Gagnon et al., 1984). Gonzalez Echeverria et al. (1984) have noted that the addition of a protein extract from hamster epididymis could increase the fertilizing ability of hamster spermatozoa. Blaquier et al. (1987) found that polyclonal antibodies to human epididymal sperm proteins bound selectively in the acrosomal cap region in fertile men, but that a large percentage of sperm bound these antibodies in a nonspecific manner in infertile men. Such data lead to the proposal that several highly specific sperm-surface protein markers of epididymal origin might be altered, not only in cases of infertility, but also in response to various drugs.

Antigenic determinants on the surface of spermatozoa change markedly as sperm traverse the epididymis (Eddy et al., 1985). Monoclonal antibodies to mouse "sperm-maturation antigens" that arise in mice during epididymal passage have been characterized, and at least one (SM4) has been shown to appear on sperm only while they are in the corpus epididymis (Eddy et al., 1985; Vernon et al., 1985). The appearance of such determinants might be due not only to the addition of a protein to the surface of spermatozoa, but also to the removal of proteins or to the enzyme-mediated exposure of pre-existing sites. This new approach to the characterization of changes that take place in spermatozoa during epididymal transit might provide some useful biologic markers, if a clear linkage between the state of maturation of a spermatozoon and the presence or absence of a given epitope can be established.

The lipid composition of spermatozoa (Dacheux, 1977; Evans and Setchell, 1979), particularly that of their plasma membranes (Nikolopoulou et al., 1985), changes dramatically during epididymal transit: the amount of cholesterol decreases, and the amounts of desmosterol and cholesterol sulfate increase (Legault et al., 1979; Inskeep and Hammerstedt, 1982; Nikolopoulou et al., 1985); the relative distribution of the different polar lipids also changes, some increasing while others decrease (Nikolopoulou et al., 1985). The change in lipid composition has been proposed as the cause of the increased sensitivity of caudal spermatozoa to cold (Nikolopoulou et al., 1985). The relationship between the state of maturation of spermatozoa and the lipid makeup of their cell membrane has not yet been determined, but there might well be a significant relationship, because of the constraints put on membrane fluidity in sperm-egg interaction. If such a relationship were established, measurements of lipid composition or membrane fluidity of spermatozoa might become useful markers.

The changes in metabolic activity of spermatozoa during epididymal transit are numerous (Voglmayr, 1975; Brooks, 1981). Increases in the glycolytic and respiratory activity of spermatozoa during epididymal transit have been reported (Dacheux et al., 1979; Voglmayr and White, 1979). The increase in cAMP associated with spermatozoa as they mature (Del Rio and Raisman, 1978; Amann et al., 1982) has been attributed to both an increase in the synthesis of this second messenger and a decrease in its hydrolysis (Purvis et al., 1982). There are so many indices of metabolic activity that it is not clear which, if any, would be useful markers of sperm maturation (Cooper et al., 1988). Research in this subject could yield important results, but for the present it must be viewed as a "fishing expedition."

Functional Changes

Content of Epididymal Spermatozoa

One of the most powerful available biologic markers of epididymal function is the number of spermatozoa in different regions of the epididymis (Amann, 1981). Because the heads of spermatozoa are very tightly condensed and resistant to homogenization, it is possible to homogenize a segment of epididymis, filter the homogenate, and count the number of condensed sperm heads with a hemacytometer (an instrument usually used for counting blood cells). The method requires the destruction of the tissue, but it is sensitive and

has the same degree of variability as any other method that depends on hemacytometric determinations; because of interference by debris, it has not been possible to adapt electronic cell-counting methods to this application. Results from hemacytometric determinations not only give an accurate index of sperm content in different epididymal regions where various functions take place (Robaire et al., 1977; Trasler et al., 1988), but also can be used to indicate the sperm reserve within a tissue (Amann, 1981).

Transport of Spermatozoa

In spite of the very large range in sperm production rate in different species, there is great consistency in how long it takes for spermatozoa to traverse the epididymis: approximately 10 days (see Table 4-1) (Robaire and Hermo, 1987). However, the transit time for spermatozoa in the human epididymis is variable (Rowley et al., 1970; Amann, 1981; Orgebin-Crist and Olson, 1984). The time spent by spermatozoa in the cauda epididymis is longer and more variable than that spent in any other segment. That is presumably because sperm transit in the caput and corpus epididymis is independent of ejaculatory frequency, whereas that in the cauda epididymis depends on this frequency (Amann and Almquist, 1962; Swierstra, 1971; Kirton et al., 1967). The rate of luminal flow in different segments of the rat epididymis decreases from 210 mm/h in the initial segment to 32 mm/h in the distal caput and 12 mm/h in the cauda epididymis and vas deferens (Jaakkola, 1983).

The mechanisms responsible for driving the luminal contents through the efferent ducts and epididymis include hydrostatic pressure (Johnson and Howards, 1976; Pholpramool et al., 1984), muscular contractions (Jaakkola and Talo, 1983), and ciliary action (Markkula-Viitanen et al., 1979).

Norepinephrine (Pholpramool and Triphrom, 1984), acetylcholine (Pholpramool and Triphrom, 1984), and vasopressin (Jaakkola and Talo, 1981) have been proposed as regulators of the muscular contractions and ciliary action. In addition, epididymal tubular contraction can be regulated by prostaglandins and by drugs that affect their synthesis (Cosentino et al., 1984). The relative contributions of hydrostatic pressure, electric activity, and ciliary action in driving luminal flow are still unresolved; it is unlikely that any factor accounts fully for luminal flow in all mammals.

Assessment of transit time of spermatozoa through the epididymis and the rate of luminal flow are unlikely to become practical markers for studies in humans. However, numerous chemicals can probably affect transit time through the epididymis by altering neuronal activity or affecting one of the other regulatory mechanisms. Studies are needed to establish which drugs can modulate such transport and whether the effects will alter the ability of spermatozoa to fertilize eggs or to produce normal, viable offspring.

Acquisition of Fertilizing Ability

In mammals, spermatozoa leaving the testis do not have the ability to fertilize eggs, whereas those in the cauda epididymis have acquired this function. It took several decades to resolve whether the role of the epididymis in that function was passive or active, but several elegant studies by Orgebin-Crist and her colleagues (Orgebin-Crist et al., 1975) and by Bedford (1966) clearly established that the epididymis was actively involved in converting immature to mature (i.e., fertile) spermatozoa. The acquisition of the potential to fertilize eggs and, separately, produce viable offspring appears not to be a simple on-off situation (Nishikawa and Waida, 1952; Orgebin-Crist, 1967b; Orgebin-Crist, 1968; Blandau and Rumery, 1964; Dyson and Orgebin-Crist, 1973; Frenkel et al., 1978; Fournier-Delpech et al., 1979, 1981). There is some species variation with respect to the site at which spermatozoa gain their fertilizing potential, but it is evident that passage through some part of the caput is essential and that no species relies on the entire length of the epididymis for acquiring fertilizing potential. (For reviews, see Orgebin-Crist, 1969; Orgebin-Crist

et al., 1975; Bedford, 1979; Turner, 1979; Courot, 1981; and Orgebin-Crist and Olson, 1984.)

Whether, and if so where, within the epididymis spermatozoa acquire not only the ability to fertilize eggs, but also to produce normal, viable offspring is the most important marker of epididymal function. To use such a marker, spermatozoa from different regions of the epididymis need to be removed and inseminated into females of known fertility, and the resulting progeny outcome must be analyzed. Such studies are expensive and tedious and require the use of large numbers of animals; however, they are the only proven means of determining whether a substance will selectively affect the site of maturation of spermatozoa, if there is any maturation at all. These tests need to be used more extensively to determine whether there are substances that alter the ability of the epididymis to mature sperm or the site where such maturation is acquired. The development of new simple approaches with great predictive value that will clearly establish whether failure of spermatozoa to mature has been caused by an epididymal defect has high priority.

Associated with the acquisition by spermatozoa of the ability to fertilize eggs is a gain in potential for motility. In studies in which different regions of the epididymis were ligated and spermatozoa were removed, it became apparent that spermatozoa could acquire the potential for motility without acquiring the ability to fertilize eggs (Bedford, 1967; Orgebin-Crist, 1967a; Cummins, 1976).

Except possibly in rabbits, the cauda fluid composition prevents movement of spermatozoa. The underlying mechanism of the acquisition of potential for motility is unknown. A number of factors have been proposed as regulators or mediators, including forward-motility protein (Hoskins et al., 1978; Acott et al., 1979), acidic epididymal glycoprotein (Pholpramool et al., 1983), albumin (Pholpramool et al., 1983), carnitine (Hinton et al., 1981; Inskeep and Hammerstedt, 1982), cAMP (Hoskins and Casillas, 1975; Amann et al., 1982), sperm-motility inhibiting factor (Turner and Giles, 1982), sperm-motility quiescence factor (Carr and Acott, 1984), and immobilin (Usselman and Cone, 1983).

Only in the past few years has quantitative objective assessment of spermatozoal motility become available. The difficulty in monitoring sperm motility lies in the variability in the degree and type of motion of a large number of spermatozoa. The method routinely used in most laboratories was the subjective determination of the percent of spermatozoa that were motile and the type of motility on a 0-4 or 0-10 scale, with 0 being immotile and with increasing numbers reflecting more progressive motility (shaking, circular, forward, forward progressive). The advent of high-speed videomicrography made it possible to record spermatozoal motility accurately. Such recordings can be analyzed with computer imaging techniques to provide accurate, objective assessment of the distribution of spermatozoal speed, extent and angle of circular movement, lateral head displacement, beat angle of the flagellum, and so on. These methods have been developed for human semen analysis (Ginsburg et al., 1988; Mahony et al., 1988) and are only beginning to be used to characterize the acquisition of sperm motility during epididymal transit (Working and Hurtt, 1987). Whether and how sperm motility can be altered by chemicals that affect the epididymis has yet to be determined. Various measures of sperm motility have the potential of being powerful markers of one of the major functions of the epididymis, but fulfilling the potential will require much more research, which is now technically feasible.

Storage of Spermatozoa

The major site for storage of spermatozoa in the mammalian duct system is the cauda epididymis. Although normal transit time of spermatozoa through the cauda is some 3-10 days (Robaire and Hermo, 1987), they can be stored in this tissue for periods of over 30 days (Orgebin-Crist et al., 1975). On storage in the cauda, a loss in fertilizing ability was found to occur before a loss in motility (Martin-Deleon et al., 1973; Cummins, 1976).

An important difference between humans and a number of other mammalian species is that other mammals have a high sperm production rate so that the number of stored spermatozoa available for ejaculation is 3-5 times greater than the daily sperm production rate and 2-3 times greater than that found in a "typical" ejaculate (Amann, 1981). In contrast, humans have a sperm production rate well below that of most other mammals and a sperm reserve available for ejaculation that is only about equal to the number of sperm in an ejaculate, whether the person has been at sexual rest or not.

To assess the ability of the cauda epididymis to store spermatozoa, one either would have to count the sperm in the tissue after homogenization (a method clearly limited to animal experimentation) or would have to obtain serial ejaculates (minutes to hours apart) to assess the size of the sperm reserve in the cauda epididymis. This marker of epididymal function might provide some useful information about the ability of the cauda epididymis to store spermatozoa, but it is not very practical and has not yet been used to monitor damage to the epididymis other than that caused by heat (Bedford, 1977; Bedford, 1978a,b).

EPIDIDYMAL LUMINAL FLUID

The epididymal lumen contains water, ions, small organic molecules, proteins and glycoproteins, spermatozoa, and other particulate matter of undefined origin. From the efferent ducts all the way through to the vas deferens, numerous changes take place in the makeup of this complex of substances. The only available means of monitoring changes in the composition of the luminal fluid, without removing or irreversibly damaging the tissue, is analysis of the epididymal contribution to semen. Such analysis has been used extensively to monitor several organic molecules and proteins as markers of epididymal function, but is of no value in assessing changes in ionic makeup along the epididymis. Micropuncture of epididymal tubules has provided valuable information, but does not allow for serial studies and cannot be used in humans (Hinton and

Howards, 1982). We therefore focus here on luminal markers that either can easily be measured in semen or are particularly closely related to the epididymis but need to be measured in isolated tissues.

There is a precipitous decrease (by nearly 100 mM) in the concentration of chloride between the efferent ducts and the caput epididymis, whereas a similar decrease in sodium is observed between the caput and cauda epididymis (Crabo, 1965; Levine and Marsh, 1971; Jenkins et al., 1980). The increase in potassium ion (more than 30 mM) does not account for the change in osmolarity, and it has therefore been proposed that the epididymis is involved in the secretion of organic ions (Levine and Marsh, 1971). Phosphorus, whose serum concentration is usually around 2 mM, becomes higher than 90 mM in the corpus epididymis; this high concentration is accounted for, in part, by the incorporation of phosphorus into glycerylphosphorylcholine and phosphocholine, which make up approximately 50 mM and 20 mM, respectively, leaving an inorganic phosphorus concentration of more than 10 mM in the cauda epididymis and vas deferens (Hinton and Setchell, 1980a). It would be of interest to determine whether the very high inorganic phosphorus concentration is maintained by the action of parathyroid hormone or vitamin D in this tissue.

Carnitine (Hinton and Setchell, 1980b), glycerylphosphorylcholine (Hinton and Setchell, 1980a), phosphocholine (Hinton, 1980), inositol (Hinton, et al.,1980), sialic acid (Arora et al., 1975), glycerol (Cooper and Brooks, 1981), and steroids (Ganjam and Amann, 1976; Turner et al., 1984) are dramatically concentrated in the lumen or change radically in concentration as one moves down the testicular duct system. The physiologic role of high concentrations of carnitine has not yet been resolved, but one of the proposed functions of carnitine is as a precursor to acetylcarnitine, which is used as a source of energy by spermatozoa or is used to promote the maturation of spermatozoa during epididymal transit (Casillas and Chaipayungpan, 1979). Why the epididymal lumen should accumulate specifically

inositol and no other sugars is most intriguing, in that no component of intermediary metabolism of either spermatozoa or epididymal epithelium seems to have a preference for inositol as a source of energy; such high concentrations are also unlikely to be required to mediate hormone action via the phosphatidyl-inositol phosphate system. Testosterone is the most abundant androgen entering the epididymis, and dihydrotestosterone becomes the major androgen in the caput epididymis, increasing by more than 200-fold between the rete testis and the caput epididymis in bulls and rats (Ganjam and Amann, 1976; Turner et al., 1984).

All the above chemicals have been measured in semen as potential indicators of epididymal function; the large collection of studies has not provided a clear answer as to their value as markers (Mann and Lutwak-Mann, 1981; Cooper et al., 1988). That stems in part from the fact that many of the clinical studies were poorly controlled and in part from the lack of basic physiologic knowledge about the epididymis, which would permit rational interpretation of the results. The possibility that chemicals or enzymes secreted by sex accessory tissues can alter the concentrations of substances secreted by the epididymis should also be kept in mind.

Some specific proteins, separable by electrophoresis, are present in the lumen of the epididymal duct system of mammals. Using micropuncture methods, Koskimies and Kormano (1975) and Turner (1979) demonstrated not only that the disk gel electrophoretic patterns of proteins of different segments of the epididymis and vas deferens differed from those of serum or rete testis fluid, but also that they differed from each other—i.e., there were gradual changes in electrophoretic pattern from the caput to the corpus epididymis and vas deferens.

The identities of most of the proteins have not yet been established. However, in the past few years, a few molecules found in the epididymal lumen with specific physicochemical characteristics or identifiable biological activity have been described. Albumin, α_2-macroglobulin, transferrin, and androgen-binding protein have been found in epididymal luminal fluid (Amann et al., 1973; Skinner and Griswold, 1982; Turner et al., 1984). Other proteins—such as acidic epididymal glycoprotein (Lea and French, 1981), dimeric acidic glycoprotein (Sylvester et al., 1984), forward-motility protein (Brandt et al., 1978), angiotensin-1-converting enzyme (Vanha-Perttula et al., 1985), and α-lactalbumin-like protein (Hamilton, 1981)—are also likely to be in the mammalian epididymal lumen, but evidence of their presence in the luminal compartment is indirect. Many of the proteins are beginning to be used as markers of epididymal function; no clear picture has yet emerged regarding their ability to reflect epididymal functions, but such results are likely to become available soon.

It is important to note that 90-95% of the fluid coming from the testes is resorbed in the efferent ducts and the initial segment of the epididymis (see, e.g., Robaire and Hermo, 1987). This water resorption will be a factor in the concentration measured for all the chemicals mentioned above.

EPIDIDYMAL EPITHELIAL FUNCTION

Histology

Presence and Relative Distribution of Various Cell Types

Detailed descriptions of the appearance, at the light and electron microscopic levels, of the epididymis in a number of animals and humans are available (Benoit, 1926; Hamilton, 1975; Connell and Donjacour, 1985), and have been reviewed by Robaire and Hermo (1988). The regional changes in epididymal histology have been described (e.g., for rat, see Reid and Cleland, 1957).

The epididymal epithelium is pseudo-stratified and is made up of principal, basal, clear, and halo cells. The major cell type is the tall, columnar principal cell, which makes up about 80% of all epididymal epithelial cells (Robaire and Hermo, 1987). These cells are involved in resorption of fluid and organic material and in secretion of small organic molecules, as well as proteins and glycoproteins. The next most important cell type (about 15%) is the basal cell, flat elongated cells that are found throughout the epididymis in contact with the basement membrane; their function has not yet been ascertained. Clear and halo cells are more sparsely distributed along the epididymis. In the species in which they are found, clear cells might be involved in resorbing the elements of cytoplasmic droplets that are shed by spermatozoa as they mature. These cells are not found in all mammals, e.g., the ram epididymis does not have any. Halo cells are believed to be part of the immune system and have been described as either monocytes or lymphocytes (Robaire and Hermo, 1987).

The quantitative distribution of these cell types as one moves from one segment of the epididymis to the next has recently been described not only in intact rats, but also in animals that had been treated chronically with the antitumor and immunosuppressive agent cyclophosphamide (Robaire and Hermo, 1987; Trasler et al., 1987b). There were marked changes in the proportion of various cell types from the initial segment of the caput to the cauda epididymis and changes in the relative cell surface area of a particular cell type in selected segments of the epididymis at specified times after initiation of treatment.

For several reasons, epididymal biopsies have not become routine. Because the epididymis is a single long convoluted tubule, a biopsy would result in a disruption of the tubule and hence prevent normal transport of spermatozoa through it. Relatively little is known about the functional histology of the tissue, so the practical consequences of an altered histologic appearance would not be evident. Thus, although it is clear that biopsies of epididymal tissues should not be routine, it is also evident that detailed analyses of the relative distribution of various cell types and of histologic appearance can provide valuable information on the actions of some drugs on this tissue and accordingly can act as useful markers.

Blood-Epididymis Barrier

The anatomic and functional existence of a blood-testis barrier is well established (Setchell and Waites, 1975), but only since the late 1970s has there been substantial evidence of a blood-epididymis barrier (Friend and Gilula, 1972; Howards et al., 1976; Suzuki and Nagano, 1978; Cavicchia, 1979; Greenberg and Forssmann, 1983; Hoffer and Hinton, 1984), as reviewed by Robaire and Hermo (1988). Given the presence in spermatozoa of proteins that are recognized by the body as foreign, it stands to reason that there should be a continuation of a functional barrier beyond the testis.

There is extensive functional evidence of such a barrier: the large differences in the concentrations of inorganic and organic compounds between the luminal fluid and the blood (Crabo and Gustafsson, 1964; Jenkins et al., 1980; Turner et al., 1984; Hinton, 1985). The barrier is reportedly resistant to various treatments, such as high-dose administration of estradiol (Turner et al., 1981), the application of gossypol (Hoffer and Hinton, 1984), vasectomy (Turner and Howards, 1985), or the presence of varicocele (Turner and Howards, 1985), but little is known about its role in protecting spermatozoa from immunoglobulins and toxicants. Indeed, the ability of the barrier to maintain its tightness under conditions of stress might be pivotal in allowing the epididymis to sustain its functions. There is no evidence that environmental toxicants or drugs can disrupt the blood-epididymis barrier. However, if it were disrupted, the potential consequences could include immobilization of spermatozoa by antibodies and alteration of the sperm genome by chemicals. It is therefore important to develop simple markers that will allow for the monitoring of the integrity of the blood-epididymis barrier without disrupting epididymal function.

Spermiophagy

The fate of spermatozoa that are not ejaculated has been and remains controversial. There is now morphologic evidence, especially for the vas deferens, that spermiophagy does occur in a number of species under normal conditions or after mechanical or chemical manipulation. Such a process might involve the epithelial cells that line the duct system or the presence of luminal macrophages. In either case, spermatozoa in various stages of degeneration have been reported in epithelial cells and in luminal macrophages (Cooper and Hamilton, 1977; Holstein, 1978; Murakami et al., 1982a,b; Murakami et al., 1984), and it has been suggested that the spermatozoa are undergoing lysis. Most observers have reached the conclusion that the epithelial cells of the epididymis and the vas deferens are not involved in such activity under normal conditions (Bedford, 1975; Orgebin-Crist, 1984). However, mechanical or chemical manipulation of the epididymis and vas deferens has been shown to cause epithelial cells of the different segments of the duct system to become phagocytic and to engulf and digest spermatozoa (Glover, 1969; Flickinger, 1972; Alexander, 1973; Hoffer et al., 1973; Hoffer and Hamilton, 1974; Hoffer et al., 1975; Neaves, 1975). Also, spermiophagy by luminal macrophages has been shown to occur under abnormal conditions (Neaves, 1975; Phadke, 1975; Bedford, 1976; Flickinger, 1982).

The ability of the epithelial lining of the duct system to become active in spermiophagy apparently depends on either the presence of excess spermatozoa or the presence of "abnormal" spermatozoa in the lumen; the mechanism is not clear. However, histologic observation and measurement of the extent of spermiophagy by the epididymal or vas deferens epithelium might provide a good indication of the extent of sperm damage. This potential marker has not been tested systematically for any family of drugs.

Biochemical Markers

Hormone Receptors and Regulation of Epididymal Function

Androgens—especially 5α-dihydrotestosterone (DHT), the 5α-reduced metabolite of testosterone—are the primary modulators of epididymal function. However, it has become apparent that many other regulatory molecules play specialized roles in maintaining normal epididymal function. The factors that regulate epididymal function—i.e., those for which there are specific binding proteins—do not reach the epididymis only through the circulation (endocrine), but also through direct input from the testis (paracrine).

Androgens. The dependence of the epididymis on androgens has been established for over 60 years (Benoit, 1926). Many studies during the past 3 decades have tested the effects of castration and testosterone replacement on a large array of characteristics (Orgebin-Crist et al., 1975; Brooks, 1979a), and several general conclusions can be drawn. The epididymis atrophies and treatment with physiologic concentrations of androgen only partially maintains epididymal weight; the remainder attributable to spermatozoa and luminal fluid (Karkun et al., 1974; Robaire et al., 1977; Brooks, 1979b). In that respect, changes in epididymal weight could be a marker of changes in testosterone.

Androgens regulate intermediary metabolism (Brooks, 1979a), the transport of ions across the epididymal epithelium (Wong and Yeung, 1977), the transport of inositol and carnitine across the membranes of epididymal epithelial cells (Böhmer et al., 1977; Brooks, 1980; Pholpramool et al., 1982), and the synthesis and secretion of a number of epididymal glycoproteins as well as the activity of several enzymes, e.g., glutathione S-transferase (Rastogi et al., 1979; Jones et al., 1980; Moore, 1981; Mayorga and Bertini, 1982; Robaire and Hales, 1982; Brooks, 1983). Whether any of these effects are directly mediated by androgens or require the synthesis of new mRNA and therefore the synthesis of new proteins is being investigated.

Acquisition of fertilizing ability and storage of spermatozoa both depend directly on androgens (Benoit, 1926; Bedford, 1975; Cohen et al., 1981). The presence of an inhibitor of 5α-reductase caused decreases in the number of motile spermatozoa, in the percentage of oocytes fertilized, and in the number of blastocysts found—but only when the inhibitor was given with testosterone.

There is no debate that androgens regulate epididymal function, but the mechanism of their action is far less clear. Two molecules in the epididymis have high affinity for DHT; these have been named "androgen receptor" and "androgen-binding protein." It must be stressed, however, that the classical conditions necessary to demonstrate that a given molecule is the androgen receptor in the epididymis have not been fulfilled.

Exogenously administered testosterone is transformed to DHT and is found bound to a cytosolic protein in the epididymis of a number of animals (Blaquier, 1971; Ritzen et al., 1971; Danzo et al., 1973; Younes and Pierrepoint, 1981; Carreau et al., 1984). Epididymal cytoplasmic receptor seems to have many features in common with prostatic androgen receptor (Tindall et al., 1975; Younes and Pierrepoint, 1981). The binding of DHT to the cytoplasmic receptor is blocked by antiandrogens, such as cyproterone acetate and flutamide (Tindall et al., 1975; Danzo and Eller, 1975).

Androgen-binding protein (ABP) has been found in many animals (Danzo et al., 1977; Fabre et al., 1979; Carreau et al., 1980) and humans (Hsu and Troen, 1978). It has a number of physicochemical properties that differ markedly from those of the cytoplasmic androgen receptor (Hansson et al., 1975). ABP is synthesized by Sertoli cells and enters the epididymis via the efferent ducts (Attramadal et al., 1981; Musto et al., 1982). Its functions are still unresolved, but it has been proposed that it acts as an androgen sink in seminiferous tubules (Hansson et al., 1975), transports androgens to the epididymis (Ritzen et al., 1971), and acts as a regulator of epididymal 5α-reductase (Robaire et al., 1981). In a rat model, a correlation between the amount of epididymal ABP and the fertilizing ability of spermatozoa has been established (Anthony et al., 1984).

Several biologic markers have been used to assess androgenic action on the epididymis. The first is the serum concentration of testosterone. Although of limited value (it does not specifically reflect epididymal activity), it is a useful gross indicator. The concentrations of ABP in semen and in epididymal tissue fractions or luminal fluid have been used as a combined indicator of Sertoli cell and epididymal functions. Androgen-receptor and 5α-reductase activities have been measured as markers of the androgenic status of the epididymis; although these two markers are highly specific and are probably among the most useful of the markers discussed in this section, they require removal of tissue, which is possible only under experimental conditions. If noninvasive techniques could be devised to monitor these aspects of androgen-related epididymal activity, they would be extremely useful.

Estrogens. The administration of estrogens can affect the male reproductive system in general (Gay and Dever, 1971; Swerdloff and Walsh, 1973; Karr et al., 1974; Verjans et al., 1974; Ewing et al., 1977) and the epididymis in particular (Meistrich et al., 1975; Orgebin-Crist et al., 1983). Although it has been generally accepted that mediation of estrogen action takes place via the hypothalamo-pituitary-gonadal axis (Swerdloff and Walsh, 1973; Verjans et al., 1974; Robaire et al., 1979), substantial evidence of the direct action of estrogens on androgen target tissues has accumulated over the past 2 decades. Specific, well-regulated, high-affinity cytosolic and nuclear estrogen-binding proteins, presumably receptors, have been identified in the epididymis of mice (Schleicher et al., 1984), rats (Van Beurden-Lamers et al. 1974), rabbits (Danzo and Eller, 1979), dogs (Younes et al., 1979), and humans (Murphy et al., 1980). It is proposed that stromal (as opposed to epithelial) cells (Cunha et al., 1980) act as primary target sites because estrogen administration to young dogs (Connell and Donjacour, 1985) or rabbits (Orgebin-Crist et al., 1983) leads to increases in epididymal weight that were due almost entirely to stromal hyperplasia.

Autoradiographic localization of tritiated DHT and estradiol in the adult mouse epididymis revealed differences in distribution of grains in different cell types, as well as in different segments of the epididymis (Schleicher et al., 1984). That finding brings up the intriguing possibility that endogenous circulating, or potentially locally synthesized, estradiol serves a specific function in the epididymis, e.g., modulation of clear cell function. The difficulty in identifying estradiol receptors and aromatase activity in adult mammalian epididymis might be due to the localization of activity to cells (e.g., clear cells) that are not abundant in this tissue. The inherent obstacles in establishing the role of estradiol and of its receptor as indicators of specific epididymal functions suggest that the development of these markers should be given a low priority.

Aldosterone. The concentrations of ions change along the duct system, but the mechanism responsible for controlling these ion fluxes has not yet been elucidated. Because of the similarities in ion fluxes between the epididymis and the kidney, Wong and Lee (1982) and Jenkins et al. (1983) have proposed similar regulatory mechanisms. Indeed, specific binding sites for aldosterone, the adrenal mineralocorticosteroid normally considered to have the kidney as its target site, have been found autoradiographically to be selectively disposed over clear cells (Hinton and Keefer, 1985) and have been shown to be involved in regulating the concentration of spermatozoa in the epididymis (Turner and Cesarini, 1983). More studies are required to elucidate the role of aldosterone in epididymal function and to determine whether specific markers can be developed to reflect its epididymal activity.

Prolactin. In light of the possible coregulation of gonadotropin and prolactin release, the apparent presence of prolactin "receptors" in the epididymis (Aragona and Freisen, 1975; Orgebin-Crist and Djiane, 1979), and the proposal that prolactin regulates ion transport (Shiu and Friesen, 1980), it is plausible that this hormone plays a major role in regulating some facets of epididymal function.

Prolactin and its binding protein might eventually become useful markers of epididymal function, but the available data suggest that this subject be given low priority for research.

Vitamin D. The known primary function of vitamin D (a sterol hormone) is the homeostasis of calcium and phosphorus; its major sites of action are bone, intestine, and kidney (De Luca, 1978). A number of complementary observations suggest that the epididymis can also be a target for its action. First, specific binding proteins for $1,25\text{-}(OH)_2$ vitamin D have been found in the rat epididymis (Walters et al., 1982). Second, the epididymis was found to take up 25-(OH) vitamin D_3 and metabolize it to both $1,25\text{-}(OH)_2$ vitamin D_3 and $24,25\text{-}(OH)_2$ vitamin D_3 (Kidroni et al., 1983). Third, the epididymis has a higher concentration of $24,25\text{-}(OH)_2$ vitamin D_3 than any other tissue studied, and the cauda has three times more of this metabolite than the caput (Kidroni et al., 1983). One of the ions known to be regulated by the metabolites of vitamin D is phosphorus, which, in both organic and inorganic forms, is found in extraordinarily high concentrations in the cauda epididymis (Hinton, 1980). More studies are required to assess the importance of vitamin D, or of one of its metabolites, as a marker of specific epididymal function.

Vitamin A. Vitamin A (retinol) is a fat-soluble vitamin that is essential for life. It can be substituted for by retinoic acid in most tissues; known exceptions are the retina, the testis, and the epididymis. Selective binding proteins for both retinol and retinoic acid have been identified in most male reproductive tissues (Porter et al., 1985). The functional significance of the binding proteins is still unclear. Given that they are present in such high concentrations and that their localization is so finely regulated, vitamin A probably plays an important role in the regulation of epididymal function. It is worth noting that the 13-*cis* form of retinoic acid, isotretinoin, is now commonly used for the treatment of severe acne; its potential effects on epididymal function have not yet been elucidated.

Steroid Biosynthesis and Metabolism

The epididymis does not appear to be able to synthesize testosterone de novo (Benoit, 1926; Karkun et al., 1974; Robaire et al., 1977); however, conflicting data have been published (Frankel and Eik-Nes, 1970; Hamilton and Fawcett, 1970; Amann, 1987). Over the past 15 years, many studies in a wide variety of animals (from mouse to human) have measured steroid content and conversion in tissue homogenates (Gloyna and Wilson, 1969; Inano et al., 1969) and in cell and organ cultures (Robaire and Hermo, 1987). Some common features of the metabolism of testosterone by the mammalian epididymis have emerged: the importance of the conversion of testosterone to DHT, changes in 5α-reductase and 3α-hydroxysteroid dehydrogenase activities along the epididymis, alterations in 5α-reductase activity during development and aging, and further metabolism of 5α-reduced steroids in epididymal tissue.

Because of the importance of DHT in mediating androgenic action in the epididymis and the well-established gradients of various androgens in this tissue, it is important to monitor the activities of the enzymes that metabolize testosterone in the epididymis. Those activities are pivotal markers for our understanding of epididymal function, but present technology requires the isolation of the tissue for their assessment. The need for noninvasive methods of monitoring is evident.

Intermediary Metabolism

The relative importance of glycolysis, the tricarboxylic acid cycle, and the pentose cycle in epididymal intermediary metabolism has been studied by a number of investigators (Turner and Johnson, 1973; Brooks, 1979b, 1981). The rate-limiting step in glycolysis is not the activity of the enzymes in the pathway, but the availability of glucose (Brooks, 1981). Indeed, evidence has been gathered to support the existence of a specific membrane-transport system for glucose (Turner and Johnson, 1973; Brooks, 1981; Hinton and Howards, 1982). The transport system is situated on the basolateral membrane, cannot transport other hexoses, and can be inhibited by 3-O-methylglucose, a nonmetabolizable analogue that is recognized by the transport system. Oxidative metabolism is high right after birth in the rat epididymis, decreases for the first 2 weeks of life, and then increases in parallel with the increase in circulating androgens (Delongeas et al., 1984).

A number of other enzymatic processes, such as prostaglandin biosynthesis and metabolism, have been described for the epididymis (Robaire and Hermo, 1987). We limit our discussion here to glutathione, because of its potential usefulness as a marker of epididymal function. Several enzymes involved in the biosynthesis, metabolism, and conjugation of glutathione are present in the epididymis. They are particularly important, because free glutathione is viewed by many toxicologists as a mechanism of "mopping up" electrophiles that might damage cellular components. The family of epididymal enzymes involved in the conjugation of glutathione with electrophilic chemicals, the glutathione S-transferases, can be resolved into six peaks, each with a characteristic isoelectric point and substrate specificity (Hales et al., 1980); the most acidic peak is much higher in the epididymis than in other tissues and might be specific for it. Results of studies on the longitudinal distribution of these enzymes in the epididymis indicate a general trend toward decreasing activity for each of the substrates from the caput to the cauda epididymis (Hales et al., 1980). Thiol oxidase, an enzyme that converts $2R\text{-}SH + O_2$ into $R\text{-}S\text{-}S\text{-}R$ and H_2O, has been reported in rat and hamster epididymis (Chang and Zirkin, 1978); enzyme activity was higher in the cauda than in the caput epididymis. It was proposed that such an enzyme could protect spermatozoa from endogenous free sulfhydryls. Exceptionally high concentrations of γ-glutamyl-transpeptidase have also been report-ed in the rat epididymis (DeLap et al., 1977). That enzyme is involved in the transport of γ-glutamyl amino acids and can act as a glutathione oxidase, converting reduced to oxidized glutathione (Tate and Orlando, 1979).

EPIDIDYMALLY MEDIATED TOXIC DRUG EFFECTS

A growing number of compounds have been shown to affect the epididymis, e.g., α-chlorohydrin, 6-chloro-6-deoxyglucose, and possibly gossypol (Robaire and Hermo, 1987). The epididymis, however, has not usually been studied as one of the target tissues for the toxic effects of drugs or environmental toxicants, so little is known about the importance of this tissue as a target for toxic drug effects.

For some compounds, such as dibromochloropropane (DBCP), it has clearly been shown that there is an effect on the epididymis, e.g., on epididymal weight or sperm reserves (Amann and Berndtson, 1986); but it is still not clear whether the deleterious effects of this drug, and others like it, on reproductive outcome are directly linked to their action on the epididymis or epididymal spermatozoa.

Clear evidence has emerged from some recent studies that—after a week of treatment with such drugs as cyclophosphamide, an anticancer and immunosuppressive drug (Trasler et al., 1985, 1986); methylnitrosourea, an alkylating agent (Nagao, 1987); or methyl chloride, an industrial gas (Chellman et al., 1986)—a sharp increase in preimplantation and postimplantation loss (dominant lethal mutations) could be found. Such male-mediated adverse effects on reproductive outcome were associated, in the case of cyclophosphamide, with specific changes in the distribution of cell types in the epididymal epithelium and an increase in the number of spermatozoa with abnormal flagellae in the rat epididymis (Trasler et al., 1988).

6

Biologic Markers of Accessory Sex Organ Structure and Function

This brief review on markers of accessory sex organ function shows that few non-invasive or minimally invasive markers exist for the assessment of accessory sex organ function in human males exposed to toxicants. That is probably because little is known about the function of these glands, because they are inaccessible, and because they vary between mammalian species in their presence, size, and structure. Those difficulties are exacerbated by the mixing of the accessory sex organ secretions at ejaculation, by the variation in the volume and chemical makeup of the secretions between ejaculates even of the same individual, and by the presence in semen of enzymes that alter the chemical composition during clotting, liquefaction, and storage. However, it has been demonstrated that many exogenous chemicals are trapped in accessory sex organ secretions. Therefore, research in appropriate animal models will be important.

The male accessory sex organs characteristic of mammals are the prostate, seminal vesicles, ampullae of the vas deferens, bulbourethral (Cowper's) glands, urethral (Littre's) glands, and preputial glands (Fig. 6-1). The prostate is purported to be the only accessory sex organ present in all mammals (Coffey and Isaacs, 1981). Species vary in the presence, size, and structure of the male accessory sex organs,

but the organs share some important characteristics: they all contain secretory epithelium that is magnified greatly by villous infoldings or a compound tubulo-alveolar structure, and they all depend on androgen for differentiation, growth, and secretory function (Coffey, 1986).

Little is known about the function of these organs beyond their obvious secretion of fluids that mix with sperm at ejaculation. It has been suggested that the accessory sex organ secretions are bacteriostatic and can protect the male genital tract from bacterial infection. (Refer to reviews Coffey (1986) and Mann and Lutwak-Mann (1981) for detailed descriptions of the structures and functions of the male accessory sex organs.)

The growth and structure of accessory sex organs and their secretion of specific chemicals into seminal plasma constitute biologic markers of exposure to or effects of toxicants. This chapter discusses primarily the prostate and to a lesser extent the seminal vesicles. It does not discuss the ampullae, the bulbourethral glands, or the preputial glands. There are several reasons for the exclusion: the prostate is probably the only accessory sex organ that occurs in all mammals; the human prostate has been studied extensively, not because of its response to toxicants, but because prostatitis, benign prostatic hyperplasia, and prostatic cancer are

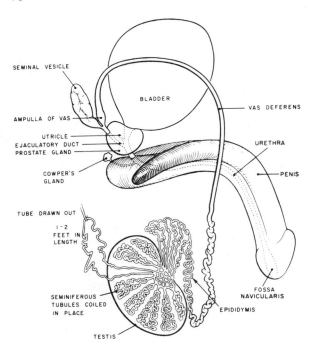

FIGURE 6-1 Diagrammatic representation of testis, showing duct system and relation of ducts to accessory sex glands and penis. Source: Reprinted with permission of Macmillan Publishing Co. from *The Human Body: Its Structure and Physiology*, 2nd ed., Grollman, 1987. Copyright 1987 by Sigmund Grollman.

major human diseases; and general concepts established for the prostate probably will have application to other accessory sex organs. It is impossible to restrict our comments to the human, because data on potential biologic markers are fragmentary. Therefore, we include animal data—frequently obtained from the dog, whose prostate has many similarities with the human prostate. The dog is a particularly valuable animal model, because it has no seminal vesicles or bulbourethral glands, and prostatic secretion therefore constitutes more than 95% of its seminal plasma.

PHYSICAL MARKERS

Size is an important consideration in the clinical evaluation of the prostate in humans (Coffey, 1986) and dogs (Blum et al., 1985). Accessory sex organ weight in general and prostate weight in particular are used widely as androgen bioassays, because androgen is required for growth and secretion. Although useful in animal experiments, prostatic weight is not useful in humans as a marker, because changes in weight are nonspecific (growth does not necessarily reflect secretion) and imprecise and its measurement requires autopsy.

Estimates of prostatic volume obtained by measuring length, width, and height in situ with calipers at laparotomy are highly correlated with prostatic weight in the dog (Walsh and Wilson, 1976; Deklerk et al., 1979). That relationship is not as applicable to irregularly shaped structures, such as the seminal vesicles or multilobular prostates. Even with the dog prostate, there are practical limits to the frequency and number of such measurements, because of the long recovery time associated with laparotomy, the formation of prostatic adhesions, the risk of infections, and the tedious, time-consuming effort associated with repeated surgery.

Berry et al. (1985) recently developed a method of generating a three-dimensional representation of the dog prostate from a pair of orthogonal radiographs of the prostate in situ to which six metal markers are surgically attached. The externally estimated prostate weight is highly correlated with actual prostate weight ($r \approx 0.90$). The technique affords precise estimates of changes in weight within a prostate over time. Although it allows continuous monitoring of a dog's prostate during the course of an experiment, it requires extensive surgery and access to sophisticated x-ray equipment. A simple, noninvasive technique of imaging the prostate is still needed.

Transrectal and transabdominal ultrasound have been used to assess prostatic volume in the human (Henneberry et al., 1979; Bartsch et al., 1982). There is a statistically significant relationship between the ultrasound estimate and the size of the prostate (Henneberry et al., 1979; Bartsch et al., 1982). Blum et al. (1985) recently compared prostatic size measured by in vivo ultrasound with actual volumetric measurement of the dog prostate at autopsy; they also reported a statistically significant correlation between the two measures. The advantages of using ultrasound to determine prostatic size are the ease and rapidity of the technique, the involvement of only minimal trauma, and the avoidance of prostatic scarring. However, greater precision and accuracy are required.

STRUCTURAL MARKERS

The human prostate, a complex organ, is divided into five zones: anterior fibromuscular stroma, peripheral zone, central zone, preprostatic tissue, and transition zone (Coffey, 1986), containing stromal and epithelial cells. Efforts are under way to identify these various zones with magnetic resonance imaging in situ (Iverson al., 1988; Sommer et al., 1988; Allen et al., 1989).

The complex and highly differentiated cytoarchitecture of the prostate has been used at the light microscopic level to diagnose a variety of clinical prostatic disorders. But it has not been used widely to study the effect of toxicants on prostatic function in humans or experimental animals, because it requires invasive biopsy techniques or autopsy specimens; because diagnostic histopathology requires judgment by a highly trained specialist and is qualitative, rather than quantitative; and because it is difficult

to determine whether a toxicant acts directly on a specific cell type in the epithelium, indirectly via an effect on a stromal element, or even more indirectly via the hypothalamo-hypophyseal-gonadal axis.

Stereologic techniques have been used to quantify the volume of prostatic lumen, epithelial cell, and stromal elements in the human (Bartsch and Rohr, 1977; Bartsch et al., 1979) and dog (Zirkin and Strandbergh, 1984) at the light microscopic level. The same approach has been extended to the ultrastructural level to quantify the volume of cytoplasmic organelles in specific cell types in dog prostate (Zirkin and Strandbergh, 1984). Although invasive and tedious, these are sensitive and precise biologic markers of prostatic structure and will be particularly useful in understanding mechanisms of toxicity.

FUNCTIONAL MARKERS

Accessory sex organs secrete fluids that are emitted into the urethra and later ejaculated or otherwise excreted. The average human ejaculate contains approximately 3 ml; 1% of the volume is spermatozoa and 99% is seminal fluid derived from accessory sex organ secretions (Coffey, 1986). In the human, about 1.5-2.0 ml of the seminal plasma in an ejaculate is derived from the seminal vesicles and about 0.5 ml from the prostate. The volume of seminal plasma and the chemical constituents in seminal plasma can be used as markers to monitor the effect of toxicants on accessory sex organ secretion.

Human seminal plasma can be obtained noninvasively and repetitively by masturbation. The volume of the human ejaculate is only a crude measure of the secretory capacity of the accessory sex organs, because it is influenced by frequency. Also, the volume is nonspecific, in that it reflects the accumulation of secretions not only of the prostate and seminal vesicles, but also of the epididymis, ampullae of the vas deferens, bulbourethral glands, and urethral glands. In contrast, the dog ejaculate reflects prostatic secretion more directly, because the dog lacks seminal vesicles and bulbourethral glands. Efforts have been made to obtain more specific information in humans by mechanically splitting ejaculates obtained through masturbation into temporal fractions, the first of which are rich in sperm and prostatic secretions and the latter in seminal vesicular secretions (Coffey, 1986). That approach has met with only limited success, because of individual variability and the difficulty in collecting split ejaculates. Inaccuracies and imprecision in the volumetric measurement of human seminal fluid are also caused by interindividual variation and by intraindividual variation over time.

More specific information regarding accessory sex organ secretory function can be derived by measuring the concentrations of specific endogenous chemicals in seminal fluid. It is beyond the scope of this report to review exhaustively the 100 or more endogenously produced chemicals that appear in mammalian seminal plasma, and the reader is referred to the reviews by Coffey (1986) and Mann and Lutwak-Mann (1981). We limit our discussion here to a few well-known chemical markers of individual human accessory sex organ function and later discuss some potential markers discovered in experimental animals.

Individual human accessory sex organs have been shown to be the principal sources of specific chemicals in seminal fluid. For example, fructose is derived from human seminal vesicles (Mann and Lutwak-Mann, 1981), and zinc, acid phosphatase, and spermine from the human prostate (Coffey, 1986). Measurement of those chemicals in seminal fluid should reflect the exposure and the effects of exposure of the prostate or seminal vesicles to toxicants (Bygdeman and Eliasson, 1969). However, such data will not prove that a toxic chemical acted directly on a specific accessory sex organ. Repetitive samples must be collected from the same person over time to offset the variability inherent in the volume and chemical makeup of a single sample of seminal fluid. And the collection and preservation of semen must be controlled rigorously, because clotting and liquefaction can take place, because enzymatic activities in seminal fluid

alter its chemical composition, because the concentration of enzyme inhibitors is variable, and because such seminal fluid constituents as fructose are metabolized by spermatozoa (Rui et al., 1986). Perhaps the existence of those characteristics explains why the technique has not been used widely as a biologic marker. Research and development might make it more useful.

Numerous drugs and toxic chemicals have been identified in human semen (Reeves et al., 1973; Stamey et al., 1973; Malmborg, 1978; Cohn et al., 1982). Examples of exogenous chemicals transported into and accumulating in semen of various mammals are carcinogens, antibiotics (including basic macrolides and sulfonamides), and methadone. In general, these chemicals are assumed to cross cell membranes of the accessory sex organs by nonionic diffusion. Several factors probably are important in trapping these chemicals in accessory sex organ secretions, including lipid solubility, pK of the chemical, binding of the chemical to proteins, and pH of sex organ secretions. Chemicals are absorbed intravaginally in the human female (Benziger and Edelson, 1983). It has been shown that thalidomide in the rabbit (Lutwak-Mann, 1964) and cyclophosphamide in the rat (Hales et al., 1986; Trasker et al., 1987) cause paternally mediated adverse effects on the progeny. Therefore, it is possible that toxic agents in human semen absorbed by the mother can affect the conceptus. Alternatively, the same chemicals might affect the quantity or viability of or the genomic information in spermatozoa. Clearly, the subject warrants high priority for research.

Human seminal plasma contains a variety of enzymes, metalloproteins, flavoproteins, and mucoproteins (Coffey, 1986; Mann and Lutwak-Mann, 1981), many of which occur in other bodily compartments and therefore are not specific markers. However, some proteins exhibit organ specificity; for example, serum acid phosphatase concentration has been used as a marker of prostatic carcinoma (Coffey, 1986; Mann and Lutwak-Mann, 1981). It is surprising that macromolecules in semen have not been exploited by toxicologists as markers of individual accessory sex organ function.

More specific and therefore more promising biologic markers have been identified. Kistler and colleagues (Ostrowski et al., 1979, 1982) purified an androgen-dependent rat seminal vesicular protein IV. Recently, they isolated and characterized a genomic clone for rat seminal vesicular protein IV (Kandala et al., 1983). French and coworkers (Lea et al., 1979) purified a major secretory protein of rat ventral prostate (prostatein). French and colleagues also have purified two androgen-dependent secretory proteins of rat dorsal prostate and coagulating gland that, although anatomically distinct, synthesize similar proteins in response to androgens (Wilson and French, 1980; Wilson et al., 1981).

A single (unnamed) protein accounts for over 90% of the total protein in dog seminal plasma (Isaacs and Shaper, 1983). Clearly, a primary function of the dog prostate is to synthesize and secrete that protein in response to androgens (Isaacs and Shaper, 1985). Therefore, its measurement in seminal fluid is a sensitive and specific marker of androgen-dependent prostate function in the dog. The protein has been shown to have many characteristics of a glandular kallikrein (Isaacs and Coffey, 1984). Kallikreins are proteolytic enzymes that have the capacity to process precursor molecules into biologically active peptides. It was recently discovered that the human prostate secretes a kallikrein into seminal fluid (Fink et al., 1985).

Those findings suggest that macromolecules secreted into seminal plasma have characteristics that make them desirable biologic markers of exposure to or effect of toxic chemicals. First, individual accessory sex organs synthesize and secrete specific macromolecules into seminal fluid. Second, the process is regulated by androgen in some instances and therefore reflects the adequacy of the production of androgen. Third, many small molecules in seminal fluid might result from diffusion into accessory sex organ secretions from the blood; that would limit their usefulness as biologic markers, but this is unlikely to be a problem with macromolecules synthesized in accessory sex organs under the influence of androgens. Fourth, specific antibodies can be generated against a purified protein and thus permit the study of their cellular origin, the development of immunoassay measurement techniques, and the development of a cDNA clone to ensure production of adequate amounts of the protein for molecular biologic analysis. Much research remains to be done to identify specific proteins in appropriate animal models and in humans. Once that is achieved, it will be imperative to test the effects of a series of toxic agents on specific markers in a standardized protocol.

7

Biologic Markers of Human Male Reproductive Health and Physiologic Damage

NEEDS FOR BIOLOGIC MARKERS OF HUMAN MALE REPRODUCTIVE HEALTH

The uncertainties associated with risk extrapolation from animal data argue strongly for the development of validated and sensitive methods for measuring germinal and reproductive damage directly in people. In males, reproductive damage can be divided into two broad types: pathophysiologic and genetic. This chapter reviews the markers of reproductive health suitable for detecting physiologic damage. Physiologic damage may reduce the chance of successfully fertilizing an egg. Human markers of germinal genetic toxicity and heritable mutations are reviewed in Chapter 9.

Currently available methods used in the evaluation of the reproductive health of human males are in three broad classes:

- Personal history.
- Physical examination.
- Laboratory analyses.

The roles of personal history and physical examination in fertility assessment was described earlier (Chapter 3).

Laboratory analyses include testicular biopsy, hormonal analyses, and semen analyses. Testicular biopsy and hormonal analyses were described earlier with emphasis on animal studies (Chapters 4-6).

Table 7-1 lists the categories of biologic markers of physiologic damage to male reproduction discussed in this chapter. These markers are grouped by the source of tissue and data required: testicular tissue, semen, blood, surveys and medical records, and maternal urine. They differ in the numbers and kinds of assays available to measure them, in the degree of quantitation attainable so far, in the extent to which underlying mechanisms are understood, and in their feasibility for human studies. The development and validation of markers of human male reproductive health require a multidisciplinary approach that includes basic research in animal and human reproductive biology, engineering and statistical development of automated and quantitative procedures, clinical studies of human factors that affect variation and of the predictive value of individual markers, and epidemiologic studies of populations exposed to xenobiotic agents.

This chapter begins with a short review of the kinds of epidemiologic studies that have been performed to evaluate fertility effects of human males exposed to reproductive toxicants. It then discusses in detail markers currently available as well

TABLE 7-1 Biologic Markers of Physiologic Damage to Human Male Reproduction[a]

Tissue or Data Required	Markers of
Testis (or biopsy)[b]	Histopathology
Seminal sperm	Sperm number
	Structure[c]
	Motility[c]
	Double F bodies
	Viability
	Agglutination
	Penetration and egg interaction:
	cervical mucus
	hamster eggs
	nonliving human eggs
	Internal and surface domains
	Chromatin structure
Other seminal parameters	Physical characteristics
	Immature germ cells
	Non-germ cells
	Chemical composition:
	normal and xenobiotic constituents
	Sertoli cell, Leydig cell, and accessory gland function
Blood[b]	Hormone levels
Survey and medical records	Fertility status:
	standardized fertility ratio
	time to conception
Maternal urine	Indicators of early pregnancy

[a]Available markers that have been used to detect effects of ionizing radiation or chemicals in exposed men and markers on which human baseline data are available or under investigation. See Table 7-2 for further details.
[b]Discussed in Chapter 4.
[c]Automated methods are under development.

as the status of new areas of research that may lead to new, useful biologic markers. Several features of human semen are evaluated including some physical characteristics of the ejaculate, presence of nonsperm cells, sperm number, sperm motility, sperm viability, sperm structure, and various aspects of sperm function and penetration. In addition, the chemical composition of semen has been evaluated to a limited extent as an indicator of reproductive function and toxicity.

EPIDEMIOLOGIC STUDIES OF HUMAN SPERM PRODUCTION AND FERTILITY

Interview data, medical records, and demographic birth records, as well as semen and blood hormone analyses have been used to evaluate the effects of radiation or chemicals on human fertility and spermato-genesis. Although surveys are not generally considered markers of reproductive health, they are a means of evaluating the fertility status of well-characterized groups and thus provide important benchmarks in the development and validation of markers related to fertility.

Epidemiologic Surveys of Fertility Status

Interviews and the use of medical records are indirect approaches to assessing fertility; epidemiologically, they measure the lack of an event (the event being the birth of a healthy child). Reduction in fertility is measured by comparing birth rates or intervals between births or pregnancies to assess couples' ability to procreate. The male component is evaluated by comparing the average fertility of a reference group with the fertility of a group of couples in which all males have a common exposure to an environmental or occupational agent.

Wong, Levine, and coworkers (Wong et al., 1979; Levine, 1983) developed the standardized fertility ratio (SFR) method, which compares the numbers of births observed with the numbers expected on the basis of the number of person-years of observation, considering such factors as age, race, marital status, and parity. Statistically and epidemiologically, the SFR has advantages and limitations in interpretation (Tsai and Wen, 1986). The SFR method was used to evaluate several male occupational exposures, including exposures to ethylene dibromide and DBCP (Levine et al., 1980; 1983).

The interval between births or pregnancies is a second indirect measure of fertility (Baird et al., 1986). Several factors affect this measure, including use of contraceptives and birth order. Birth order needs to be controlled, because the interval between births increases with increasing parity. Births within a family are not independent events, because of the presence of social, genetic, and health factors. Statistical methods that account for these factors must be used in analyses. The intervals between births and between pregnancies were increased in smoking mothers (Baird and Wilcox, 1985). Further work and additional approaches are needed to improve the quantitation of survey-based indicators of human fertility status.

Markers of Early Pregnancy

Early pregnancy monitoring may provide a very sensitive method for measuring male-mediated effects on fertility and pregnancy. It has been difficult to detect the early days of human pregnancy, and most women don't suspect that they are pregnant until a menstrual period is missed. However, early pregnancy is a time of elevated frequencies of embryo loss, although neither the precise nor the portion that can be attributed the human male (i.e., via the sperm) are known. Sensitive methods are being developed to detect early pregnancy; for example, the immunological detection of β-chorionic gonadotropin

in the maternal urine can detect pregnancy by about 10 days after conception (Canfield et al., 1987). Methods such as these might be useful for evaluating male fertility and for identifying male factors that may affect the frequency of early embryo loss. Further research is encouraged to determine human baseline variations for early pregnancy loss and of male factors that might alter these rates and to develop methods that detect even earlier pregnancies.

Cohort Studies with Semen and Blood Samples

Sperm number, motility, and structure have been used to evaluate the effects of exposures to physical and chemical agents. Detrimental effects of increased scrotal temperatures and ionizing radiation and exposure to over 50 therapeutic, occupational, and environmental chemicals have been identified (Wyrobek et al., 1983a). Cohort studies or case reports of exposure-related changes in sperm characteristics provide no direct information on fertility effects, because specific "sperm-characteristics versus fertility" relationships and within-laboratory standards for normal fertility are usually not presented in these studies.

Primarily, two types of study designs have been used with semen analyses: cross-sectional and longitudinal. Cross-sectional studies require the analyses of single semen samples from exposed men and from men of at least one reference group (Wyrobek et al., 1982). The reference group can consist of unexposed men sampled concurrently or historical within-in-laboratory values from unexposed men. In rare cases, workers can be stratified by exposure to assess dose-effect relationships, as exemplified by Lancranjan et al. (1975) for lead workers. They reported that proportions of men with oligospermia, asthenospermia, and teratospermia increased with increased peripheral blood lead concentrations. More typically, exposure is difficult to quantify, and studies are limited to group comparisons of exposed and unexposed men. The statistical evaluation of confounding factors is crucial in cross-sectional studies; abstinence time, illness, smoking habits, and other chemical or drug exposures are common confounding factors.

Among-person variations in sperm number, motility, and structure scores can be used to estimate statistical power of studies and to calculate sample sizes required to detect changes with the cross-sectional design. For example, one laboratory (Wyrobek et al., 1982) calculated that detection of a 20% change in mean between the exposed and unexposed groups required analyses of semen from approximately 800 men for sperm concentration (400 exposed and 400 controls) and 80 men for sperm structure. Considering realistic participation rates, more men would have to be identified in each group to achieve the required numbers of semen samples.

The insensitivity of the cross-sectional design plagues human semen studies, because it is difficult and expensive to find, gain cooperation of, and evaluate large numbers of men. The large sample requirements also limit the kinds of factories and agents that can be investigated. The design is nevertheless commonly used, probably because it permits rapid collection and analysis of semen samples (in comparison with the longitudinal design described below). Studies of occupational exposures to DBCP (Whorton et al., 1977; Whorton et al., 1979; Babich et al., 1981), carbaryl (Wyrobek et al., 1981), and wastewater treatment (Rosenberg et al., 1985) and others (Wyrobek et al., 1983a) have used this design.

The longitudinal design, requires analysis of at least two sequential semen samples from each man (Sherins et al., 1977). The samples should be separated by at least several months (and more than two repeat samples are preferred). Sample collection times are selected in relation to exposure. For example, the first sample might be collected before exposure begins, and the second about 6 months after it begins (this would allow for two to three spermatogenic durations between sampling). The advantage of this design is based on the observation that within-person variation of sperm concentration, motility, and structure is generally smaller than among-person variation. Thus, smaller numbers of men are required for detection of a specific exposure-induced change in group mean. For example, a longitudinal study to detect a 20% change in group mean would require approximately 80 men for sperm concentration, but only about 10 men for sperm structure (A. Wyrobek, Lawrence Livermore National Laboratory, unpublished data, 1989). Furthermore, this design does not, in principle, require an unexposed cohort for reference, although an unexposed cohort might be helpful in controlling factors closely correlated with exposure. The longitudinal design with semen analyses has been used primarily to evaluate the effects of exposure to drugs, e.g., AMSA (da Cunha et al., 1982), colchicine, various chemotherapeutic agents, and antifertility drug candidates (see review by Wyrobek et al., 1983a). Further studies are needed to evaluate the relative utilities of longitudinal and cross-sectional designs in exposed populations.

Blood hormone levels have been evaluated as surrogate measures of spermatogenesis. The levels of follicle stimulating hormone (FSH) are elevated in men with very low or zero sperm numbers, as seen with DBCP and certain cancer chemotherapeutic compounds (see review by Sever and Hessol, 1985). However, this method is "insensitive," and hormone measurements have not been used commonly in studies of environmental exposures.

HUMAN SPERMATOGENESIS AND DEVELOPMENT OF SEMEN-BASED MARKERS OF MALE REPRODUCTIVE HEALTH

Human sperm production is unique among animals in several ways:

- In most animals (laboratory, as well as domestic), sperm are produced in great excess over what is needed for adequate fertility. Even in mice, among the smallest mammals, males produce some 10 times the sperm required for fertility. In comparison, the mean sperm concentration man is a smaller proportion of levels associated with subfertility.

- Individual variation in semen quality is much greater in men than in laboratory and domestic animals. For example, approximately 10% of men within a normal fertile group might be at or below 20 million sperm per milliliter (MacLeod and Wang, 1979; Whorton and Meyer, 1984). Such variation would be virtually nonexistent among fertile laboratory and domestic animals.

- There is more variation among the types of cells and cell structure in the typical human ejaculate than in animals' ejaculate.

- Humans are genetically heterogeneous, so responses (i.e., pharmacokinetics and metabolism) to particular chemicals can differ unpredictably.

- The minimal number of sperm required for adequate and minimal fertility might depend on species. For humans, the number for adequate fertility is generally considered to be 20 million sperm per milliliter of ejaculate or more. Some reports suggest that the minimum can be any number greater than zero (Barfield et al., 1979; Clark and Sherins, 1986). Little is known of how animals and humans compare on this point.

- Human seminiferous epithelium is unique among mammals in several ways (Hellerand and Clermont, 1964; Clermont, 1972), including kinetics of stem cell renewal, capacity to regenerate after toxic insult, effects of chromosomal abnormalities on spermatogenic differentiation, cellular association patterns, and the duration of differentiation.

In addition to the above features of human spermatogenesis, semen is an unusual body fluid—a person's ability to produce it has no direct bearing on his health or longevity.

Semen is also a complex fluid. Although sperm are considered its primary constituent, it also contains numerous other germinal and somatic cell types, chemical constituents from germinal and supporting somatic tissues, hormones, and probably xenobiotic agents.

Unlike most cells found in other body fluids, seminal sperm are not terminally differentiated in one major respect. That is, from the perspective of the human germ line and its generational cycles, seminal sperm are in a more or less central location of each cycle, which begins with the germinal stem cells in an adult and ends with the germinal stem cells in offspring. Thus, in principle, sperm could provide two types of markers: those which are retrospective, in the sense that they monitor differentiation events that occurred earlier in the testis and epididymis, and those which are predictive of the fertility status of the adult and of the genetic health of his offspring. The fertility status of an adult is judged by his ability to produce sperm that successfully fertilize a female egg; markers of fertility status are a major topic of this chapter. (Markers for assessing genetic damage in sperm and genetic health of offspring are reviewed in Chapter 9.) Both retrospective and predictive sperm markers are needed for evaluating the reproductive health of males.

Semen analysis is a common component of fertility examinations (see Chapter 3). It has had a very long history; sperm are thought to have been one of the first kinds of cells analyzed under the microscope by its inventors, Leeuwenhoek and his student Hamm. The biology of the human spermatozoa has been the topic of several comprehensive reviews (Zaneveld, 1979; Olson, 1982)

Three applications of semen markers have been proposed: as markers of sperm production and function, as indicators of fertility status, and as indicators of exposure to a reproductive toxin.

In this chapter, we distinguish among these three applications because we still lack the data to understand the quantitative relationship among them. With sperm count, for example, there remains uncertainty in how to interpret a 10% reduction in group mean value even though all would agree that a 100% reduction (i.e., no sperm in the ejaculate) assures infertility.

The use of semen markers to discriminate the effects of environmental, therapeutic, and occupational exposures does not necessarily require that a marker be associated with fertility status. As discussed earlier in this chapter, semen from exposed and unexposed cohorts can be compared with each other and with historical controls to identify and evaluate the effects of hazardous agents.

Markers of sperm production and function reflecting both testicular and accessory organ function are needed for understanding the molecular mechanisms of human spermatogenesis and will provide candidate markers for fertility and toxicologic evaluations.

No semen marker has emerged as a definitive indicator of male reproductive health. It is generally agreed that a battery of sperm characteristics should be measured jointly to assess male reproductive health, and repeat measurements are very useful. The underlying hypothesis is that male fertility is multifactorial, requiring the normal function of numerous molecular and physical aspects of sperm and seminal fluid (see, for example, Amann and Berndtson, 1986).

Table 7-2 assigns the biologic markers of male reproductive toxicology to the following groups: (1) markers that have been successfully used to detect the human male reproductive effects of exposure to radiation or chemical agents and (2) markers for which no toxic effects data are available but for which human baseline data have been obtained or are under investigation. Also listed are new research concepts based on modern cellular, molecular, and recombinant DNA techniques that promise the next generation of biologic markers of human male reproductive health. For all markers, baseline data in normal individuals usually are considered prerequisite for detecting abnormal events and for determining the sample sizes required to detect changes. Markers that already have been used to detect the effects of toxic exposure are not necessarily further in their development than markers for which baseline data are under investigation. Both categories of markers need continued research to improve their sensitivity, investigate underlying mechanisms, determine their quantitative relationship to changes in human fertility, and other aspects as detailed in the next sections.

PHYSICAL CHARACTERISTICS OF THE HUMAN EJACULATE

Human semen coagulates shortly after ejaculation, and prostatic enzymes liquefy it within about 30 minutes, although in some men this may take hours. The coagulation and increased semen viscosity are due to prostatic and seminal vesicle components added during ejaculation. The specific relationship between semen viscosity and fertility remains unclear. The color of the ejaculate varies among shades of white, yellow, and gray. The white cloudy component is thought to be concentrated sperm, but the relevance of semen color to fertility remains obscure. The pH of normal semen is between 7 and 8 (low pH might be due to obstruction of the ejaculatory ducts, a rare disorder in men), but the relationship of subtle changes in semen pH to fertility also is unclear. No reports are available on the effects of exposure to xenobiotic agents on coagulation or color (Wyrobek et al., 1983a). Recently, Welch et al. (1988) reported a very slight elevation in semen pH of shipyard painters exposed to ethylene glycol ethers.

The volume of human ejaculate is usually about 2 to 5 ml (Hargreave and Nilsson, 1983), although samples of less than 1 ml and greater than 6 ml are not uncommon in large surveys. Semen volume can be affected by abstinence time and has been reported as increasing by 0.4 ml/day and reaching a plateau by about 5 days (Clark and Sherins, 1986). Human semen volume is usually not affected by exposure to xenobiotics (Wyrobek et al., 1983). However, methadone treatment has been reported to lead to an apparent increase in sperm concentration by decreasing semen volume (Cicero et al., 1975).

PRESENCE OF NONSPERM CELLS IN SEMEN

Four types of cells other than sperm can be found in human semen: microorganisms, white blood cells, duct cells, and immature germ cells (Amelar and Dubin, 1977; Belsey et al., 1980; Amann, 1981; Eliasson, 1981; Alexander, 1982). White blood cells and pathogens, such as bacteria, are indicative of reproductive tract infections, which are usually treated accordingly. Duct-lining cells and immature germ cells are normal components of the human ejaculate. Immature cell types can be distinguished morphologically on smears with Papanicolaou's stain or Harris-Shorr technique (Auroux et al., 1985). However, immature germ cells are not commonly scored as markers of reproductive health, because their identification is highly subjective and their relevance to fertility status is not established. Immature germ cells are an unexplored and promising area for future marker development.

SPERM NUMBER

Sperm number refers to the number of sperm in the ejaculate, expressed as total sperm or numbers per milliliter of semen. Objective measurements are easily made with a hemocytometer. Electronic scoring with a Coulter counter is an alternative to the use of a hemocytometer, but has reduced accuracy at low sperm concentrations (under 10 million per milliliter) (Gordon et al., 1965, 1967). Also, debris can clog the measurement orifice, and other cells or fragments can produce electronic measurement artifacts.

As Marker of Sperm Production

The presence of sperm in semen is compelling evidence of active sperm production. Its absence can reflect the inactivity of the seminiferous epithelium or a post-testicular tubular obstruction that can be resolved by testicular biopsy and vasography. Clinically, men without sperm in their ejaculates are termed azoospermic; men with some sperm, but fewer than 20 million per milliliter are generally termed oligospermic; and men with higher concentrations are termed normospermic. Men who cannot produce semen are termed aspermic.

To allow a continuum of sperm production that can reach approximately 100 million per day, there are probably 4 to 6 million divisions of stem cells to form new stem cells and committed spermatogonia each and every day. Sperm concentrations vary markedly among men; group-average values in fertile men are typically about 60-100 million per milliliter; about 1% of men are azoospermic and 10% are oligospermic (MacLeod and Wang, 1979; Whorton and Meyer, 1984). Sperm concentration also varies among ejaculates of a given person. Katz et al. (1981) evaluated the sources of variation in sperm concentration and reported 73% of the total variation to be due to "among-donor" effects and only 27% to be due to "within-donor" effects. A large source of the variability is due to the interval of sexual abstinence.

In mice, the number of mature epididymal sperm is correlated with the number of spermatogenic stem cells (Meistrich, 1982), but a similar relationship has not been established for human sperm.

As Marker of Fertility Status

There is little agreement about the specific relationship between sperm number and fertility status, nor is there agreement on whether total number per ejaculate or sperm concentration (number per ml) are more relevant. Using group data from several infertility clinics, Meistrich and Brown (1983) evaluated the mathematical relationships between sperm concentration and likelihood of infertility. On the basis of group averages, men with sperm counts below approximately 20 million per milliliter show an increased likelihood of infertility. However, at higher sperm concentrations, there was no apparent correlation between concentration and fertility status. Very high sperm concentrations (over 200 million per milliliter) also might increase the likelihood of infertility (Niendorf, 1964). The biologic basis for the increased likelihood of infertility at both extremes of sperm concentration is not well understood. (The reason that millions of sperm are required to fertilize a single human egg is unknown.)

The relevance to individuals of relationships based on group data is uncertain, for several reasons. First, total semen volume and sperm number differ among people, irrespective of their fertility (MacLeod and Gold, 1951; Smith et al., 1977; Zukerman et al., 1977; Homonnai et al., 1980a). Second, sperm number fluctuates within each person; for example, a man with an initial sperm concentration of 100 million per milliliter could have a true mean sperm concentration, based on six subsequent samples, of 50-230 million per milliliter (Schwartz et al., 1979). Third, the sperm number threshold for subfertility remains uncertain and might depend on the couple. Meistrich and Brown's (1983) analyses suggest that a person's fertility status is of a probabilistic nature below the value of approximately 20 million per milliliter; the lower the sperm count, the less likely a man would be to impregnate his partner. Similarly, the duration required to achieve fertilization might increase with lower sperm counts (Bostofte et al., 1982a; Collins et al., 1983). However, the numerical relationship for individuals is unknown.

Because of the large variation, some laboratories suggest that a person's true mean sperm concentration can be assessed only from repeated semen samples (three to six, or more) collected over periods of many months. In spite of these efforts, sperm concentration might have only minor utility as a marker of a person's fertility status, although it seems to be useful for comparing group effects.

As Indicator of Exposure to Toxic Agents

Germ cell killing is a common consequence of exposure to agents that reduce fertility or produce germinal mutations. The relationships among testicular exposure, time-course of sperm-concentration changes, and time-course of fertility changes are understood best for ionizing radiation (Searle and Beechey, 1974; Oakberg, 1975). In irradiated mice, approximately 10-fold reductions in sperm concentration are required for a noticeable reduction in fertility, and the duration of the induced sterile period is dose-dependent (e.g., 8-10 Gy led to a 10- to 11-week-long sterile period) (Oakberg, 1975). Human data relating dose to sperm

concentration changes show that men might be more sensitive to germ cell killing than mice and that the effects are longer-lived; acute exposures as low as 15 rads induced 4-fold reductions, and 4-6 Gy led to azoospermia for 5 or more years (Rowley et al., 1974; Clifton and Bremner, 1983). As stated earlier, induced reductions in sperm concentration might be more critical in men than in most animals. In men, a 4-fold reduction in mean sperm concentration would lower it into the range of oligospermia; larger reductions are required to do that in most animals. Also, as stated earlier, sperm concentration varies greatly from person to person. Therefore, even small reductions would bring some men into the oligospermic range and could increase their likelihood of becoming infertile. In addition, as we see with DBCP, chemical exposure of a group of men can shift the sperm concentration distribution to lower values and increase the proportion of azoospermics.

Of the indicators listed in Tables 7-1 and 7-2, sperm number has been used most often in studies of human male reproductive toxicity; 87 of the 89 chemical exposures surveyed in the last major review on this topic (Wyrobek et al., 1983b) and all semen studies of male reproductive effects of occupational exposure surveyed included it. Exposure to any of 57 agents led to detrimental effects on human sperm production. However, very few of the studies used rigorous study designs, including power calculations and sample size estimates (see the discussion of epidemiologic studies above).

Other Factors That Affect Sperm Number

Several factors not related to chemical exposure decrease sperm concentration, including short abstinence period, some illnesses, some viral infections, increased scrotal temperature, and some genetic and chromosomal disorders. Abstinence period is one of the best-understood factors (Schwartz et al., 1979; Baker et al., 1981; Mortimer et al., 1982). As abstinence period increased from 1 day to 1 week, sperm concentration increased by 10-15 million per milliliter per day, and total sperm count increased by 50-90 million per day (semen volume increases by 0.4 ml per day). The progressive increase in sperm concentration is thought to be due to accumulation of sperm within the epididymis and vas deferens, which are referred to as the extragonadal reserves. The capacity of these reserves is considerable (Amann and Howards, 1980; Johnson et al., 1980a; Tyler et al., 1982). Frequent (daily) ejaculation reduces sperm concentration to about 25% of normal; that suggests that the 4-fold higher sperm numbers seen after a week of abstinence are recruited from extragonadal reserves. After 2 weeks of abstinence, there might be up to a 10-fold increase in sperm concentration, compared with the values after frequent ejaculation (Johnson, 1982).

In addition, the means and location of semen collection may affect its quality. It is generally agreed that collection of the sample in the physician's office or clinic is superior to bringing the sample from home. On the other hand, the volume and total number of sperm in the sample might be influenced by the degree of sexual arousal that could be greater for samples collected at home, though there are few studies directly addressing this point.

SPERM STRUCTURE

Sperm structure is the study of sperm shape and size. Numerous systems have been developed for classifying human sperm into categories based on head, midpiece, and tail features. Scoring methods generally rely on observer judgment and visual criteria (e.g., Hotchkiss et al., 1938; MacLeod, 1964; Wyrobek et al., 1982; Eliasson, 1983). Sperm with differing features are assigned to shape-abnormality classes, such as double, tapered, narrow, and irregular (the specific numbers and types of categories depend on the classification system used).

Sperm structure has been used as a marker of sperm quality and fertility status and to monitor exposure to reproductive toxins in animals and humans. However,

its value in assessing sperm quality and fertility status remains controversial. In part, that is due to the lack of consistency among scoring methods and the associated difficulties in making interlaboratory comparisons (Freund, 1966). The problem is dramatized by a comparison of the distributions of the proportions of abnormal sperm among groups of men attending major fertility clinics in Europe (Fredricsson, 1979); the mean proportions of abnormally shaped sperm ranged from approximately 20% to 70% among clinics. Some laboratories have high within-laboratory reliability by relying on one highly trained scorer or only a few, by using decision-tree logic in assigning sperm to structural categories, or by using reference slides for setting and maintaining quantitative standards. However, the problems associated with visually determined sperm structure remain a major impediment to large-scale evaluation of the utility of sperm structure as a marker of male reproductive health, and further work is needed to automate these measures.

As Marker of the Quality of Sperm Production

Sperm structure is a measure of the quality of sperm produced by the testis and of alterations occurring in the efferent ducts and accessory glands. The effects of genetic and environmental factors on sperm nuclear structure have been investigated in detail in laboratory animals (see review by Wyrobek et al., 1983b). In inbred and hybrid lines of mice, adult males produce proportions and types of sperm-head shape abnormality that are characteristic of their genotype. The proportions of sperm with shape abnormalities in unexposed mice are determined by multiple genetic loci, including both Y-chromosomal and autosomal factors. Specific recessive mutations and chromosomal abnormalities have also been associated with abnormal sperm-head shapes, and the length of midpiece also depends on the strain of mouse studied. Abnormally shaped sperm are less likely to reach the oviduct and site of fertilization (Nestor and Handel, 1984).

TABLE 7-2 Status of Biologic Markers of Physiologic Damage to Human Male Reproduction

Tissue or Data Required	Markers Used to Detect Radiation or Chemical Effects in Exposed Humans	Markers Used to Obtain Human Baseline Data	Promising New Concepts and Research Topics
Testis (biopsy)	Histopathology of germ cells	Histopathology of Sertoli and Leydig cells	Computer-assisted image analysis of cells and tissue sections; markers of differentiation and function with recombinant DNA and monoclonal antibodies; cloning male reproduction genes
Seminal sperm	Sperm number, motility, morphology; double F-bodies; SDS sensitivity	Penetration of cervical mucus, zona pellucida, hamster eggs; computer-assisted image analysis of motility and structure; viability tests for dye exclusion and hypoosmotic shock	Monoclonal antibody mapping of sperm components: surface domains, acrosome, etc.; markers of differentiation with antibodies and recombinant probes; markers of sperm function: motility, capacitation, fertilization, penetration, etc.
Other seminal components		Immature germ cells; nongerm cells	Antibody detection of nonsperm cells, biochemicals and xenobiotics; semen markers of somatic tissue function: (Sertoli and Leydig cells, ducts, accessory sex glands)
Surveys, records, and maternal urine	Standardized fertility ratio	Time to pregnancy; early pregnancy marker	Detection of very early pregnancy

Human data on the genetic component controlling sperm-shape abnormalities are incomplete, but available evidence is consistent with animal findings. Men with some genetic diseases and chromosomal disorders show increased proportions of sperm-shape abnormalities (see review by Wyrobek et al., 1983a). Some cases of human sterility have been associated with specific types of sperm-shape abnormalities, such as round-headed sperm. Human sperm structural characteristics also might have a familial component.

Men show little change in their sperm structure distributions over periods of many years (Wyrobek et al., 1983a). At the average site in the seminiferous epithelium, it is estimated that approximately 23 spermatogenic stem cell renewals occur per year. Thus, the constancy seen in the proportion and types of sperm abnormalities over time might reflect genetic determination like that reported for mice. Further research is required to elucidate the genes and protein functions responsible for human sperm shaping.

As Marker of Fertility Status

Evidence from both animals and men links increased proportions of abnormal sperm forms with reduced fertility. Human sperm with abnormal head shapes are less motile in vitro than normally shaped sperm, and sperm structure has been correlated with poor hamster-egg penetration (Shalgi et al., 1985). In general, as the proportion of morphologically abnormal sperm increases, fertility decreases. The relationship appears to be nonlinear but has not been well described. There are case reports of specific sperm-shape abnormalities associated with sterility (Weissenberg et al., 1983). Although it is not common to find a single type of sperm defect indicative of infertility, this has been seen in bulls (e.g., Dag effect, where the head and tail are separated). In addition, Bostofte et al. (1982b) found a correlation among time to pregnancy, number of children, and proportion of morphologically abnormal sperm. Partners of men with higher proportions of abnormal sperm took longer to get pregnant, and the men generally had fewer children.

As Indicator of Exposure to Toxic Agents

Sperm structure has been used to assess the effects of ionizing radiation and chemicals on animal and human sperm production. For mice and several domestic animals exposed to nonsterilizing doses of ionizing radiation, two-component time responses were generally observed—a large transient increase in the proportion of morphologically abnormal forms shortly after exposure in mice treated with x-rays, beginning within 1 week after treatment, peaking at week 5 to 6, and returning to near background by 11 weeks after end of treatment. This is followed, depending on genotype, by a smaller but persistent increase above background (e.g., Bruce et al., 1974). In mice, both the transient and persistent increases were dose-dependent and have been observed also after chemical treatments (Wyrobek et al., 1983b; Meistrich et al., 1985).

The induction of sperm-shape abnormalities is highly indicative of exposure to a male reproductive toxin. A sperm morphology assay was developed and used to evaluate the germ cell effects of ionizing radiation and chemicals (Wyrobek et al., 1983a,b). Classifying sperm by their structure has statistical characteristics that make it well suited for use in both cross-sectional and longitudinal study designs.

Earlier studies in mice reported correlations between agents' abilities to induce sperm-shape abnormalities and their germ cell genotoxicity, as measured by tests for dominant lethality, heritable translocations, and seven specific-locus mutations (discussed in the next chapter of the report and reviewed by Wyrobek et al., 1983b). Because these correlations were based on small numbers of highly selected agents and the molecular mechanisms underlying them have not been identified, more data are required to resolve this point. Furthermore, for several agents known to induce sperm-shape abnormalities in mice, no mutational damage has been detected. Also, in a retrospective human study of 534 pregnancies, Homonnai et al. (1980b) found no relationship between sperm quality and adverse pregnancy outcome.

The available evidence suggests that the transient induction of sperm-shape abnormalities after exposure represents physiologic damage, rather than mutational damage to germ cells. The correlations observed between sperm structure and mutagenicity in mice might reflect the fact that chemical mutagens generally also damage cells physiologically. Two aspects of the sperm-structure response deserve closer scrutiny in regard to mutational damage: evaluations in the proportion of sperm-shape abnormalities that persist long after exposure, and the multigenerational inheritance of sperm-shape defects after mutagen treatment of male mice (e.g., Timourian et al., 1983).

The effect of ionizing radiation on human sperm structure has not been as fully investigated as the effect on sperm concentration. The results indicate that ionizing radiation induces sperm-shape abnormalities in exposed men. In addition, 44 of the 89 (49%) chemical exposures in the above-mentioned survey (Wyrobek et al., 1983a) used sperm structure as one of the markers of spermatogenic damage. Where data are available, the time course of induced sperm structural changes approximate the time course of reduction in sperm number. The chemical exposures studied included occupational, therapeutic, and environmental chemicals.

Other Factors That Affect Sperm Structure

As in other species, human sperm structure is relatively unaffected by abstinence time and several of the technical factors that influence sperm concentration and motility (Amann, 1981). However, sperm structure is sensitive to testicular temperature. Febrile diseases and severe allergic reactions might lead to increased sperm-shape abnormalities (MacLeod, 1964). Increased temperatures—such as those found in saunas or hot baths and those experienced by professional truck drivers—could also induce sperm-shape abnormalities (see review by Wyrobek et al., 1983b). Although major increases in testicular temperature are clearly associated with increased sperm abnormalities, the effects of small and irregular increases (as in the occasional use of hot tubs or in common illnesses, such as seasonal colds and influenza) have not been well studied, and their impact on human sperm production and fertility remains uncertain. Any factor that raises the core temperature of the testis or reduces heat dissipation is a candidate for heat-induced injury to spermatogenesis. Also, sperm structure is usually assessed on stained sperm smear and thus might be affected by slide preparation and staining conditions. However, the lack of consistent scoring criteria is by far the largest single factor affecting the assessment of sperm structure; efforts are under way to rectify this based on image analyses.

SPERM MOTILITY

Sperm motility is a measure of the "movement" characteristics of sperm. Sperm are flagellate cells propelled by tails equipped with contractile proteins whose movements relative to each other control tails' characteristic wavelike motion. The contractile proteins are arranged in longitudinal organelles within the sperm tail and include the coarse outer fibers, subfilaments, and microtubules. Although sperm motility is crucial in fertilization, sperm motility probably is not the primary method for transporting sperm within the female reproductive tract. That seems to be accomplished primarily by muscular contractility and ciliary activity of the female reproductive tract. Sperm motility is important for traversing several key junctures within the female tract—the cervix and the uterotubular junction—and might be essential for sperm penetration of cumulus cells and zonae pellucidae.

Sperm motility is highly sensitive to extracellular conditions both within and outside the body. Sperm are immotile while in the lumen of the seminiferous epithelium in the efferent ducts and in the proximal and middle portions of the epididymis; they do not attain their motility potential until they reach the distal epididymis. It is hypothesized that the sperm cell membrane contains specific chemical receptors. These sites could regulate the repetitive depolarization and repolarization cycles and might function to coordinate the molecular events of the beating of the tail.

Within ejaculates there is great diversity in the speed, direction, and type of sperm motion, which may be in part due to postejaculation technical factors, such as control of temperature (Phillips, 1972; Makler et al., 1979a). Little is known about the motility selection processes that exist during transport within the female; only a few hundred highly motile sperm reach an ovum, and probably only one activates it. Once these processes are better understood on the molecular level, diversity in motility of ejaculated sperm could become more useful for identifying fertile and infertile sperm and semen. The visual procedures for measuring human sperm motility used in most clinical laboratories are imprecise and subjective (Sherins and Howards, 1986). Typically, laboratory normal values depend heavily on method and scorer. The subjectivity of motility scoring has plagued interlaboratory and interscorer comparisons. Efforts are under way to use computer-assisted image analyses to provide objective measures of sperm motility.

As Marker of Sperm Function

The presence of motile sperm in the ejaculate is strong evidence of normal sperm production and function. However, immotile sperm are not uncommon and arise because of a variety of technical and biologic factors. Clinically, persons who produce only immotile sperm are termed asthenospermic. Defects in sperm structure and motility might be correlated, inasmuch as abnormally shaped sperm have poorer sperm motility than do normally shaped sperm (Katz et al., 1982).

As Marker of Fertility Status

The importance of sperm motility in fertility has been well established; numerous studies have demonstrated a correlation between motility and fertility (Freund, 1968; Eliasson, 1975; Hargreave and Etton,

1983). Motility is probably not directly coupled with fertilizing capacity. For example, freezing causes sperm fertilizing capacity to be lost before motility is lost. Although sperm motility might be a good correlate of fertility, a man with highly motile sperm could be infertile. However, men who repeatedly produce immotile sperm are generally not fertile. The subjectivity of most clinical motility measurements has made it difficult to establish standardized criteria for motility and to investigate possible quantitative relationships between specific aspects of motility and fertility.

As Indicator of Exposure to Toxic Agents

Exposure of humans to some toxic agents can affect their sperm motility. In a survey of the effects of chemical exposures on human semen (Wyrobek et al., 1983a), 59 of 89 agents were evaluated for sperm motility; 22 showed significant decreases in exposed men. None of the studies measured sperm motility quantitatively, and most did not use control or comparison groups. Furthermore, none of the agents surveyed reduced sperm motility without also decreasing sperm number (Wyrobek et al., 1983a; Ratcliffe et al., 1987).

Sperm motility is difficult to evaluate in human field studies. In fertility clinics, men are usually motivated to provide semen samples on site: but participants in environmental or occupational exposure studies often do not show such willingness. More typically, such participants prefer to collect samples at home and to deliver them to the laboratory at their convenience. That requires careful protocol preparation, patient instruction, adequate equipment for transporting the semen to the laboratory, and attention to temperature fluctuations and to the time between collection and analyses. Video-equipped microscopes have been used to collect moving images in the field for subsequent analyses in the laboratory, thus reducing time between semen collection and motility measurement.

Other Factors That Affect Sperm Motility

Sperm motility is relatively insensitive to duration of abstinence. However, several endogenous male factors have been reported to affect sperm motility, including donor age, extent of sperm maturation, and surface-active agents (antibodies and agglutination factors). Motility is also sensitive to exogenous factors, including viscosity, osmolality, pH, temperature, ionic composition, nature of the suspending fluids, and presence of chemical modifiers, such as inorganic ions, hormones, cyclic nucleotides, kinins, prostaglandins, and immunologic agents. Compared with sperm count and structure, sperm-motility measurement is especially sensitive to postcollection factors, especially time and temperature. Collection procedures can vary among clinical laboratories and those factors can affect motility in vivo (within the male and female tracts) and during semen handling and analysis. Thus all interlaboratory comparisons of visually determined motility data are problematic. As discussed in the next section, automation itself does not circumvent the variability caused by postejaculation factors, and controlling them will continue to be important as clinics turn to more automated sperm-motility measurements.

SPERM VIABILITY

Viability refers to the membrane integrity of sperm. Dyes that are excluded by live cells but incorporated by dead cells permit the determination of the proportion of live cells (Hargreave and Nilsson, 1983). That determination is particularly useful for distinguishing between live immotile cells and dead cells. A method for evaluating the membrane integrity of sperm is based on measurement of a cell's resistance to hypo-osmotic shock (Jeyendran et al, 1984). Such methods are sensitive to postejaculation factors, and baselines in normal men and the effects of intrinsic and extrinsic factors have not been fully evaluated.

SPERM FUNCTION

Sperm number, structure, and, to some extent, motility measure the physical aspects of human sperm, but indicate little about sperm function (i.e., ability to travel the female tract and to fertilize the ovum). Numerous attempts have been made to assess the functional aspects of sperm. The following are two in vitro measures of sperm function.

Cervical-Mucus Penetration

The ability to traverse cervical mucus is one of the first major requirements of fertile sperm within the female tract (Moghissi, 1976). Several laboratory methods have been proposed to evaluate sperm in cervical mucus (Kremer, 1965; Ulstein, 1972; Katz et al., 1980; Bergman et al., 1981). Usually, the ability of sperm to enter the mucus, and their motility within the mucus, are measured as well as their viability after penetration. The postcoital test is included in this group. The data provided by these tests are highly variable, owing in part to natural variations in the quality of human cervical mucus. Normal changes occur in cervical mucus during the menstrual cycle and test results are affected by the time at which the mucus is collected. In addition, the mucus of different women differs in other ways that make it difficult to establish objective criteria for mucus penetration and to assess the quantitative relationship between mucus penetration and fertility (Hargreave and Nilsson, 1983).

Sperm-Oocyte Interaction

For fertilization to occur, one of the several hundred sperm that reach the periphery of an ovum must traverse the cumulus cells and zona pellucida to penetrate it successfully. Ideally, living human ova would be used to assess the ovum-penetration ability of human sperm, but ethical considerations bar these types of analyses for diagnostic purposes. Two alternate techniques have been developed: the zona-free hamster egg penetration test, and the nonliving, human egg penetration test.

The zona-free hamster egg penetration test scores the proportion of enzymatically denuded hamster eggs that are successfully penetrated by human sperm (Yanagimachi et al., 1976; Barros et al., 1978; Rogers et al., 1979; Hall, 1981; Chang and Albertson, 1984). Penetration is scored as the proportion of eggs that contain sperm undergoing nuclear decondensation. However, there is large variability in results, and only samples with very low penetration rates (e.g., less than 10%) are considered potentially abnormal. Even repeat samples from the same men show large variability. The utility of this test as an indicator of fertility remains highly controversial. Furthermore, the underlying molecular mechanisms are unknown, and it remains unknown exactly what the hamster assay is measuring.

Using nonliving human eggs and human zona pellucida circumvents some of the disadvantages of using hamster eggs (Overstreet et al., 1980). However, this technique is not widely available, because of the difficulties in obtaining a supply of human eggs.

OTHER SPERM MEASUREMENTS

Sperm Agglutination

Techniques for detecting antisperm antibodies in blood, semen, and cervical mucus are available (Rumke and Hellinga, 1959; Halpern et al., 1967; Rumke, 1968; Haas et al., 1980; Mathur et al., 1981). They are used clinically where incompatibility between a man and a women is suspected (for example, poor cervical mucus penetration). The presence of sperm antibodies is a concept closely related to sperm domains, which is discussed later in this chapter.

Nuclear Chromatin

The nucleus of the differentiating germ cell undergoes dramatic changes during spermatogenesis in chromatin structure and in the constitution of basic proteins. The mammalian sperm nucleus is highly compact, rigid, and of high specific gravity; its DNA is genetically inactive and relatively inert in response to xenobiotic chemicals. During meiosis, several histones peculiar to meiosis appear in the nucleus with residual variants common to somatic cells. After meiosis, the meiotic histones are replaced by a series of transition proteins that are finally replaced with the basic sperm protein called protamine. Protamine is thought to facilitate the dense compaction of the sperm nucleus, making it genetically inactive and relatively inert to chemical exposure and imparting nuclear rigidity required for fertility. A single protamine is found in sperm of most mammals. In some mammals—including men, mice, and hamsters—sperm also contain a second protamine that is more variable in length and sequence than that of the first protamine. In human sperm, about 15% of the nuclear protein is histones. It is not known whether the protamine and histone content of human sperm is related to nuclear structure or function. Sperm nuclei naturally decondense on entry into the ovum in fertilization.

The following are two approaches for evaluating the ability of sperm to decondense in vitro:

• *Chemical decondensation of sperm nuclei in vitro.* Sperm nuclei can be decondensed in vitro with invertebrate egg extract, in high-salt solutions, with reducing agents, with detergents, or with special salts (Huret, 1986). The susceptibility of ejaculated human sperm nuclei to these agents in vitro might be indicative of the extent of chromatin condensation or nuclear protein cross-linkage. However, that has not been confirmed experimentally. Wildt and coworkers (1983) suggested that lead-exposed workers have an increased susceptibility to decondensation with SDS.

• *Colorimetric measurements of sperm chromatin.* Evenson and colleagues (1980) developed a flow-cytometric method for quantifying the ratio of single- and double-stranded sperm DNA. The method measures the fluorescent dye acridine orange, which distinguishes strandedness of DNA. In mutagen-exposed mice, the fluorescence ratio was highly correlated with the degree of induced sperm-shape abnormalities. Infertile bulls seem to have higher fluorescence ratios than fertile bulls. Data on humans are insufficient for evaluation. The method requires further study.

Double F Bodies

When stained with the fluorescence dye quinacrine and viewed in a fluorescent microscope, human sperm fluoresce over their entire nucleus. Some sperm have a brighter spot within the nucleus, thought to be the Y chromosome (Barlow and Vosa, 1970). The Y chromosome is known to fluoresce brightly in stained somatic nuclei and metaphase spreads of male cells. However, contrary to expectation, the proportion of sperm with one spot seldom reached the 50% value expected if the spots truly represented all sperm with Y chromosomes. Also, Kapp and associates (1979) suggested that the rare sperm with two spots could be due to Y-chromosomal aneuploidy and proposed that the "YFF" test be an indicator of induced aneuploidy. However, no data support that contention; in fact, mass measurements and comparisons with chromosomal analyses of human-hamster hybrid chromosomes suggest that sperm with two bright spots do not necessarily represent aneuploid sperm. The proportion of sperm with double fluorescence (double F bodies) increases with exposure to DBCP, adriamycin, and several other clinical agents (Kapp et al., 1979) and might be an indicator of exposure. However, the molecular and cellular mechanisms underlying this observation are unknown.

CHEMICAL COMPOSITION OF SEMINAL FLUID

Human seminal fluid is a biochemically diverse solution containing peptide, lipid, carbohydrate, glycoprotein, and salt components derived from the testis, efferent ducts, epididymis, and accessory glands. Normal seminal fluid constituents were described in detail in Chapters 4-6. Considered a research tool, chemical analysis of semen is not commonly used in fertility evaluation.

Semen can contain xenobiotic agents, including metals and chlorinated organic substances (Dougherty et al., 1981). However, it is not easy to determine whether xenobiotics in semen represent testicular or accessory sex organ exposures. It is also unclear how xenobiotics from accessory glands can affect sperm and their fate in the female tract; the events are likely to be agent-dependent. Because fertile sperm are removed from the seminal fluid early during their transit to the fertilization site, it is likely that only agents that are tightly bound to or internalized in sperm will be present at the fertilization site (e.g., pesticides that may be solubilized in the plasma membrane). Alternatively, xenobiotics in semen might affect fertilization and development by indirect exposure via the mother's circulatory system. However, exposure to the egg and embryo via maternal circulation is likely to be very small. The role of seminal fluid components on fertilization and development warrants further basic research. Reliable markers of the normal and xenobiotic constituents of semen are needed.

PROMISING RESEARCH CONCEPTS

The following are selected research concepts (see Table 7-2) that have promise of yielding semen-based markers of male reproductive health. Most are still too early in their development to evaluate their utility (1) for assessing male fertility, (2) as indicators of sperm production and function, or (3) as indicators of exposure to agents that interfere with male reproduction. Some approaches, however, are already so advanced that human baselines are being established (e.g., computer-assisted image analyses of motility and structure).

Automated Sperm Measurements

Automation is a general concept that may eventually be considered in the development of all sperm markers. At present, it is best exemplified by machine-based measurement of sperm motility and structure. However, a sizable gap remains between the capabilities of research instruments and their commercial availability.

Machine-Based Sperm-Motility Measurements

Considerable progress has been made toward developing automated methods for measuring sperm motility (Walker et al., 1982). With early quantitative approaches, sperm motion was observed directly under the microscope or was determined from photomicrographs of stroboscopically illuminated sperm. The most widely used quantitative approaches include multiple-exposure photomicrography (Makler et al., 1979b) and videomicrography. Some of these methods have been used on thousands of men (e.g., Katz and Overstreet, 1981; Overstreet et al., 1981) as part of routine clinical evaluations to determine percent sperm motility, mean swimming speed, and percent progressive motility. With videomicrography, microscopic images of multiple sperm fields can be permanently recorded for later analysis with manual techniques, such as the placement of graduated overlays on the video image (Overstreet et al., 1981), or with automated methods.

Modern automated methods generally rely on image analysis to locate the sperm and to quantify their motion. Edge-contrasting methods convert the visual images of sperm into binary masks from which the central coordinates (centroid) of each head are determined. By following the trace of the centroid in time, one can compute both rectilinear and curvilinear movements to evaluate sperm progressiveness. However, sperm motility is complex, and only a few aspects of sperm movement have been evaluated so far. Clearly, additional descriptors of sperm head and flagellar movement are needed. Furthermore, most automated methods provide distributional data and, statistical analyses should include descriptors of central tendency, dispersion, as well as distributional outliers.

Automation of sperm-motility measurement is a young field. To date, there have been few detailed evaluations of any automated method or motility characteristic.

Also, there is still no systematic study of normal values for any sperm-motility characteristic in a well defined population. It is important to emphasize that, for each new motility measure proposed, its range for normal values, sources of variation, relationship with fertility, and utility as an indicator of exposure must be established anew as part of the validation process for diagnostic purposes or exposure monitoring.

At least three automated motility-measuring systems are available commercially. Extensive validation is needed before their utility can be thoroughly evaluated.

Quantitative Sperm Structure

As with sperm motility, subjectivity plagued all attempts at interlaboratory comparisons of visual sperm structure assessments (Freund, 1966). Disagreement persists regarding the precise shape and size of fertile sperm, in part because different laboratories use different classification criteria (Fredricsson, 1979), sperm sizes are known to be donor-dependent, and no one has actually measured the structure of fertilizing sperm. In addition, fertile donors can differ markedly in the proportion of sperm in various shape classes.

Several decades ago, MacLeod and Gold (1951) developed a widely recognized visual scoring system; his success was due in large part to the high level of internal standardization he obtained by scoring smears himself. Later efforts to improve visual sperm classification have included complex assignment schemes that consider head, midpiece, and tail structure separately (David et al., 1975) and the use of decision-tree logic and reference slides (Wyrobek et al., 1982).

Recently, objective methods have been proposed for measuring sperm size and shape. Katz and Overstreet (1981) classified human sperm with graduated overlays on previously recorded video images. Schmassman et al. (1982) used an image-analysis system to measure sperm nuclear size and reported that the sperm of infertile men had larger nuclear area than the sperm of fertile men. In another evaluation of image analyses, 27 aspects of nuclear size, shape, orientation, and stain content were measured for each sperm nucleus and were evaluated statistically to identify parameter groupings that most accurately assigned sperm to 1 of 10 visually based shape classes (Moruzzi et al., 1988). An 86% overall classification accuracy was obtained with measurements, including basic morphometric characteristics, indicators of stain content, and measures of nuclear inhomogeneity. Those preliminary examples of the utility of automated sperm-structure analysis are encouraging, but much work is needed before these methods can be validated for clinical use. Additional research is needed to improve cell-staining procedures for image analysis, to "train" computers to recognize specific shape classes, and to develop methods for classifying sperm in real time. Cooperation will be required among researchers and clinicians to define individual sperm-shape classes with morphometric characteristics rather than visual criteria.

Alternatively, morphometric characteristics could be used to describe sperm samples independently of visual classification systems. For example, new classification systems could be based on distributional data for such measures as nuclear area, length, and width. Evaluations could be multivariate and whenever possible should consider descriptors that represent central tendencies (e.g., average nuclear area), dispersion (e.g., variance of nuclear area), and distributional outliers (e.g., proportions of cells above or below threshold areas).

General Considerations Regarding Automated Methods

Automation is very attractive because it provides a degree of precision and objectivity in measuring sperm motility and structure not attainable with previous visual methods. However, there are several factors that should be considered in the application of automated methods for any sperm parameter, not limited to sperm motility or structure:

1. Automated methods do not in themselves circumvent the fundamental problems of semen collection, handling, and preparation that especially affect sperm motility.

2. Machines do not make diagnoses; they simply provide quantitative data. The nature of data depends on the sophistication of the machine, the program language used, and the ingenuity of the programer and operator. These instruments typically provide voluminous data, and strategies for data handling must be developed concurrently.

3. Quantitative methods require reference or normal values for interpretation and statistical methods for making comparisons.

4. Each new machine method and especially each new proposed sperm measure will require new investigations of normal values among fertile men and of the effects of confounding factors. There are few data to establish normal values for any of the quantitative sperm measurements in well-defined populations.

5. Also, the relationship of each characteristic to fertility must be evaluated before its clinical relevance can be ascertained. At present we can only guess which characteristics will be useful for predicting fertility status and which will be sensitive to environmental exposures. Until methods and sperm characteristics are validated, they must be considered research tools, rather than clinical instruments.

These features of automated sperm analyses do not diminish the benefits of these technologic developments. Rather, they should serve as caution against overinterpretation and as guidelines for future needs.

Markers of Sperm Function

Sperm progression through the female tract is a multistage process, and our understanding of the underlying molecular components that control sperm capacitation and sperm penetration through the cervical junction, uterotubular junction, cumulus cells, zona pellucida, and plasma membrane of the egg is growing slowly. Proposed markers of sperm penetration are either mucus-based or chemical-based; both show promise. Change in androgen binding also deserves further study as a possible marker of sperm-penetration potential.

There are few promising approaches for measuring sperm capacitation directly in sperm, because underlying mechanisms are not clear. Some sperm surface components are lost during capacitation, including antigens (Vernon et al., 1985) and an acrosomal stabilizing factor (Eng and Oliphant, 1978). After capacitation, an antigen initially detected only on the posterior sperm tail becomes localized predominantly on the midpiece (Myles and Primakoff, 1984). Further research is encouraged to define the key molecular aspects of capacitation.

Fertile sperm require normal functioning of many enzyme systems. In vitro assays of enzyme functions related to energy metabolism, sperm motility, sperm penetration, and fertilization would provide valuable insight into the biology of reproduction and might have clinical application in establishing the biochemical basis of infertility. An example of a promising approach is the recent observation that carboxymethylase activity is decreased in the semen of men with immotile sperm (Gagnon et al., 1982; 1986).

Available tests of sperm-egg interaction typically require a ready supply of fresh eggs and are plagued by both poor reproducibility and poor quantitation. Improvements and alternatives are needed. Optimistically, surrogate methods based on understanding specific molecular aspects of normal sperm-egg interaction might yield markers that are sperm-based but do not require mammalian eggs.

Immunologic Reagents for Studies of Semen and Spermatogenesis

Antibodies are highly specific for detecting single antigenic determinants, can be available in virtually unlimited quantities, and are usable in a wide variety of physiologic, morphologic, biochemical, and molecular studies. They recognize specific antigenic determinants on proteins and glycoproteins present in single domains on living cells, and they can be measured with sensitive and highly specific enzyme-linked immunosorbent assays (ELISA) and radio-immunoassay (RIA) procedures. Immunologic reagents can be used to investigate the molecular arrangement of both surface and internal components of human sperm, and they provide an approach for elucidating male reproductive abnormalities on a biochemical and molecular level. This section is organized by sperm components and well as surface domains (acrosome, tail and midpiece, and nucleus). In addition, the need of immunologic reagents for identifying and measuring natural and foreign components of seminal plasma, for determining hormonal status (not discussed in this section), and for use with recombinant-DNA gene-expression vectors is emphasized.

Antibodies can be used to study either cells or molecules by ELISA and RIA techniques. For applications requiring the analysis of single cells, cytometry can provide detailed measurement of antibody distribution on individual cells and cell-to-cell variations with a precision not possible through microscopic inspection. Flow cytometry and image analysis provide powerful complementary analytic capabilities. In addition, flow cytometry can be used to sort individual cells for microscopic and biochemical analysis (Van Dilla and Mendelsohn, 1979). Typically, cytometric analysis of antibody-labeled cells requires well-controlled fluorescence staining techniques.

Sperm Surface Domains

The sperm plasma membrane is divided into distinct regions or domains, each of which contains characteristic marker antigens, overlies a major structural component of the sperm, and has well-defined boundaries (Myles et al., 1981). The plasma membrane of the sperm head includes the anterior acrosome, equatorial segment, and postacrosomal domains. A variety of probes have been used to characterize the domains, but monoclonal antibodies have been used most effectively to map the location of sperm surface domains, identify antigens in particular domains, define the origin of such antigens, and determine the role of specific sperm surface antigens in reproductive processes (Eddy, 1987).

For example, during the acrosome reaction, the plasma membrane of the anterior acrosome domain fuses with the underlying acrosomal membrane, releasing enzymes from the acrosome and allowing the sperm to penetrate investments of the egg. A monoclonal antibody to a protein in the anterior acrosome domain of mouse sperm blocks the acrosome reaction and prevents fertilization (Saling, 1986). In addition, when the sperm reaches the egg surface, the membrane of the posterior part of the sperm head fuses with the membrane of the egg, allowing the sperm to enter the egg cytoplasm. In the mouse, a monoclonal antibody to a protein in the equatorial segment domain blocks this process and prevents fertilization (Saling et al., 1985). A result of the acrosome reaction is that most of the plasma membrane is lost from the anterior acrosome and equatorial segment domains. At the same time, part of the acrosomal membrane becomes continuous with the plasma membrane at the anterior edge of the equatorial segment, forming a new domain of the plasma membrane over the anterior sperm head. On fertilization, the sperm plasma membrane is inserted into the egg plasma membrane. Some sperm surface components appear to remain in a small patch in the egg plasma membrane (Gabel et al., 1979); others diffuse over the entire egg surface (Gaunt, 1983).

Monoclonal antibodies have been used to identify some of the antigens added to the sperm surface in the epididymis (Orgebin-Crist and Fournier-Delpech, 1982; Saling, 1982; Eddy et al., 1985). One such antigen is secreted by the epithelium in a discrete region of the mouse epididymis and attaches to the midpiece and distal tail portion, apparently by binding to specific acceptor sites in these domains (Vernon et al., 1982). Surface alterations also occur during ejaculation, when proteins secreted by the accessory glands of the reproductive system bind to sperm (Irwin et al., 1983; Isaacs and Coffey, 1984).

Additional changes take place in the plasma membrane as sperm undergo capacitation in the female reproductive tract. That process must occur if sperm are to complete the acrosome reaction and fertilize the ovum. Some sperm surface components are lost during capacitation, including antigens (Vernon et al., 1985) and an acrosomal stabilizing factor (Eng and Oliphant, 1978) that are acquired in the epididymis. The boundaries and contents of domains are also modified. An antigen initially detected only on the posterior tail domain of guinea pig sperm becomes localized predominantly on the midpiece domain after capacitation (Myles and Primakoff, 1984). Another antigen migrates from the postacrosomal segment domain into the acrosomal domain after the acrosome reaction (Myles and Primakoff, 1984).

Such studies indicate that sperm plasma membrane domains are dynamic, undergo structural and functional changes throughout the life of the cell, and contain antigens that serve vital roles in reproduction.

The establishment of domains requires synthesis of specific components in appropriate quantities, delivery of the components to the cell surface, and segregation of the components to specific regions of the sperm surface. Toxic agents could perturb domain formation, composition, or maintenance by acting directly on those processes or indirectly by altering spermatogenesis, Sertoli cell function, or endocrine processes. Later modifications of the sperm surface in the male reproductive tract depend on the normal metabolic and secretory functions of the epididymis and accessory glands. Toxic agents might affect those modifications directly (by interfering with processes at the sperm surface required for the addition, removal, or alteration of domain components) or indirectly (by perturbing biochemical or physiologic activities of the ducts or glands of the male reproductive tract). Such effects of toxic agents might be expected to alter the composition and distribution of domains and to be detrimental to male fertility.

It is recommended that studies to determine the usefulness of monoclonal antibodies to sperm surface components for detecting toxic effects on the male reproductive system be accorded high priority. They have considerable potential to identify the site and mechanism of action of toxic agents. These studies can be initiated with antibodies that are now available, but larger numbers of antibodies to rodent and human sperm should be prepared specifically for this purpose. If toxic agents are found to have detectable effects on sperm surface domains in experimental animals, monoclonal antibodies can probably be used for detecting toxic effects on the human male reproductive system.

Components in the Sperm

Several biochemical components of the human acrosome have been identified (Abyholm et al., 1981; Eliasson, 1982), including acrosin, hyaluronidase, corona-penetrating enzyme (CPE), ATPase, acid phosphatase, aspartyl amidase, and beta-glucuronidase. These components aid in sperm penetration of the zona pellucida (acrosin), in penetration of cervical mucus (acrosin) (Beyler and Zaneveld, 1979), in sperm penetration of the cumulus oophorus (hyaluronidase), in sperm penetration of the corona radiata (CPE), and in the sperm capacitation process (acrosin, CPE, and ATPase). The roles of acid phosphatase, aspartyl amidase, and beta-glucuronidase remain unknown.

A monoclonal antibody for human acrosin recently was described (Elce et al., 1986). The production of acrosome-specific monoclonal antibodies would be instrumental in characterizing and measuring the acrosomal components of "fertile" sperm and in identifying abnormalities.

At the center and along the length of the tail is an arrangement of microtubular doublets, similar to that in cilia. Further research is warranted to catalog the structural constituents (including dynein and the tubulins) and to decipher the molecular mechanisms of energy conversion and the roles of motion-related proteins and carbohydrates of the sperm tail. Antibodies to such sperm molecules will help in this research and could lead to the development of molecular markers of sperm tail

and motility dysfunction.

Monoclonal antibodies also could be valuable for investigating the nuclear constitution of sperm. Monoclonal antibodies specific for human protamines (Stanker et al., 1987a) and some sperm histones have been developed and characterized with ELISA and Western-blot methods. These antibodies require further study in well-characterized populations of fertile and infertile men for evaluating the relationship between sperm nuclear constitution and fertility. In addition, sperm antibodies against transition proteins, meiotic histones, and other components of the sperm nucleus are needed for basic studies of sperm biochemical structure, identification of candidate reagents for clinical study, and evaluation of the sperm-nuclear effects of environmental exposures.

Use of Antibodies for Studying Gene Expression

With specially engineered phage and bacterial hosts (e.g., Riva et al., 1986), antibodies can be used for screening clones that contain genes producing testis-specific proteins recognized by specific antibodies. For example, human testis cDNA libraries have been constructed into expression vectors, such as lambda gt11 and resulting bacterial plaques can be induced to produce human testis proteins. Isolated clones can be used as probes for the gene and to monitor sperm differentiation. Such probes also permit molecular characterization of chromosomal location, haploid copy number, and genetic controlling elements. In principle, any human protein or peptide (provided that it is antigenic) can be used to prepare monoclonal antibodies to screen for and clone the corresponding human gene. Thus, the development of antibodies against sperm components provides the useful first step for identifying and characterizing the genes that control that aspect of human sperm production.

Antibodies for Characterizing Semen

In addition to sperm, the human ejaculate

contains several other cell types described in Chapter 4. Monoclonal antibodies specific for each cell type would facilitate the objective characterization and measurement of these seminal components. In principle, antibody methods could measure the chemical composition of semen. Antibodies are needed for nearly all the normal seminal constituents, including hormones, enzymes, proteins, carbohydrates, and lipids. Once developed, the antibodies might be used to determine the concentrations of those normal constituents in groups of fertile and infertile men and in men exposed to reproductive toxins, to evaluate their utility. Semen is a complex mixture derived from several glands; coupled with animal research, these antibodies could be used to determine the glandular source of each constituent. Research on monoclonal antibodies for detecting and measuring normal chemical components of semen should receive high priority.

In addition, antibodies could be used to detect chemicals not normally found in semen, such as environmental and occupational chemicals and drugs. Antibody methods have advantages over spectrophotometric and chromatographic methods: they cost less to perform, are highly specific, and can have excellent sensitivity. The sensitivity of the antibody approach for detecting xenobiotics is illustrated by the recent demonstration that antibodies can detect small chlorinated carbon molecules at concentrations of less than 1 ppm in soil samples (Stanker et al., 1987b; Vanderlaan et al., 1987). Such methods have not yet been applied to semen.

SEMEN MARKERS OF SERTOLI CELL AND LEYDIG CELL FUNCTION

Semen provides a natural window for evaluating retrospectively the function of the major somatic cells that support sperm production: Sertoli and Leydig cells, for evaluating retrospectively some aspects of epididymal function, and for evaluating prostatic and seminal vesicle function. As discussed in Chapter 4, those cells and organs contribute specific constituents to semen. For example,

semen is made up of a minute amount of sperm-dense epididymal fluid mixed at ejaculation with the secretions of the accessory glands (Lilja et al., 1987). The major structural protein in the coagulated component of the ejaculate is a high-molecular-weight protein from the seminal vesicles termed HMW-SV protein, or semenogelin. Its transformation into three subunits during liquefaction results in a series of basic low-molecular-weight proteins. The seminal gel liquefies through proteolysis by prostatic kallikreinlike serine protease (also known as prostate-specific antigen). Fibronectin, an "adhesive" glycoprotein, is also part of the seminal gel where it is linked to semenogelin. The seminal vesicles are thought to excrete lactoferrin, a metal-chelating protein that adheres to sperm. Easy and reliable quantitative methods are needed to monitor semen for those and other specific secretory products. The molecular function of most of those products, as well as of the factors that modulate seminal amounts and activities, remains to be evaluated. In addition, the normal baselines as well as the relationship to fertility of semen markers of the Sertoli cell, Leydig cell, and accessory organ function remain to be determined.

RECOMBINANT-DNA METHODS FOR STUDY OF HUMAN SPERMATOGENESIS AND SEMEN

As reviewed by Hecht (1987a), spermatogenesis is an ideal differentiating system for investigation of the control of gene expression as related to normal protein and cell function. Its morphology and kinetics have been well described, there is only a single cell product, and our understanding of biochemical mechanisms is rapidly increasing. The availability of DNA probes for spermatogenic proteins derived from both messenger RNA and genomic DNA is increasing rapidly. As a result, the genetic factors that control sperm differentiation are beginning to unfold. These recombinant-DNA probes are discussed in detail in Chapters 4 and 9.

Recombinant-DNA probes promise at least two types of markers that might become useful for assessing male reproductive health.

First, some recombinant-DNA probes might provide a means for evaluating male infertility at the gene level. For example, in conjunction with other technologies, such as cellular staging under the microscope and antibodies against spermatogenic proteins and other cell markers, recombinant probes could be useful for determining whether transcription is altered in infertile men and, if so, what spermatogenic cell stages are involved. In addition, analysis of genomic DNA from carefully selected men with similar spermatogenic arrest patterns could lead to the identification of characteristic genetic lesions that cause infertility.

Second, as described by Hecht (1987b), recombinant probes could also provide a means to analyze the DNA of individual sperm. That could be useful in measuring the proportion of sperm with specific genetic lesions and investigating the lesions in detail. However, this application has many additional technologic hurdles, because methods for probing DNA of single cells are still in their infancy. Although chromosome-specific repetitive DNA probes have been used to assess the ploidy of individual cells, methods are not yet sensitive enough to detect unique sequences in single cells.

In spite of present limitations, research and development of recombinant-DNA probes for spermatogenic genes are progressing rapidly, and they are expected to have broad applications for assessing male reproductive health. Recombinant techniques are also applicable to the study of Sertoli cell, Leydig cell, and accessory gland function, and continuing research in this field is strongly encouraged.

8

Assessing Transmitted Mutations in Mice

This chapter reviews laboratory tests for assessing exposure or heritable genetic effects of exposure in laboratory animals. Genetic damage can occur in somatic cells and in germ cells. Induced genetic damage in germ cells can lead to alterations in cell functions or cell death. Alternatively, induced genetic damage can be transmitted to the next generation, in which case the conceptus might suffer no ill effects or might have undesirable manifestations (including death) during some or all stages of life (prenatal and postnatal).

Mutagenic chemicals and radiation of various kinds are widely distributed in the environment. The mutagens that attract the most attention are products of the chemical industry, but some exist naturally in the environment (NRC, 1982). Many genetic systems are available for mutagenicity screening. For the purpose of this report, only end points of mutational damage to mammalian germ cells will be considered. When a chemical to which humans are exposed causes mutations in a laboratory mammal, such as the mouse, the genetic risk associated with human exposure to the chemical becomes a matter of serious concern. During the last decade, many chemicals under development for possible use in or as drugs, cosmetics, and food additives have undergone mutagenicity

testing. A dilemma arises when a chemical, either under development or already in the human environment, is of value to at least some segments of society and has been shown to be mutagenic. It is difficult to determine the largest human exposure that poses no substantial harm to human health. Determination of such an exposure is referred to as genetic risk assessment.

Some environmental chemicals cause genetic damage to germ cells of experimental mammals—e.g., ethylene oxide, trimethyl phosphate, dibromochloropropane (DBCP), acrylamide, bisacrylamide, and many cancer chemotherapeutic drugs. It is assumed that these chemicals will be mutagenic to human germ cells under appropriate conditions. Direct study of chemically induced transmitted genetic effects (mutations) in humans is virtually impossible, so the risk must be estimated from a variety of experimental test systems. The systems are usually categorized into two groups—mammalian germline (MG) and nonmammalian germline (NMG). Several ways of evaluating genetic risk have been proposed; they differ not only in how MG data are used, but also in the emphasis placed on NMG data.

There is no consensus on how to assess genetic risk associated with environmental chemical mutagens, and acceptable strategy and guidelines are crucially needed (NRC, 1982, 1983). In the United States, no chemical has ever been regulated on the basis of its potential for increasing the mutation load of later generations, nor has genetic risk evaluation contributed to the regulatory decision-making process (OTA, 1986). Regardless of the specifics accepted for genetic risk assessment, data on transmissible genetic effects in laboratory mammals will be indispensable—not only as a measure of end points, but also to form a standard for evaluating the usefulness of results of NMG tests as indicators of genetic risk to humans.

Assessment of the genetic risk associated with exposure to a chemical includes several components:

• Defensible evidence that the chemical in question has the potential to induce genetic damage to human germ cells.

• Identification and quantification in experimental systems of the types of mutations that are expected to be produced and transmitted to the next generation.

• Extrapolation of experimental results to humans (i.e., quantification of the increase expected for each class of mutation associated with likely human exposures).

• Estimation of the expected total contribution to the human genetic load.

• Estimation of the impact of the expected mutational increase on society.

The list is formidable. If progress is to be made in practical genetic risk assessment, a simple concept that makes use of carefully selected biologic markers needs to be adopted.

Evaluation of genetic risk of a chemical follows a three-step process—detection of mutagenicity, measurement of genetic effects, and extrapolation of results. Chemical mutagens can vary in the manner in which they react with various cellular and chromosomal components. Consequently, the genetic damage that they produce can vary, and no test system can measure every conceivable type of genetic damage. Methods for measuring some of the end points have been established; methods for measuring others are still under development. Obviously, it is impractical, as well as expensive, to use all the established tests for transmissible genetic effects for every chemical that needs to be evaluated. Therefore, a simple concept must be developed for the purpose of practical genetic risk assessment.

ASSESSING MARKERS IN LABORATORY ANIMALS

Our understanding of the mechanism and effects of interaction between xenobiotic substances and mammalian DNA comes predominantly from in vitro and in vivo studies of animal cells. The susceptibility of the male parent to induced mutations in reproductive cells was demonstrated by Müller with irradiated *Drosophila* males 6 decades ago (Müller, 1927). Since the 1940s, laboratory mice have been intensively studied for spontaneous and induced gene mutations and chromosomal abnormalities. Radiation and over half the approximately 20 chemicals tested so far in mice induced heritable mutations in mouse male germ cells (as measured by the specific-locus-mutation and heritable-translocation tests). As yet, there is no validated murine test for measuring chemically induced germline mutations directly in the germ cells of exposed males, and all germinal mutagenicity in mice is inferred from heritable-mutagenicity tests. Generally, there has been good agreement between results in somatic cells in vivo and heritable effects of treated differentiating male germ cells. However, the induction of somatic mutations has not been predictive of mutagenicity in spermatogenic stem cells. Only a subset of agents that induce mutations in somatic cells also induces mutations in the spermatogenic stem cells. Continuing studies in mice have attempted to reveal the nature of the selective immutability of stem cells by some agents, and their results have suggested that these cells have high repair capability. Studies in mice might also be used to investigate the molecular aspects of the different somatic and germinal mutational responses. This work might identify specific genetic lesions for which there is high somatic-germinal concordance. Studies in mice have been and will continue to be the cornerstone of our understanding of the basic aspects of spontaneous and induced mutations. And, the mouse will continue to play a key role in the quest to understand the molecular nature of mutations and their effect on phenotype and health.

An accepted tenet of toxicology and carcinogen testing is that we should not rely on one species for risk extrapolations to human beings. It is well known that metabolic activation is required for the toxicity, carcinogenicity, and mutagenicity of some chemicals and that animals and people can differ in their metabolism. For example, the germinal effects of exposure to DBCP differ markedly among species (Wyrobek et al., in press). It kills spermatogenic cells in most species, including rats, hamsters, rabbits, and humans; and it induces dominant lethal mutations in treated male rats and might induce spontaneous abortions in the spouses of exposed men. Mice are the only animals known whose male germ cells are essentially nonresponsive to DBCP, showing neither toxic nor mutational effects. Thus, for DBCP, the mouse would be a poor choice as a test species for estimating human germinal toxicity and mutational risks. Mouse-human discrepancies have been observed also with the germinal toxicity of some cancer chemotherapeutic agents, such as adriamycin (Meistrich et al., 1985). Possible solutions for the problem associated with interspecies extrapolation would be the use of molecular dosimetry (e.g., DNA or protein adducts) to develop quantitative methods for extrapolation, the development of a second laboratory species for measuring germinal and heritable mutations, and the development of methods for detecting germinal and heritable mutations directly in people.

MARKERS OF EXPOSURE

Some measure of exposure is necessary, not only to establish whether a chemical or its active metabolite reached the germ cells, but also to relate a genetic response to specific molecular target sites qualitatively and quantitatively.

When the genetic response is clearly positive, the question of whether the chemical reached the target cell is academic. Absence of a genetic response can mean that the chemical did not reach the germ cells, that the chemical is a nonmutagen, or that the chemical is a mutagen but the test system is insensitive or has an inherently effective repair capability. Thus, markers of exposure are necessary for proper interpretation of results.

The markers that indicate exposure of male and female germ cells were discussed in detail recently by Russell and Shelby (1985). They can be classified into the following categories: cytotoxicity, cytogenetic effects, cellular biochemical responses, molecular binding, and cellular morphologic responses.

Cytotoxicity

Cytotoxicity to some germ cells implies that the test chemical reached the gonads and supports the assumption that the surviving cells were also exposed. Cytotoxic effects might be determined directly by histologically examining the seminiferous tubules (or of the ovary for female animals) at an appropriate interval (usually days) after exposure, allowing for the manifestation of cellular degeneration or for the disappearance of affected cells. When specific germ cell stages are scored separately, the method is sensitive. Very low levels of cell-killing that might not result in a demonstrable effect on fertility might be detected. Often, reproductive performance can be affected; without histologic verification, however, that is unreliable as a measure of germ cell exposure, because fertility can also be reduced by nongerminal means.

Cytogenetic Effects

Demonstrable chromosomal damage is direct evidence of exposure. The cytogenetic end points used widely are chromosomal aberration, sister-chromatid exchange (SCE), and micronucleus formation. In all cases, scoring is done in descendants of exposed cells. In males, chromosomal aberrations can be scored in spermatogonial metaphases, in meiocytes, and in the zygotic metaphase; micronucleus information in spermatogonia, in spermatids, and in two-cell embryos; and SCE in spermatogonial and meiotic metaphases. In females, chromosomal aberrations can be scored in the metaphase-II and in the zygotic metaphase stages and micronucleus in two-cell embryos. (SCE induction in female germ cells has not been reported.) SCEs can occur in the presence or absence of demonstrable chromosomal aberrations and point mutations. Micronucleus formation is generally believed to result from the chromosomal elimination that follows chromosomal breakage or misdivision.

Biochemical Responses

Introduction of exogenous substances into the cell elicits enzymatic responses. In the case of mutagens, DNA damage (an indication of exposure) can trigger unscheduled DNA synthesis (UDS) in some male and female germ cell stages. Spermatocytes, spermatids, and oocytes do not normally undergo DNA synthesis; however, when chemicals bind to DNA, these germ cells respond by repairing some altered sites. If the germ cells are provided with radioactive thymidine during repair, the amount of repair activity (thus, the amount of DNA damage) can be measured.

Molecular Binding

One of the most direct measures of germ cell exposure is the demonstration of molecular binding. In the context of mutagenesis, the most important molecular target sites are the chromosomal DNA and proteins (histones and protamines). Various techniques of molecular dosimetry can be used to measure very low frequencies of adduct formation through the reaction of the test chemical or its metabolite with germ cell DNA. If the germ cell stage studied in molecular dosimetry and in mutagenesis is the same, it is possible to relate the magnitude and quality of DNA binding to mutation induction. This is the most sensitive marker of exposure so far, although adduct formation in germ cells has been studied only in males.

Cellular Morphologic Responses

During spermatogenesis, cells undergo changes that culminate in spermatozoa with the morphologic characteristics of their species (see previous chapters for extended discussion). Changes in sperm structure that result from chemical exposure of the male indicate toxicity either directly to the maturing spermatogenic cells or indirectly through damage to other systems. The distinction between the two types of toxic response is difficult to make.

TESTS IN MICE TO DETERMINE TRANSMITTED GENETIC EFFECTS

Chemical mutagens react with various cellular and chromosomal components in different ways, so they produce different types of genetic damage. Generally, mutations are of two types: either gene (or point) mutations and small deficiencies or chromosomal aberrations (changes in chromosomal structure or number). The tests of induced transmitted mutations in mice are summarized in Table 8-1. The genetic tests that are involved with these markers have been discussed in detail by Russell and Shelby (1985); what follows here is a brief summary.

Specific-Locus Test with Visible Markers

This is the most widely used system for detecting induced point mutations and small deficiencies. The test makes use of genetic information on up to seven loci that affect visible characteristics of the animal. Animals of one strain, which has normal (or wild-type) alleles at all seven loci, are exposed to the test agent and then mated with animals of a tester stock, which is homozygous for recessive alleles at all the loci. Normally, all the progeny would resemble the wild-type parent. However, if mutations are induced at any of the loci, the type or distribution of visible characteristics—such as coat pigment, eye color, hair structure, or structure of the external ear—might be affected. The test can be used to study

TABLE 8-1 Procedures for Detecting Transmitted Mutations in Mice[a]

Genetic Test	Biologic Marker	Primary Genetic Lesions Detected to Date	Sex Studied	Status of Test
Specific-locus tests with visible markers	External features	Intragenic mutations, small deficiencies	Male, female	Developed; in selective use
with biochemically detectable markers	Electrophoretic pattern; enzyme activity	Intragenic mutations, small deficiencies	Male	Developed; in limited use
with immunologic markers	Skin-graft rejection	Intragenic mutations, small deficiencies	Male	Undeveloped; dormant
Recessive lethals at nonspecific loci	Intrauterine death; absence of marked class	Deficiencies, gene mutations	Male	Undeveloped; dormant
Heritable-translocation test	Reproductive disturbance	Reciprocal translocation	Male, female[b]	Developed; in selective use
Inversion test	Cytologic aberrations	Inversions (paracentric or pericentric)	Male	Developed; in limited use
Dominant-lethal test	Early embryonic death	Breakage, rearrangement, missegregation	Male, female	Developed; in extensive use
Cytogenetic analysis of zygotes	Abnormal pronuclear complement	Breakage, rearrangement, missegregation	Male, female	Developed; in limited use
Sex-chromosome-loss test	External features	Breakage, rearrangement, missegregation	Male, female	Developed; in limited use
Nondisjunction tests	Chromosomal hyperploidy, external features	Missegregation	Female	Under development
Dominant-mutation tests	Skeletal changes, cataracts, altered behavior, misdivision, phenotypes, congenital defects enzyme aberrations, sperm anomalies in F_1 males	Various, usually unknown	Male	Generally undeveloped; in limited use[c]

[a]Adapted from Russell and Shelby, 1985.
[b]Results of female studies, which have been limited to ionizing radiation, indicate very low induction rates.
[c]Both the dominant skeletal and dominant cataract procedures are developed.

mutational response in both male and female germ cells.

Specific-Locus Test with Biochemical Markers

Induced point mutations and small deficiencies can change protein structure and thus alter the electrophoretic mobility of proteins or, in the case of enzymes, alter the magnitude of enzyme activity. Electrophoresis and enzyme activity assessments have therefore been used on a limited scale to screen for induced biochemical mutations in male germ cells. No study in females has been reported. A heritable altered electrophoretic pattern is a good marker for mutation in the structural gene, but alteration of enzyme activity can also arise as a result of genetic change at sites other than the structural gene. Because there are many more biochemical markers than visible markers, these tests have the advantage that they are more likely to detect a toxic effect. However, the biochemical methods have the distinct disadvantage that tissues need to be removed by biopsy or processed from blood. Electrophoresis and enzyme activity assessments have been developed and used widely, but improvement in tissue processing and electrophoretic techniques can be expected. For additional references on these methods see Feuers and Bishop (1986), Lewis and Johnson (1986), J. Peters et al. (1986), and Pretsch (1986).

Specific-Locus Test with Immunologic Markers (H Test)

This test is based on the large number of H genes that control cell surface antigens that induce histocompatibility responses. There are about 40 H genes in the mouse. Mutations might result in a new antigenic form or the loss of an antigen. Detection of the mutants is based on skingraft rejection patterns in a transplantation scheme that involves exposed mice and tester strains. The test detects primarily intragenic changes and small deficiencies in the H loci. It has been used only in males. The use of the test has been restricted because transplantation of skin grafts is a surgical procedure whose outcome can be influenced by nongenetic factors.

Recessive Lethal Test

Induced mutations in this class are lethal in the homozygous or hemizygous state, that is, when both copies of the gene are mutant alleles or when the single copy of the sex-linked gene in males is in the mutant form. Most of these lethal mutations are small deficiencies and small intragenic changes. However, induced reciprocal translocations that are lethal in homozygous conditions are observed occasionally; in these situations, a translocation of genetic material from one chromosome to another alters the expression or structure of the gene in such a way that the condition is lethal in the homozygous animal.

Two methods for detecting the small mutational changes have been attempted or proposed. In one method, the entire genome is screened. That requires three successive generations of specific mating patterns. Daughters in the final generation are mated to their sires. The presence of mutations is detected as an increase in intrauterine death rate due to homozygosity for the mutations among the conceptuses. In the other proposed method, only specific chromosomal segments, either autosomes or X chromosomes, are screened; this makes use of inversions with genetic markers. The second method requires two or three successive generations of specific mating patterns. If a recessive lethal mutation is present in the inverted segment or is closely linked to it, resulting conceptuses that are homozygous for the inverted segment containing the mutation have an increased risk of fetal death. Therefore, these genetically marked progeny are either absent or reduced in number in the final generation.

Dominant Lethal Test

Dominant lethal mutations cause death among first-generation progeny. Generally, death occurs either during an early cleavage stage (in which case the affected embryo fails to implant) or around the time of implantation (in which case the affected embryo stimulates decidual reactions and the formation of a resorption body in mice). Dominant lethal mutations reflect primarily chromosomal breaks. Embryonic lethality occurs because of the resulting deletions and asymmetric exchanges.

The dominant lethal test has been the most widely used in vivo mammalian mutagenicity test and has been used to study genetic effects in male and female germ cells. Increased embryonic lethality after exposure of male parents indicates clastogenicity (that is, induced chromosomal breakage) in male germ cells. However, increased embryonic lethality after exposure of female parents can result from genetic or nongenetic causes (that is, uterine damage). The effect might reflect chromosomal breakage or a phenomenon acting through the maternal environment. Observation of chromosomal aberrations in the pronuclear metaphase or of micronuclei in two-cell embryos would strongly favor genetic causation.

Effects of dominant lethal mutations can also be measured with an in vitro technique. The biologic marker in this case is the ability of two-cell embryos to develop to the trophectoderm outgrowth stage when cultured in vitro. The relative frequency of successful development of embryos of the experimental group, compared with the control group, is used to determine the dominant lethal mutation rate. This in vitro procedure allows distinction between preimplantation loss due to dominant lethal mutations and reduced implantation due to reduced fertilization.

Heritable-Translocation Test

Reciprocal exchange of genetic material between nonhomologous chromosomes is much more readily inducible in male than in female germ cells. With chemicals, therefore, this test has been used solely in studies involving exposed males. When a reciprocal exchange is induced in a male germ cell, the resulting progeny are heterozygous for the translocation.

The two methods used to screen for translocation carriers are both based on meiotic and segregation properties of the heterozygous offspring. The first method is fertility testing for semisterility (i.e., reduced number of living conceptuses). Semisterility is expressed when translocation-heterozygous offspring are mated to normal animals. From these matings, slightly more than half (on the average) of conceptuses produced have unbalanced chromosomal constitutions; that is, about half the fetuses have duplication and deficiency in the region that has undergone translocations. Inasmuch as some types of exchanges cause blockage in early spermatogenesis, this method also screens for completely sterile translocation carriers. Chromosomes of meiotic or somatic cells can be analyzed for verification of translocations. The other method is direct cytologic examination of meiotic cells for multivalent chromosomal association; that is, the translocated chromosomes will form quadrivalent, instead of the normal bivalent, associations. Breeding tests in females often require an extra generation, and cytogenetic analysis of oocytes is hampered by the limitation in the number of oocytes that can be analyzed and by the relative complexity of the procedure. Therefore, screening is restricted to progeny of exposed males.

Inversions

Inversions are chromosomal rearrangements that involve a segment within one chromosome. The segment's orientation of the transcription process is inverted. Because all chromosomes in the standard mouse are telocentric in nature (that is, the centromere is near the end of a chromosome), inversions are paracentric (do not include the centromere). Detection of progeny carrying newly induced inversion is done cytologically. Crossing-over within the inversion produces a dicentric chromatid, which results in the formation of a bridge in the first anaphase. This cytologic test is more easily applied to male, than to female, progeny. The use of inversions as a biologic marker of induced chromosomal breakage and rearrangement is not likely to be important because of the low induction rate and the cytologic scoring procedure, which requires technical expertise and is time consuming.

Sex-Chromosome Loss

In the mouse, the XO condition (only one X chromosome and no other sex chromosome) results in viable females, and the YO condition (a Y chromosome and no other sex chromosome) is lethal. Theoretically, these conditions arise either through chromosomal breakage and elimination or through nondisjunction (improper separation of chromosome pairs). So far, however, the induced XO condition is only the result of chromosomal breakage. Detection of mutation is based on the differential expression of X-linked markers in hemizygous (XO) and heterozygous (XX) female progeny. The XO condition is verified cytologically or by breeding tests. Sex-chromosome loss has been screened for in offspring of male and female exposed parents, although only to a limited extent.

Nondisjunction Test

Nondisjunction leads to unequal distribution of homologous chromosomes in progeny cells; that is, a progeny cell will contain either too many or too few chromosomes. In the standard mouse, when autosomes are involved, the animal dies. Autosomal monosomic animals (i.e., with only one copy of a particular autosome) die early in embryonic development, whereas autosomal trisomic animals (i.e., with three copies of a particular autosome) might survive up to late fetal and early postnatal stages. Except for animals with the YO condition, all other monosomic and trisomic products of sex-chromosomal nondisjunction are viable. Because monosomy can also be produced via chromosomal breakage, trisomies are the most reliable biologic markers of transmitted nondisjunctional products.

Cytologic evidence of nondisjunction induced in germ cells has been reported, but there is no clear evidence of a transmitted induced aneuploidy. Trisomy of the sex chromosomes appears to be the most promising biologic marker in experimental aneuploidy, because these offspring are viable. Trisomies among progeny of exposed parents can be detected either visually or by reproductive tests. The former method makes use of X-linked markers that determine visible characteristics; the latter is based on the finding that XXY and XYY males are sterile. In both cases, aneuploidy can be verified cytologically.

One proposed method of testing for inducible nondisjunction uses genetic markers on autosomes in high-nondisjunction tester stock. The high-nondisjunction tester stock produces a high frequency of gametes that either lack the marked chromosome (nullisomic) or have an extra copy of the marked chromosome (disomic). If the mutagenic treatment of the exposed mice produces nullisomy and disomy of the same chromosome, the complementing combinations from the matings of the exposed mice and the tester stock (i.e., pairing of nullisomic and disomic gametes) should result in viable conceptuses. Some of the nondisjunction progeny are detected on the basis of external genetic markers expressed in viable complementing types. This method is complex, and it is not clear how useful it will be in large-scale testing.

Aneuploidy can be scored in zygotic pronuclear metaphases. At this stage of the conceptus, the male and female contributions are still separate, and often they are distinguishable from one another. False-positive test results might occur, because loss of chromosomes can result from the cytologic procedure. Therefore, a more reliable indicator of aneuploidy is the presence of at least one extra chromosome; i.e., the presence of an extra chromosome is a better criterion than the presence of too few chromosomes. However, the reliability of the cytologic data remains doubtful unless they are matched with similar findings in later embryonic and postnatal stages.

One method worth exploring is the late-fetal-death method. In mice and rats, dead implants are expressed primarily as resorption bodies. Midgestation and late fetal death are uncommon. Most autosomal trisomies cause lethality during the second half of gestation, so induced trisomies can be scored by uterine examination in late gestation. For additional references, see Russell (1985) and Searle and Beechey (1985).

Cytogenetic Analysis of Zygotes

The pronuclear metaphase stage has been analyzed for numerical and structural chromosomal anomalies after exposure of male and female germ cells to an agent before, at, or after fertilization. Structural aberrations that can be scored include deletions, exchanges, and chromosomal fragmentation. This method is useful in followup studies of suspected chromosomal effects that lead to embryonic mortality (dominant lethals), particularly when the exposed parents are female.

Tests for Dominant Mutations

Mutations that have dominant or semidominant effects are the most important class of mutations for the next generation, because the mutations will affect about half of these progeny. Dominant mutations vary from small intragenic changes to gross chromosomal exchanges. Several methods have been used to detect mutagen-induced increases in dominant mutations. The markers of first-generation effects that have been used to date are also varied, affecting either specific organs, organ systems, or function or any visually detected unusual phenotype. They are miscellaneous phenotypes detected postnatally, congenital anomalies in fetuses, abnormal enzyme activities, sperm anomalies, behavioral changes, cataract development, and skeletal changes. The last two have been the most useful for risk evaluation, because the bases for measuring mutation rates and risk are the best established. Further improvement in methods of measuring rates of induction of dominant mutations is needed. Common among these methods are the problems associated with incomplete penetrance, variable expressivity, and expression of variant phenotype caused by nongenetic factors. As more data become available, a well-defined set of phenotypic markers and loci might be developed for mutant detection and counting. The skeletal and cataract systems are progressing in this direction.

NEEDED RESEARCH ON GENETIC DAMAGE IN LABORATORY ANIMALS

Base changes, DNA deletions, gene transposition through chromosomal rearrangements, and chromosomal misdivision are the genetic changes generally recognized as the major mechanisms of induced mutagenesis. Integration of transposable elements to new sites is also emerging as a mechanism.

Male and female germ cells and the various germ cell stages differ in many ways, including ability to repair DNA lesions, length of cell-cycle time, and interval between S phases. In somatic cell systems, each cell is autonomous with respect to the fixation of aberrations. In the case of male meiotic and postmeiotic germ cells, however, the fixation of chromosomal breaks and exchanges is a joint venture between the fertilizing sperm and the egg. The sperm brings in the premutational lesion, and the fertilized egg either repairs it or processes it into a break and exchange. Chemical mutagens can differ from one another in the degree to which each chromosomal target site reacts, but no known mutagen binds to only a single molecular entity. Finally, mutagens bind not only to DNA sites, but also to chromosomal and extrachromosomal proteins. Taken together, all these factors exemplify the complexities involved in understanding the mechanisms, from the initial step of the mutation process (adduct formation) to the expression of mutation in conceptuses. We still have only a minimal understanding of these mechanisms.

Relationship Between Molecular Target Sites and Production of Various Types of Mutations

To understand this relationship, one must keep in mind not only the reaction properties and molecular nature of the mutation, but also the biologic properties of various germ cell stages. Thus, in addition to studies of the relation between the different adducts formed with germ cell chromosomal DNA and protein and the types of transmitted genetic effects produced, studies of repair of specific DNA adducts are also essential. It is generally assumed that base adducts are the important reaction products in DNA. But the oxygen of the phosphate backbone (forming phosphotriesters) is also a target, and it is the primary site of alkylation for some alkylating mutagens, such as isopropyl methane sulfonate and ethylnitrosourea. The questions of whether phosphotriesters have mutational consequence and whether mammalian germ cells have corresponding specific repair enzymes await detailed studies.

Alkylation of protamines has been hypothesized to lead to chromosomal breakage. The extent to which the hypothesis is true needs to be investigated further.

Molecular Nature and Expression of Mutations

Transmitted genetic damage that has dominant or semidominant expression is especially important in genetic risk considerations, because it usually shows up in the first generation. This class of mutation includes both gene mutations and chromosomal rearrangement. The molecular nature of mutagen-induced genetic damage, the way in which deleterious effects are expressed, and why some mutations have incomplete penetrance or highly variable expression are important problems—not only for risk considerations, but also for basic genetics. With the rapid development of DNA experimentation, solutions of these problems are now accessible.

One clue to the molecular nature of genetic damage might come from reciprocal translocations. The question has been raised whether some human genetic disorders that have been assumed to result from single gene mutations could instead be associated with chromosomal rearrangements. It has been generally believed that balanced reciprocal translocations do not involve loss or gain in chromosomal

components. However, an increasing number of clear associations between balanced exchange and deleterious effects suggest that the breakpoint might be in a structural gene or that it might affect the activity of genes in the immediate vicinity. Stocks of mice are available for DNA sequencing and for gene-expression studies.

More than 400 sites of autosomal dominant mutations in the human genome are recognized; many of them involve serious disorders. Most human genetic disorders fail to yield simple Mendelian ratios, and these disorders are sometimes referred to as irregularly inherited. Results of studies of induced dominant skeletal mutations in mice suggest that many irregularly inherited disorders might also result from single dominant mutations with incomplete penetrance. Furthermore, these mutations have pleiotropic effects when they are expressed. Stocks of mice are available for studying the types of DNA damage that cause dominant skeletal defects and how the mutations influence development to cause variability in expression and pleiotropic manifestations.

Mutagen-Caused Induction of Integration of Endogenous Transposable Elements in New Sites

That the insertion of transposable gene elements in the vicinity of, or onto, any given gene might cause a variant phenotypic expression of the gene—originally suggested by Barbara McClintock in the late 1940s—is now a well-established genetic phenomenon. For example, many spontaneous mutations in the white locus of *Drosophila* species are caused by the insertion of transposons, such as the copia element; the dilute locus of laboratory mice is associated with the insertion of an ecotropic murine leukemia virus (MuLV) genome; and virus-induced oncogenesis involves the insertion of retroviral regulatory gene elements at a proto-oncogene locus (converting it to an oncogene). Although it is not yet known, it is generally believed that there are many endogenous transposable elements in the mammalian genome. If exposure of germ cells to mutagens can induce integration of endogenous transposable elements in new sites, the fundamental and risk implications are great. The issue must be resolved through exhaustive studies.

Mouse Mutants as Models of Human Genetic Diseases

Many laboratories in the United States have rich collections of spontaneously occurring or mutagen-induced mouse mutations. Many of these mutations are useful in studying the development of genetic disorders that are similar to those found in humans. Three examples of modeling are described here.

Deficiency in the enzyme ornithine carbamoyltransferase is known in human and mouse mutants and results in urinary defects. The mouse sparse-fur mutant has, besides its hair abnormality, the tendency to produce kidney or bladder stones that are composed primarily of orotic acid. The sparse-fur locus is on the X chromosome, and a quantitative measure of X inactivation can be studied by examining the amount of enzyme present in animals in which the X chromosome has been fragmented in translocation mutants. The location of the presumptive X inactivation center on the X chromosome might thus be determined. A biologic marker in this case causes a physiologic defect that is used as a tool to study the basic problem of the natural inactivation of one of the two X chromosomes.

An electrophoretic variant of the enzyme pyruvate kinase known in the mouse results from an alteration of the gene on chromosome 9. This locus occurs in the region where many chromosomal deletions have been isolated. By determining whether the enzyme variant is present in F_1 progeny of crosses that involve the deletion mutant and the pyruvate kinase electrophoretic mutant, one can further define the boundary of the deletion. The human pyruvate kinase can also be distinguished from the mouse enzyme electrophoretically; following this enzyme in mouse-human hybrid cells makes it possible to identify the chromosomal location. The biologic marker in this case is useful for genetic mapping studies.

Synthesis of the neurotransmitters serotonin and norepinephrine and conversion of phenylalanine to tyrosine are carried out by enzymes that all use a common cofactor, tetrahydrobiopterin (BH_4). The biosynthesis of BH_4 proceeds from guanosine triphosphate through a pathway that involves three or four enzymes. Human and mouse mutants that result in a reduced concentration of BH_4 have recently been identified. In the human case, a mental deficiency, atypical phenylketonuria, is the phenotypic expression of the mutation; in the mouse, several behavioral abnormalities are manifested. The biologic markers that need to be examined are the individual enzymes, so that the regulation and normal function of BH_4 can be understood and means of alleviating the defects in such mutants can be devised.

Nondisjunction

Aneuploidy resulting from chromosomal missegregation constitutes an important fraction of transmitted human genetic anomalies. The extent to which it is inducible by chemicals in male or female cells is not clear, mainly because no chemical has been clearly established as an inducer of nondisjunction in these cells. Conceivably, chromosomal missegregation results from damage to the spindle and kinetochore and their precursors or via chromosomal rearrangement, which, in turn, could affect normal pairing and segregation of homologous chromosomes during meiotic stages. For the latter possible mechanism, it is essential to know when synapsis actually takes place during spermatogenesis. A provocative hypothesis stating that synapsis and recombination occur during the last premeiotic *S* phase, rather than later during zygotene and pachytene, respectively, has been raised. The issue needs to be resolved.

DNA Methods

Failure to detect increases in mutation rates in supposedly mutagenized human populations has triggered interest in the use of molecular methods that entail direct analysis of human DNA. A recent publication of a workshop report (Delehanty et al., 1986) identified six DNA methods to detect human heritable mutations (see also Chapter 9). It cautioned, however, that none of the methods was ready for field application but that refinements in DNA experimentation, would soon permit the analysis of mutation in human populations. Because of the dynamic nature of DNA experimentation, it is assumed that better methods will eventually become available for use in human genetic epidemiology.

The report (Delehanty et al., 1986) enumerated several properties as essential for new methods to be successful. They must be able to examine 10^{10} base pairs; detect a well-defined and wide spectrum of mutational end points; have extremely low error rates; use easily accessible samples; conserve time, people, and resources; cope with the complexity of the human genome; and recognize recombination, polymorphism, physically variable genes, somatic mutations masquerading as heritable mutations, and false paternity. Those properties also constitute one of the main reasons why appropriate mutagenesis studies in laboratory mice are necessary. Research in mice with chemical mutagens would not only contribute to method development, but also provide the basis for interpreting human results.

9

Markers for Measuring Germinal Genetic Toxicity and Heritable Mutations in People

Germinal genetic toxicity is an integral part of reproductive toxicity. The induction of new germline mutations by exposure of either parent to mutagens is of serious concern, because it would increase the frequencies of genetic diseases in the parent's children and in later generations. Also, certain mutations would result in early embryo loss and thus contribute to the inability of couples to have children. The contribution of chromosomal abnormalities to human infertility is well documented (Chandley et al., 1975; Chandley, 1984). Thus, accidental, occupational, environmental, or therapeutic exposures to ionizing radiation or some chemicals might cause germinal mutations that will reduce a person's ability to produce normal, healthy children.

When dealing with animal data, the evaluation of genetic toxicity involves the detection of genetic damage, measurement of genetic effects, and extrapolation of results. This chapter discusses markers of germinal genetic toxicity and mutagenesis in exposed human males and their offspring and thus focuses only on detection and measurement. Some background knowledge of genetic toxicology and research methods used in laboratory animals that are described in Chapter 8 is assumed. Extrapolation from one species to another, one dose to another, or one genetic effect

to another is a complex process that deserves separate attention.

Mutations are long-lasting changes in the genetic information carried in DNA, which, when inherited, can cause severe diseases and disabilities in affected organisms. Inherited genetic diseases are incurable (in the sense that DNA changes cannot be reversed), and very few can be treated effectively. Genetic defects account for a substantial portion of human chronic disease and contribute to the human burden of infertility, developmental and neonatal mortality, and mental retardation (Vogel and Rathenberg, 1975). Several of the most common genetic disorders are known to be caused by mutations (e.g., hemophilia and Duchenne-type muscular dystrophy). Increased risks of some chronic conditions, such as some forms of heart disease and cancer, also have a strong genetic component.

Approximately 80% of the single-gene mutations that manifest themselves in genetic disease in some people have been transmitted from generation to generation (UNSCEAR, 1982). Each mutation is thought to have arisen in the germ cells of an ancestor and then passed through the germ line to the present generation. The remaining 20% of single-gene mutations and most chromosomal abnormalities are thought to have occurred sporadically in the gonadal cells of one of the parents of the affected person. The frequency of newly occurring human germline mutations depends on several factors, including the sex and age at the time of the child's conception.

The susceptibility of rodent germ cells to the induction of heritable mutations by ionizing radiation and some potent chemicals has been well demonstrated and characterized (Livingston, 1984; Russell and Shelby, 1985). Several groups of people exposed to agents known to induce heritable mutations in laboratory animals have been under intense scrutiny, but to date, researchers have been unable to detect an increase in the frequency of heritable mutations in any of these groups. The two largest continuing investigations are of the Japanese atomic bomb survivors (Sankaranarayanan, 1982; Satoh et al., 1982) and the survivors of cancer therapy (Mulvihill and Byrne, 1985). As reviewed later in this chapter, other groups on which data are being gathered are the offspring of survivors of attempted suicide by chemical means and the offspring of men exposed to chemicals occupationally.

Our present inability to detect induced human heritable mutations is thought to be due not to the nonresponsiveness of people compared with mice, but rather to the limitations of available detection methods and the insufficient sizes of the groups studied. For example, the largest groups studied to date (the Japanese atomic bomb and cancer-therapy survivors) are too small to detect an increase in point-mutation frequency of less than a factor of about 5 or 10. Studies in irradiated mice predict only a factor of 2 increase at the doses experienced in the Japanese bombings. In some cases, however, it is not possible or practical to study larger cohorts, because larger numbers of exposed people are not available or the expense is too great. More sensitive markers for detecting human germline mutations are needed.

The development of markers of genetic damage in the human line is proceeding in three interrelated directions. The directions involve research and development leading to markers of exposure of human germ cells to genotoxic agents, to markers of germinal mutations (i.e., mutations in the germ cells of mutagen-exposed persons), and to markers of heritable mutations (i.e., mutations that are inherited by the offspring of mutagen-exposed people). These markers are characterized by whether they provide data on genotoxicity, mutation, or mutant frequencies and by which tissue or cell type of the human germ line is used for analysis. The markers reviewed in this chapter are listed in Table 9-1, organized by the source of tissue and the type of data used in performing the analyses.

Markers of exposure includes both germ-cell-based and record-based methods that might be useful for identifying high-risk persons or exposures. Markers in this category are generally qualitative indicators of genotoxic exposure only and do not provide quantitative mutation data. Germ-cell-based methods in this category, which are all limited to males, include measurements of sperm-DNA adducts, sperm-protein adducts, and sperm-DNA alkaline elution. One example of a record-based method in this category, which is potentially applicable to exposures of both male and female parents, is a spontaneous-abortion study in which the parents under investigation were exposed only before conception.

Markers of germinal and heritable human mutations, on the other hand, are designed to provide quantitative data suitable for assessing mutation or mutant frequencies. They are distinguished from each other by the germline tissue and cell stages sampled (Fig. 9-1) and by the type of information obtained.

Heritable mutations can be detected only by monitoring for offspring with traits not found in the somatic cells of either parent. The analysis of induced heritable mutations applies to situations in which only the male parent, only the female parent, or both parents were exposed before conception of their offspring.

Germinal mutations, are mutational changes that occur in the germ cells of persons irrespective of whether they have offspring. In principle, they can be detected in two ways: by direct measurement

TABLE 9-1 Potential Markers of Genetic Damage and Heritable Mutations in the Male Germline Reviewed in This Chapter

Tissue	Marker
Testis	Cytogenetic analyses of cells in mitosis, meiosis I, and meiosis II
Semen	
Sperm	Sperm cytogenetics
	Sperm DNA and protein adduction
	Gene mutations in sperm
	Sperm aneuploidy
Immature germ cells	Spermatid micronuclei
	Cytogenetics of ejaculated meiotic I cells
Questionnaire and medical records	Sex ratio
	Spontaneous abortion
	Offspring cancer
	Sentinel phenotypes
Offspring tissue	Cytogenetics
	DNA sequencing
	Protein mutations
	Restriction-length polymorphism of DNA
	RNAase digestion
	Subtractive hybridization of DNA
	Denaturing gel electrophoresis of DNA
	Pulse-field electrophoresis of DNA
Mother's urine	Detection of early fetal loss
Somatic cell surrogates	
In white blood cells	HGPRT mutations
In red blood cells	Hemoglobin mutations
	Glycophorin A mutations

of mutated germ cells of adults or by inference from offspring studies of heritable mutations. Men produce large numbers of gametes throughout their adult lives, and methods are being developed to detect germinal mutations directly in their germ cells. It is not yet possible to measure germinal mutations directly in women, because the numbers of oocytes present after puberty are too small, and they are inaccessible by noninvasive methods. Thus, germinal mutations in women must be inferred through the analysis of heritable mutations.

Germinal and heritable mutations also differ in the selection steps that new mutations have encountered before analysis. As illustrated in Figure 9-1, germinal mutations induced early in gametogenesis must survive numerous possible selection steps to be manifest as phenotypic traits in a liveborn infant. This is best understood for cytogenetic mutations, and the frequency of some cytogenet-

ic lesions is known to change markedly as cells progress through spermatogonial mitosis, meiosis, first cleavage of the fertilized egg, and development to birth. Although the frequencies and characteristics of gene mutations in germ cells and those in offspring can differ, little is known of the selection pressures exerted on gene mutations during their progression through the germ line from one generation to the next. Detection methods for both germinal and heritable mutations must be developed to investigate the induction and persistence of germline mutations from gonocytes to birth.

The following sections review the current methods for measuring human heritable and germinal mutations and for detecting genetic damage in human male germ cells. This chapter reviews the available methods for measuring human heritable mutations, briefly discussing the findings of studies with mutagen-exposed populations. Because human heritable mutations are dif-

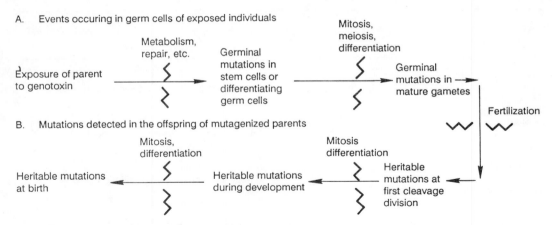

FIGURE 9-1 Schematic representation of human germline, including some of the cell types in which genetic damage can be measured. ⌇ represents possible selection sites. Part A brackets the events occurring within the gonad and efferent ducts of the exposed individual. Part B includes the postconception events until birth. Source: A. Wyrobek.

ficult to detect, this chapter also reviews the surrogate approaches with human blood cells as well as several promising molecular approaches. The remainder of the chapter reviews the current methods and promising approaches for measuring germinal mutations and genetic alterations directly in human male germ cells.

CURRENT METHODS FOR MEASURING HUMAN HERITABLE MUTATIONS

Spontaneous heritable mutations in humans have been studied by three methods: those based on the incidence of specific genetic diseases (sentinel phenotypes), those based on changes in chromosomal structure or number, and those based on changes in structure or function of blood proteins.

Sentinel Phenotypes

Sentinel phenotypes are a special class of severe clinical disorders that occur sporadically, probably as the result of single-gene mutations. They are manifest at birth or within the first few months of life. The low fertility of carriers of these diseases suggests that

they represent de novo dominant mutations in germ cells, rather than the transmission of existing mutations across generations. Mulvihill and Czeizel (1983) compiled a list of sentinel phenotypes—including autosomal dominant and X-linked disorders (Table 9-2). The incidence of each specific genotype is very small, from 1 per 10,000 to 1 per 10,000,000 newborns, with an arithmetic mean of approximately 2 per 100,000 genes per generation (Vogel and Rathenberg, 1975).

Surveillance of sentinel phenotypes in large populations is useful for estimating the background rate of mutations that lead to dominant disorders. As a method for detecting the effects of exposure to a potential mutagen, it is probably not useful, unless very large populations and international efforts are involved.

Chromosomal Abnormalities

The adverse effects of chromosomal abnormalities on human health are well established; at least 10 large studies have screened infants to determine the spontaneous rate of abnormal chromosomes and to evaluate their effects on health. Chromosomal abnormalities are of two broad classes: numerical (involving extra or

TABLE 9-2 Candidate Sentinel Phenotypes

	Inheritance[a]		Inheritance[a]
Phenotypes identifiable at birth			
Achondroplasia	AD	Whistling face (Freeman-Sheldon	
Cataract, bilateral, isolated	AD	syndrome	AD
Ptosis, congenital, hereditary	AD	Acrocephalosyndactyly type V,	
Osteogenesis imperfecta type I	AD	Pfeiffer syndrome	AD
Oral-facial-digital (Gorlin-Psaume)		Spondyloeipphyseal dysplasia	
syndrome)	XD	congenita	AD
Incontinentia pigmenti, Bloch-			
Sulzberger syndrome	XD	**Phenotypes not identifiable at birth**	
Split hand and foot, bilateral			
atypical	AD	Amelogenesis imperfecta	AD
Aniridia, isolated	AD	Exostosis, multiple	AD
Crouzon craniofacial dysostosis	AD	Marfan syndrome	AD
Holt-Oram (heart-hand) syndrome	AD	Myotonic dystrophy	AD
Van der Woude syndrome (cleft		Neurofibromatosis	AD
lip and/or palate with mucous		Polycystic renal disease	AD
cysts of lower lip)	AD	Polyposis coll and Gardner	
Contractural arachnodactyly	AD	syndrome	AD
Acrocenphalosyndactyly type I,		Retinoblastoma, hereditary	AD
Alpert's syndrome	AD	Tuberous sclerosis	AD
Moebius syndrome, congenital		von Hippel-Lindau syndrome	AD
facial diplegia	AD	Waardenburg syndrome	AD
Nail-patella syndrome	AD	Weidemann-Beckwith (EMG)	
Oculodentodigital dysplasia		syndrome	AD
(ODD syndrome)	AD	Wilms' tumor, hereditary	AD
Polysyndactyly, postaxial	AD	Muscular dystrophy, Duchenne type	XR
Treacher Collins syndrome,		Hemophilia A	XR
mandibulofacial dysostosis	AD	Hemophilia B	XR
Cleidocranial dysplasis	AD		
Thanatophoric dwarfism	AD		
EEC (ectrodactyly, ectodermal			
dysplasia, cleft lip and palate)			
syndrome	AD		

[a]AD refers to autosomal dominant inheritance, XD refers to X-linked dominant inheritance, XR refers to X-linked recessive inheritance. Source: Mulvihill and Czeizel, 1983.

missing whole chromosomes) and structural (involving, for instance, translocations, deletions, or insertions of parts of chromosomes).

Fetuses with numerical chromosomal abnormalities are at increased risk of being aborted spontaneously; numerical abnormalities contribute to a major part of spontaneous abortions and stillbirths. In liveborn, the presence of an extra sex chromosome or the absence of a sex chromosome is often associated with physical, behavioral, and intellectual impairments. The presence of an extra autosome is usually more detrimental, causing severe mental and physical retardation.

The relationship of structural abnormalities of chromosomes to human health is more variable and less well understood. Some rearrangements might have no apparent phenotypic effects, whereas others might be associated with mental retardation, physical malformations, and various malignant diseases. Balanced translocations have an added fertility consequence. In carriers of such translocations, meiosis produces a majority of gametes with unbalanced chromosomal complements, and resulting conceptuses might die in utero because of chromosomal duplications or deficiencies. The frequency of spontaneous abortions is also increased in translocation carriers.

Numerical and structural chromosomal abnormalities have been detected in approximately 60% of recognized spontaneous abortions, 6% of perinatal deaths, and 0.6% of liveborn infants (Hook, 1981; Boué et al., 1985). Prenatally and postnatally, chromosomal aberrations can be identified on the basis of karyotype. However, most conceptuses with chromosomal abnormalities are aborted spontaneously before pregnancy is recognized, and the frequency of mutation at birth is known to represent only a small fraction of the frequency of these mutations at conception (see Fig. 9-1). A developing method for scoring chromosomal abnormalities in human sperm is beginning to provide data to bracket the uncertainty (see discussion later in this chapter). As determined with this procedure, approximately 7% of sperm have some numerical or structural abnormality, most of which would be expected to result in an aborted pregnancy. In addition, broad variability was found among people noted with this method.

Chromosomal abnormalities are useful for measuring mutation rates, because such an abnormality is generally a new mutation: the parents could not have survived or remained fertile if they carried the defect systemically or laboratory analysis showed that neither parent carried the same defect. However, it is difficult to compare the mutation rate based on chromosomal abnormalities with that based on analysis of sentinel phenotypes, because chromosomal aberrations are multigenic, whereas sentinel phenotypes involve only one gene, and because there is strong evidence of postconceptional selection against most chromosomal aberrations, whereas selection against single-gene defects is less certain (see Fig. 9-1).

Prevalence of Chromosomal Abnormalities at Birth

Although presenting only a partial picture (an underestimate) of the true incidence of cytogenetic abnormalities in germ cells or fertilized eggs, analyses of newborns show a range of frequencies of 2 in 100,000 for inversions to 121 in 100,000 for trisomy 21, the most common numerical chromosomal abnormality at birth (Table 9-3). These data are based on analyses of

67,014 offspring from six countries (Sankaranarayanan, 1982). Cases of numerical abnormality are more frequent at birth than cases of structural abnormality. However, the latter could be artifactually deflated by the use of techniques that are insensitive to balanced translocations, a class of chromosomal rearrangements that are least likely to lead to cell death and pregnancy loss. Methods that use fluorescence hybridization with chromosome-specific DNA probes are being developed to improve the sensitivity of detection and ease of scoring of both translocations and numerical abnormalities.

Mutant Proteins

Sentinel phenotypes can be used to measure the incidence of disorders related to dominant mutations. Biochemical studies of protein variants in asymptomatic people can be used to determine the cumulative rate of recessive mutations. In principle, any protein in the body can be studied; for reasons of accessibility, blood proteins are used most often. Three major techniques have been used: electrophoresis, study of enzyme-deficiency variants, and analysis of unstable hemoglobin. Only the first is discussed here for illustration.

Electrophoresis separates proteins on the basis of net molecular charge. A mixture of proteins is loaded onto a starch or acrylamide gel, and an electric current is applied. Each protein moves according to its charge for a predetermined time, after which the proteins are stained for visualization. For the study of heritable mutations, the protein pattern of a child is compared with those of its parents. The identification in a child of a variant protein not present in either biologic parent is evidence of a new mutation. Theoretically, approximately one-third of amino acid substitutions produce a change in net molecular charge. It is estimated that electrophoresis probably detects about 50% of base-pair substitutions within the coding region of a gene, including base-pair substitutions that affect mobility through changes in molecular conformation (Neel et al., 1986).

TABLE 9-3 Prevalence of Chromosomal Abnormalities at Birth

Chromosome Abnormality	Total Population	Number of New Mutants[a]	New Chromosome Abnormalities per 100,000 Newborns per Generation
Numerical anomalies			
Autosomal trisomies:[b]			
Trisomy 13	67,014	3	5
Trisomy 18	67,014	8	12
Trisomy 21	67,014	81	121
Male sex chromosome anomalies:			
47,XYY	43,048 males	43	100
47,XXY	43,048 males	42	98
Female sex chromosome anomalies:			
45,x	23,966 females	2	8
47,xxx	23,966 females	24	100
Balanced structural rearrangements			
Robertsonian translocations:[c]			
D/DD[d]	67,014	48(2/29) = 3.3	5
D/G	67,014	14(2/11) = 2.5	4
Reciprocal translocations and insertions	67,014	60(13/43) = 18	27
Inversions	67,014	12(1/8) = 1.5	2
Unbalanced structural rearrangements			
Translocations, inversions, and deletions	67,014	37(7/16) = 16	24

[a]See discussion in text for proportion of new mutants among all those identified with structural rearrangements
[b]Trisomies refer to conditions in which there are 47 chromosomes (instead of the normal 46)
[c]Rearrangements in which the long arms of two chromosomes fuse
[d]D and G refer to groups of chromosome numbers 13-15 and 21-22.
Source: Sankaranarayanan, 1982.

Several large-scale studies have been conducted to determine the spontaneous rate of electrophoretic variants (Harris et al., 1974; Neel et al., 1980; Atland et al., 1982; Neel et al., 1983; Neel et al., 1986); the summary results of 4 mutations in 1,255,246 loci tested have shown Poisson-dominated fluctuation and an overall rate of approximately 3 mutations per million loci.

The ready availability of human blood makes the electrophoretic approach feasible for large-scale studies. Furthermore, electrophoresis is applicable to animal studies and thus allows for interspecies comparisons. It is especially important, because it evaluates the mutation rate of a set of genes that code for functional proteins and, in principle, permits detailed study of the relationship between gene mutation and protein expression. However, it says little about the mutability of the noncoding part of human DNA and is limited to DNA changes that affect protein electrophoretic mobility. It also is labor-intensive and requires large populations for the detection of even moderate effects that might be due to mutagen exposure.

The electrophoretic method described above has been extended to two dimensions in a technique (Neel et al., 1984) referred to as two-dimensional polyacrylamide gel electrophoresis (2D PAGE). With this technique, proteins are separated first by isoelectric focusing on the basis of their molecular charge, and then by electrophoresis on the basis of their molecular weight. Some 1,000 protein spots might be visible in a single gel, and about 100 sufficiently delineated to detect rare variant proteins recognized by spots that have changed position. To search for new mutations in proteins, as for one-dimensional gel electrophoresis, blood samples must be drawn from each child and both its parents and analyzed. The spot patterns of the trio are compared by trained eye or computer, to identify children with variants not present in either parent.

An advantage of 2D PAGE is that it is possible to use with current technology (although at some expense). However, it is technically demanding, and few proteins can be scored practically. As with other techniques described above, only a certain spectrum of mutations is detectable with 2D gels. The technique has great sensitivity for detecting point mutations and deletions in DNA that change the size or charge of a protein. However, it is not sensitive to DNA changes in the gene that do not cause these electrophoretic alterations, to chromosomal rearrangements, or to small insertions or deletions. Also, null mutations cannot be detected reliably. It is estimated that 2D PAGE can detect approximately 30% of all spontaneous gene mutations in the proteins that can be clearly identified. Several technical aspects of 2D PAGE of blood proteins are under development, including improved computerization and improved strategies for detecting null mutations.

In summary, computer-based analysis allows the application of 2D PAGE to large populations. However, only a small number of people have been analyzed with this technique, compared with one-dimensional protein separation. The identity of most of the proteins that can be visualized in 2D gels is unknown. Further research is required to improve the methods to recover proteins from the gels and to determine their amino acid sequences. In the future, these sequences might be used to develop nucleic acid probes to locate and characterize the corresponding DNA in the human genome.

RESULTS OF EPIDEMIOLOGIC STUDIES OF HUMAN HERITABLE MUTATIONS IN EXPOSED POPULATIONS

Atomic Bomb Survivors

The three heritable-mutation-detecting methods noted above (sentinel phenotypes, chromosomal aberrations, and mutant proteins) were applied to the offspring of Japanese atomic bomb survivors to determine whether exposure of parents to ionizing radiation induced heritable mutations. Although exposed people showed increased frequencies of some diseases, such as mental retardation (especially in those exposed in utero) and cancer, no detectable inherited effects have been measured on the basis of sex-ratio alterations, spontaneous abortions, chromosomal aberrations, or electrophoretic mutations of blood proteins (Schull et al., 1981a,b; Awa et al., 1981; Neel et al., 1986). It is generally agreed that the lack of effects is due to the inefficiency of the methods used, rather than to an inherent resistance of humans to genetic damage, and larger sample sizes would be needed to detect effects with the current methods (Miller, 1983).

Survivors of Cancer Chemotherapy

Cancer therapy often includes agents known to be germline mutagens in laboratory animals. Researchers have used epidemiologic methods to investigate the effects of cancer treatments on the incidence of genetic diseases and abnormal reproductive outcome in pregnancies in which one of the parents was treated before conception. Several thousand pregnancies have been investigated, but no treatment-related heritable effects have been documented (Mulvihill and Byrne, 1985). Those results underscore the inefficiency of the survey approach. Also, other factors might account for the negative findings. First, studies with mutagen-treated male mice indicate that stem cells are relatively resistant to the chemical induction of events leading to dominant lethality, which would predict very little or no increase in human spontaneous abortions for conceptions arising from sperm that were exposed to mutagens as stem cells (Russell and Shelby, 1985). Second, the studies usually do not discriminate between mutagenic and nonmutagenic chemotherapy. Thus, for either reason, the mutation rate might appear lower than it is.

Other Exposure Groups

Several other human groups have been investigated for induced heritable mutations. Czeizel (1982) studied a group of survivors of suicide attempts that used chemical agents, but found no effects on genetically inherited traits or on spontaneous abortions. However, their cohort was small, and the exposures were variable and involved complex mixtures. Further studies that include more sensitive markers of genetic damage and larger numbers of suicide-attempt survivors would be beneficial.

Several studies have reported abnormal reproductive outcome after occupational exposure of the father to DBCP, wastewater treatment chemicals, lead, and anesthetic gases (Narod et al., 1988). Although each of these studies has shown an increase in abnormal reproductive outcome (usually spontaneous abortion), occupational exposures are inherently poorly documented, and fathers often continue to be exposed during pregnancies. Thus, one cannot be confident that the spontaneous abortions associated with the fathers' exposures occurred via mutational mechanisms. The epidemiologic evidence of increased spontaneous abortions in wives of men exposed to chemicals occupationally and the evidence of increased frequencies of childhood cancers have not withstood statistical scrutiny or have not been reproduced, or the exposures were not limited to male parents before conception.

HUMAN SOMATIC MUTATION METHODS

Markers are needed for identifying people who have been exposed to mutagens and for identifying specific mutagens with an eye to controlling or preventing human exposure. For those purposes, mutation tests must be designed around the cells of exposed people (without requiring cells of their offspring). Such tests must be essentially noninvasive, be able to detect background mutation with small amounts of tissue, and be able to detect exposures to a wide variety of mutagens.

The following describes two examples of somatic cell mutation tests under development for measuring human somatic mutations in vivo. A major uncertainty in the development of these tests is the relevance of somatic mutagenicity to germinal mutagenicity– specifically, the extent to which the rates and kinds of somatic mutations have any predictive relationship with the rates and kinds of germinal mutations. Further development and testing of somatic and germinal methods are required and encouraged, so that these relationships can be elucidated. However, even without knowledge of the details of the relationships among somatic, germinal, and heritable mutagenicity, somatic mutation tests are important in their own right. Primarily, they can provide an early warning that people are being exposed to mutagens; in addition, they are relevant to studies of aging and carcinogenesis.

Current somatic mutation methods have been developed for readily accessible cells. The most accessible human tissue is blood, and all the somatic tests under development have used red or white blood cells. Instead of examining DNA directly, they are designed to detect rare cells with marker phenotypes from among large numbers of mostly normal cells through the use of selective growth conditions or mechanical methods. As these methods become established, mutation assays should be developed for other somatic cells so that tissue differences in response can be compared.

Detection of Somatic Mutations in White Blood Cells

Groups led by Albertini (Albertini et al., 1982) and Morley (Morley et al., 1983) have developed assays to detect mutations that affect the phenotype (i.e., drug resistance) of lymphocytes. Mutant cells are detected by their inability to produce the enzyme hypoxanthine-guanine phosphoribosyl transferase (HGPRT). Cell culture conditions are set so that normal cells use that enzyme to metabolize 6-thioguanine to a toxic chemical that kills the cells; mutated cells survive, because without the enzyme they are incapable of producing the toxic metabolite. Two methods have been developed to identify T lymphocytes in human blood samples that are mutant with respect to HGPRT: an autoradiographic method that detects the 6-thioguanine-resistant cells by their ability to incorporate thymidine and replicate in the culture conditions that kill normal cells and a clonal method in which surviving cells grow to form visible colonies, which takes about 10-14 days. The clonal assay allows characterization of mutations found in the cells, whereas the autoradiographic method permits only the determination of their frequency.

Studies of normal people have identified several sources of variation. Group-average mutant frequency in normal people shows a 15-fold variation, ranging from about 1 to 15 mutant cells per million (OTA, 1986). Variations among normal persons within each group were even larger; Albertini (1985) noted a 100-fold variation in one group of 23 normal people with a range of 0.4-42 mutants per million cells. Variations among normal persons might be explained in part by donor age and smoking status.

Exposure to physical and chemical agents can increase the frequency of mutant T lymphocytes, as indicated by the results of studies of cancer patients who received chemotherapy and radiotherapy (Dempsey et al., 1985) and of blood from atomic bomb survivors (Albertini et al., 1988). However, the responses to exposure are highly variable; some of the cancer patients had frequencies as high as 150 mutants per million white blood cells, whereas some had frequencies indistinguishable from normal. Research being conducted with this method is aimed at characterizing the sources of measurement variation, applying it to people exposed to and not exposed to mutagens, and investigating the molecular nature of the mutations.

Detection of Somatic Mutations in Red Blood Cells

Unlike white blood cells, mature red blood cells contain neither nuclei nor DNA. Two mutational assays based on glycophorin and hemoglobin peptide changes are under development. High-speed automated microscopy or flow cytometry are used to detect mutant cells and sort them from normal cells. The glycophorin-based method is described here as an example.

This assay was designed to detect expression-loss mutations of the glycophorin-A gene. Mutant cells are identified by the absence of cellular protein corresponding to the gene. Glycophorin-A is a transmembrane glycoprotein present on the surface of red blood cells. Two variant forms exist normally; they are called "M" and "N," and they differ from each other by 2 of a total of 131 amino acids. The M and N serotype have no known biologic function, and each is expressed independently and codominantly in heterozygous persons. Monoclonal antibodies have been developed and can be tagged with fluorochromes to distinguish M and N proteins with high specificity. In this manner, each red cell of heterozygous persons would be labeled with both colors. Mutant cells are detected with internal control, i.e., as cells that have lost the color for one serotype and retained the color for the other. A flow cytometer with dual-beam excitation and sorting capability is used for detection and visual verification of variant cells.

Studies in normal people have indicated a spontaneous frequency of 1 to about 20 mutant cells per million; age and smoking are known to contribute to the frequency (Langlois et al., 1986; Jensen et al., 1987). Certain chemotherapy increases the fraction of mutant cells in exposed people (Bigbee et al., in press). Most dramatically, it has been shown that atomic bomb survivors have dose-dependent fractions of mutant cells, now over 40 years after exposure to bomb radiation (Langlois et al., 1987). Current research with this method includes investigations of sources of measurement variation, studies of people with ataxia telangiectasia (Bigbee et al., 1989) and Bloom's syndrome (Langlois et al., 1989), and new recombinant-DNA approaches to evaluate the variant cells on a molecular level.

NEW MOLECULAR APPROACHES FOR DETECTING HUMAN HERITABLE MUTATIONS

New developments in molecular biology have suggested the possibility of examining human DNA directly for mutations. Until recently, human mutations could be studied in fine detail only in particular genes, such as the HPRT gene, and there was no way to study mutations in regions not expressed as proteins. This section summarizes several new methods for examining mutations in large regions of genomic DNA. Most of these were introduced as potential methods at a December 1984 workshop cosponsored by the International Commission for Protection Against Environmental Mutagens and Carcinogens (ICPEMC) and the U.S. Department of Energy (Delehanty et al., 1986). The approaches summarized here are described in more detail elsewhere (OTA, 1986, Chapter 4) and in general are very early in their development phase.

DNA Sequencing

The most comprehensive analyses of new mutations would be comparisons of the complete DNA sequences of a child and its parents. This method would detect the entire mutational spectrum, including DNA changes in exons, introns, and repeat sequences, as well as chromosomal rearrangements of all types. The complete DNA sequence of even one person would provide an invaluable reference for all laboratories to locate their DNA sequences, to identify polymorphisms, and to evaluate new mutations. Assuming the spontaneous occurrence of 1 mutant base pair per 10^8 base pairs per generation, one would expect about 30 new mutations per haploid genome per generation. That is, each child would be expected to carry about 60 new base-pair changes that did not occur in the somatic cells of either parent.

Maxam and Gilbert (1977) and Sanger and coworkers (1977) developed a theoretical basis for sequencing DNA. Later developments have made it possible to cut genomic DNA into small pieces with restriction enzymes and then to use chemical methods to determine the nucleotide sequences within each fragment (Church and Gilbert, 1984). Although these methods are in common use, a recent estimate determined that only 4×10^6 total base pairs had been sequenced by this technique—the equivalent of slightly more than 0.001 of the human haploid genome. With current technology, it might cost $3 billion or more to sequence even one genome (Freeman, 1985), and considerably more effort in method development is required before this approach becomes feasible.

Restriction-Fragment-Length Polymorphism (RFLP)

A less complete survey of mutations than genomic sequencing involves the use of gel electrophoresis and DNA-cutting enzymes to detect specific nucleotide substitutions, as well as small DNA insertions and deletions. With this method (described in Figs. 9-2 and 9-3), genomic DNA is isolated from somatic cells (usually white blood cells) and treated with restriction enzymes that recognize specific DNA sequences and cut the DNA at specific base-pair sites in relation to these sequences. The total number of fragments and their lengths are characteristics of the given restriction enzyme. The RFLP method examines DNA in both expressed and unexpressed regions, but it requires detailed additional studies to determine whether a specific fragment represents introns, exons, or other DNA regions.

Each restriction enzyme is site-specific, and it is not practical to use enough different enzymes to examine all possible DNA sequences. Typically, small sets of enzymes are used to increase the number of nucleotides surveyed. Two approaches are available to analyze the fragments: a direct procedure and one involving cloning.

- *Direct analysis.* On gel electrophoresis, DNA fragments of the human genome

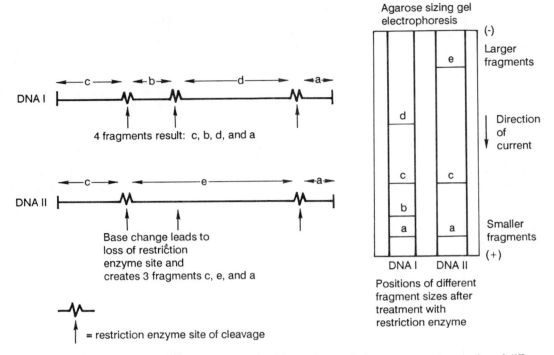

FIGURE 9-2 Production of restriction-length polymorphisms using restriction enzymes and separation of different DNA fragment sizes by Agarose gel electrophoresis. Source: OTA, 1986.

form smears of indistinguishable bands. For visualizing individual bands with the direct procedure, the Southern blotting procedure is used to transfer the DNA onto a paper matrix. The DNA, naturally double-stranded, dissociates into single strands of DNA and is then allowed to reassociate with small added radiolabeled pieces of specific genes or other small pieces of DNA. The added fragments, or probes, hybridize with homologous genomic DNA, and the bands are detected with autoradiography.

• *Cloning*. In contrast with the direct method in which probes are used to detect the DNA fragments, the cloning method first replicates the fragment by cloning in bacterial phage. Human DNA fragments are incorporated into the DNA of the bacterial virus lambda. Lambda replicates itself and the inserted human DNA while in *E. coli*, and large amounts of each human fragment are produced. Each clone is treated with a restriction enzyme to produce fragments, which are subjected to electrophoresis and form visible bands.

For both direct analysis and cloning,

the DNA bands of the child are compared with those of its parents; the presence of a band in the child's DNA that is not present in the DNA of either parent indicates a nucleotide substitution, deletion, or insertion, which suggests that a heritable mutation has occurred.

Ribonuclease Cleavage

Myers and coworkers (1985) proposed a mutation detection method that uses ribonuclease A (RNases), an enzyme that cuts double-stranded RNA/DNA heteroduplexes where specific mismatch of base pairs occurs. RNaseA cleaves the RNA/DNA molecule where a cytosine in RNA erroneously occurs opposite an adenine in DNA (a normal pairing would be cytosine with guanine or adenine with thymine). The method is applicable to one-twelfth of the possible base-pair mismatches. Approximately 50% of substitutions are detected, if probes for both strands are used. Further study is required to evaluate the efficiency of the RNaseA method when the mismatch occurs among differing sequences, and

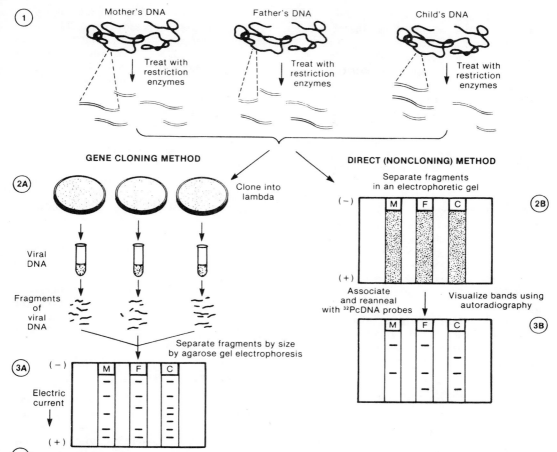

① Isolate geonomic DNA from white blood cells. Using restriction enzymes, cut DNA into double-stranded fragments of various lengths.

Gene Cloning Method:

②A Clone each fragment into the bacterial virus lambda, infect into *E.coli*, grown in petri dishes, and allow the virus to replicate within the bacteria. Isolate large quantities of the viral DNA, which now contains segments of human DNA, and cleave it with restriction enzymes.

③A Separate DNA fragments using agarose gel electrophoresis. Visualize bands corresponding to different sizes of fragments using fluorescent staining. Fragments found in the child's DNA samples and not in the parents' DNA may contain heritable mutations. These bands can be removed and the DNA analyzed for specific mutations.

Direct (noncloning) Method:

②B Apply samples of DNA fragments to an electrophoretic gel, producing smears of indistinguishable bands.

③B Dissociate DNA fragments into single strands and incubate with radioactive, single-stranded, ^{32}P-labeled DNA probes for specific human genes. The probes hybridize with complementary sequences in the DNA and label their position in the gel. Visualize the position of bands using autoradiography. Changes in the position of bands in the child's DNA compared to the parents' DNA suggest possible new mutations. Bands can be isolated and DNA analyzed for sequence differences.

FIGURE 9-3 Restriction-length polymorphism. Source: OTA, 1986.

further RNaseAs needed to be developed for the other mismatches theoretically possible.

In this method, genomic DNA is isolated and mixed with radiolabeled RNA probes, and the mixture is heated and cooled so that DNA dissociates and later hybridizes with labeled RNA. RNA and DNA sequences that are homologous can form heteroduplexes, and the presence of a single base-pair mismatch does not prevent heteroduplexes from forming. Heteroduplexes are treated with RNaseAs and analyzed with agarose gel electrophoresis, which separates molecules on the basis of size. Perfectly paired molecules are unaffected by RNaseA and produce bands on the gel that correspond to their original size. Heteroduplexes with cytosine:adenine mismatches are cut at those sites and form two smaller fragments with greater mobility on the gels.

For mutation analysis, the parental DNA and child DNA are run separately, and the gel patterns are compared for the presence of bands due to mismatches.

Subtractive Hybridization

Church described an interesting method of finding DNA sequences in a child that are not present in either parent (Delehanty et al., 1986). The basis of the method is the consideration of the human genome as a group of unique sequences small enough to be studied experimentally. He calculated that a sequence of 18 nucleotides (18-mer) is long enough to be unique, because any sequence of that length is expected statistically to occur only once per haploid genome. Church's approach proposes that every possible 18-mer be synthesized (there are 4^{18} or 70 billion possibilities); the task is feasible experimentally.

To detect mutational events, DNA from child and parents is isolated and cut with restriction enzymes into pieces of 40-200 base pairs long. These are dissociated into single strands and mixed with the 18-mers under conditions that permit the formation only of perfect hybrid molecules, i.e., those with perfect base-pair matching along the entire 18-mer sequence. The

18-mers that fail to find a match among the DNA of both parents are separated from the hybrid molecules and are used to investigate the child's DNA. That is done by mixing these 18-mers with the child's DNA to identify 18-mers that hybridize perfectly with the child's DNA. Thus, any 18-mer that hybridizes perfectly with a sequence that is present in the child's DNA will have identified a DNA sequence present only in the child, not in the DNA of either parent.

Thermodynamic calculations (E. Branscomb, Lawrence Livermore National Laboratory, unpublished data, 1989) indicate, however, that such "surrogate oligo" approaches probably cannot be made to work, given the amount of purification required. Also, oligomer melting curves are too broad compared with the amount that single-base mismatches displace them in temperature. Alternative approaches that retain the idea of comparing parental genomes with those of offspring by hybridization have been proposed and may yet prove feasible.

Pulse-Field Gel Electrophoresis

When DNA is cut with restriction enzymes and run in gel electrophoresis, it forms so many bands that it appears as a smear of DNA. If DNA were cut with restriction enzymes for which there were only several hundred sites on the entire genome, the resulting pieces of DNA would be too large to separate with conventional electrophoresis. A new technique, pulse-field gel electrophoresis (PFGE), is being developed to allow separation of large fragments of DNA. Such fragments could be used to detect submicroscopic chromosomal mutations (such as small chromosomal deletions and insertions) that are intermediate in size between chromosomal aberrations that are visible microscopically with conventional cytogenetic methods and single-base-pair changes.

With this method, it is important to avoid random breakage of the long DNA fragments that occur during normal isolation procedures. That is avoided by suspending whole cells in agarose, which forms into a gel, and using enzymes to digest protein and RNA. Rare-cutter restriction enzymes are then used to cut the DNA at specific sites into fragments of 50-1000 kilobases. The fragments are separated on gels in which electric current is pulsed in perpendicular directions at a frequency that maximizes separation. Conceptually, the pulse time is selected so that fragments are constantly untangling and reorienting. The gels are then screened with the conventional Southern blotting technique and selected radiolabeled probes. Each labeled fragment appears as a visible band after the gel is developed. A chromosomal mutation appears as a shift in the position of the involved fragment when child DNA and parental DNA are compared. Studies with the DNA of lower organisms have suggested that the DNA change must be at least 5% as large as the original fragment to be visible with this method. Further work is required to optimize the pulse-controlling strategies and to adapt the procedures to human chromosomes.

One-Dimensional Denaturing Gel Electrophoresis

Fischer and Lerman (1983) developed a modification of the standard electrophoretic procedure that separates DNA on the basis of size, as well as nucleotide sequence (Fig. 9-4). Separation by sequence is based on the fact that DNA dissociates into single strands when exposed to denaturing chemicals, such as formamide or urea, and the nucleotide composition determines the concentration of denaturing chemical required for dissociation. A gradient of increasing strength of such chemicals is incorporated into an electrophoresis gel. When DNA is subjected to electrophoresis on these gels, it migrates in the electric field as a double-stranded molecule separating from other molecules on the basis of its size. Depending on sequence, each fragment begins to dissociate at a characteristic concentration of the denaturing agent in the gel. Dissociation of DNA into two single strands causes the split molecule to get stuck in the gel pores and cease to migrate further. It has been estimated that a difference of only one base pair between two otherwise identical pieces of DNA of 250 base pairs is suffi-

cient to separate two fragments into two distinct bands.

In this method, total genomic DNA is isolated from a child, denatured, and mixed with selected, single-stranded radiolabeled DNA probes. The molecules are formed into heteroduplexes, treated with restriction enzymes, and subjected to electrophoresis. The presence of a single mismatch between a nucleotide in the reference probe and the sequence obtained from a child with a mutation will cause the heteroduplexes to travel to different gel positions. Gels are dried and radiographed. The banding patterns of the child are compared with those of its parents to determine whether mutations have occurred.

Two-Dimensional Denaturing Gradient Gel Electrophoresis

Lerman (1985) proposed an additional technique that uses two-dimensional electrophoretic separation of child and parental DNA to discriminate three types of DNA: sequences common to all three, polymorphisms that are inherited from one parent, and sequences with new mutations. The method separates first on the basis of sequence length and then on the basis of sequence composition. Size separation is carried out in standard agarose gels. The gel strips are then laid across the top of a gel that contains a gradient of a denaturing agent to separate on the basis of sequence composition. To study new mutations in children, two DNA mixtures are made—one sample containing DNA of both parents and the second sample containing DNA from both parents and the child. These are subjected individually to electrophoresis in both dimensions, as described above. The denaturing gel is then cut into strips that contain DNA sequences that melt at common denaturing conditions, i.e., isomelting DNA. The DNA in the strips is fully dissociated and reannealed to convert the homoduplex DNA into heteroduplex DNA, i.e., with individual single strands from different persons. Each strip is placed on another denaturing gel and subjected to electrophoresis. With increasing denaturant concentration, three

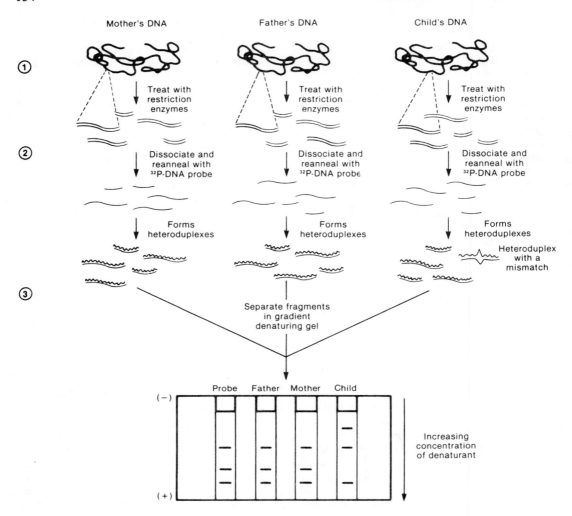

① Isolate genomic DNA from white blood cells. Using restriction enzymes, cut DNA into (double-stranded) fragments of various lengths.

② Dissociate double-stranded DNA fragments into single strands, and reanneal in the presence of radioactive ^{32}P-labeled, single-stranded DNA probes. Heteroduplexes form between probe and sample DNA even if the base sequences are not perfectly complementary; if mutations are present, some of these heteroduplexes will contain mismatches.

③ Separate heteroduplex fragments in a denaturing gradient gel. Fragments with mismatches denature in a lower concentration of denaturant than fragments that are perfectly complementary. Visualize the position of the fragments with autoradiography. Fragments containing the child's DNA that denature sooner than the parents' fragments can be analyzed for new mutations.

FIGURE 9-4 One-dimensional denaturing gradient gel electrophoresis. Source: OTA, 1986.

classes of DNA are recovered. Two of the classes do not contain recognizable mutations; sequences that are perfectly homologous will travel to the highest concentration of denaturant before dissociating (farthest down in the gel), and sequences that travel slightly slower are polymorphisms that are present in only one parent and are seen as bands of spots. The third class moves the shortest distance in the gel and contains more than one polymorphism per fragment. These mismatches, which are seen as a few spots at the top of the gel, are candidates for mutation analysis. By comparing the spot patterns in the gels derived from the two mixtures, new spots can be observed. These would represent single-base changes or very small deletions or insertions in the child's DNA that are not present in either parent. Few data are available to evaluate the feasibility of this method. It does have the advantage that it can be scaled up to inspect large portions of the genome by analyzing multiple isomelting regions of DNA.

The above molecular approaches for measuring heritable mutations are based on comparisons of children's DNA with that of their parents. These methods are under development and not sufficiently validated for detecting induced mutations in people.

The following approaches are being designed to detect genetic damage or germinal mutations directly in human male germ cells. They include testicular markers of cytogenetic damage as well as sperm markers of cytogenetic damage, DNA strand breakage, specific locus mutations, and aneupolidy.

TESTICULAR MARKERS OF HUMAN GERMINAL CYTOGENETIC DAMAGE

Several approaches are available for the direct measurement of cytogenetic damage directly in germ cells of men exposed to a potential germline mutagen. As indicated in Table 9-4, either testicular tissue or semen is required. Although the methods that require testicular tissue are inherently difficult to apply in human studies, they provide important bench-

marks for the development of the more practical semen-based assays.

In principle, cytogenetic analysis of metaphase chromosomes is possible during three periods of spermatogenesis: during spermatogonial mitosis, meiosis I (MI), and meiosis II (MII). In animals, exposure to some chemicals or to ionizing radiation induces cytogenetic abnormalities detectable in spermatogonial metaphase chromosomes, and both chromosomal aberrations and sister-chromatid exchanges have been described (Hsu et al., 1979; King et al., 1982). However, human counterparts of spermatogonial analyses have not yet been established, and such research is encouraged.

Two distinct types of cytogenetic abnormalities can be detected in meiotic cells: chromosomal rearrangements during MI and aneuploidy during MII. Hulten and colleagues (1985) studied MI and MII metaphases in normal fertile men, while McIlree et al. (1966) (Koulischer and Schoysman, 1974) investigated subfertile men. Although the total numbers of metaphases analyzed were small, very low frequencies of abnormal metaphases were found. In spite of the technical difficulties, it is expected that these human cytogenetic data bases will continue to increase in size and will gain importance in establishing spontaneous rates of chromosomal abnormalities in the germ line of the human male.

The most convincing, and in fact the only direct, evidence of the sensitivity of human germ cells to agent-induced genetic damage has been the demonstration by Brewen et al. (1975) of a dose-related increase in MI chromosomal rearrangements in testicular preparations of men exposed to graded testicular doses of radiation up to 6 Gy. Their results indicate that men are about as sensitive as marmosets and twice as sensitive as mice to this type of damage. A dose-dependent response was observed in the frequency of reciprocal translocations with a frequency of 7.0 $\pm$ 1.3% in men who received acute stem-cell irradiation of 2 Gy. That method's main limitation is that it requires testis-tissue samples; this dramatically restricts its application and makes it an

TABLE 9-4 Status of Human Biologic Markers of Genetic Damage to Male Germline

Tissue Required or Data Source	Markers Used to Detect Effects of Chemical or Radiation Exposure	Markers Used to Determine Human Baselines	Promising New Concepts
Testis	Cytogenetic analysis of meiotic I cells	Cytogenetic analysis of meiotic II cells	None
Semen Sperm	Sperm cytogenetics	None	Sperm DNA and protein adduction Gene mutations in sperm Sperm aneuploidy
Immature germ cells	None	None	Spermatid micronuclei Cytogenetic analysis of ejaculated meiotic I cells
Questionnaire or medical	None	Sex ratio Spontaneous abortion Offspring cancer Sentinel phenotypes	None
Offspring tissue	None	Cytogenetics Protein mutations	DNA sequencing Restriction-length polymorphism Subtractive hybridization Denaturing gel electrophoresis Pulse-field electrophoresis
Mother's urine	None	Detection of early fetal loss	

impractical marker of induced human germline mutations. It has not been applied to the study of other men exposed to mutagens, such as men receiving cancer radiotherapy or chemotherapy.

SEMEN MARKERS OF HUMAN GERMINAL MUTATIONS AND GENETIC TOXICITY

Of all the germ-cell types in men, sperm are by far the easiest to obtain in large numbers. Sperm methods obviously are limited to males; to date, there are no direct methods that can be applied to assess genetic damage in the germ cells of females.

As stated earlier, damage inherited via female germ cells must be assessed by survey or by analysis of mutations using offspring methods.

Several approaches have been proposed or are under development for measuring genetic alterations in the sperm of exposed persons, including detection of cytogenetic abnormalities in sperm, gene-mutation analysis in sperm, detection of DNA adduction in sperm, detection of aneuploidy in sperm, and alkaline elution of sperm DNA. In addition, the human ejaculate differs from that of laboratory and domestic animals in that it can contain both spermatocytes and spermatids that were presumably exfoliated from the seminiferous epithelium before completing differentiation into mature sperm (Auroux et al., 1985). Thus, additional markers of genotoxicity might be possible if these cells are used, such as meiotic chromosomal analysis of seminal spermatocytes and scoring of micronuclei in seminal spermatids.

The semen-based methods can be grouped by whether they can be applied as qualitative indicators of germinal genotoxicity or as quantitative markers that yield data useful for estimating mutation frequencies.

The markers that might provide qualitative indications of human germ-cell genotoxicity include DNA and protein adduction in sperm, alkaline elution of sperm DNA, and meiotic chromosomal analyses of seminal spermatocytes and micronuclei in seminal spermatids. The markers designed to be quantitative markers suitable for estimating chromosomal and genic mutation frequencies in germ cells include the markers of cytogenetic abnormalities, gene mutations, and aneuploidy in sperm. As discussed below, none of these methods has been sufficiently developed or validated for use in large-scale studies of people exposed to mutagens.

Cytogenetics of Human Sperm

After the pioneering work of Rudak et al. (1978), several laboratories confirmed that artificially capacitated human sperm can fuse with enzymatically denuded hamster eggs (see Fig. 9-5) to yield first-cleavage sperm chromosomes that can be evaluated by standard cytogenetic techniques (Yanagimachi, 1984). Brandriff et al. (1984), with the largest published data base, have prepared over 5,000 metaphases from 20 healthy men and found donor-specific variation in the frequency of cells carrying spontaneously occurring cytogenetic aberrations from about 1% to 15%. Overall, they reported 2.1% aneuploid cells and 6.9% cells with structural cytogenetic abnormalities (Brandriff et al., 1988a). Another laboratory (Martin et al., 1983; Martin et al., 1987) has a suitable data set for comparison; it found 4.7% of aneuploid cells and 6.2% of cells with structural aberrations (10.4% abnormal sperm complements) among 1,582 cells from 33 men.

The major advantage of this method is that it provides the only available means of preparing human sperm chromosomes for standard cytogenetic analyses. The integrity of sperm chromosomes is important as an indicator of events that occurred during spermatogenesis (including effects related to paternal age, male physiology, and history of exposure to physical and chemical mutagens) and as a predictive marker of the likelihood of success of early cleavage and embryo development that would result from fertilization with these sperm. The major limitation is that the method is not yet fully developed or validated. Further studies are needed to investigate discrepancies between the data of the two laboratories and to determine the effects of age, smoking, etc., on the background rate of sperm cytogenetic abnormalities. Recent studies of men receiving radiotherapy found increased proportions of cytogenetically abnormal cells (Brandriff et al., 1986; Jenderny and Röhrborn, 1987; Martin, 1988). The procedure warrants further validation in studies of mutagen-exposed people, but requires simplification before routine application is practical.

Sperm-DNA Alkaline Elution

The alkaline elution method is an indirect measure of DNA damage in sperm. It has been developed for mouse and rat germ cells and sperm (Skare and Schrotel, 1984; Bradley and Dysart, 1985; Sega et al., 1986). This method might be a useful indicator of DNA breakage in sperm and of exposure to DNA-breaking agents. Application to human sperm requires methods of DNA-break detection not based on incorporated radioactivity. Sega and coworkers are evaluating fluorescence procedures to monitor human sperm-DNA breakage (G. Sega, Oak Ridge National Laboratory, unpublished data, 1988).

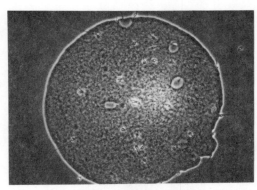

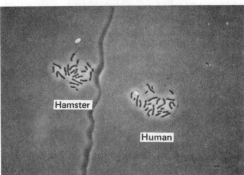

FIGURE 9-5 Decondensing human sperm (top panel) and human sperm chromosomes (bottom panel) revealed in the cytoplasm of hamster eggs. Source: Brandriff, Lawrence Livermore National Laboratory, unpublished.

Specific-Locus Mutations in Sperm

Sperm are highly structured cells that contain many sperm-specific proteins organized into specific internal and surface compartments. Several approaches have been proposed to use these proteins and their compartmentalization for the detection of mutant sperm. Malling and associates (Burkhart et al., 1985) used polyclonal antibodies against rat $LDHC_4$, a testis-specific isozyme of lactate dehydrogenase, to search for mutated sperm in mice that became "ratlike" in their antigenicity, but this method was not reproducible.

An attempt is under way to develop a gene-expression-loss assay that uses fluorescently labeled monoclonal antibodies against each of the two human protamines (Stanker et al., 1987a,b) to label sperm with null mutations affecting the human protamine gene. Flow cytometry is used to identify and score the mutants in a manner similar to that used with the glycophorin-A gene-expression-loss assay

of red blood cells described earlier in this chapter.

Other approaches are needed, with emphasis on well-characterized epitopes and binding sites of monoclonal antibodies or lectins, use of sperm-specific proteins, and understanding of the molecular aspects of the genetic target being measured.

DNA Adduction in Sperm

Many chemicals that break chromosomes and induce gene mutations do so via DNA adduction that interferes with DNA replication and repair. Specific DNA adducts have been measured in the testicular germ cells and sperm of mutagen-exposed mice (Sega and Owens, 1983), but no DNA adduction procedure has yet been developed or validated for human sperm. Work by Sega and coworkers suggested the importance of protamine adducts in the formation of heritable mutations. The suggestion was based on correlation (as yet not understood) in mutagen-exposed mice between protamine adduction, the induction of dominant lethality, and the increased alkaline elution of spermiogenic DNA (Sega et al., 1986).

The ability to measure DNA adducts in human sperm would provide a direct means to quantify damaged germline DNA for identifying human exposures that are genotoxic and for identifying exposed persons. However, adduction is a functional description, and individual adducts differ markedly in their chemical structure. Sega has identified four different protamine adducts so far in his studies with mice: methyl, ethyl, hydroxyethyl, and an adduct derived from acrylamide (Sega and Owens, 1987). However, DNA adducts were not detectable with acrylamide. Thus, detection methods must be tailored to the specific adducts to be measured. More research and development are encouraged in this regard.

Aneuploidy Detection in Sperm

Aneuploidy contributes substantially to the occurrence of spontaneous abortion, fetal death, and genetic defects (e.g., Down's syndrome). Approximately one-fourth of cases of Down's syndrome are thought to be due to fertilization with a sperm carrying two chromosomes 21. Data from live births and from human sperm cytogenetics (Fig. 9-4 and text) indicate that the frequency of aneuploid sperm might be approximately 2%. However, there is no practical method for measuring the rate of aneuploidy among cells of individual ejaculates.

More than 15 years ago (Barlow and Vosa, 1970), a method based on the bright fluorescence of the Y chromosome of quinacrine-stained sperm was proposed for detecting sperm with two Y chromosomes (referred to as the YFF test). However, it was shown that it was probably not a valid measure of aneuploidy, because mass measurements and comparisons between YFF frequencies and sperm cytogenetic analyses indicated that YFF overestimates the frequency of sperm with two Y chromosomes (Wyrobek et al., 1984).

Recently, there has been active development of chromosome-specific DNA probes (Rappold et al., 1984), including probes for sex chromosomes and autosomes. In principle, these probes provide a means of detecting sperm that are aneuploid for sex chromosomes or autosomes (Joseph et al., 1984; Wyrobek and Pinkel, 1986). Applied to somatic and germ cells of the same person, this technique could provide a means for comparing induction and persistence of chromosomal aneuploidy in somatic and germinal cells of mutagen-exposed men.

Cytogenetic Analysis of Meiotic Cells in Semen

The human ejaculate is unusual among mammals, in that it contains immature germinal cells at concentrations of up to several percent in healthy men and perhaps higher in some infertile men. Templado and coworkers (1986) have developed a technique for using meiotic cells in the ejaculate for analyses of multivalent chromosomes at MI. Egozcue et al. (1983) analyzed the chromosomes of meiotic cells from the semen of 501 men and found an incidence of 4.3% of cells with meiotic anomalies, including univalents at metaphase I, pairing anomalies in prophase I, desynapsis, and meiotic arrest. This method is limited by the small numbers of cells available for analysis, and further efforts are warranted to develop enrichment techniques and to improve reproducibility.

Recently, Templado and associates improved their procedure to provide analyzable meiotic cells in 58.1% of 50 consecutive cases. Continuing efforts are under way to improve the preparative procedures and to evaluate the utility of seminal meiotic cells for assessing cytogenetic damage in the male germ line. Also, it remains unknown whether meiotic cells in the semen are representative of normal meiotic cells in the seminiferous epithelium. Additional studies are required to determine the concordance between the cytogenetic analyses of meiotic cells and sperm in semen and to relate within the same subjects these measurements to aberrations in somatic cells.

Micronuclei in Seminal Spermatids

Tates and deBoer (1984) and Laehdetie

and Parvinen (1981) showed that exposure of rats to germinal mutagens increases the proportions of spermatids with micronuclei. In somatic cells, micronuclei are thought to represent acentric chromosomal fragments that failed to segregate normally in prior cell divisions. Assessment of micronuclei in spermatids has not yet been applied to men or to human semen, but the presence of spermatids in some human ejaculates suggests that further studies are warranted to develop this technique. Once it is developed, research will be required to determine the concordance between seminal spermatids with micronuclei and sperm with cytogenetic abnormalities. Seminal micronucleus analysis also permits comparison with micronuclei in peripheral blood cells on a person-by-person basis, providing another means of comparing somatic and germinal responses in people.

SUMMARY

Table 9-3 summarizes the status of the mutation methods evaluated in this chapter. Other than the analysis of meiotic testicular cells of irradiated men (Brewen et al., 1975) and the preliminary sperm cytogenetic data (Jenderry and Röhrborn, 1987; Brandriff et al., 1988b; Martin, 1988) with radiotherapy patients, no method has given conclusive direct evidence of agent-increased genetic damage in the germ line of mutagen-exposed people. A limited number of methods have been used to establish a human baseline (Table 9-4, column 3), and these, if applied in appropriate studies, might provide evidence of germinal genotoxicity and induced heritable mutations in mutagen-exposed people. However, as emphasized in this chapter, the available methods for detecting human germline mutations are inefficient and inadequate for most human exposures. Future research should emphasize and be aimed toward the promising new cellular and molecular approaches using semen and offspring tissue (Table 9-4, column 4).

10

Conclusions and Recommendations

There is a paucity of information on the prevalence of naturally occurring reproductive abnormalities and the prevalence of reproductive abnormalities induced by physical agents or toxic chemicals in the human male. Two major reproductive health outcomes are of concern when human males are exposed to toxic agents: *pathophysiologic changes* (e.g., reduction in spermatozoal concentration, decrease in motility, increase in proportion of sperm with abnormal forms), which might or might not be associated with fertility status, and *heritable genetic damage*. Chromosomal and genic abnormalities in parental germ cells may lead to reduced fertility, early or late pregnancy loss, congenital malformations, or other defects and diseases in the offspring, some of which may not have a health effect until later in adult life. The biologic markers used to assess pathophysiologic changes in human males in response to exposure to toxic chemicals are potency, fertility, serum concentrations of gonadal steroids and pituitary hormones, and semen characteristics. Blood serum concentration of gonadal steroid and pituitary hormones and several of the semen markers, especially sperm count, motility, and morphology, have been used to identify human male reproductive hazards.

Currently available markers of heritable genetic damage in the human male genome in response to exposures can include pregnancy outcome, presence of sentinel phenotypes (e.g., that of achondroplasia), chromosomal damage, and physicochemical changes in macromolecules of offspring (e.g., occurrence of electrophoretic variants of red blood cell enzymes). These genetic markers have thus far been used to measure the spontaneous rate of human germinal mutations but have proved ineffective for assessing exposure to or effects of suspected mutagenic chemicals in humans.

One of the major problems with the use of animal models is the difficulty in extrapolation of their results to human beings. Some of the reasons follow, and research to address these questions is encouraged:

- Few toxicologic studies in laboratory animals have been directed at male reproductive health, relative to the large number of potentially toxic chemicals to which people are exposed.
- Studies vary in their experimental design, species used, dose and route of administration of toxic chemical, and choice of markers. That makes it difficult to compare results of different studies and results from different laboratories.
- Few animal studies have yielded infor-

141

mation on the correlation of markers with fertility or reproductive outcomes.
• Techniques for extrapolating data from laboratory animals to human males are rudimentary and not validated.

New biologic markers of reproductive and genetic toxicity in the laboratory human male are needed. The male reproductive system consists of several organs that interact in a complex manner with each other and with the neural and endocrine systems. A medical history and a physical examination, although important, are unlikely to detect exposure to a variety of toxic chemicals or effects on male reproduction. However, case reports that tentatively link infertility or abnormal reproductive outcome with particular occupations, chemicals, exposures, or drugs might be valuable indicators that lead to identification of new human male reproductive toxicants. Limiting the analysis to individual markers might be an oversimplified approach. Instead, a battery of markers that reflects a wide array of reproductive functions and heritable genetic damage is required, in combination with a thorough medical history and physical examination. Finally, it must be remembered that some changes in markers could be so subtle that important alterations in response to toxic insult will be seen only in a large population. The subcommittee makes the following general recommendations:

• Extensive basic research in laboratory animals must continue to identify additional markers of physiologic function and heritable genetic damage.
• Markers of exposure to toxic chemicals must be correlated with markers of effect and with changes in reproductive health in the human male.
• Mechanisms of toxicity must be investigated in laboratory animals and in vitro culture systems and related to the human response.
• Controlled reproductive toxicologic studies must be completed on several species to validate markers.
• Risk assessment procedures must be developed, to allow data from laboratory animals exposed to toxic chemicals to be extrapolated to human males.
• Human markers that measure the effective magnitude of toxic exposure or effect on germ cell, epididymis, and other reproductive organs must be identified.
• New markers must be evaluated with cross-sectional and longitudinal data for individuals and for groups (the latter stratified by regions, occupations, races, etc.), to establish baselines.
• Develop assays in semen, saliva, and urine or use noninvasive externally derived signals, such as from ultrasound, that can be applied to the screening of large populations.
• Continue to develop more invasive techniques limited to use in subgroups, to gain insight into mechanisms of toxicity.

The subcommittee focused its specific recommendations, which follow, on the identification of markers that could be used to detect exposures or their effects on male reproductive function and heritable genetic damage, criteria for the development of markers of male reproduction, and the development of strategies for testing the effects of toxic chemicals on markers of male reproduction.

IDENTIFICATION OF MARKERS OF ABNORMAL PHYSIOLOGIC FUNCTION

Most of the potentially useful biologic markers of male reproductive function are in the developmental stage in the laboratory. Research needs to be continued in various subjects, as follows, in an effort to identify additional markers of specific physiologic functions.

Testes

• Evaluate the use of noninvasive physical measurements, such as magnetic resonance imaging and ultrasound, for providing signals that can be used to assess such criteria as size, consistency, blood flow, and function.
• Investigate the capacity of the testis to metabolize xenobiotics.

• Investigate the pharmacokinetics of toxic chemicals and their metabolites with respect to transport into the interstitium and seminiferous tubules.

• Carry out basic research on spermatogenesis, with special consideration of stem cell numbers, division, and clonal nature.

• Test the effect of toxic agents on markers of Leydig cells, Sertoli cells, and specific germ cells; special attention must be given to the time course of toxic effects and the potential recovery period.

• Develop assays for inhibin, androgen-binding protein, Müllerian inhibiting substance, etc., as biologic markers of Sertoli cell function.

• Identify and characterize additional molecules that are produced by Leydig, Sertoli, and germ cells and secreted into the blood and semen.

• Carry out basic research and develop probes for testicular RNA, DNA, and macromolecules; these can be used to develop assays for quantifying specific steps in the development of germ cells.

• Continue basic research on mechanisms of intratesticular communication among Leydig, Sertoli, and germ cells with the goal of identifying additional markers of testicular function.

• Compare the toxic responses of human and laboratory animal germ cells, to identify useful laboratory animal models of human effects.

Epididymis

• Carry out basic research to determine which facets of spermatozoa change during epididymal transit (e.g., specific surface antigens).

• Evaluate the effect of toxic chemicals on spermatozoal epididymal transit time, site of acquisition of motility, and fertilizing potential.

• Evaluate the effect of toxic chemicals on epididymal structure, integrity of the blood-epididymis barrier, and function of specific cells.

• Evaluate the capacity of the epididymis to metabolize xenobiotics.

• Continue basic research on the functional interaction between the immune system and the epididymis.

• Develop molecular probes of epididymal function applicable for blood or semen evaluation.

Accessory Sex Organs

• Identify specific molecular markers and their cDNA probes for individual accessory sex organs (e.g., prostatein).

• Continue basic research on the regulation of the biosynthesis and secretion of those molecular markers, especially their composition in seminal plasma.

• Develop ultrasound and other physical measurements to provide externally derived signals that can be used to assess site and function of specific accessory sex organs.

• Investigate the pharmacokinetics of toxic chemicals in individual accessory sex organs.

SEMEN MARKERS OF ABNORMAL PHYSIOLOGIC FUNCTION

In principle, semen can be used to evaluate the cellular product of spermatogenesis, the function of the somatic supporting testicular cells, the function and integrity of the efferent duct system (including the epididymis), the function of the accessory glands, hormonal state, and perhaps the xenobiotic exposure of the male organism. Several available markers have been evaluated (Table 7-2) and some have shown detrimental effects in people exposed to radiation or chemicals. For some markers, baselines have been established, but they have not been evaluated in exposed people. Most markers remain inadequate for quantitatively assessing male reproductive health at this time. More research is required for elucidating underlying molecular mechanisms, for evaluating the predictive value of individual markers in relation to fertility, and for identifying the utility of markers as indicators of exposure to xenobiotic agents.

Chapter 7 highlighted several promising concepts (Table 7-2), including automation of sperm measurements; markers of sperm function; immunologic reagents for molecular studies of spermatogenesis; recombinant probes of spermatogenic genes; and semen markers of Sertoli-cell, epididymal, prostatic, and seminal vesicle function. The list is not intended to be complete; rather, it gives examples of research subjects with early promising results. In the near future, we hope to have markers of cell differentiation and function that are specific for events at the molecular level.

The following are specific recommendations for semen studies:

• Develop in vitro sperm assays that measure the capacity of animal and human spermatozoa to fertilize oocytes.
• Develop seminal plasma molecular markers of individual accessory sex organ function.
• Develop monoclonal-antibody assays of specific functional domains on spermatozoal plasma membranes.
• Develop computer-based and automated techniques for making quantitative sperm measurements and for sorting sperm, e.g., on the basis of numbers, motility, morphology, domains, and enzyme function.
• Study the pharmacokinetics of toxic chemicals and their metabolites in seminal plasma and the capacity of semen and sperm to carry toxic agents to the female and the site of fertilization.
• Develop semen-based assays for Sertoli cell and Leydig cell function.

NEED FOR IMPROVED MEASURES OF FERTILITY STATUS AND EXPOSURE

In addition to the need for improved semen markers, improved approaches for assessing their validity are needed. The evaluation of a semen marker's utility for fertility assessment requires sensitive quantitative measures of fertility for comparison. Further work is required to quantify fertility, including further evaluation of standardized fertility ratios and time required to conception. Another measure of a male's fertility might be the immunologic detection of β-chorionic gonadotropin in his mate's urine, which indicates pregnancy about 10 days after conception (Table 9-3). This measure of fertility has disadvantages—it is insensitive during the first 10 days of pregnancy, and it obviously requires that cou-

ples plan to have children. At present, it appears to be the most sensitive indicator of early pregnancy, and, on further evaluation, could become the fertility reference of choice for determining the predictive value of selected semen markers.

The assessment of a semen marker's utility as an indicator of exposure to xenobiotic agents requires sensitive quantitative exposure measures for each agent to be evaluated. Although a crucial aspect of the validation process, the ability to quantify human exposure varies dramatically with agent and circumstances. Several approaches are possible, including biologic, physical, and chemical dosimetry and exposure history. However, with the exception of very few agents for which exposure can be determined with reasonable certainty (e.g., therapeutic ionizing radiation and chemotherapeutic drugs), human exposure assessments are usually little more than guesses. It might not be possible generally to determine the quantitative dose-response relationship between a semen marker and human exposure, as required for validation. However, several solutions are possible: select a single agent to model the human dose-response relationship and extrapolate to other human exposures, develop improved biologic dosimetry (with sensitive methods for detecting adduction of DNA and proteins), and perform detailed dose-response evaluations in animal models and extrapolate the results to humans.

IDENTIFICATION OF MARKERS OF GERMINAL GENETIC TOXICITY AND HERITABLE MUTATIONS

Recommendations for research needs in the development of markers of germinal genetic toxicity and heritable mutations include studies with human and animal tissues:

• Study whether integration and excision of transposable elements constitute one mechanism for inducing germinal mutations.
• Determine the extent to which transposable elements exist in humans.

- Carry out basic research on selective pressures for and against chromosomal defects during spermatogenesis.
- Compare the consequences of protein and DNA adducts in various types of germ and somatic cells in mice.
- Study the extent and conditions of induction of aneuploidy in male germ cells.
- Investigate and compare the molecular nature of stem cell and post-stem-cell mutations, both spontaneous and induced, in mice.
- Develop sperm-based markers of genetic damage in mice, such as alkaline elution, cytogenetic abnormalities, aneuploidy, and gene mutations analogous to those in humans.
- Evaluate the effect of toxic chemicals on the capacity of epididymal spermatozoa, as opposed to testicular germ cells, to produce genetically normal offspring.
- Develop additional animal models for studying the role of toxic chemical metabolism, the mechanism of toxic-chemical damage, and repair mechanisms in the induction of germinal and heritable mutations.
- Develop and validate human semen markers of genetic toxicity and induced mutations, including methods for detecting gene mutations, aneuploidy, chromosomal aberrations, and DNA adducts in mature sperm and in immature germ cells.
- Develop and validate DNA markers of human heritable mutations including Lerner gels, restriction-fragment-length polymorphisms (RFLPs), subtractive hybridization, and RNAse digestion.

CRITERIA FOR DEVELOPMENT AND VALIDATION OF MARKERS OF MALE REPRODUCTION

Identifying biologic markers that represent exposure to toxic chemicals or the effects of exposure on specific male reproductive tract functions and heritable genetic damage is only the first step. The subcommittee has compiled a series of steps that should be followed when a potentially useful marker is to be validated.

- Establish normal baseline values and distribution of each marker in laboratory animals and humans.

- Evaluate the sensitivity and specificity of each marker to predict a health outcome (e.g., fertility) and heritable genetic damage.
- Understand in detail the dose-response relations and time course of response of each marker to a given toxic chemical, with special attention to the recovery process.
- Evaluate new markers in studies, including existing proven markers of pathophysiologic functions and heritable genetic damage, and understand the relative value of each marker to others in the battery.
- Develop a strategy and consensus for the use of multiple species in toxicologic studies.
- Test the effect of toxic chemicals in several species, including mice and rats; mice are particularly useful for genetic studies, because they have been widely studied for mutagenesis, but murine metabolism of some chemicals might differ markedly from human metabolisms, in which case species more closely resembling humans must be studied.
- Encourage the development of animal markers with human correlates so that risk extrapolation models can be developed and evaluated.

STRATEGY FOR TESTING EFFECT OF TOXIC CHEMICALS ON MARKERS OF MALE REPRODUCTION

It is suggested that a task force be established to develop and carry out a strategy for evaluating biologic markers in laboratory animals. The task force might have the following functions:

- Select a battery of markers of male reproductive functions and heritable genetic damage.
- Select a number of toxic chemicals and nontoxic analogs.
- Select a single source for production of highly purified chemicals.
- Design important aspects of a protocol, such as species to be used and dosage, duration, and route of chemical administration.

- Select specific laboratories that have the expertise to carry out the measurements required for testing a specific marker. No single laboratory would carry out all the measurements at this stage, because unique methods might be required.
- Serve as a clearinghouse for data analysis and evaluation of a marker of specificity, sensitivity, precision, and accuracy in reflecting exposure to toxic chemicals and the effects of exposure on reproductive function or heritable genetic damage in laboratory animals.

II

Biologic Markers in
Female Reproductive Toxicology

11

Introduction

Less information is available concerning reproductive toxicology for females than for males. This lack of formal data applies to work with laboratory animals and humans and reflects substantial differences between males and females in the chronology of gametogenesis, the number and accessibility of germinal cells, and probable differences in neuroendocrine milieu.

This chapter identifies biologic processes that may be sensitive to environmental influences and hence may be particularly ripe for development and application of markers. It is arranged as an overview of reproductive biology and covers oogenesis, development of the female reproductive tract, maturation, maintenance of reproductive function, fertilization, implantation, and reproductive senescence. In utero assessments are limited to female reproductive tract development, and the use of human chorionic gonadotropin (hCG) as an assessment of fecundity in women. The introductory material is followed by four chapters that discuss biologic markers of germ cell damage; development and aging; cyclic ovarian function; and conception, implantation, and early embryonic loss. This review is not comprehensive but focuses on major substantive aspects of female reproductive toxicity.

The subcommittee's conclusions and recommendations regarding research opportunities and directions for a program on biologic markers of female reproduction are presented in Chapter 16. A summary table lists and categorizes biologic markers according to whether each marker may be used in large-scale human studies or only in studies of special populations and whether it needs further development or needs to be applied in animal studies.

The physiologic processes and consequent landmarks of female reproductive function are discussed in this chapter. Identification of landmarks that describe normal functions is a necessary first step for identifying biologic assessments that might be used as markers of exposure or effect. Hypothetical and documented toxic effects are discussed for each stage of female reproductive function. Specific biologic markers are discussed in later chapters.

OOGENESIS

Primordial germ cells in humans are first detectable in the yolk sac at 3 weeks of development. These oogonia (approximately 1,700) migrate to the gonadal ridge and undergo a period of active mitosis that peaks—with an increase to 5-7 x 10^6 cells—around month 5 of gestation. All the oogonia enter the initial stage of meiosis by month 7, at which point they become known as primary oocytes. Approximately 60% of the store of germ cells is lost to physiologic atresia between month 5 and birth; these cannot be replaced (Fig. 11-1). Shortly after birth, the primary oocytes are arrested in late prophase of meiosis I. With the onset of puberty, the increasing production of gonadotropins stimulates the resumption of meiosis. The first meiotic division is completed just before ovulation (Fig. 11-2), and the second division is completed only when fertilization ensues.

As the oogonia enter into meiosis, human cells contain a diploid number of chromosomes and four copies of DNA. The two divisions of meiosis entail halving the number of chromosomes (meiosis I) and then halving the amount of DNA (meiosis II). The female pronucleus is formed at fertilization and ultimately combines with the male pronucleus to reestablish the diploid state.

Several aspects of oogenesis are noteworthy. First, exposure of female conceptuses to germ cell toxicants is of particular importance, because damaged oocytes never can be replaced. The stock of gametes is large, however, and some destruction undoubtedly can be tolerated without re-

ductions in fertility or length of reproductive life. Second, mitosis in females ceases in utero, so the incidence of replication-dependent mutations should be lower than in the male, whose stem cells divide continuously. Third, exposures or physiologic deterioration sustained during the long resting phase of oocyte maturation may account for the increased sensitivity of female germ cells to meiotic errors, such as nondisjunction. Whether a blood/follicle barrier exists to protect the oocyte from toxic substances is controversial, and more research on this question is needed. The mature oocyte appears to be capable of repairing its own damaged DNA (Pederson and Mangia, 1978) and, after fertilization, damaged sperm DNA (Generoso et al., 1979).

DEVELOPMENT OF THE FEMALE REPRODUCTIVE TRACT

The female reproductive tract develops on the urogenital ridge early in fetal life. The paired Müllerian ducts form the fallopian tubes, and fuse to form the uterus and cervix. The vagina develops partly from the Müllerian duct and partly from the urogenital sinus, whereas the external genitalia evolve from the genital tuber-

FIGURE 11-1 Changes in total population of germ cells in human ovary with increasing age. Reprinted with permission from Baker, 1971. Copyright 1971, C.V. Mosby Co.

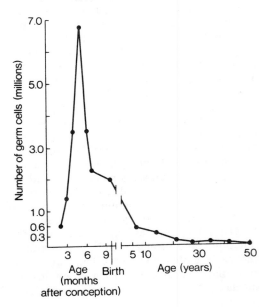

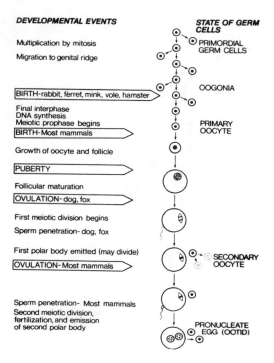

DEVELOPMENTAL EVENTS

Multiplication by mitosis

Migration to genital ridge

BIRTH-rabbit, ferret, mink, vole, hamster

Final interphase
DNA synthesis
Meiotic prophase begins
BIRTH-Most mammals

Growth of oocyte and follicle

PUBERTY

Follicular maturation

OVULATION- dog, fox

First meiotic division begins

Sperm penetration- dog, fox

First polar body emitted (may divide)

OVULATION-Most mammals

Sperm penetration- Most mammals
Second meiotic division,
fertilization, and emission
of second polar body

STATE OF GERM CELLS

PRIMORDIAL GERM CELLS

OOGONIA

PRIMARY OOCYTE

SECONDARY OOCYTE

PRONUCLEATE EGG (OOTID)

FIGURE 11-2 Life cycle of a female germ cell. Reprinted with permission from Austin and Short, 1982. Copyright 1982, Cambridge University Press.

cle, labioscrotal folds, and labioscrotal swellings.

The hypothalamus and anterior pituitary, which are important in regulating reproduction, form from the developing central nervous system and the oral ectoderm, respectively. Sex-based structural and functional differences in the hypothalamus of animals depend on the hormonal milieu early in development (Gorski, 1968; Brawer and Naftolin, 1979). A similar sexual differentiation is believed by some to occur prenatally in the human hypothalamus; this may influence later psychosexual orientation and reproductive function (Ehrhardt and Meyer-Bahlburg, 1979).

Prenatal exposure to the synthetic hormone diethylstilbestrol (DES) is the best-documented example of toxic influence on the developing female reproductive tract. Anomalies of vagina, cervix, uterus, fallopian tubes, and mesonephric duct remnants found in humans with hysterosalpingography and usually detected after reproductive maturity seem to originate from disturbances in embryologic development (Newbold and McLachlan, 1982). Among women affected, these abnormalities of

structure may compromise reproductive performance (Herbst et al., 1980). Whether prenatal exposure to DES also alters the hypothalamus or pituitary is being investigated (for example, in a study of sexual activity and functioning in DES daughters (Meyer-Bahlburg et al., 1985)).

MATURATION

Hormonal Changes

Assumption of adult female reproductive function involves major increases in ovarian and adrenal production of steroids that induce maturation of cells that are responsive to sex steroids. The steroid production changes—at least those due to increased ovarian activity at puberty—are controlled by the hypothalamic-pituitary axis.

During childhood, the hypothalamus is very sensitive to sex steroids, and low concentrations suppress gonadotropin-releasing hormone (GnRH) (Fig. 11-3). This sensitivity decreases with the onset of puberty, and increasing concentrations of steroids are required for suppression.

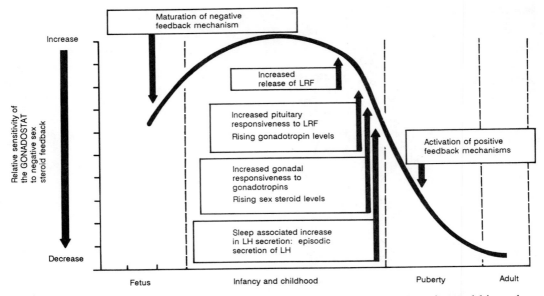

FIGURE 11-3 Schematic illustration of development of hypothalamic-pituitary-gonadotropin-gonadal interrelationship in relation to onset of puberty. From Grumback et al., 1974.

As a consequence, a new set point is reached, and pulsatile GnRH releases are sufficient to raise circulating luteinizing hormone (LH) and follicle-stimulating hormone (FSH) to the point where follicular activity ensues. The resulting increased ovarian steroid production has negative and positive feedback effects on gonadotropin release. The underlying hypothalamic mechanism that initiates these hormonal changes at puberty is thought to be the augmentation of pulsatile GnRH secretion under the control of an oscillator in the arcuate nucleus (Knobil, 1980). Late in puberty, positive feedback effects of estradiol (E_2) lead to a preovulatory LH surge. Terasawa (1985) showed that these patterns usually cannot be evoked at earlier ages in the absence of the hormonal milieu of a sexually mature adult. Maturational changes in the control of GnRH secretion occur in monkeys ovariectomized before puberty and thus are probably independent of ovarian steroids. Similar conclusions are drawn from the pattern of gonadotropin secretion in women with Turner's Syndrome (a form of gonadal dysgenesis).

The nocturnal pattern of LH secretion also changes dramatically during puberty (Boyar et al., 1973) (Fig. 11-4). The frequency of pulsatile LH release over 24 hours and the response to E_2 are sensitive indicators of hypothalamic maturation, but cannot be readily applied as general assays to evaluate effects of toxicants on humans. In adult primates, the maturation of ovarian follicles depends on the frequency of LH pulses and is impaired by slight slowing of these pulses (Pohl et al., 1983).

Menarche

A major landmark of maturation is menarche, the beginning of the menstrual function. This usually occurs between the ages of 9 and 16 years, with a mean of 13 years (Tanner, 1981). That precocious menarche can occur as early as the age of 1 month indicates that gonadal cells that respond to hormones have differentiated to the stage of hormonal competence in neonates and can respond to endocrine signals at any age. Menarche and other characteristics of puberty can be delayed indefinitely, as in persons with Turner's syndrome and other types of gonadal dysgenesis.

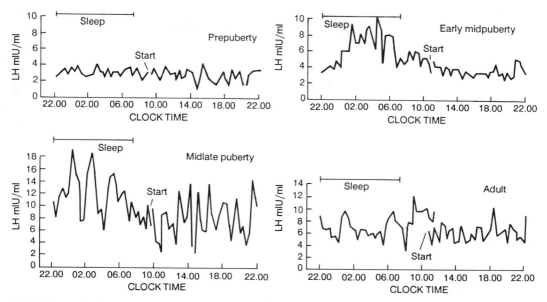

FIGURE 11-4 Daily plasma LH patterns in a representative prepubertal girl (9 years), early pubertal (15 years), late pubertal (16 years), and young adult (23 years) males. Sleep stage pattern is depicted for each nocturnal sleep period. Source: Reprinted with permission from Weitzman et al., 1975.

The first menstrual cycle signals maturation of the hypothalamus and the reproductive tract. Initial cycles often are irregular and can be anovulatory (Treloar et al., 1970; Tanner, 1981) (Fig. 11-5). The length of the anovulatory phase varies and can be several years (Ashley-Montague, 1957; Ojeda et al., 1980).

Little is known about how particular toxicants influence menarche and the onset of fertility. But it is known that puberty can be delayed by exogenous factors such as stress, nutritional deficiencies, and emotional distress. (Girls with emotional distress also usually have deficiencies of gonadotropins that imply hypothalamic dysfunctions.) Environmental stress delays puberty in laboratory rodents. The onset of puberty is associated with achievement of a critical body size and fat content (Frisch, 1980; Tanner, 1981). The restricted diets often chosen by athletes and dancers can delay menarche until the age of 20 years (Frisch, 1985); similar effects occur during anorexia nervosa. Loss of more than one-third of body fat at any age causes reversible amenorrhea (Frisch, 1985).

Although the neuroendocrine basis of the effects of stress and diet on puberty is unclear, some toxicants that interact with the stress-mediating hypothalamic-pituitary and sympathoadrenal systems might influence puberty. Such indirect effects could also result from toxicants that influence appetite or nutrient absorption.

An example of a toxicant that affects puberty in rodents is the insecticide DDT. Exposure of neonatal rats to DDT causes major changes in neuroendocrine functions, including early puberty and a syndrome of delayed-onset, persistent estrus in association with a polyfollicular ovarian status (Heinrichs et al., 1971). This permanent reproductive impairment in female rodents resembles the neonatal masculinization of the hypothalamus and the polyfollicular, anovulatory ovarian syndrome caused by exogenous steroids (Gorski, 1971; Mobbs et al., 1984). Whether such effects of DDT are limited to a critical period during development is unknown. Exposure of rodents just before or after birth to estrogens and other steroids has profound effects

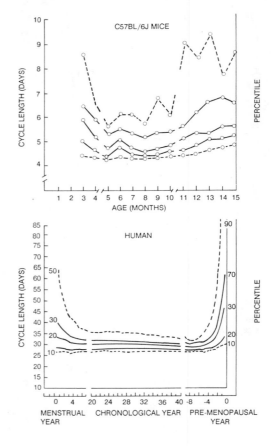

FIGURE 11-5 Comparisons of age-related changes in ovulatory cycle length distribution from longitudinal studies of C57BL/6J mice and humans. The ages are scaled to midlife. Sources: (Mice) Reprinted with permission from Nelson et al., 1982. Copyright 1982, Society for the Study of Reproduction. (Human) Reprinted with permission from Treloar et al., 1970. Copyright 1970, Allen Press, Inc.

on adult reproductive functions that can be manifested at puberty or can induce precocious cessation of fertile cycles (delayed anovulatory syndrome). The mechanism by which DDT causes persistent estrus may involve an estrogenic action, since DDT has a uterotrophic effect (Bitman et al., 1968) and also can bind to cytosolic E_2 receptors (Robison et al., 1985). This example shows how environmental toxicants can interact with neuroendocrine maturation. No analogous phenomenon in humans has been found. Women exposed in utero to DES have normal age at menarche and appear to be fertile; menstrual cycles are reported to be regular in some studies (Barnes, 1979) and irregular in others (Herbst et al., 1980).

Adrenarche

Adrenarche is the onset of menstruation and other physiologic changes of puberty induced by hyperactivity of the adrenal cortex. The concentration of the weak adrenal androgen dehydroepiandrosterone (DHEA) increases after age 6, several years before the increased secretion of estrogens, gonadotropins, and prolactin in pubertal girls (Hopper and Yen, 1975; Ojeda et al., 1980; Cutler and Loriaux, 1980; Reiter and Grumbach, 1982). DHEA appears largely as a sulfate, DHEA-S, in human blood. Another androgen, Δ-4-androstenedione, increases at slightly later ages. Accelerated growth and appearance of pubic and axillary hair are associated with increased blood concentrations of adrenal steroids. The hormonal mechanisms underlying adrenarche are not known, but seem to be independent of mechanisms that influence gonadal development (Cutler and Loriaux, 1980; Reiter and Grumbach, 1982); children with gonadal dysgenesis can have normal adrenarche (Boyar et al., 1973).

Adrenal insufficiency in humans may delay puberty (Boyar et al., 1973), but this association is not found in all cases (Reiter and Grumbach, 1982). Adrenalectomy of rats delays puberty, whereas corticoid replacement restores the normal onset (Ramaley, 1978). There is general agreement that adrenal steroids are not obligatory for maturation of hypothalamic controls over the gonad in either sex.

Thelarche

Thelarche (the beginning of breast development) is an estrogen-dependent pubertal event and can be evaluated by well-established criteria (Marshall and Tanner, 1969). Its absence by the age of 13 years is diagnostic of delayed puberty (Ojeda et al., 1980). A number of hormones are involved, including estrogens, progesterone, corticosteroids, prolactin, and growth hormone. However, prolactin and E_2 are considered pivotal for thelarche (Kleinberg, 1980).

If children of either sex are accidentally exposed to estrogens, early stages of thelarche occur. Premature and abnormal breast growth in children who were exposed to an estrogen-containing ointment has been documented (Halperin and Sizonenko, 1983). An outbreak of precocious thelarche in Puerto Rican children is still unexplained (Haddock et al., 1985; New, 1985).

CYCLIC OVARIAN FUNCTION

The ultimate purpose of cyclic ovarian function is to provide viable oocytes for fertilization by sperm ascending through the female genital tract. Ovarian hormones control the oviductal environment for fertilization and gamete transport and the endometrial environment for implantation and embryonic development. Synthesis of the steroids responsible for this sequence of events must be timely and requires maturation of the follicle, release of the oocyte, a normally functioning corpus luteum, and ovarian responsiveness to the presence of a conceptus.

Unlike that in men, reproductive function in women is characterized by cyclic fluctuations in pituitary gonadotropins and sex steroids (Fig. 11-6). These hormones have a dynamic relationship with positive and negative steroid feedback on release of gonadotropins from the pituitary. Other nonsteroidal regulatory factors of ovarian origin are also involved in this dynamic system.

Cyclic ovulatory function starts with sexual maturity and continues until the fifth decade of life. This cyclicity depends on continual maturation of ovarian follicles, which is stimulated by a normally functioning hypothalamic-pituitary axis. The process begins with the recruitment of a cohort of competing follicles that are able to respond to gonadotropins. Once a dominant follicle is selected, the remaining follicles in the cohort are suppressed.

Approximately 65 days before ovulation, the primary oocyte begins maturing within a primordial follicle (Gougeon, 1982). The early stages of this process appear to be independent of gonadotropin stimulation. The primary oocyte is surrounded by a single layer of cells—presumably forerunners of granulosa cells—and little is known of the physiology of early follicular development.

Only the final 14 days of follicular maturation appear to be influenced cyclically by gonadotropins in primates. Women with surgically absent or irradiated pituitaries can be made to ovulate with a relatively short course of human menopausal gonadotropin (hMG), with hCG as a surrogate midcycle LH surge (Schwartz and Jewelewicz, 1981). A typical course of hMG requires 10-14 days to develop preovulatory follicles, according to sonographic follicular size and estrogen production.

The follicular phase in primates lasts approximately 14 days. Ablation experiments in nonhuman primates have suggested that, from the midpoint of the follicular phase through the late luteal phase, a dominant cyclic structure—a preovulatory

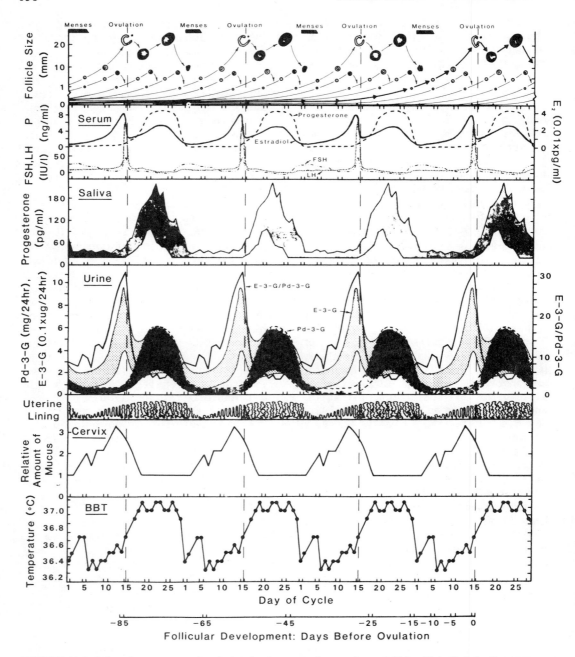

FIGURE 11-6 Selected events occurring during development of an ovulatory follicle. Note that development occupies three full cycles. Parameters include size of developing follicles measured by ultrasound or during laparoscopy; gonadotropin concentrations in serum; estrogen and progestogen concentrations in serum, saliva, and urine; ratio of urinary steroids; state of endometrial proliferation; relative cervical mucus volume; and basal body temperature. An average 28-29 day cycle was used as a model to which were fitted several results. Confidence intervals for salivary progesterone include ± standard deviation of the mean; for urinary steroids, the 80% confidence interval is shown. Source: Reprinted with permission from Campbell, 1985.

follicle or corpus luteum—is present. Removal of this structure at any time results in recruitment of another dominant follicle and ovulation in approximately 14 days (Goodman et al., 1977). The best explanation of this is that follicles are always maturing to a critical point at which either a dominant follicle is selected or follicular atresia results. FSH appears to be the gonadotropin responsible, in that LH does not rise early in the follicular phase coincidental with the process of follicle selection, as FSH does, and FSH can induce follicular development by itself.

Asymmetrical steroid secretion of the ovary indicates the presence of a dominant cyclic structure by day 7 of the menstrual cycle, as determined by estrogen and progesterone concentrations in the venous drainage of the ovaries (diZerega et al., 1980).

How primates release only a single oocyte per cycle is unknown, and the mechanism by which other developing follicles are suppressed remains controversial. Available data point to a local effect, inasmuch as removal of the corpus luteum results in ovulation from the contralateral ovary. When the cyclic structure (either a follicle or corpus luteum) is excised, FSH concentrations are 3 times higher than in a cycle in which the structure is not removed (Goodman and Hodgen, 1977). Despite this relatively high FSH content, a single ovulation still occurs 14 days after ablation, with steroid production at appropriate magnitudes for a single follicle.

Steroid production by preovulatory follicles parallels their health and states of differentiation, and estrogen production is the hallmark of follicular health (McNatty et al., 1976). The cellular source of estrogen secretion is controversial and involves a complex biosynthetic process that requires interaction between follicular compartments. Some studies indicated that the theca (on the follicular sheath) is the major source of estrogen biosynthesis (Channing and Coudert, 1976; Younglai and Short, 1970), whereas others suggest that thecal and granulosa compartments are needed (Falck,

1959). This "two-cell" idea is strengthened by the observation that granulosa cells are unable to synthesize androgens in appreciable amounts (Short, 1962; Bjersing and Carstensen, 1967), although the granulosa cell compartment processes greater aromatase activity than the thecal compartment (Schomberg, 1979). Thecal cells have not been demonstrated to have FSH receptors (Richards et al., 1976), whereas granulosa cells have FSH-inducible aromatase activity (Dorrington et al., 1975); therefore, FSH stimulation of aromatase activity might be a key regulatory step in follicular estrogen production.

Estrogen appears critical to follicular health. It is mitogenic to granulosa cells and inhibits premature biochemical luteinization, and high concentrations are found in the fluid of healthy follicles (McNatty and Baird, 1978). Development of FSH-induced aromatase activity could be critical in preventing atresia. That is probably the mechanism by which therapeutic agents like clomiphene citrate (which raises FSH content) initiate follicular maturation that leads to ovulation. Attainment of granulosa cell aromatase activity appears to be critical in controlling the number of follicles maturing in each cycle.

As preovulatory estrogen content rises, gonadotropin release is inhibited and FSH is suppressed differentially. Paradoxically, a sustained estrogen content for approximately 24-36 hours has a positive feedback effect and results in a surge of both gonadotropins. That causes a series of events within the follicle, including intrafollicular prostaglandin synthesis, terminal oocyte maturation, a shift in steroidogenesis from estrogen to progesterone by the granulosa cells, morphologic luteinization, and, ultimately, rupture of the follicle and release of the oocyte with its investment of cumulus cells into the peritoneal cavity (LeMaire et al., 1973).

Even before ovulation, preparation for luteinization by granulosa cells is well under way. The high estrogen content in follicular fluid is responsible for the dramatic increase in the number of granulosa cells, as well as for the appearance of LH receptors. A decrease in serum E_2 after the gonadotropin surge is consistent with blocked aromatase activity. Serum and follicular fluid progesterone increase concomitantly and mark the initiation of the luteinization process (McNatty et al., 1976). Neovascularization of the granulosa compartment begins as vessels penetrate the basement membrane and invest the luteinized granulosa cells. Recent observations in which circulating steroids around the midcycle were carefully measured showed a rise in progesterone a few hours before follicular rupture (Horning et al., 1981).

Without sustaining factors secreted by a conceptus, a corpus luteum undergoes a programmed demise, with a life span of approximately 14 days. As noted earlier, peripheral progesterone begins to increase with the initiation of the LH surge and continues to increase until the midpoint of the luteal phase, when it begins a gradual decline that results in menses.

In humans, hCG maintains the corpus luteum during early pregnancy. Circulating hCG can be detected a few days after luteal progesterone peaks and prevents the programmed involution of the corpus luteum (Saxena et al., 1974; Catt et al., 1975).

Excision of the corpus luteum or the ovaries at 7-8 weeks of gestation does not interrupt pregnancy; that suggests that the fetoplacental unit takes over the critical hormone production function necessary to maintain pregnancy (Csapo et al., 1972). hCG content is high at this point in pregnancy and gradually declines after 13 weeks of gestation. No further role of the ovary is apparent in the development of human pregnancy.

Screening for ovarian toxicity in vivo is difficult, because of the interrelationships with the hypothalamic-pituitary axis and the fact that clinical problems arise only when there are major functional changes. The observation of disruption of the reproductive cycle does not permit discrimination of ovarian or central nervous system action. More detailed hormonal studies over some period are required.

Effects of exposure to toxicants on cyclic ovarian function have not been investigated widely, but several observations

suggest that occurrence of menses is sensitive to environmental influences. Nutritional factors, stress, and vigorous exercise are well-established risk factors in anovulation (Warren, 1982, 1983; Green et al., 1986). Cycle length was examined in women exposed to inorganic mercury vapor, with some indication that higher levels of exposure increased the risk of oligomenorrhea (DeRosis et al., 1985). Hormones and central nervous system toxicants, such as metals, can disturb menstrual function. Amenorrhea and altered menstrual hormone levels are parts of an ovarian failure pattern observed in women treated with antineoplastic agents, such as cyclophosphamide (Chapman, 1983). Polymenorrhea has been found in women working with synthetic hormones (Harrington et al., 1978), and menstrual dysfunction has been suggested by some studies of prenatal exposure to DES (Bibbo et al., 1977; Barnes, 1979; Peress et al., 1982). Solvents have been associated with menstrual dysfunction (WHO, 1986). However, a recent study of women working with styrene found no increase in menstrual disorders (Lemasters et al., 1985).

The mechanisms of female reproductive toxicity vary. For example, a toxicant might mimic the action of naturally occurring reproductive hormones. Many chemically unrelated compounds—including stilbene derivatives (e.g., DES), industrial chemicals (e.g., PCBs), and pesticides (e.g., DDT)—exhibit estrogenic activity in bioassays, and thus have the potential to alter the normal estrogen feedback relationship between the gonad and the brain and so disrupt ovulation in a manner analogous to that of oral contraceptives. Environmental agents might alter hormone synthesis, storage, release, transport, or metabolism. For example, compounds with estrogenic activity can cause luteolysis and inhibit the production of progesterone; as a consequence of these and other considerations (such as alterations in fallopian tube and endometrium functions), DES has been used as a morning-after pill to victims of sexual assault. Environmental agents also might cause gamete cytotoxicity. In a generally monotocous species, such as the human, substantial loss of oocytes can be tolerated without disrupting the menstrual pattern. The injury is apparent only long after exposure, when premature ovarian senescence occurs, which makes identification of the responsible agent particularly difficult.

FERTILIZATION

Fertilization in the course of a reproductive cycle involves several processes, including:

- Sperm-egg attachment.
- The fertilized oocyte's blocking of penetration by additional sperm.
- Completion of meiosis, with extrusion of the second polar body.
- Development of the two pronuclei.

Fertilization failure or delay might occur if exposure to toxicants interferes with timely ovulation or gamete transport. Toxicants also have the potential to alter oocyte biochemistry, affecting oocyte activation and pronuclear formation. During the peri-implantation period (weeks 1-3 in humans), the fertilized egg moves down the oviduct to the uterus, where it implants approximately 8 days after conception, and then blastulation and gastrulation occur. Insults that occur during this period can result in repair through compensatory hyperplasia, in embryo lethality, or in diverse malformations that suggest genetic damage to the early zygote. Generoso et al. (1987) reported that early exposure of mouse zygotes to mammalian germ-cell mutagens induces a variety of congenital defects, as well as death. In this part of the report, one marker of pregnancy and early loss that has been used in epidemiologic studies is discussed. Biologic markers related to pregnancy are discussed in detail in Part III of this report.

REPRODUCTIVE SENESCENCE

Menopause is the cessation of menstruation. The last menstrual cycle usually occurs at the age of 42-58 years, with a median of about 50 (MacMahon and Worcester, 1966; Gosden, 1985). Menopause is preceded by gradual cycle lengthening and irregularity over 5 years or more (Fig. 11-4). Perimenopausal cycles may be prolonged and anovulatory and are associated with decreased E_2 and progesterone and increased LH and FSH (Sherman et al., 1976; Metcalf et al., 1981). Foreshortened, possibly anovulatory, cycles also occur in association with abbreviated follicular phases. Menopause is assumed to have occurred if amenorhea has persisted for at least 6 months.

Decreasing fertility (ability to produce offspring) is a hallmark of approaching menopause and follows remarkably similar trends in modern populations of industrial countries, as well as in groups, such as the Hutterites, that try to maximize fertility and live offspring (Fig. 11-7). Marked decreases in fertility can occur 5 years before cycles become obviously irregular. Therefore, menstrual cycle regularity is not a reliable assessment of fertility. Observed age-related decrease of fertility in populations results from an increased number of women who have become sterile for many different reasons, as well as a reduced rate of conception among those who are still fertile (Federation CECOS, 1982; Menken et al., 1986).

Down syndrome and some other chromosomal abnormalities in offspring increase sharply after age 30 (Ferguson-Smith, 1983). About 50% of all first-trimester spontaneous abortions in humans are aneuploid (Porter, 1986), so the doubling in spontaneous abortion from the age of 20 to the age of 40 could reflect an increase in chromosomal abnormalities. These abnormalities often are lethal, and the evidence does not support weakened selection against abnormal fetuses with increasing maternal age (Gosden, 1985). A promising animal model, the rat, shows that delayed ovulation can increase embryo abnormalities (Page et al., 1983), thereby suggesting a role of age-related cycle lengthening to increased birth defects. However, no relation has been shown between cycle regularity, early menopause, and Down syndrome (Sigler et al., 1967). The inadequate understanding of this major phenomenon

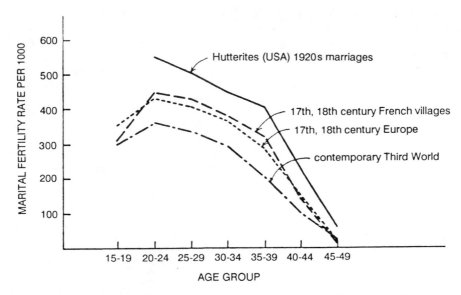

FIGURE 11-7 Age-specific fecundity in selected populations that do not practice contraception: rates per 1,000 married women. Source: Based on data in Leridon, 1977.

also hampers studying the interactions of toxicants, age, and birth defects.

The major cause of menopause is depletion of the ovarian stock of oocytes, which declines from birth onwards in all mammals (Fig. 11-8). Individual variations in the onset of menopause are thought to result from different rates of ovarian oocyte loss or different sizes of the initial stock (Nelson and Felicio, 1987; Richardson et al., 1987). Even within the same inbred strain, mice show extensive individual differences in initial oocyte stocks and in the numbers of oocytes remaining at the approach to acyclicity (Gosden et al., 1983). Rodent models also show age-related ovarian depletion and lengthening of fertility cycles (Fig. 11-5).

As ovarian oocytes and growing follicles become depleted, circulating concentrations of estrogens and progesterone decrease to those found in ovariectomized women. Among the major changes associated with decreased steroids are onset of hot flushes, increases in blood LH and FSH, and atrophy of most organs that respond to sex-steroids. These changes vary widely, but replacement of ovarian steroids usually prevents or attenuates them. Ovar-

ian steroids have some effect on postmenopausal osteoporosis; oophorectomy in young women can precipitate premature bone loss, and steroid therapy appears to reduce the risk of age-related fractures. However, inasmuch as bone loss begins about 10 years before measurable deficits in blood estrogens, other factors that do not depend directly on changes in blood estrogen concentrations must also be involved. DHEA-S decreases progressively after the age of 30 years and is not related to menopause (Orentreich et al., 1984).

Menopause might be influenced by toxicants and other environmental factors. For example, smokers of 14 or more cigarettes per day have menopause as much as 2 years earlier than nonsmokers and smokers of fewer than 14 cigarettes per day (Jick and Porter, 1977; Van Keep et al., 1979; Lindquist and Bengtsson, 1979; Kaufman et al., 1980). Benzo[a]pyrene, a carcinogen that occurs in tobacco smoke and urban air, can kill oocytes in mice (Mattison and Thorgeirsson, 1977) and may be an active causal agent in premature ovarian oocyte loss and early menopause. In addition, cigarette smoke has been shown to be antiestrogenic (Michnovicz et al., 1986).

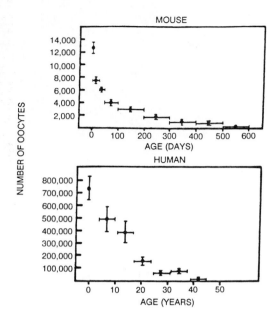

FIGURE 11-8 Loss of ovarian oocytes during aging in the mouse and human. Sources: (Mouse) Redrawn from Jones and Krohn, 1961; (human) redrawn from Block, 1952.

Other environmental influences on menopause are less certain. Several reports have indicated a trend of increasingly later ages at menopause during the past century (Frommer, 1964), but other reports have not confirmed this trend (Van Keep et al., 1979; Gosden, 1985). Economically underprivileged populations may also have slightly earlier menopause (MacMahon and Worcester, 1966; Soberon et al., 1960), but no general effect is established (Gosden, 1985).

Toxic environmental estrogens are important, because they resemble some aspects of reproductive senescence. Sheep fed on phytoestrogen-containing clovers develop a permanent infertility syndrome with hypothalamic histopathology (Adams, 1976, 1977). These changes are intriguingly similar to the changes in aging acyclic rodents (Schipper et al., 1981; Finch et al., 1984) that can also be induced by chronic exposure to estrogens (Brawer et al., 1983; Mobbs et al., 1985). In addition to neural damage, sheep with clover disease have impaired luteal function and sperm transport (Adams, 1983; Adams et al., 1981; Adams and Martin, 1983; Lightfoot et al., 1967). Phytoestrogens interact with environmental trace metals, cobalt intensifies the effects of phytoestrogen, and selenium blunts the effects of cobalt (Gardiner and Narin, 1969).

12

Biologic Markers of Genetic Damage in Females

Female reproduction can be adversely affected through a variety of mechanisms, including damage to germ cells from prenatal or postnatal exposure. Resulting damage encompasses oocyte destruction, as well as genetic alterations that may be transmitted to offspring. This chapter focuses on markers of genotoxic damage to somatic cells and to germ cells and briefly discusses ovotoxicity. Biologic markers are discussed under the general headings of exposure to toxicants, oocyte toxicity, markers of genotoxic damage or repair, and markers of mutational events.

Direct measurements of genetic and other germ cell damage in humans are rare, because female germ cells are few and access for investigation is difficult. Work with oocytes in other mammals also is scarce, for the same reasons. Laboratory and human studies have relied heavily on reproductive end points, such as infertility and fetal loss, as indicators of germ cell damage. However, changes in reproductive function may occur for reasons unrelated to toxic exposures. Therefore, findings from these studies may be difficult to interpret.

Among the materials available for assessing germ cell effects on female reproduction are gametes, other ovarian cells and tissues, follicular fluid, cervical and vaginal secretions, and tissue from the conceptus (amnion, chorion, placenta, embryo, and fetus). (Markers of genetic damage in the conceptus are discussed in Part III—Biologic Markers of Pregnancy.) Some of these materials usually can be obtained only through invasive procedures and are not widely available to researchers. Some expendable tissues, such as placentas and abortuses, can be used for a few research purposes, but until recently, the only access to ovarian materials has been through surgery (e.g., oophorectomy and hysterectomy) and organ-donor programs. In vitro fertilization and embryo transfer (IVF/ET) centers present an important research opportunity. In some countries, investigators involved with in vitro studies now have available gametes, embryos, and other materials from IVF programs. Any study based on difficult-to-obtain materials should include a comparison with other, more widely studied, tissues and fluids, so that the findings of the study can be interpreted properly.

Further systematic studies of exposed women and their progeny that use markers of genotoxicity (e.g., DNA adducts) or of mutational events (e.g., micronuclei) should improve measurement precision in evaluating associations between exposure and outcome. Clinical relevance of markers should be validated. Studies in New York State (Hatcher and Hook, 1981a,b) used routine samples of biologic material (e.g., cord blood and blood from heel sticks) and routine birth records to examine the relations among sister-chromatid exchanges (SCEs), chromosomal aberrations, and birthweight, and efforts of this type should be extended.

MARKERS OF EXPOSURE

Biologic assessments of human genotoxic exposure typically are accomplished through chemical analyses or bacterial mutagenesis assays of body fluid—usually blood and urine, because of their availability and expendability. These tissues can be obtained from either men or women, so the assessments can be conducted in both. From men, semen also can be assessed. Recently, follicular fluid has been used to assess exposure in females (see below).

Chemical analyses of body fluids with gas chromatography and mass spectrophotometry (GC/MS) or immunoassays give a direct measure of exposure to specific compounds or metabolites and the resulting internal dose. Mutagenesis assays, however, are indirect markers of exposure to mutagens and are useful when exposure is unknown or complex or when analytic standards are lacking. Moreover, such assays demonstrate biologic activity, rather than simply toxicant presence.

A widely used mutagenesis assay with potential applications in female toxicity is the Salmonella/microsome mutagenicity test (Ames et al., 1975), which has been well validated and is quite sensitive to DNA damage. It uses tester strains that cannot synthesize the amino acid, histidine, but revert and begin to grow in the presence of mutagens. Mutagenesis assays are generally less sensitive than such analytic methods as GC/MS or immunoassays, particularly for some types of compounds, including hormones. When the mutagens are present in a low concentration, large sample volumes may be needed to yield enough mutagen to be detected. Storage and extraction procedures can distort results, and genotoxic activity may be compounded by body concentrations of extraneous agents (e.g., foods). Nonethe-

less, cost and performance time compare favorably with those of chemical analyses. New assays, such as the *E. coli* multitest system (Toman et al., 1985), require smaller biologic samples and are used to evaluate several genetic and nongenetic end points but have not been applied to human samples.

Follicular fluid is important in oocyte maintenance and ovulation. Because the blood-follicle barrier is permeable, toxicants in this milieu could affect oocyte integrity, meiosis, fertilization, and implantation. Follicular aspirates from laparoscopies in an IVF program have been used to measure several common chlorinated hydrocarbons, including DDT, PCBs, and hexachlorobenzene (Trapp et al., 1984; Baukloh et al., 1985). In a majority of the 47 women sampled, most of the pollutants sought were found. In these studies, the investigators noted that the oocyte recovery rates and subsequent embryo cleavage rates were inversely related to the hydrocarbon concentrations.

MARKERS OF OOCYTE TOXICITY

Species apparently differ substantially in oocyte sensitivity to toxicants, molecular target within oocytes, and susceptible stages of development. During human fetal life, oogenesis is a sensitive period for exposure to ovotoxicants, because germ cells damaged during this period can not be replaced and the cells are metabolically active. A later period of vulnerability occurs during the preovulatory stage of the menstrual cycle, when oocyte maturation resumes and the cells are again metabolically active (Mattison, 1982).

Impaired fertility as a result of in utero oocyte exposure to toxic substances has not been proved. In mature females, chemotherapy is associated with ovarian failure and appears to involve damage to growing oocytes (Chapman, 1983). Earlier onset of menopause in current smokers than in nonsmokers (McKinlay et al., 1985) has been viewed as evidence that exposures can increase oocyte atresia rates.

Other markers of ovodepletion need to

be developed. Two possibilities are alterations in gonadotropic hormones (in that oocyte atresia is controlled by the pituitary) and ultrasound evaluation of ovarian size to detect gross changes. Some of the new DNA technology may potentially provide useful tools for detecting subtle changes in ovarian function; for example, ovarian DNA probes enable the measurement of gonadal peptides (Mason et al., 1985).

MARKERS OF GENOTOXIC DAMAGE OR REPAIR

Much of the development and application of genotoxic markers has been in the field of carcinogenesis. Actual or potential markers in humans include DNA adducts, unscheduled DNA synthesis, and SCE. These are not direct markers of female reproductive toxicity because, to date, they have been used only in somatic cells. But they hold promise as indirect markers.

DNA Adducts

DNA extracted from human tissue can be analyzed to detect adducts formed by covalent binding of a genotoxicant with a DNA base. The significance of an adduct appears to differ according to its size, structure, and site of binding. For example, bulky adducts are more likely to interfere with DNA replication than are smaller adducts (Brusick et al., 1981). Studies correlating various adducts with specific-locus mutations have indicated that the O^6-guanine adduct is critical in mutagenesis (van Zeeland, 1986).

Agent-specific and generic methods for measuring adducts are available (Perera et al., 1986; Wogan, 1988). Agent-specific methods permit the extent of binding of a known agent or metabolite to be determined. Generic methods are especially useful when the exposure or the operative agent in the exposure is unknown (see Chapters 9 and 18).

Methods for detecting chemical-specific adducts rely on rapid, reproducible immunoassays that use antibodies against modified DNA, nucleotides, or nucleosides. Binding of the antibody of interest is measured after binding of a radiolabeled or enzyme-linked second antibody. These assays require approximately 300 μg of DNA (or about 300 mil of blood) and detect adducts at a concentration of about 1 per 10^8 bases (Perera et al., 1986). The postlabeling technique (Gupta et al., 1982) is a generic method to detect binding by aromatic compounds that requires a DNA sample of approximately 10 μg (1 ml blood) and that has a detection level of 1 per 10^8-10^{10} bases. The assay procedure entails digesting DNA to nucleotide derivatives, labeling the nucleotides with phosphorus-32, removing normal nucleotides by thin layer chromatography, and performing autoradiography of the ^{32}P-labeled nucleotides. Adducts can be characterized by comparing chromatographic patterns with those of adducts formed by known compounds.

Assays that measure hemoglobin alkylation have been proposed as a surrogate for immunoassays that use white-blood-cell DNA (Ehrenberg and Osterman-Golkar, 1980); this promising approach requires very little blood.

DNA adducts might be measured in granulosa cells found in follicular fluid or in oocytes; no reports of this were found in the literature on humans or experimental animals. In the mouse, fetal adducts have been measured to assess transplacental DNA damage from known mutagens administered to the mother (Lu et al., 1986). Adducts were found in all fetal organs—generally at concentrations lower than in maternal tissues—but the tissue distribution of adducts differed between mother and fetus; therefore, fetal concentrations could not be predicted from maternal concentrations. In human placental tissue (Everson et al., 1986), maternal smoking was found to be strongly related to an adduct detected by the postlabeling assay (16 of 17 smokers compared with 3 of 14 nonsmokers). The adduct has not been characterized, but appears to correlate with the birthweight of the offspring (Everson et al., 1988).

DNA adducts hold promise as dosimeters for in utero exposure to genotoxicants that might be detrimental to developing fetal oocytes, as well as to somatic cells. Adducts as direct measure of germ-line exposure are untested in humans. In the laboratory, only males have been studied for adduct formation in germinal cells, primarily because of the inaccessibility of female germ cells. Background concentrations of DNA adducts and age and sex influences on adduct formation are suggested by recent data (Randerath et al., 1986; Reddy and Randerath, 1987). More information is needed to guide design and interpretation of studies that use DNA adducts.

Unscheduled DNA Synthesis

Measuring excision repair of adducts as unscheduled DNA synthesis (UDS) has been proposed as an indicator of exposure to DNA-damaging agents (Williams, 1977). UDS assays require the addition of tritium-labeled thymidine to nondividing cultured cells. UDS is measured as the extent of label incorporated during repair of single-strand gaps that result from excision of damaged nucleotides. UDS has been validated with a wide range of direct-acting chemicals, as well as agents that require activation (Santella, 1987).

Oocytes appear to have effective mechanisms for repairing damaged DNA. UDS is a sign of gene repair and has been observed in oocytes of mice exposed to UV radiation (Pedersen and Mangia, 1978). UDS could prove useful as a marker of reproductive damage in humans, although no such applications have been reported.

Sister-Chromatid Exchange

SCE is the reciprocal interchange of DNA between chromatids at one locus and does not result in alteration of chromosomal structure. SCEs reflect repair of several types of lesions; they are more efficiently induced by compounds that form DNA adducts or otherwise intercalate into DNA (e.g., alkylating agents) than by agents that break the DNA backbone (e.g., radiation).

SCEs are not mutational events, nor have they any known health consequences (Carrano et al., 1980). They are a manifestation of repair of damaged DNA and correlate with specific-locus mutations in Chinese hamster ovary cells (Carrano et

al., 1978), but they do not correlate well with adduct formation—not even with the O^6-guanine adduct thought to be important in mutagenesis (Tice et al., 1984). SCEs are increased in patients with Bloom's syndrome (Evans, 1982), and they generally are accepted as indicators of potential genetic hazards (Archer et al., 1981; Latt et al., 1981).

SCEs are quicker and easier to score than chromosomal aberrations and will often be detected even when other short-term assays are negative (Carrano et al., 1980). The measurable background concentration of SCEs is variable, but standardized protocols give reproducible data. Cells are cultured from about 10 ml of blood. Measurement of SCEs requires two cycles of DNA replication in the presence of bromidine deoxyuridyl, followed by microscopic analysis of stained metaphase cells.

SCE analysis of fetal cells has been used to detect transplacental mutagens in the mouse (Kram et al., 1979); significant increases in fetal SCEs are seen at doses of known mutagens well below their teratogenic doses, although fetal SCE frequencies generally are lower than maternal frequencies (Kram et al., 1980). Induction of SCEs in fetal cells decreases during gestation for direct- and indirect-acting compounds. That may reflect reduced placental transport, stage-dependent influences on SCE formation rates, or differing rates of cell turnover.

That SCEs can be induced by such reproductive toxins as antineoplastic drugs (Norppa et al., 1980) and ethylene oxide (Yager et al., 1983) indicates their utility as exposure markers. Conflicting findings suggest that SCEs are uncertain indicators of in utero exposure or markers of fetal damage.

MARKERS OF MUTATIONAL EVENTS

In considering mutational events, chromosomal lesions and point mutations are of interest. Because of the inaccessibility of female germ cells, work to date has relied heavily on observations in other species, mainly the mouse.

Laboratory Animals

The work in rodent systems is described in detail in Chapter 8 (see also Russell and Shelby, 1985). Tests for genetic damage are conducted less often in female than in male animals, because the yield of mutations is lower. The dominant-lethal, heritable-translocation, and specific-locus tests are the most extensively used laboratory methods; in all of these, risk is evaluated in F_1 progeny. The dominant-lethal test measures early embryonic death induced by a variety of chromosomal abnormalities, primarily numerical, rather than structural (Brusick, 1980). The heritable-translocation assay measures transmissible structural chromosomal lesions, and the specific-locus test can measure recessive point mutations. Dobson and Felton (1983) have underscored the importance of accounting for species differences in target and cell killing when extrapolating genetic risk in laboratory animals to human females.

A protocol for cytogenetic analysis of meiosis II oocytes and first-cleavage zygotes recently was proposed as a sensitive assay for aneuploidy in meiosis I and II cells, as well as for structural aberrations after DNA synthesis (Mailhes et al., 1986). A measure of aneuploidy in female germ cells is of great interest because more than 80% of aneuploidy in humans originates there (Hassold, 1985).

Human Oocytes

Although analysis of human oocyte chromosomes has been attempted with material from ovariectomies or biopsies (Jagiello and Lin, 1974; Jagiello et al., 1975), early cytogenetic studies of meiotic cells were hampered by technical difficulties in chromosome preparation. More recently, new chromosome preparation techniques have been applied to develop estimates of chromosomal aberrations with oocytes from stimulated cycles of infertile women (Wramsby et al., 1987). The resulting data place the background frequency of oocyte chromosomal anomalies at 50%, but the estimate is limited by the select nature and small size of the population studied and by the unknown effect of ovulation induction. However, those limitations do not preclude using IVF materials to examine cytogenetic characteristics of the egg in relation to successful fertilization and implantation. In addition, it should be possible to compare results in infertile women with those obtained with oocytes from unstimulated cycles of fertile women.

Successful efforts also have been made to analyze chromosomal anomalies in early human embryos obtained from IVF (Angell et al., 1983).

Human Somatic Cells

Chromosomal Aberrations

Alterations in chromosomal structure are unequivocal markers of genetic damage. The basic lesion thought to underlie all structural anomalies is a break in the chromatin fiber. Aberrations in somatic cells are considered undesirable, but not directly predictive of health effects. Chromosomal aberrations are generally less sensitive measures of chemical exposure, which tend to induce aberrations only when the cell is in the S-phase (i.e., chromatid aberration detectable as SCEs) (Morgan and Wolff, 1984). In addition, scoring aberrations can be time-consuming.

Significant increases in spontaneous abortions or heritable chromosomal abnormalities have not been found among Japanese who survived the atomic bomb blasts, despite somatic cell chromosomal damage (Schull et al., 1981a); however, the observed trends of increasing frequency with increasing dose received were predicted (see also discussion in the Chapter 9). An early study of chromosomal aberrations in newborns suggested an association with low birthweight (Bochkov, 1974), but the relationship was not confirmed in a later study that used matched, contemporaneous controls (Hatcher and Hook, 1981a).

Micronuclei

Micronuclei are microscopically visible DNA-containing bodies in the cytoplasm of a cell with no structural connection to the nucleus. The presence of these extranuclear bodies is considered to indicate chromosomal damage (aneuploidy, as well as breakage). Metaphase analysis is not required, so the test is relatively easy and inexpensive, and it is amenable to automation. The background rate of micronuclei is very low (less than 1%), so many cells are required for analysis.

A major limitation of this test is that the human spleen eventually removes blood cells with micronuclei. Early applications of the micronucleus test used exfoliated surface cells, but more recent assessments used bone marrow cells, which can be obtained only through invasive procedures. Efforts are under way to apply the method to more accessible cell populations, such as peripheral lymphocytes. A technique is needed to pick up cells at the first mitoses; a proposed method uses an inhibitor to prevent cytoplasmic division after nuclear division (Fenech and Morley, 1986).

The micronucleus test has been used in rodents with fetal liver erythroblasts as a model system for transplacentally induced chromosomal damage. Because rodent fetal liver is metabolically more active than adult bone marrow, the transplacental test is generally more sensitive to genotoxic agents (Cole et al., 1983), as well as being better suited for assessing risks to the fetus from maternal exposures (King and Wild, 1979). No reports of analogous applications of micronuclei in humans were found.

Specific-Locus Mutations

Standardized tests to detect specific-locus mutations in somatic cells are only beginning to be available. Mutation rates may be lower in germ cells than in somatic cells, because of differences in repair rates. (In addition, female germ cells undergo fewer doublings than male germ cells.) Nonetheless, these tests are likely to prove useful, especially to elucidate differences in species.

The hypoxanthine phosphoribosyl transferase (HPRT) specific-locus test was one of the first to detect spontaneously occurring mutants in cultured human lymphocytes. In this test, mutants are detected as cells resistant to azaguanine or thioguanine in the culture medium. A high frequency of thioguanine-resistant cells has been reported in patients treated with known mutagens (Archer et al., 1981).

Recently, a test for glycophorin A loss was applied to red blood cells of Japanese atomic bomb survivors; mutations were linearly related to the estimated radiation dose (Langlois et al., 1987). Jensen and Thilly (1986) reported that characteristic mutation spectra are produced in human B lymphocytes by specific chemicals, and gradient denaturing gel electrophoresis might be usable for generating a mutation spectrum from as little as 10 ml of blood (Liber et al., 1985).

The tests mentioned here are relatively new and have not been broadly applied.

13

Biologic Markers
of Reproductive Development and Aging

Toxicants and other environmental factors can influence female reproductive function during development, as well as during later age-related changes. This chapter discusses biologic markers of female reproduction across the life span, including markers of neuroendocrine function that are potentially pertinent to reproduction. Some of these markers might be used to assess effects of toxicants.

Little information is available on critical periods or ages of particular vulnerability for effects of toxicants on postnatal development and aging. Some effects of toxicants could be equivalent to accelerated senescence.

Cryptic damage—damage not immediately manifested—might interact with other insults. All-or-none effects might not be associated immediately with a particular exposure. Daughters exposed to DES in utero manifested an increased incidence of genital tract cancer as adults (Herbst et al., 1974). Exposure of neonatal rats to DDT causes major impairments of female reproductive functions that emerge after puberty (Heinrichs et al., 1971). Chronic exposure of mice to endogenous ovarian steroids secreted during young adult life is a cause of age-related estrous cycle lengthening (Felicio et al., 1986) or premature loss of estrous cycles after the

estrogen treatment ends (Mobbs et al., 1984; Kohama et al., 1986).

These rodent phenomena have no human analogues. However, the likelihood of hot flushes at menopause when estrogen concentrations decrease appears to depend on exposure to ovarian steroids during puberty. Women with Turner's syndrome, whose estrogen concentrations are the same as postmenopausal concentrations, do not have hot flushes; however, withdrawal from estrogen treatment will cause hot flushes in these women (Yen, 1977). That important result indicates that the adult human nervous system has memory mechanisms for exposure to steroids such that the effects can be manifested decades after the exposure (Finch et al., 1984). In analyzing interactions of estrogenic toxicants with the nervous system, we should anticipate potential cryptic effects on the numerous neurons throughout the brain that contain receptors with high affinity for estrogens and other steroids.

Many examples demonstrate cryptic brain damage during early adulthood. The relation between viral encephalitis (Von Economo's disease) and parkinsonism is a classic case (Poskanzer and Schwab, 1961; Finch, 1976; Calne et al., 1986). Parkinsonism induced by ingesting the neurotoxicant methyl(phenyl)tetrahydropyridine (MPTP), which often contaminates synthetic heroin, might not occur immediately; some persons without neurologic symptoms have depressed metabolisms in their dopaminergic systems, as determined by positron emission tomography analysis (Calne et al., 1985).

Those examples constitute ample precedent for considering possible long-term adverse effects of toxicants on neuroendocrine loci that might be incurred occupationally (e.g., by metal workers).

Most biologic markers of toxicant-related effects on female reproductive functions are physiologic or morphologic. Among the best-characterized markers used in epidemiologic surveys and individual case studies are infertility and length of menstrual cycles, which appear to be sensitive to many of the same environmental influences in humans and rodent models. However, human neuroendocrine reproductive functions appear to differ in important ways from those of rodents; the hypothalamus might be less crucial in regulating the preovulatory surge in humans than of rodents (Knobil, 1980). Nonetheless, human female reproductive functions clearly are susceptible to neurogenic influences from stress (Peyser et al., 1973) and perhaps from pheromones (Russell et al., 1980).

Transgenerational toxic effects can arise in several ways. The oocyte complement of an adult female is attained in utero before the midpoint of gestation. Toxicants might reduce the number of oocytes and, if mutagenic, affect later generations. Furthermore, toxicants in the maternal environment that affect fetal brain development could influence the maternal physiology and behavior of female offspring, which then affects the next generation. For example, environmental influences that extend to the F_2 generation have been demonstrated in rats (Zamenhof et al., 1972). Malnutrition during pregnancy reduces the number of brain cells in rats for at least two generations, despite crossfostering of pups with normal surrogate nurses (Zamenhof et al., 1972). The concept of transgenerational environmental effects is well known to develop-mental biologists, but has not been discussed widely as an aspect of toxicology.

MARKERS OF MATURATION

Adrenarche

Increases in plasma DHEA-S (a metabolite of DHEA) precede increases in estrogens by several years in humans and indicate maturation of adrenal cortical function. The stimuli for increased secretions of DHEA and other adrenal steroids before puberty are poorly understood and do not seem to involve direct action of ACTH or gonadotropins. Although readily measured by radioimmunoassays, the increase in the plasma metabolite DHEA-S is linked only circumstantially to functional changes in maturation.

Ovary and Uterus

Size changes in the uterus and ovary can be followed by ultrasonography (Orsini et al., 1984). The data base is modest for ultrasound measurements of these morphologic changes, but such measurements can be obtained in conjunction with other common clinical markers of puberty.

Menarche

Onset of menstrual bleeding is the most obvious sign of active ovarian steroid secretion. The onset is triggered by increases in E_2 and progesterone, followed by decreases in progesterone within a few days. Ovulatory cycles usually are established months after menarche and vary considerably in timing and hormonal characteristics.

Thelarche and Pubic Hair

The ages of thelarche and the appearance of pubic hair are used widely to judge whether puberty is precocious or delayed, and an extensive data base is available. Five morphologic stages of breast development generally are accepted (Marshall and Tanner, 1969). The five stages can vary extensively in duration and can revert to earlier stages. Four pubic hair stages also show extensive individual variations. Breast and pubic hair stages often are asynchronous and by themselves do not precisely indicate the rate of development.

MENSTRUATION

Menstrual cycles constitute the most accessible and noninvasive biologic marker of female reproductive function in humans. The cycle lengths can vary (average, 28 days); determination of amenorrhea or other acyclic conditions requires a daily menstrual record for at least some 3 months in humans and higher primates. Because the menstrual cycle can fluctuate as a result of nutrition, stress, use of oral contraceptives, and other influences, detailed personal and medical histories must be collected from study subjects.

Laboratory rodents do not have menstrual cycles, but have estrous cycles, and their cycle status usually can be established from cell changes in daily vaginal smears. Acyclicity is defined as 14-30 days without evidence of ovulation in a proestrus smear. The onset of acyclicity can be studied as part of the aging processes or as a response to experimental intervention. Strings of cornified smears for 14 days or more indicate a polyfollicular anovulatory condition (persistent vaginal cornification), which is the most common initial acyclic state during aging (Finch et al., 1984). Similar vaginal smear patterns after acute or chronic exposure of young rodents to suspected environmental toxicants indicate severe disruption of the reproductive neuroendocrine system. Whether acyclicity is reversible depends on the agent, age at exposure, and duration of exposure. Permanent acyclicity has been induced in rats by DDT (Heinrichs et al., 1971).

Menstrual or estrous cycles can be analyzed for cycle length distributions and for lengths of consecutive cycles. The most detailed analyses have been done on rodents during aging. The frequency distribution of estrous-cycle lengths (e.g., 4-day, 5-day, or 6-day cycles) and the frequency of length transitions (e.g., 4-day to 4-day or 4-day to 5-day transitions) are sensitive indicators of maturation and aging as well as for effects of estrogen toxicity in mice (Fig. 13-1) (Nelson et al., 1982).

Available longitudinal menstrual records should be analyzed in more detail (see Treloar et al., 1970). Many powerful statistical time-series analyses might detect random and structured effects of toxicants on cycle length. For example, a digital filtering technique that removed atypical frequency variations (Orr and Hoffman, 1974) detected cyclic increases in gonadotropins in premenarchal girls (Hanson et al., 1975). Several groups have analyzed pulsatile LH in rats (Ellis and Desjardins, 1984) with the iterative approach of the Cycle Detector Program (Clifton and Steiner, 1983). Phase relationships and couplings among cycle length distributions should be investigated in depth.

LOSS OF FERTILITY AND FECUNDITY

Using infertility to assess the effects of toxicants is difficult, because of the common use of oral contraceptives and the importance of male-specific factors (e.g., oligospermia). Proper diagnosis of infertility usually requires endocrine and gynecologic data.

Resorption of most abnormal rodent fetuses is reflected in reduced litter size; the age-related decrease in litter size does not result from the shedding of fewer ova at ovulation (Holinka et al., 1979). Age-related increases in number of stillborn pups contribute to reductions in litter size and are associated with an increase in length of gestation (Holinka et al., 1979). In humans, there is also an age-related decrease in number of offspring long before the approach to menopause, in which behavioral factors also are important (see Fig. 11-7). The age-related frequency of increasing fetal malformations causes increased rates of spontaneous abortion.

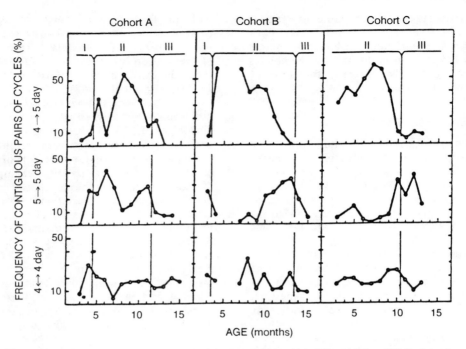

FIGURE 13-1 Frequency profiles of estrous-cycle length transitions (4-4, 4-5, and 4-5 + 5-4 days) in three cohorts of aging virgin mice. Phase designations are described in Figure 11-5. Reprinted with permission from Nelson et al., 1982.

PRECOCIOUS MENOPAUSE

Early menopause can have various causes, including hereditary influences associated with either parent, autoimmune destruction of ovarian tissue, mumps oophoritis, and exposure to ionizing radiation (Mattison et al., 1983; Gosden, 1985; Finch and Gosden, 1986). Premature onset of infertility and menopause usually is attributed to premature depletion of the ovarian follicular stock (Gosden, 1985).

Rodents are susceptible to premature infertility syndromes in association with neuroendocrine damage, as when steroids are administered to neonates in submasculinizing doses (Mobbs et al., 1984). Natural variation of rodent infertility occurs due to in utero factors; female fetuses flanked by males eventually become infertile several months before fetuses flanked by females; such effects probably are limited to a critical period during development (Vom Saal and Moyer, 1985).

OVARIAN OOCYTE DEPLETION

Oocytes become depleted throughout the lifetime of an individual female. Assays of the oocyte stock require laborious histologic analyses of excised ovaries. Very few human ovaries have been sectioned serially to determine oocyte numbers. Biopsies or ultrasound examinations can show whether the ovaries are cystic or atrophied (Orsini et al., 1984), and continued improvements in image analysis systems could make large-scale histological analyses of human oocyte stock plausible in the near future.

More information about ovarian oocyte stock might come from administering exogenous gonadotropins for controlled hyperstimulation in new reproductive technologies. Mouse ovaries produce ova with endogenous or exogenous gonadotropic stimuli until almost immediately before the stock is exhausted (Gosden et al., 1983), as do human ovaries (Sherman et al., 1976).

HORMONES

Gonadotropins and steroids can be assayed accurately in small blood samples-0.01-5 ml. Most assays can be scaled down for mice and rats with appropriate immunoreagents. In addition, human urine and saliva are sources for measurements of major changes.

Gonadotropins

Pregnancy-related increases in chorionic gonadotropins in urine can be detected by a highly sensitive assay as early as 8 days after fertilization. The assay can help to identify groups with inapparent spontaneous abortion in early pregnancy (Wilcox et al., 1985) that might be due to effects of toxicants. Daily urine samples are needed, in addition to records of intercourse frequency and menstrual cycles. A large epidemiologic study is described in Chapter 15.

Measurement of LH and FSH to monitor the complex changes during puberty and menopause is problematic because of individual variations. However, postmenopausal LH and FSH increases are detected easily, and their absence after menopause might be an indirect biologic marker of hyper-prolactinemia, since prolactin suppresses LH secretion at the pituitary (Cheung, 1983). Stable, increased LH and FSH constitute a biologic marker of premature ovarian exhaustion or atrophy. More refined analyses of daily fluctuations in women of reproductive age require multiple daily blood sampling over several months, which is difficult to do on a large scale. Resolution of toxic effects on high-frequency LH surges is even more difficult, and requires sampling every 5 to 15 minutes.

Prolactin also might be important for monitoring toxicants and drugs that promote growth of lactotropes (pituitary acidophilic cells that secrete prolactin). Rodents are susceptible to prolactinemia and the spontaneous lactotrope-containing pituitary tumors that commonly arise during acyclicity; these can be induced by chronic exposure to estrogens in some genotypes (Finch et al., 1984).

Although lactotrope adenomas in humans have not been linked to estrogen exposure (El Etreby, 1980), the possibility is still being considered. Some widely prescribed drugs, such as haloperidol and reserpine, increase prolactin. Monitoring prolactin may not require serial daily blood sampling. Fluctuations in prolactin across menstrual cycles (Guyda and Friesen, 1973) are much smaller than increases often seen in hypersecreting pituitary tumors.

Steroids

E_2, progesterone, and DHEA-S are major age-related biologic markers from menarche through menopause. All are best measured from blood, but all have small daily fluctuations and can be assayed in small blood samples (5-10 ml). Urine also can be used to assay estrogens (Thijssen et al., 1975), progesterone metabolites (Teitz et al., 1971; Speroff et al., 1983; Rebar, 1986; Shackleton, 1986), and etiocholanolone, an androgen related to DHEA-S (Bulbrook et al., 1971). DHEA-S changes little across the menstrual cycle (Guerrero et al., 1976) or diurnally (Rosenfeld et al., 1975). Daily sampling for progesterone is needed to identify changes in corpus luteum functions, but a single mid-luteal sample suffices to document corpus luteum formation.

Cortisol is also of interest, because of its close relation to stress, which can affect cyclicity, and therefore, fertility. However, extensive diurnal fluctuations and response to activity and ambient temperature contraindicate cortisol as a useful general biologic marker of reproductive function.

NERVOUS SYSTEM

A major issue in interpreting the impact of chronic exposure to toxicants on brain functions is the extent of underlying age changes. Rodents become more sensitive to some agents with age, as judged by greater depletion of dopamine in 28- versus 4-month-old rats exposed to the same dose of the neurotoxin 6-hydroxydopamine (Marshall et al., 1983).

These age changes might be caused by altered clearance or weakened detoxification mechanisms, of which there is ample evidence (Vestel and Dawson, 1985). Some rodent studies have shown age-related decreases in extracellular space, which might increase the efficacy of toxicants (Bondareff, 1973). The effects of age on toxicant stability and metabolism in vivo need more study.

Alternatively, toxic effects might be greater with age because of age-related reductions in "neuronal reserve" (Fig. 13-2). Particular support for that idea is found in Huntington's chorea, an autosomal dominant disease that has onset in persons 30 to 60 years old and is associated with the selective neuronal deterioration. Asymptomatic carriers have shown neurologic abnormalities after drug challenge (Klawans et al., 1972), as though their resistance to the drug is diminished or their capacity to function under stress has diminished (Finch, 1980). Thus, progressive, age-related changes in the brain might approach a dysfunction threshold so that the margin of safety for a range of insults is smaller, through either loss of neurons or erosion of the dendritic arbor (Finch, 1976; Calne et al., 1986). Those phenomena are suspected to occur in the hypothalamus with advancing age, but few data are available.

Dopaminergic losses do not produce neur-

ologic symptoms until critical threshold is reached (cross-hatched region) leading to parkinsonian symptoms (see Fig. 13-2). Ultimately, lesion summates with normal age trend for loss of nigro-striatal dopaminergic activity to exceed putative threshold (Finch, 1976). Greater induction of reversible parkinsonian symptoms in older patients by antipsychotic dopaminergic antagonists (Ayd, 1961) is also consistent with this model. Parkinson's disease thus could afford a "window" on aging changes. Huntington's chorea (Finch, 1980) and Alzheimer's disease (Calne et al., 1986) are among other age-related neurologic diseases with distinctive ages of incidence that might be viewed similarly (from Finch, 1981).

The extent to which age-related manifestations are time-dependent and cumulative is also uncertain; for example, the incidence of skin cancer in mice is related more closely to duration of exposure to benzopyrene than to age (Peto et al., 1975). A general representation of that concept is the experience space of Figure 13-3, which shows how duration of exposure and strength of the stimulus (e.g., toxicant or drug) could interact to determine different trajectories that approach the hypothetical dysfunction threshold at different rates.

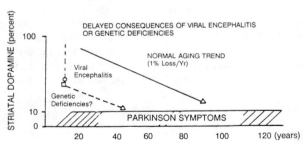

FIGURE 13-2 Late-midlife onset of some types of parkinsonism might result from early viral lesion (encephalitis lethargica) that causes loss of substantial nigral dopaminergic neurons (Poskanzer and Schwab, 1961) or from genetic deficiency (Mjones, 1949; Myrianthopoulos et al., 1969) of nigral dopaminergic function. Dopaminergic losses do not produce neurologic symptoms until critical threshold is reached (cross-hatched region), leading to parkinsonian symptoms. Ultimately, lesion summates with normal age trend for loss of nigro-striatal dopaminergic activity to exceed putative threshold (Finch, 1976). Greater induction of reversible parkinsonian symptoms in older patients by antipsychotic dopaminergic antagonists (Ayd, 1961) is also consistent with this model. Parkinson's disease thus might afford a window on aging changes. Huntington's chorea (Finch, 1980) and Alzheimer's disease (Calne et al., 1986) are among other age-related neurologic diseases with distinctive ages of incidence that might be viewed similarly. Source: Finch, 1981.

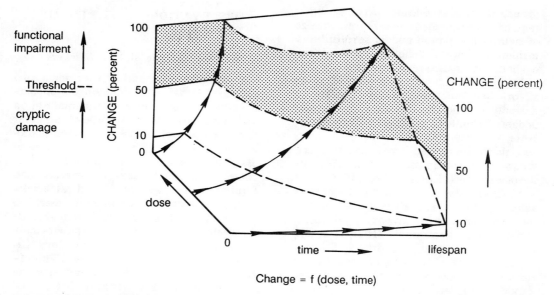

Change = f (dose, time)

FIGURE 13-3 Age-related phenomena might be represented on a 3-dimensional experience surface whose axes are time, dose (strength of a cause of age-related change, e.g., estradiol in ovary-dependent neuroendocrine syndrome of rodents), and change (impairment as function of dose and time). In many cases, change might be cryptic functional consequence) until some threshold is reached (stippled background). Three trajectories at different doses are shown (arrows). Source: Finch, 1987.

Behavior

Rodents have a robust repertoire of sex-steroid-dependent behavior that can be used to assess the effects of environmental toxicants—such as lordosis and open-field activity. Hormonal influences on sexual behavior in women have been difficult to prove, and no markers have been generally accepted. Increasing evidence shows that sexual interest might be linked to concentrations of plasma androgens (Morris et al., 1987).

Cell Populations

A few reports suggest age-related loss of hypothalamic neurons in the human female (Sheehan, 1968; Swaab and Fliers, 1985). The data are scarce, however, and do not establish any change in neuron populations as a suitable marker of reproductive neurotoxicity. However, studies in rodents indicate that this may be a useful biologic marker.

In rodents, at least three markers of age-related neuron damage are available. Glial hyperactivity in the arcuate nucleus increases during female reproductive senescence (Schipper et al., 1981) and can be induced in young rats by estrogen exposure (Brawer et al., 1983). N-Methyl-d-aspartate causes premature cycle lengthening and also kills 30% of arcuate neurons (May and Kohama, 1986). Binding of lead-210 by the hypothalamic median eminence in autoradiographic studies (Stumpf et al., 1980) and the effect of lead on neurotransmitters (Silbergeld, 1983) are additional possible CNS markers of toxicity in rodents. Postmortem human brain tissue might be used for similar analyses, but few studies have been attempted.

In humans, research is needed to determine whether changes in hypothalamic cell populations can be measured noninvasively and whether these changes correlate with reproductive toxicity.

Important information must be obtained by brain-imaging techniques, such as 2-deoxyglucose uptake, which is sensitive to neuronal degeneration in other brain regions. Sex-related differences in size of neuron population and in susceptibility to damage from endogenous hormones or exogenous agents also are of interest.

Neuron damage often is viewed as an all-or-none outcome, with cell death as the only end point. However, a spectrum of damage should be considered—from reversible to irreversible. Neurotoxicants and their interactions with hypoxia and sugar in hippocampal neurons illustrate the possibilities (Sapolsky, 1985, 1986).

Molecular approaches should be applicable, e.g., by following changes of mRNA that are induced by hypoxia (Pulsinelli, 1985) and corticosteroids (Nichols et al., 1986). Other molecular markers of intermediate stages of neuron damage during early and late responses to toxicants should be sought.

Neuron loss from acute or chronic exposures to toxicants cannot be evaluated without a greatly extended data base on possibly age-related neuron loss. Data are available from postmortem material that is meager for ages under 65 years. Most studies have been conducted on fewer than 5 brains from persons within any 20-year age span. Consequently, the normative range of neuronal numbers for different brain regions and pathways in healthy young adults is unknown. The absence of these data severely limits future studies on neurotoxicant-age interactions in the hypothalamus. More information will emerge slowly from studies of nerve cell loss during Alzheimer disease, since a range of control groups is being sought.

Feasibility studies are needed to determine which of the toxicant-sensitive neuron populations can be counted most reliably by automated image analysis to determine loss. Normative age-related profiles could be established later from postmortem specimens of neurologically normal brain donors with known health status enrolled in the Alzheimer's Disease Research Centers. Such tissues would provide a basis for investigating groups that might be at risk of toxicant-related neuron damage, e.g., from occupational exposure to metals or from dietary exposure to phytoestrogens.

OBSERVATIONS ON THE USE OF THESE MARKERS

Need for Multiple Biologic Markers

Onset, changes, and loss of menstrual cycles and their duration are major biologic markers of changes during maturation and menopause. Fertility depends on other functions as well; therefore, other markers of reproductive status are needed to assess the impact of environmental toxicants. Multiple assays of ovarian and pituitary hormones are needed—especially during transitional periods—to establish the stability of basal values and characterize the frequency and amplitude of episodic changes, such as preovulatory LH surges and pulsatile LH release. The effects of lead in reducing progesterone concentrations and lengthening luteal phases and menstrual cycles of young adult monkeys (Franks et al., 1986) illustrate the value of these biologic markers and the need for data on menstrual cycle length and hormonal status. Accounts of irregular menstrual cyclicity in association with phytoestrogens from tulip bulbs eaten in the Netherlands during World War II (Burroughs et al., 1985) cannot be evaluated clearly without supporting data on nutritional and endocrine status.

Differential Susceptibility of Individuals and Populations

Biologic markers are needed to identify individuals or populations whose reproductive functions are particularly susceptible to toxic effects. Human individuality might arise from genotypic differences or from a broad spectrum of environmental influences throughout life. A wide range of examples of genotypic and environmental influences on responses to toxicants has been modeled in laboratory animal studies, genetic influences on the cytochrome P-450 drug metabolizing enzymes (Gonder et al., 1985; Koizumi et al., 1986). Human genetic polymorphisms in responses to toxicants have not been identified, but studies of twins and drug clearance indicate greater concordance between monozygotic twins than between dizygotic twins (Vesell et al., 1971). The well-known prevalence of lactase deficiency in adult Orientals (McKusick, 1975) that causes intolerance to the lactose in milk and unfermented milk products also supports the presence of genetic polymorphisms that could influence responses to environmental toxicants.

Analysis of Menstrual Cycle Lengths

The frequency of menstrual cycles generally has been characterized in statistical terms. The large longitudinal data bases of cycle lengths from hundreds of women collected by Treloar et al. (1970) and others could be used to develop new statistical descriptors of changes in cycle frequency, for example, the frequency of consecutive cycles of particular lengths. Many sophisticated approaches might detect structured or random toxic effects on cycle length and on lengths of consecutive cycles. That information also might bear on the increase in birth defects with maternal age. Pilot studies are needed to evaluate the applicability of existing approaches and needs for further development. Primate and rodent responses to toxicants that influence cycle frequency could be used to evaluate the sensitivity of these approaches.

Long-Term Neurologic Consequences of Toxicant Exposure

In view of the many examples of cryptic neuron damage that results in neurologic disorders years after toxicant exposure, pilot studies should be established to track subjects exposed to neurotoxicants early in life. The recently established followup studies of MPTP exposure (Calne et al., 1985) are a precedent for this approach that might be extended to long-term effects from occupational exposures to lead, manganese, and other neurotoxic agents.

Longitudinal neurologic and psychiatric studies at the Alzheimer Disease Research Centers compare various normal and disease-afflicted groups with dementia patients and might be helpful in determining long-term effects. Additional groups for antemortem or postmortem studies could be added easily.

14

Biologic Markers
of Nonconceptive Menstrual Cycles

Use of biologic markers to evaluate the effects of environmental toxicants on the normal menstrual cycle is a recent concept. Biologic monitoring can use a variety of body fluids (including fluids specific to reproduction) and tissues (Tables 14-1 and 14-2). This section discusses biologic markers that can be evaluated during the nonconceptive menstrual cycle.

To produce menstrual dysfunction, environmental agents have to have systemic toxicity that is manifested in the reproductive system or specificity for the reproductive system. Agents with activity peculiar to the reproductive tract might be difficult to identify because of the following:

- The complex nature of the reproductive process.
- The presence of spontaneously occurring disease with identical symptoms.
- The act of procreation is largely voluntary and, therefore, lack of conception might be a matter of choice rather than toxic effect.
- The lack of a reliable epidemiologic data base on reproductive events or continuing surveillance of reproductive markers in the general population.

The only reproductive tract fluid so far studied for the purpose of biologic markers is follicular fluid collected from patients undergoing in vitro fertilization/embryo transfer (IVF/ET). Several xenobiotics have been identified in follicular fluid (Trapp et al., 1984), but associations with adverse effects have not been investigated systemically. That is of particular concern, because follicular fluid is in direct contact with the oocyte and the steroidally active granulosa cells. Toxic agents in the follicular fluid have the potential to alter granulosa-luteal cell function, as well as the developing early embryo.

The menstrual cycle is a normal physiologic event, and the fundamental markers of female reproductive function are the phenomena of the reproductive cycle, which are the following (in order of increasing complexity):

- Characteristics of the normal menstrual cycle—such as interval, regularity, duration, and character of menses and the presence of premenstrual molimina (see below).
- Biophysical changes associated with the reproductive cycle—such as basal body temperatures, cervical mucus

TABLE 14-1 Human Body Fluids Potentially Useful in Measuring Biologic Markers

Fluid	Availability[a]	Comments
Nonreproductive tract fluids		
Blood	+ + +	Significant temporal fluctuations in hormone concentrations assessed in blood
Urine	+ + + +	Provides good indication of cumulative exposure and includes accumulated metabolites
Saliva	+ + + +	An ultrafiltrate of plasma
Cerebrospinal fluid (CSF)	+	Measurement of neurotransmitters limited by CSF/brain barrier
Reproductive tract fluids		
Vaginal secretions	+ + +	Cycle-specific, use hampered by bacterial contamination, often used in animal studies
Cervical secretions	+ + +	Cycle-specific, needs further development
Uterine luminal fluid	+	Poorly characterized secretory products, cycle-specific, require further development
Tubal secretions	+	Poorly characterized secretory products, cycle-specific, need further development
Follicular fluid	+ +	Only in stimulated cycles in IVF/ET,[b] requires further development
Peritoneal fluid	+ +	Contains exudate of ovary, requires further development
Menstrual effluent	+ +	Limited by autolysis, requires further development

[a]Increasing number of +'s indicates more readily available. Some fluids available only from patient samples, such as from surgical patients.
[b]In vitro fertilizations/embryo transfer procedures.

changes, vaginal cornification, and sexual behavior.

• Endocrinologic characteristics of the ovulatory cycle.

A normal menstrual cycle is defined by a pattern—every 26-30 days—(Treloar et al., 1970) of an ill-characterized constellation of symptoms termed premenstrual molimina (e.g., breast tenderness, bloating, and mood swings followed by vaginal bleeding). Ovulatory menses, typically, is associated with some degree of lower abdominal cramping due to the effect of prostaglandins on the myometrium. The menstrual effluent comprises autolyzed endometrium, tissue fluid, and blood that has undergone clotting and lysis. Passage of excessive blood clots with the menstrual flow is unusual in ovulatory cycles, unless mechanical problems, such as endometrial polyps or leiomyomata, are present.

Some biophysical changes can be used to determine whether ovulation has occurred, such as basal body temperature shifts after ovulation and changes in the character and quality of cervical mucus. Other descriptors of the menstrual cycle, such as cyclic fluctuations in vaginal cytology and changes in sexual behavior, are subject to nonendocrine influences and are less useful for that purpose.

Any agent that disrupts normal cyclic menstrual function may be described as causing reproductive toxicity. The reproductive tract is susceptible to disruption by environmental agents that affect the cerebral cortex, hypothalamus, pituitary, ovaries, fallopian tubes, or uterus. Aberrant menstrual cycles can result from disruptions of a diverse group of functions, including:

• Synthesis, storage, transport, release, and metabolism of neurotransmitters, gonadotropin-releasing hormone (GnRH), gonadotropins, and ovarian regulatory peptides.
• Gonadotropin responsiveness.
• Ovarian steroidogenesis.

TABLE 14-2 Human Tissues Available for Use in Measuring Biologic Markers of Reproductive Toxicity

Tissue	Availability[a]	Comment
Hypothalamus	+	Autopsy material
Pituitary	+	Autopsy material
Uterus		
Endometrium	+ + +	Cycle specific, difficult to culture, heterogeneous cell population
Myometrium	+	Not endocrinologically active
Fallopian tube	+	Cycle-specific secretory products, unclear physiologic significance
Ovarian cells		
Granulosa-luteal	+ +	Large-culture methods possible, available in late follicular phase
Thecal	+	Organ cultures only, difficult to purify
Stromal	+ +	Minimal hormone secretion
Adipose tissue	+ + + +	Useful as internal dose marker comparable with ovary, owing to lipid composition and steroid enzyme activity, available only from surgical procedures
Cytologic specimens		All composed primarily of exfoliated cells, most of which are degenerative; most require further development
Vaginal	+ + +	Available through Pap smears
Cervical	+ + +	Available through Pap smears
Endometrial	+	Requires invasive procedure
Peritoneal fluid	+ +	Composed primarily of macrophages

[a]Increasing number of +'s indicates more readily available.

- Gametogenesis.
- End-organ response to sex steroids.

Nonreproductive tract tissues, such as the liver and adrenal glands, can affect menstrual cyclicity by altering the production of sex-hormone-binding globulins (Vermeulen et al., 1969; Forest and Bertrand, 1972; Anderson, 1974) or causing excessive synthesis of nongonadal androgens (Molinatti et al., 1964; Mahesh et al., 1968; Riddick and Hammond, 1975) Furthermore, the reproductive process is extremely sensitive to the general health of the woman, and metabolic stress (such as weight loss, hypothyroidism, excessive exercise, and glucocorticoid excess) could stop ovulation (Warren et al., 1975; Vigersky et al., 1977; Smith, 1980; Warren, 1980; Shangold et al., 1981). Severe psychologic stress can alter the reproductive process by suppressing gonadotropin products and inducing amenorrhea (Fries et al., 1974; Rabkin and Struening, 1976; Sommer, 1978; Henry, 1980).

Healthy women occasionally experience anovulatory menstrual cycles that may be accompanied by altered menstrual cycle length and character but without long-term changes in fertility. Episodic toxic exposure might be analogous in effect to short-term use of oral contraceptives, which temporarily prevent ovulation but have no lasting anovulatory effect (Golditch, 1972; Jacobs et al., 1977; Tolis et al., 1979; Henzl, 1986). Continuous exposure—e.g., through chronic ingestion of contaminated food or water, occupational exposure, or long-term pharmacologic treatment—might have a profound influence on reproduction. Women undergoing anesthesia is a case of short-term exposure (Soules et al., 1980). Women working in operating rooms and chronically exposed to anesthesia have an increased number of miscarriages with increasing duration of exposure to anesthetic chemicals.

Adverse effects might be overlooked if they are not life-threatening, rare, or specifically monitored in the population at risk. For example, the drug spironolactone was used for many years to treat hypertension, before it was noted that it had antiandrogenic activity and induced gynecomastia in males (Corvol et al., 1975; Boisselle and Tremblay, 1979; Shapiro and Evron, 1980; Cumming et al., 1982). Infertility, however, is neither life-threatening nor rare and is likely to be undetected in the general population, unless it is monitored.

SPECIFIC MARKERS

A sophisticated array of assays of biochemical events associated with ovulatory menstrual cycles is available, including assays of gonadal sex steroids (E_2, estrone, progesterone, 17-OH-progesterone, testosterone, and androstenedione), pituitary hormones (FSH, LH, and prolactin), hypothalamic neurotransmitters (GnRH and dopamine), and gonadal regulatory peptides (inhibin and activin) (Tables

14-3 and 14-4). Random measurement of these hormones to evaluate menstrual cycle normalcy in unselected women is not cost-effective; a better strategy would involve hormone measurements standardized for time in the cycle in an at-risk population with high exposures or symptoms, e.g., irregular menses. With this strategy, clinically inapparent hormonal changes in gonadal or pituitary hormones might be detected with cycle-specific hormone measurements.

Alterations of the normal cycle are nonspecific responses and can occur against a background of spontaneous disease with an indistinguishable pattern of menstrual dysfunction. Therefore, alterations do not necessarily indicate that exposure to a toxicant has occurred. Once an abnormality of the reproductive cycle is detected, characterization of a specific alteration in the hormonal profile is usually

TABLE 14-3 Potential Biochemical Markers of Reproductive Toxicity for Evaluation In Vivo

Origin	Biologic Markers	Comments
Central nervous system	Neurotransmitters	Measurement of CNS concentrations might not reflect circulating concentrations owing to blood-brain and brain-CSF barriers
Pituitary	Luteinizing hormone, follicle-stimulating hormone, prolactin, adrenocorticotropic hormone, thyroid-stimulating hormone, growth hormone	Pulsatile secretion makes adequate sampling difficult
Ovary	Steroids: estradiol, estrone, progesterone, testosterone, andro-stenedione	Cycle-specific
	Regulatory factors: relaxin, progestin-associated endometrial protein (PEP), prolactin, plasminogen activator, inhibin, oocyte maturation inhibitor, luteinization inhibitor	Poorly characterized, cycle-specific, unclear physiologic significance
Fallopian tube	Secretory proteins	Poorly characterized, specific, and the physiologic significance is unclear
Uterus	Prolactin, other secretory proteins	Poorly characterized, cycle-specific, of unclear physiologic significance
	Prostaglandins	Nonspecific, rapidly metabolized paracrine and autocrine hormones
Cervix	Mucus	Cycle-specific
Vagina	Secretory proteins	Poorly characterized, cycle-specific, easily contaminated with bacteria

TABLE 14-4 Potential Biologic Markers of Reproductive Toxicity for Evaluation In Vitro

Tissue	Biologic Marker	Comments
Adipose	Steroidogenesis	Lipid-soluble toxins might accumulate, as in ovary
Uterus		
Cervix	Rate of mucus production	Estrogen-dependent
Endometrium	Concentration of secretory proteins	Cycle-specific, unclear physiologic significance
Myometrium	Steroid hormone response	Little relationship to menstrual cycle
Fallopian tube	Concentrations of secretory proteins	Cycle-specific, of unclear physiologic significance
Ovary		
Granulosa	Rate of steroid-ogenesis; gonado-tropin-receptor number or responsive-ness; synthesis of regulatory factors	Excellent culture method
Thecal cells	Rate of steroid-ogenesis; gonadotro-pin receptor number or responsiveness	Difficult to purify

possible. The underlying course, however, often is not well understood, and unrecognized reproductive toxicity is likely.

Alterations of the normal menstrual cycle can occur either through disruption of follicular development or through onset of luteolysis. The exact mechanisms are unknown, but in most naturally occurring instances, altered gonadotropin or progesterone secretion or action is suspected. Reproductive toxicants might also alter the secretion or action of these hormones.

Cyclicity

An initial approach to exploring environmental toxicity is to evaluate whether menstrual cycles are regular. However, regardless of which organ is affected or the mechanism of toxicity, a substantial endocrinologic disruption of the menstrual cycle must occur, if irregular menses is to be clinically detectable. For example, quantitative changes in the function of the corpus luteum can occur without alterations in overall cycle length (Murthy et al., 1970; Abraham et al., 1974; Jones et al., 1974; Radwanska et al., 1976; Soules et al., 1977; Rosenfeld et al., 1980; Radwanska et al., 1981; Gravanis et al., 1984).

Character of Menstrual Flow

In a normal genital tract, menses is the inevitable result of a nonconceptive ovulatory cycle. Not all vaginal bleeding, however, is indicative of ovulation; anovulatory bleeding can occur often enough to be clinically indistinguishable from ovulatory bleeding. Typically, the flow lasts 3-7 days and is associated with some degree of dysmenorrhea. Few clots are passed with menstrual blood in an ovulatory cycle. Substantial changes in length from cycle to cycle or the presence of clots may indicate effects of external factors. When it is clinically important to distinguish ovulatory from nonovulatory bleeding, a more accurate indicator than the character of the vaginal bleeding is required.

Detection of Corpus Luteum

The hallmark of primate ovulation is a spontaneous luteal phase whose dominant feature is the production of progesterone. That steroid has various biophysical effects; hence, luteal function is detectable through a variety of clinical and laboratory measures, as noted briefly below.

Progesterone in the Peripheral Serum

Progesterone, a readily diffusible steroid, can be measured in various body fluids (Yoshimi et al., 1969; Mikhail, 1970; Yussman and Taymor, 1970; Lloyd et al., 1971; Rondell, 1974; Sherman and Korenman, 1975; Maathuis et al., 1978; Aedo et al., 1980; Donnez et al., 1982; Zorn et al., 1982; Crain and Luciano, 1983; Loumaye et al., 1985; Bouckaert et al., 1986; Koninckx et al., 1986) and reproductive tract fluids (Fowler et al., 1978; Botero-Ruiz et al., 1984), as can its glucuronidated metabolite in urine (Tietz et al., 1971; Speroff et al., 1983; Rebar, 1986; Shackleton, 1986). Urine samples are easy to collect and simple to store and do not exhibit the rapid fluctuations in progesterone concentration observed in blood. Detailed comparisons of blood, saliva, and urine must be performed for each assay to validate the relationships for that method. Each assay system might have specific storage requirements, such as serum separation, freezing, and use of antibacterial preservatives.

Only one midluteal progesterone value is necessary to document that ovulation has occurred (Israel et al., 1972). However, evaluation of an inadequate luteal phase is controversial. No method is universally accepted but several serum measurements of progesterone in the luteal phase would characterize the function of the corpus luteum.

Basal Body-Temperature Shift

Progesterone is a thermogenic hormone; when circulating concentrations rise above about 2-3 ng/ml, the basal body temperature rises by approximately 0.5°C, until the demise of the corpus luteum just before menses. Basal body temperature can be reliably measured orally or rectally. New electronic thermometers might improve accuracy and, if coupled with a recorder, might prove applicable to large-scale monitoring.

Cervical Mucus Changes

E_2, the dominant follicular-phase steroid, differs markedly from progesterone, the dominant luteal-phase steroid, in effect on cervical mucus quality and quantity. Changes in the character and quantity of cervical mucus can be useful clinically to predict hormonal status (Clift, 1945; Pommerrenke, 1946; Birnberg, 1958; Moghissi, 1966; MacDonald, 1969; Moghissi et al., 1972; Moghissi, 1973). Cervical mucus evaluation is qualitative and requires a pelvic examination.

Vaginal Cytology

The effects of E_2 and progesterone on vaginal cytologic findings also differ (Papanicolaou, 1933; Riley et al., 1955; Rakoff, 1961; Frost, 1974). Vaginal cytology is useful in monitoring the cyclicity of female rodents, but has less reliability in women because the changes from one day to the next are less evident. Self-collected vaginal specimens or measurements compatible with tampon use are feasible, but bacterial or fungal contamination might compromise individual measurements.

Endometrial Histology

Characteristic histologic changes accompany follicular phase development in response to rising E_2 concentrations. With the onset of a luteal phase, increasing progesterone concentration sequentially changes the histologic appearance in a well-defined progression to the menstrual endometrium (Noyes et al., 1950). Histology is a reliable means of detecting ovulation and is semiquantitatively correlated progesterone production. Endometrial biopsy is extremely accurate for assessing the presence and competence of a corpus luteum, but has the serious disadvantage of being an invasive, painful, and expensive procedure that carries some risk of genital tract infection.

Secretion of Progesterone-Stimulated Endometrial Proteins into Uterine Lumen

The only two well-characterized progestin-dependent secretory proteins are prolactin (Maslar and Riddick, 1979; Daly et al., 1983a) and progestin-associated endometrial protein (PEP) (Mazurkiewicz et al., 1981; Joshi, 1983). Neither has been used clinically to evaluate the luteal phase, but efforts are under way to assess their usefulness. Cyclic changes of specific uterine luminal proteins have not been characterized, except for those of prolactin (Maier and Kuslis, 1987). Aspiration of uterine luminal fluid is comparable in utility with endometrial biopsy and is an uncomfortable, invasive procedure with some hazard of infection; application as a marker must await a better understanding of uterine physiology.

Reproductive Endocrinology In Vivo

Measurements of other circulating reproductive hormones and regulatory factors, such as GnRH, the gonadotropins, testosterone, and inhibin, have been used experimentally and clinically to study the physiology or pathophysiology of the menstrual cycle. An increasing catalog of paracrine hormones and growth factors is providing a new class of regulatory substances to consider.

Measurements of reproductive hormones or regulatory factors, such as GnRH and gonadotropins, are accurate in detecting anovulation, but are of value mainly in identifying the site of the problem. Measurements of other ovarian hormones (e.g., testosterone and androstenedione) are of little value because their production is a consequence, not a cause, of anovulation.

In Vitro Markers

A variety of important biochemical events in the menstrual cycle can be evaluated with in vitro systems and used to determine the effects of putative toxicants (Table 14-4). Any adverse influence of environmental agents needs to be confirmed with in vivo testing, and in vitro systems can be used as models to screen agents for reproductive toxicity, as well as to provide details of the mechanisms. Examples of such models are:

- Gonadotropin secretion by dispersed pituitary cell cultures.
- Steroidogenesis, regulatory factor synthesis, and gonadotropin binding by ovarian cell organ or cell cultures (i.e., granulosa cells, thecal cells, and stromal cells).
- Secretion of luteal-specific proteins by endometrial cell cultures.
- Electrophysiology of myometrial cells in organ culture.
- Mucus production by endocervical cell cultures.

Follicular fluid, granulosa cells, oocytes, and other reproductive tract materials are readily available for routine study from surgical specimens.

Evaluating cellular physiology in vitro permits the effects of environmental agents on biochemical events of the normal menstrual cycle to be investigated. That approach has been applied to investigation of diseases that cause anovulatory infertility, such as polycystic ovary syndrome (Haney et al., 1986). The in vitro approach should prove rewarding in estimating the potential for adverse biologic effects of environmental chemicals, and efforts in this area should be encouraged.

BIOLOGIC RATIONALE

Clinical validation of any biologic marker is critical to its application in detecting toxicity. Spontaneous anovulation in infertility patients can serve as an example of factor that is important in the consideration of environmentally related reproductive toxicity.

Of the factors that are associated with infertility—semen, cervical mucus, the endometrial cavity, oviductal function, ovulatory function, and endometriosis—ovulation seems the most likely to be affected by environmental exposure. Defective ovulation usually is manifested by irregular vaginal bleeding. Other signs and symptoms, such as galactorrhea, the absence of premenstrual molimina, and a change in the character of menses, might be helpful, but their significance is diminished without a disruption of menstrual cyclicity. The biologic markers previously discussed are used clinically in infertility patients to detect ovulation when the vaginal bleeding pattern is not regular enough to permit accurate prediction of ovulation in nonconceptive menstrual cycles.

Environmental agents whose pharmacologic action is expected to be similar to that of contraceptives should be regarded as potential reproductive toxicants. Theoretical validation of the markers of reproductive toxicity noted above also can be accomplished by considering therapeutic agents used for contraception. For instance, ovulation can be blocked by steroid-feedback inhibition of gonadotropin release. In addition, the end-organ responses of the cervix, uterus, and oviducts to the natural ovarian steroids (Mishell, 1979) are altered by the three major classes of sex steroids—androgens, estrogens, and progestins. The development of birth control pills containing estrogens and progestins is based on contraceptive efficacy and control of vaginal bleeding. Progestins alone can

alter gonadotropin release and endometrial development; without estrogen, the endometrium is fragile, and vaginal bleeding is not well controlled. Estrogens alone are effective in preventing conception and have been used clinically after sexual assaults (Kuchera, 1974). A chemically modified androgen, danazol, also has been used to suppress gonadotropins therapeutically and create an anovulatory state (Young and Blackmore, 1977; Barbieri et al., 1977; Guillebaud et al., 1977; Rannevik, 1979; Luciano et al., 1981), but its use is limited by the high frequency of undesirable androgenic side effects.

The mechanisms of other hormones can also be demonstrated pharmacologically. Progesterone antagonists do not prevent ovulation, but alter decidualization, the endometrial response to progesterone production by the corpus luteum, enough to prevent implantation (Healy et al., 1983; Kovacs et al., 1984; Paris et al., 1984; Paris et al., 1986; Koering et al., 1986). Inhibitors of progesterone synthesis induce biochemical luteolysis, which leads to failure of endometrial development in a fashion similar to that caused by progesterone antagonists (Birgersson and Johansson, 1983; van der Spuy et al., 1985; Webster et al., 1985).

15

Developing Assays of Biologic Markers for Epidemiologic Studies: Experience with a Marker of Pregnancy and Early Loss

Human chorionic gonadotropin (hCG) is a glycoprotein hormone secreted by the syncytiotrophoblast; it appears to enter the maternal circulation at the time of endometrial implantation of a fertilized ovum (Jaffe, 1986). Its principal known role is to act at the ovarian corpus luteum to stimulate further secretion of progestins to support endometrial growth. Blood and urinary hCG concentrations rise rapidly and peak approximately 12 weeks after onset of the last menstrual period; thereafter, they slowly decline until the fetus and placenta are delivered. An additional function of hCG may involve stimulus of steroid hormone production by the fetus. The period of maximum testosterone production by the fetal testis corresponds to the peak period of placental hCG production, and Jaffe's group has demonstrated that hCG binds to receptors in the fetal testis and stimulates the production of testosterone (Huhtanemi et al., 1977). (Physiologic changes associated with the initiation of pregnancy are discussed in Chapter 20.)

This chapter describes the development and application of an assay to assess urinary hCG content as a biologic marker. The purpose of describing the problems encountered in laboratory development and early use of inappropriate assays is to make clear the need for complete development in the laboratory and preliminary studies in the field before assays are used in large epidemiologic studies. It is demonstrated that even with this widely studied hormone, there remains a need to improve the assays for use in field studies and a need for more extensive studies of normal populations.

ASSAYS OF hCG

In contrast with many other plasma proteins, hCG is excreted in the urine at a concentration approximately equal to that in blood. Marshall and colleagues (1968) demonstrated that to be the case throughout the first trimester, and Armstrong et al. (1984) verified that, even when minute quantities of serum hCG are present, the urine specimen is a ready source for assay and purification of hCG. However, it is also likely that urine specimens contain partially degraded hormone fragments, and measurement of a serum specimen might be preferred for diagnostic interpretation.

Bioassays for hCG were developed by Ascheim and Zondek (1928) (the A-Z test) as a method to detect pregnancy several weeks after the first missed menstrual period. In the 1960s, immunologic assays for hCG replaced the more cumbersome bio-assays (Wide, 1962; Wide and Gemzell, 1960); these added increased sensitivity and reproducibility, which permitted earlier diagnoses of pregnancy (Bell, 1969). However, polyclonal (usually rabbit) antibodies raised against hCG always cross-reacted with human luteinizing hormone (hLH) in the maternal serum sample or with hormone components in urine, owing to the high degree of homology between hCG and hLH. The sensitivity of these assays was limited by the inevitable presence of hLH immunoreactivity in maternal serum and also by either intact hLH or its fragments in urine.

The relationship of all the homologous glycoprotein hormones became apparent with the discovery that each comprised a single α subunit and a single β subunit (Pierce et al., 1971). The amino acid sequences of the α subunits of hLH and hCG were shown to be identical, but minor differences were found in the structures of the β subunits (Bellisario et al., 1973; Morgan et al., 1973, 1975; Birken and Canfield, 1977; Kessler et al., 1979; Fiddes and Goodman, 1981; Pierce and Parsons, 1981). When immunized with the purified β subunit of hCG, rabbits occasionally made antibodies that were directed predominantly against this hormone and had limited cross-reactivity with human hLH (Vaitukaitis et al., 1972); radioimmunoassays were developed that routinely detected hCG at 1 ng/ml in serum in the presence of circulating hLH. Radioimmunoassays with urine specimens were more complex, because of interfering substances and the presence

of degradation products. Nonetheless, this assay method was used to develop non-radioisotopic home-testing kits for hCG in urine specimens to detect pregnancy with moderate reliability (Doshi, 1986) during the first 9 days after the first missed menstrual period.

Two developments led to new hCG assays with improved sensitivity and specificity. First, monoclonal antibodies against the hCG β subunit could be produced, so that urine could be extracted efficiently by immunoaffinity adsorption. Second, preparations of the unique β COOH-terminal peptide region of hCG could be used as immunogens to make high-affinity polyclonal antibodies (Birken et al., 1982).

Several laboratories have made preparations of antibodies that recognize hCG and hLH to varied degrees (c.f., Vaitukaitis et al., 1972; Birken et al., 1982; Thau et al., 1983; Ehrlich et al., 1985). Of particular interest are antibodies that exhibit a high degree of specificity for hCG, that is, have very low reactivity with hLH.

Four separate regions in hCG give rise to antibodies that may be used in assays with selectivity for hCG over hLH (Fig. 15-1). Each region presumably reflects a conformational change resulting from structural differences.

Region IV represents an antigenic determinant (epitope) that is not readily reactive in native hCG, but is present on the free β subunit and on forms of the degraded hCG β subunit excreted in human urine. That region has been most recently identified and affords the opportunity to distinguish

FIGURE 15-1 Antibodies with specificity to α and β subunits of hCG. Shaded areas indicate regions that give rise to antibodies. Source: Canfield et al., 1987.

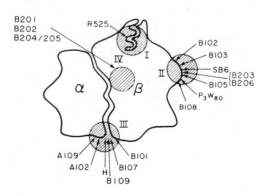

between degraded hCG β subunit fragments and the intact hormone. The complete structure of the degraded β core fragment has been determined, and monoclonal antibodies that bind to Region IV have been developed (Birken et al., 1988; Krichevsky et al., 1988).

The importance of the four discrete regions of the hCG molecule is demonstrated by immunoradiometric assays (sandwich assays) that use two antibodies selected on the basis of binding to different hCG epitope regions. An early study found that one such assay exhibited an affinity far beyond that expected (Ehrlich et al., 1982). A mathematical model was developed that accounted for that observation by predicting formation of a circular complex composed of one molecule of each antibody and two molecules of antigen (Moyle et al., 1983 a,b) (Fig. 15-2).

With those immunochemical reagents, a sensitive assay for hCG in urine was developed that used an immunoradiometric assay (Wilcox et al., 1985). The β-specific monoclonal antibody B101 is coupled to Sepharose beads, suspended in urine, spun down; the hCG binds to the antibody. To measure the amount of hCG captured, a

[125]I-labeled antibody from the rabbit (R525) is mixed with the resuspended Sepharose-B101-hCG mixture. When this is spun down, amounts of radioactivity bound to the solid phase indicate quantities of hCG attached to the capture antibody. The assay permits hCG detection at concentrations approaching 0.01 ng/ml (highly purified reference preparations of hCG have a bioassay value of approximately 13,000 IU/mg) and is 50-100 times more sensitive than any other existing method (Armstrong et al., 1984)(Fig. 15-3).

When a conventional radioimmunoassay is performed on a urine specimen, contaminating proteolytic enzymes in the urine can degrade the radiolabeled tracer during its incubation. This proteolytic artifact might lead to falsely increased measurements of hCG (Maruo et al., 1979). In addition, other interfering urinary substances can add to the background noise and decrease the signal-to-noise ratio (Ayala et al., 1978). In the immunoradiometric assay, the radioactive tracer is introduced after proteases and interfering substances have been washed away. SB-6 radioimmunoassay in urine displays poor signal-to-noise ratio with wide variation

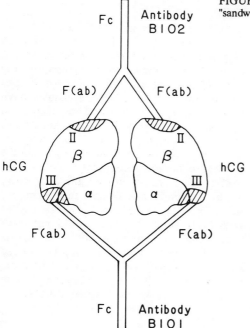

FIGURE 15-2 High-affinity circular complex that is formed in a "sandwich" assay. Source: Moyle et al., 1983b.

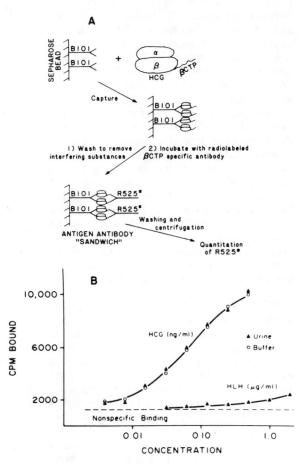

Figure 15-3 A. Schematic diagram of procedure for immunoradiometric assay of hCG in urine; B. Sensitivity of hCG detection by immunoradiometric assay. Source: Canfield et al., 1987.

between 1 and 10 ng/ml (Wilcox et al., 1985). This contrasts with data from immunoradiometric assay of same specimens, which displays greater sensitivity for hCG without widely fluctuating baseline. Those factors might have contributed to the differing results (8% to 57%) obtained in prior studies of fetal loss (Miller et al., 1980; Edmonds et al., 1982; Whittaker et al., 1983).

FIELD STUDIES OF EARLY FETAL LOSS: TESTING THE UTILITY OF THE hCG ASSAY

A small trial was conducted to ascertain the feasibility of the study design and whether new assays with improved sensitivity for hCG in urine would yield new data; specimens also were analyzed with earlier methods (Wehmann et al., 1981).

Paid volunteers in the field study collected approximately 1 oz of urine each morning and stored the containers frozen until the containers were picked up for assay (Wilcox et al., 1985). Three studies testing the utility of these assays have been completed. In the first study, 30 women collected daily urine specimens from the time they stopped contraception until they became pregnant or for 6 months if no pregnancy occurred (Wilcox et al., 1985). The study showed the feasibility of the epidemiologic design and permitted investigators to detect several cycles with early fetal loss that would not have been detected with other available methods. The second (control) group comprised women who had undergone tubal ligation; no patterns suggestive of pregnancy were seen. The study was performed blindly.

In the third study, women using intrauterine devices for contraception were evaluated to determine whether this contraceptive technique prevented implantation of the fertilized ovum. Urinary hCG was detected in only 1 of 107 menstrual cycles in this study. Wilcox et al. (1987a) concluded that implantation is infrequent among IUD users; the single pregnancy loss detected might have been a tubal implantation.

Figure 15-4 illustrates the types of findings that can be obtained in field studies. Three consecutive menstrual cycles are illustrated in which daily urine specimens were collected from one person during the second half of each cycle. During the first cycle, no conception occurred, and all hCG values were less than 0.01 ng/ml. During the second cycle, 3 weeks after the onset of the prior menstrual period, urinary hCG rose for 6 days to 0.38 ng/ml. hCG steadily declined over the next 6 days and became undetectable. With the decline of hCG came the onset of menses—27 days after the onset of the previous menses. The woman was unaware of her pregnancy or the episode of early fetal loss. However, the finding of 11 consecutive increases in urinary hCG above the normal background for that woman and the pattern of rise and fall leave little doubt

that a loss occurred. Three weeks after the apparent fetal loss, the woman again exhibited increasing urinary hCG. This time, she remained pregnant and delivered a normal, full-term infant (Wilcox et al., 1985).

FUTURE ASSAY DEVELOPMENTS

Tests for urinary hCG have evolved from relatively insensitive bioassays to sensitive and specific immunoassays. The ectopic secretion of hCG by some tumors and its slight increase in postmenopausal urine specimens (Armstrong et al., 1984; Kuida et al., 1988) are unlikely to diminish its epidemiologic value as the biologic marker of pregnancy in healthy women of reproductive age because the characteristic pregnancy-related changes in hCG concentration are not observed from tumor secretions. Other proteins that appear to be pregnancy-specific have shown less utility as markers in epidemiologic studies; furthermore, they are usually assayed in blood specimens, and their fate in urine is less well known. (An extended discussion of markers of early pregnancy is given in the Chapter 20.)

Field trials involving regular hCG testing in women attempting to conceive have proved adaptable to large-scale epidemio-

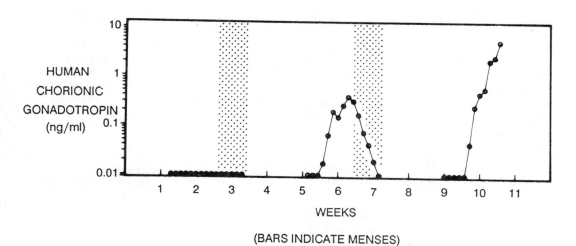

(BARS INDICATE MENSES)

FIGURE 15-4 hCG in urine during three consecutive menstrual cycles in human female unaware of pregnancy or early fetal loss. Source: Wilcox et al., 1987b.

logic studies. Within a few days of the
expected date of implantation of a fertilized ovum, hCG can be detected, as it can
at least 2 days before the onset of the next
expected menstrual period (Wilcox et al.,
1985). Early fetal loss has been detected
by urinary hCG testing; this requires an
hCG assay that is sensitive to a concentration lower than 0.05 ng/ml.

Much more needs to be done to advance
methods in this field of reproductive biology, if large-scale epidemiologic studies
are to be undertaken to search for environmental factors that alter the rate of early
fetal loss. Any change in the rate of clinically apparent abortions potentially
caused by environmental factors can be
detected readily by routine hCG testing
and by the clinical events that surround
abortion.

Improved Methods for hCG Detection

To expand the present labor-intensive
and time-consuming research methods for
immunoradiometric measurement of hCG,
several advances would be desirable.

A simpler assay method is needed that
preserves sensitivity and specificity
for hCG while reducing the needs for labor,
large urine specimens, and large quantities of antibodies. These advances
should be made in the direction of a non-
radioactive-assay format. Ideally,
many specimens could be processed accurately, rapidly, and inexpensively in the
laboratories shortly after urine collection. That would minimize the need for
extensive deep-freeze storage space,
because more than 90% of the specimens have
negative results and would be discarded.
Only urine specimens with positive screening tests that suggest an episode of fetal
loss would be shipped to a central assay
laboratory.

New Monoclonal Antibodies for hCG Detection

In the quest for an improved assay for
the urinary products of hCG, new antibodies with high affinities for the various
forms of urinary hCG will be important.
These should be monoclonal antibodies,

to ensure a continuous source of completely characterized immunochemical reagents. It is important to be able to characterize and detect not only the entire
hormone, but also its degraded forms in
urine; to maximize sensitivity; and to
ensure that unusual biologic events, such
as a selective decrease in the secretion
of one of the subunits, do not lead to incomplete or inaccurate results. New and
specific monoclonal antibody cell lines
might be required to accomplish this task.

Nonradioactive Methods for hCG Detection

Many of the most sensitive research immunoassays use radioactive iodine, but
this approach has disadvantages, including:

• Radiolabeled reagents must be freshly
prepared and characterized every few weeks
and then shipped to the users.
• An ever-increasing problem with radioactive waste disposal is of concern.
• Use of radioactive methods requires
special training and surveillance in the
laboratory.

Epidemiologic research related to
early fetal loss would be advanced substantially if a nonradioactive method to measure hCG were devised in which the reagents
had a shelf-life of 6 months to 1 year. The
most widely used nonradioactive method
is enzyme-linked immunosorbent assay
(ELISA), which requires only basic instru-
mentation (spectrophotometry) for quantification. The sensitivity of the assay
depends on the enzyme amount and activity
that can be specifically linked to the
antigen or antibody measured.

Several recent advances have been introduced to amplify the signal and increase
sensitivity, including enzyme-antibody
coupling through the avidin-biotin system
(Fuccillo, 1985), lectin-carbohydrate
coupling, and use of chimera antibodies
of double specificity (Guesdon et al.,
1983). Those techniques involve noncova-
lent linkage and therefore avoid the loss
of enzyme catalytic activity due to chemical modification.

Assays based on fluorescence or phosphorescence depend on light absorption to provide the excitation energy to produce the emission. Assays based on chemiluminescence depend on the energy from an oxidative chemical reaction to produce molecules in an electronically excited state; return to the ground state is accompanied by photon emission, which is detected photometrically. Low quantum yields and high background interference limit the sensitivity of chemiluminescent, fluorescent, and phosphorescent techniques. To avoid those problems, time-resolved fluorometric immunoassay has been developed. Lanthanide chelates have a high quantum yield and a large Stokes shift (340-nm excitation wavelength and 614-nm emission wavelength). In addition to a long decay time for europium chelates (10^3-10^6 ns), these properties permit an assay wherein a pulsed light of short duration (compared with the decay time of the lanthanide chelates) is transmitted to the sample, and the interfering rapid decay fluorescence of other serum or urinary constituents is discriminated against by activating the detection system after a delay sufficient to complete the decay of naturally occurring fluorophores.

Those techniques and others have the advantage of a nonradioactive detection system and increased reagent stability. Further investigation is required to determine which approaches would provide the most satisfactory sensitivity and specificity, which until the present have been hallmarks of radioisotope-based technology.

Requirement to Measure Other Biologic Markers

Urinary hCG is a valuable biologic marker of implantation of a fertilized ovum, but it does not detect a fertilized ovum that does not implant. New methods to test for that should be developed. One approach would be a technique to collect uterine secretions and assay for the presence of hCG secreted by the nonimplanted ovum.

Another approach involves developing adequate methods for the assay of other pregnancy markers that do not depend on implantation. Although a wide variety of pregnancy-associated proteins have been described (Horne and Nisbet, 1979), most appear to be associated with the later stages of pregnancy development. One exception is the pregnancy-specific protein early pregnancy factor (EPF) (Morton et al., 1977). EPF properties have been studied in humans (Tinneberg et al., 1985), sheep (Morton et al., 1979) and mice (Morton et al., 1976), mainly by the rosette-inhibition assay. Its appearance can be detected in maternal serum within 6-48 hours after fertilization and does not depend on implantation of the fertilized ovum for detection (Sinosich et al., 1985). EPF persists in serum throughout the first two trimesters of pregnancy (Cavanagh et al., 1982) and rapidly disappears after embryo death or surgical removal (Nancarrow et al., 1979).

In a study of 13 multiparous women (Rolfe, 1982), EPF was detected within 48 hours of fertilization (ovulation dated by progesterone determination). Of cycles studied in which intercourse occurred at the time of ovulation, EPF was detected in 18 of 28 cycles, but continued to be produced beyond 2 weeks in 4 instances; of these 4 subjects, only 2 proceeded to full-term pregnancy. In the remaining 14 of the 18 subjects, EPF became undetectable before the onset of the next menstrual cycle. The data suggest that many fertilized ova are lost before implantation. An assay less cumbersome and more sensitive than the rosette-inhibition test should be investigated. EPF should be thoroughly elucidated, and the possible presence and assay of EPF or its metabolites in urine investigated.

In epidemiologic studies of reproductive function, improved markers of ovulation need to be developed in easily collected biologic specimens—saliva or urine. These improvements are needed to develop better methods to detect the various forms of the pituitary gonadotropins and also to detect various estrogen and progesterone derivatives throughout the menstrual cycle, to estimate the time of ovulation, and to document the formation of a corpus luteum.

The requirement for new high-affinity antisera directed against hLH and FSH is based on the heterogeneity of the glycoprotein hormones from the pituitary (Franchimont et al., 1972). Isoelectric focusing data show that hLH from regularly cycling women is less acidic than that circulating in men and postmenopausal women, presumably owing to a lower sialic acid content (Wide, 1981). Differences in sialic acid content are manifested not only in physicochemical properties, but also in immunoreactivity, receptor affinity, and biologic activity, even to the extent of having a variant of hLH that is immunologically reactive but devoid of biologic effect (Axelrod et al., 1979). Multiple forms of FSH are also observed in other mammals, with as many as six immunoreactive forms present in the hamster (Ullsa-Aguirre and Chappel, 1982). The different forms have the same molecular weight, but exhibit different affinities in lectin binding and have different bioactivity-to-immunoreactivity ratios when tested by radioreceptor assay.

Those findings stress the importance of developing assays that measure the pertinent forms of excreted pituitary gonadotropins. Assays that use antisera raised against antigens derived from postmenopausal or male sources might not be optimal for assays on normally cycling women and might have contributed to the primary lack of sensitivity in ovulation detection. Recent work documented that urinary FSH patterns closely resemble those found in serum FSH patterns when measured by granulosa cell aromatase bioassay (Dahl et al., 1987). Future work in gonadotropin-assay development should include assays for the intact hormone and its subunits, as well as determination of the correlation between the immunoreactivity and bioactivity of the substances being measured (e.g., gonadotropins isolated from normally cycling women), to monitor the validity of the assay.

Two objectives are met by measuring circulating steroids or their metabolites. The first is to signal the onset of ovulation and thereby validate the accuracy of gonadotropin assays as an ovulation marker. The second is to assess the existence and adequacy of the corpus luteum. A variety of urinary estrogen metabolites have been assayed and their utility as markers of ovulation evaluated. Baker and colleagues (1979) assayed directly E_1-3-glucuronide, E_2-3-glucuronide, E_2-17β-glucuronide, E_3-3-glucuronide, and E_3-16α-glucuronide in urine and found that the E_2-17β-glucuronide assay was the most sensitive predictor of ovulation, followed closely by E_1-3-glucuronide. The authors recommended the latter, because 5 times more of it is excreted than of the former and it can be detected in a 100-fold diluted urine specimen, in which potentially interfering substances would be diluted virtually to nonexistence.

The Baker et al. study (1979) also indicated that random urine collections can be valid ovulation markers. Dividing the mass of steroid metabolite by the exact duration of collection gives the production rate in nanomoles per hour, and comparison of this pattern for first-morning voids and 24-hour collections showed excellent correlation. Results expressed as the steroid:creatinine ratio also correlated highly with the 24-hour collection results. Thus, either method may be used in lieu of 24-hour collections. The E_1-3-glucuronide:pregnanediol 3α-glucuronide ratio is independent of urine volume and proved to be another valid ovulation marker.

In a study of the urinary E_1-3-glucuronide, hLH, and pregnanediol 3α-glucuronide as markers of ovarian function, Collins et al. (1979) found a good correlation between early-morning collection and 24-hour collections and found that the ratio of E_1-3-glucuronide to pregnanediol 3α-glucuronide could be used to demarcate the duration of the fertile period.

A multicenter study (WHO, 1980b) evaluated the excretion pattern of E_1-3-glucuronide, E_2-17β-glucuronide, E_2-3-glucuronide, E_3-16α-glucuronide, E_3-3-glucuronide, pregnanediol 3α-glucuronide, and pregnanetriol 3α-glucuronide throughout the menstrual cycle, to determine which best indicated the fertile period during the cycle; E_1-3-glucuronide in the follicular phase provided a marker of ovulation 72 hours before ovulation, and assay of pregnanediol 3α-glucuronide was the best indicator of whether ovulation had occurred.

The WHO study (1980b) also confirmed that first-morning voids yielded results as reliable as those obtained from 24-hour collections.

OTHER CLINICAL OPPORTUNITIES

If research is limited to young and otherwise normal women, unusual patterns might be missed. However, women with abnormal physiology enrolled in an epidemiologic study can lead to unexpected and inexplicable findings. Thus a wide spectrum of clinical studies is important for epidemiologic research.

Fecundity decreases with age in women. The rate of early fetal loss might differ between younger women and older women, and enrollment of older women in a study might affect the outcome; but no adequate data exist on the rate of early fetal loss in women older than 35 or 40 years.

Effects of different types of prior contraceptive use such as steroidal agents, IUDs, and spermicides should be studied. Furthermore, not all women have regular menstrual periods, and those with short and long luteal phases should be studied to determine, for example, how this affects the rate of fetal loss and timing of specimen collection to detect implantation and loss. The study of luteal phase characteristics has become feasible with the development of direct assays for urinary estrogen and progestin metabolites using nonradioisotopic assay systems.

Postmenopausal women have a strong stimulus for pituitary gonadotropin secretion; to some extent, this appears to stimulate a minute degree of hCG synthesis and secretion as well (Robertson et al., 1978; Armstrong et al., 1984). Important questions to study are whether this might be a problem when studying perimenopausal women who conceive, whether this is a pattern of hCG, and whether the degree of ovarian failure that leads to a slight amount of hCG secretion occurs only when a woman becomes infertile.

The National Center for Health Statistics reports that 14% of couples in this country have a problem with infertility. Some men or women might have a genetic predisposition to a high rate of early fetal loss or failure of the fertilized ovum to implant. Artificial insemination programs provide a fine opportunity to study this problem. It is also important to determine whether frozen and fresh semen specimens lead to different rates of conception and fetal loss. Although several studies have indicated that the use of frozen semen leads to diminished fecundability, at least one study reports that there is no difference in pregnancy rate whether fresh or frozen semen is employed (Trounson et al., 1980).

Epidemiologic research provides an opportunity to identify persons with unusual reproductive patterns that might occur with low frequency in the population. Protocols should be designed that test the major epidemiologic hypotheses and also are sensitive to the occurrence of such unusual patterns.

The research described above will increase the knowledge regarding the use of hCG as a marker of pregnancy and early fetal loss; that increased knowledge will benefit not only epidemiologic studies, but also other research to detect abnormalities of human fertility.

16

Conclusions and Recommendations

Before a biologic marker is applied in a toxicologic context, its frequency and distribution in a normal, healthy population must be established. A logical next step is to validate the marker and judge its utility by using it in a high-exposure, high-risk cohort to verify that it measures what it is meant to measure.

Markers of biologic processes rely on subjective or objective assessments. Regardless of the care taken to construct measurement instruments, patient-dependent measures are likely to be highly variable; one reason is the presence of bias associated with failure of recall, lack of understanding of the question, or lack of recognition of a symptom. Objective measures are often superior, provided that they are reliable and valid.

Application of biologic markers in reproductive toxicology depends on interdisciplinary work by laboratory scientists, clinicians, and epidemiologists. Collaborations of this kind impose heavy obligations on investigators to keep abreast of and understand all the disciplines involved.

Because of the rapid pace of change in the development of new tools and bioassays, agency implementation of field studies with biologic markers of human reproduction should be guided by an oversight panel of experts conversant with the reliability and validity of such markers, measurement issues, and design and conduct of field studies. A task force might be convened periodically to reassess the status of biologic markers in use and proposed for future studies.

Biologic markers must be fully developed and studied before they are applied in epidemiologic studies as the principal indicators of events. Initially, it might be advisable to use a new biologic marker in conjunction with other assessments. Without adequate laboratory testing and cross-validation with markers having known properties, a potential marker is likely to be misused or misinterpreted.

Epidemiologic studies need to be designed carefully because of the large effort involved in collecting data, particularly if a study entails biologic assays. Nested designs might be appropriate to test questions involving intensive laboratory assessments or involving only special exposure conditions (NRC, 1985). During design of a study, careful consideration should be given to the next steps to be taken; this will help to ensure that questions can be addressed retrospectively as new information develops.

Longitudinal studies will be particularly important for the identification of toxic effects. Ways must be found to assess damage during fetal oogenesis, and identification and tracking of cohorts with relevant in utero exposures will be important. Daughters exposed to DES, for instance, soon will enter the perimenopausal period, at which time it will be possible to see whether age at menopause is indicative of oocyte damage sustained in utero. For postnatal exposure, continued monitoring of ovarian failure and recovery among women treated for cancer will be useful.

SPECIAL RESEARCH OPPORTUNITIES

Many important opportunities for collecting data relevant to the evaluation of toxicologic effects on female reproduction are at hand.

Environmental health research could be incorporated into current assessments of clinical populations and studies directed at answering biologic questions. Valuable information regarding the potential utility of biologic markers could be developed by integrating assessment of environmental exposures and tests for various biologic markers into existing work with populations undergoing in vitro fertilization, artificial insemination, prenatal diagnosis, and normal obstetric care. Such an approach would be cost effective for research funding, and could be useful in assessing effects of common exposures. Hence, the committee recommends that clinics and epidemiologic studies of women of reproductive age be encouraged to collect information relevant to patient exposure history.

Treatment of infertility problems could provide extensive information on reproductive physiology. For instance, IVF centers have contributed knowledge about factors influencing oocyte quality and the milieu for continued development of the fertilized egg. Such data are valuable for assessing and predicting the effects of various toxic exposures. Materials from IVF centers would lend themselves to systematic research and study, although ethical concerns regarding their use in research must be considered.

Fetal and placental material from spontaneous and induced abortions might provide valuable information on the pharmacokinetic properties and potential effects of common toxic exposures. Such material is not widely used to assess environmental health risks to the fetus.

Reproductive function is related to other biologic processes; for instance, ovarian hormones appear to influence the risk of cardiovascular disease and the rate of bone loss (Bush and Barrett-Connor, 1985). Thus, it might be expeditious to use current studies of osteoporosis and heart disease in women as vehicles for addressing questions about the role of toxic insult in ovarian senescence.

SPECIFIC RESEARCH RECOMMENDATIONS

The preceding chapters reviewed many biologic markers of exposure and effect. Table 16-1 summarizes the status of each marker.

Germ Cell Damage

Induced aneuploidy in female germ cells is an important subject for basic science, especially because aneuploidy in humans arises largely in maternal germ cells. Fundamental understanding of the mechanisms of chromosomal malsegregation, as well as assays to detect agents that induce aneuploidy, must be improved.

Research should be encouraged to clarify the age-related increase in oocyte aneuploidy. Effects on oocytes of substances in the follicular fluid are not well understood; what substances get through the zona pellucida and whether barriers weaken with age are unknown. Valid noninvasive methods to evaluate oocyte stock directly are needed. These important subjects involve fundamental research as well as toxicologic studies.

Sexual Differentiation

"Critical periods"—a unique feature of sex differentiation—have important implications for the subject of markers in reproduction. At the beginning of embryonic life, the gonads and other organs destined to become the genital tract and

TABLE 16-1 Status of Current and Potential Markers in Female Reproductive Toxicology

Marker	Animal Studies Needed[a]	Can Be Used in Limited Human Subsets[b]	Can Be Used in Large-Scale Human Studies[c]	Needs Further Development
Exposure markers—Chemical analysis for toxicants or metabolites, or mutagenic analysis of body fluids				
Blood, urine, saliva	+		+	
Tissues				
Intact	+			
Cytologic specimens				+
Fluids	+			
Cerebrospinal fluid	+			
Follicular fluid, amniotic fluid	+			
Placental tissue	+		+	
Peritoneal fluid	+			
Genotoxic markers—DNA adducts (chemical specific, generic)				
Oocytes, ovarian tissue			+	
Placental tissue			+	
Fetal tissues			+	
Maternal serum			+	
Fetal serum			+	
Unscheduled DNA synthesis				
Maternal lymphocytes			+	
Fetal lymphocytes	+			
SCE (sister-chromatid exchange)				
Maternal lymphocytes			+	
Fetal cells	+			
Chromosomal aberrations				
Maternal serum			+	
Abortus tissue	+			
Chorionic villi	+			
Amniotic cells	+			
Fetal serum			+	
Micronuclei				
Maternal blood	+			
Vaginal/cervical cells				+
Fetal liver cells				+
Fetal lymphocytes				+
Specific-locus mutations				+
Development/aging				
Onset of puberty				
Clinical observation, breast bud development			+	
Blood				
Melatonin			+	
DHEA-S			+	
Gonadotropin (pulsatile)	+			
Age of first menstrual bleeding	+			
Hormones: estrogens, inhibin, LH, FSH, androgens	+			+
Age of breast development			+	
Sexual behavior				+
Neurotransmitters in CSF	+			
Menstrual cycle length			+	

Marker	Animal Studies Needed[a]	Can Be Used in Limited Human Subsets[b]	Can Be Used in Large-Scale Human Studies[c]	Needs Further Development
Ovarian-oocyte stock				
Ultrasound for ovarian size				+
IVF		+		
Biopsy		+		
MRI				+
Periodic ultrasound to monitor follicular development		+		
Inhibin			+	
Premenopausal hormonal status (estrogens, gonadotropins, inhibin, LH, FSH)				+
CNS reproductive senescence	+			
Menstrual Function				
Cycle frequency and characteristics		+		
Detection of corpus luteum follicular development (ultrasound)		+		
Basal body temperature				
Thermometer			+	
Improved, self-recording, electronic thermometer			+	
Cervical mucus			+	
Sexual behavior				+
Vaginal cytology		+		
Biophysical measurements of vaginal secretions			+	
Endometrial histology		+		
Endocrinology: gonadotropins, steroids, ovulatory hormones		+		
In vitro assays				
LH-FSH				
Pituitary cells (from cadavers)	+			
Granulosa cells		+		
Luteal-specific proteins, endometrial cell cultures				+
Mucus production, endocervical cells				+
Fertilization, Implantation, and Loss				
hCG	+		+	
EPF				+
PEP				+

[a] + = not ready for application in humans.
[b] + = too invasive or too demanding of subjects for use on broad scale.
[c] + = sufficiently validated and safe for application in field studies, although might warrant further refinement and additional work with animals.

the external genitalia are bipotential. Primordial germ cells induce development of a male or female gonad, and the gonads secrete diffusible organizing substances, or hormones, that complete the process of morphogenesis of sexually relevant tissue. Bipotential tissues are inducible during only a short period; if induction or differentiation is prevented during that critical period, later exposure cannot alter morphology or function. For example, at a critical early time in the human embryo, a functional testis secretes a protein that suppresses the anlage of the Müllerian duct. The protein—called Müllerian duct-inhibiting hormone, Müllerian-inhibiting substance, and anti-Müllerian hormone—and its early secretion (a critical event) ensures that males do not retain female sex accessory organs. In the neonatal male rat, testosterone secretion on the first postnatal day permanently alters the size of hypothalamic nuclei; testosterone does not have this effect after postnatal day 5.

An environmental event might alter reproductive organs at one time—during a critical period—but not at another time, and an abnormality manifested in adults could be the result of an environmental insult that was effective only during a critical period of reproductive ontogenesis. The specific relationships between particular developmental periods and particular toxicants are not well understood. Moreover, some functional effects might not be manifested until puberty or menopause. Therefore, in utero exposures need to be documented, and exposed populations should be followed up through many periods of life.

Puberty

The onset of puberty is affected by nutritional status; hence, other environmental factors might affect puberty onset, although systematic studies are lacking. The hypothalamus presumably is insensitive to some environmental insults early in childhood. However, when hypothalamic secretions begin during puberty, exposure to some toxicants might have an effect. Studies of normal populations and of populations potentially exposed during childhood or puberty need to be conducted.

Ovulation

Probably only a small percentage of the healthy population is trying to become pregnant at any point. Consequently, the study of nonconceptive menstrual cycles is important as a way to monitor or to study exposed populations. Changes in menstrual cycles might have long-term consequences for other medical conditions, such as cancer and heart disease.

Among the components of female fertility, ovulatory function appears to be the most sensitive to environmental effects; therefore, it is of interest to investigate the influence of various agents on menstrual cycling. This should be studied in animal models, as well as human populations.

Short-term changes in the menstrual cycle induced by stress or diet differ from long-term toxicity. Markers are needed that can distinguish between these events. A battery of assessments should be developed that could be used to evaluate ovulatory function in current epidemiologic studies and in special studies of exposed human populations. Because anovulatory bleeding can be confused with regular ovulatory cycles, a more accurate detection of ovulation would be useful. These assessments should be sensitive to alterations induced by low concentrations of toxicants and should be designed so that they can be readily applied in field studies. Field studies to monitor cyclic ovarian function with urinary assays of steroid and gonadotropic hormones have been conducted successfully, although not in the context of toxicology. These studies suggest that corpus luteum formation can be documented in a practical and valid way.

Evaluating cellular physiology in vitro allows pathophysiology of naturally occurring diseases that cause anovulatory infertility to be investigated. Whether this in vitro approach will prove profitable in evaluating biologic effects of environmental chemicals remains to be seen. Future efforts in this field will expand the understanding of biologic markers.

Clinically Inapparent Loss

Early spontaneous abortions might be related to ovulatory dysfunction or to genetic damage, and studies of clinically inapparent losses should yield important toxicologic information. Such studies must be carefully considered. The assay used must be sufficiently sensitive to detect early pregnancy and loss and must be applied appropriately with regard to timing of specimen collection and choice of fluid to sample. With appropriately sensitive assays, consistent estimates of the frequency of implantation and early loss have been obtained; such estimates are important for use in studies evaluating environmental exposures.

Ovarian Senescence

An accurate way to assess the oocyte pool, either with imaging techniques or with measures of regulatory factors or gonadotropins, is needed. Presumably, rates of loss vary. Groups with premature senescence should be assessed, to identify differences from normal populations.

NATIONAL DATA BASE

A national data base should be developed to obtain information on human reproductive biologic markers in normal and exposed populations. Several major clinical centers could supply information to help to develop national reproductive profiles. The National Center for Health Statistics might conduct a survey of a probability sample of women that involved periodic assessments (including biologic assessments) of reproductive measures. The Agency for Toxic Substances and Disease Registry might appropriately develop comparable data on exposed populations and coordinate a data base to link information on the two groups. Infertility and irregular menstruation are neither life-threatening nor rare, and are likely to be undetected in the general population unless they are monitored.

EXPERIMENTAL STUDIES

The relevance of data obtained in nonhuman primates or nonprimate mammals often is questioned—principally because the 28-day human menstrual cycle does not appear to resemble the 4-day rodent estrous cycle, the spontaneous ovulation of the primate is quite different from coitus-induced ovulation in the rabbit, and the human does not show the clear seasonality of cycles manifested by Rhesus monkeys, sheep, and rodents (such as hamsters). Nevertheless, the basic processes of female reproduction are probably similar in all mammals. Species differences are apparent in some phenomena, such as luteal function; for example, the unmated rat does not have a functional corpus luteum of the cycle, whereas the primate does. The rat and mouse are probably good models for oogenesis in primates, but not for masculinization of the hypothalamus, in that testosterone administered perinatally in female rats permanently prevents estradiol from inducing an LH surge, but does not do so in primates. The Rhesus monkey is an excellent model for human follicular and luteal phases, but is expensive. More work on luteal events probably should be carried out in the guinea pig—the only small, readily available laboratory mammal with a spontaneous luteal phase; the potential for xenobiotics to alter luteal function or cause luteolysis might be better assessed in this species. Careful consideration needs to be given to selecting appropriate laboratory models for different components of female reproduction.

III

Biologic Markers of
Toxicity During Pregnancy

17

Introduction

This section of the report assesses markers of maternal, embryonic, fetal, and placental physiologic and biochemical processes and ways to develop biologic markers for clinical utility in humans. Validation of such markers in humans is a critical concern.

A biologic marker is useful to confirm a potential risk, as well as to document that an adverse event has occurred. However, the use of biologic markers for risk assessment during pregnancy is only beginning. The goal is to develop markers that can establish that a mother or conceptus might be at risk for a toxic response before expression of that response, to permit intervention to prevent the toxic response.

Successful pregnancy in mammals involves the progression through processes of fertilization, implantation, organogenesis, fetal development, and parturition. Three principal compartments must be coordinated during this progression; the maternal host, the placenta, and the embryo or fetus. Interactions that mark the progression of a conceptus through gestation are attended by specific biochemical and physiologic responses. These interactions occur at various levels of biologic organization, and data on them are not uniform. The differences are most obvious when data on events that take place before and around implantation are compared with those on events that take place during the fetal period. Few markers of cell-, tissue-, or stage-specific reactions, which make it possible to detect critical periods, are available in early development. Such markers are products of laboratory investigation; as pregnancy proceeds, different methods for analyses can be used, including clinical diagnostic procedures. Data on developmental events during the fetal period (i.e., after 8 weeks of gestation) are more numerous.

Research into the early events of pregnancy has identified many markers. But because the understanding of these early events is provisional and fragmentary, essentially no information regarding toxic exposure has been gathered. Interpretation of markers often is plagued by the absence of critical assays or by inappropriate application of or failure to apply available diagnostic methods. For example, amniocentesis and chorionic villus biopsy have not been used to document potential adverse effects of xenobiotic exposure. Nevertheless, these tests might yield valuable information regarding doses of toxicants in the tissues or genotoxic effects of exposure.

A few assessment tools are peculiar

to pregnancy, such as amniocentesis, chorionic villus biopsy, Doppler blood flow velocity measures, fetal blood sampling, fetoscopy, products of conception, real-time ultrasound, and incidence of pregnancy loss. However, many of the properties of the biologic markers derived with those tools are not well established, even for normal pregnancies. This part of the report discusses the possible use of those biologic markers to study exposures to therapeutic or environmental agents, but those markers have not been used to study whether exposures to specific xenobiotic agents are associated directly with pharmacodynamic events.

This chapter is a brief review of the biologic processes and changes that occur to the mother, conceptus, and placenta during pregnancy. This is followed by chapters that focus on disciplines in biology that are developing markers of pregnancy. Advances in molecular biology, immunology, cell biology, physiology, and pharmacology are discussed. The subcommittee's conclusions and recommendations regarding opportunities and directions for a program on biologic markers of pregnancy are presented in Chapter 24. A summary table lists and categorizes biologic markers according to whether each marker may be used in large-scale human studies or only in studies of special populations and whether it needs further development or needs to be applied in animal studies.

THE EVENTS OF PREGNANCY

Normal ovulation in the human occurs approximately 14 days before the onset of the next menses. The period during which the ovum then can be fertilized is estimated to be 18-24 hours. Fertilization of the ovum normally takes place in a fallopian tube. Entry of a spermatozoon into the ovum prevents the entry of additional spermatozoa and is followed by fusion of the spermatozoal nucleus with the nucleus of the fertilized ovum, which results in a zygote. The zygote continues its transport through the fallopian tube, undergoing a series of cell divisions. Approximately 6 or 7 days after ovulation, the embryo attaches to the apical surfaces of the endometrial epithelial cells.

By the time the developing embryo enters the uterus, a cavity is formed—the blastocoele. The outer cellular layer of the blastocyst surrounds the blastocoele and the embryo (also referred to as the inner cell mass). The placental contribution from the embryo—the trophoblast—attaches to the uterine epithelium; that attachment initiates migration of the trophoblast through the epithelium and its basal lamina. Trophoblast cells in each mammalian species differentiate according to a species-specific series of morphologic and functional changes before interacting with the maternal vasculature and establishing a definitive placenta.

During the course of these changes, the embryo initiates rapid cell division, growth, and differentiation that culminates in gastrulation. The gastrula includes two primary germ cell layers—the ectoderm and endoderm. Further development yields the third primary germ layer—the mesoderm—and is followed by regional differentiation of the embryonic disk. Each step of placental and embryonic differentiation is a possible point of adverse action of a xenobiotic agent. The concept of critical windows of exposure must be considered for specific alterations for organ development in the embryo.

MATERNAL PHYSIOLOGY

The embryo-placental unit (and later the fetal-placental unit) must alter maternal responses without jeopardizing the mother. The prodigious production of polypeptide and steroid hormones by the embryo-placental unit results in physiologic adaptations of virtually every maternal organ system.

Maternal weight increases an average of 25 lb during pregnancy. The pulse rate increases by about 20%, and blood volume per heartbeat (stroke output) also increases. The net result is an increase in cardiac output of some 30% by the end of pregnancy. The respiration rate is unchanged, but the tidal volume increases by 30-40%. Those changes might mean that an internal dose of certain airborne toxicants would be greater in a pregnant woman than in a nonpregnant woman.

Gastric emptying time increases by as much as 50%. Renal flow also increases by as much as 50% during the first trimester. The glomerular filtration rate increases early and then levels off. Changes in filtration and elimination rates might mean that toxic materials remain in the circulation of a pregnant woman longer than in that of a nonpregnant woman.

Hormonal concentrations are changed as a result of alterations in pituitary, adrenocortical, thyroid and parathyroid gland, and pancreatic function. Concentrations are affected further by altered clearance rates that result from increased glomerular filtration and decreased anion excretion in the mother and modified clearance of steroids and protein hormones by the placenta.

The many changes in the organ systems of the pregnant woman can influence the exposure concentration, metabolism, and elimination of a xenobiotic agent. These processes together affect the pharmacokinetic properties of a substance, so pregnant women might have different responses or magnitudes of response to exposure from similarly exposed nonpregnant women.

EMBRYONIC/FETAL CHANGES

In utero development is a time-sensitive process during which all mechanisms of possible interactions between and among cells leading to cellular proliferation or degradation result in modified structural and functional changes. Individual processes indicate not only cellular sensitivity but critical windows of development in which specific agents may induce damage, for example:

• Thalidomide appears to produce its major effects on development in humans—limb abnormalities—when exposure occurs between 25-40 days of development. Such exposure correlates with the development of the upper and lower limb buds in the human (Newman, 1985).

• Diethylstilbesterol (DES) is a developmental toxin that can alter reproductive tract development when administered between 8-18 weeks of gestation in the human. This is the period for development of the reproductive tract in the human. Not only are vaginal tumors noted from DES exposure, but also ovarian, uterine, and vaginal/cervical malformations (Herbst and Bern, 1981).

• Retinoids (isotretinoin) produce major alterations in cranio-facial, thymus, cardiac, and otic development, apparently due to early embryonic effects on neural crest cell migration and function (Lammer et al., 1985; Teratology Society, 1987).

• Methylmercury also is noted to be a human teratogen, yet its principal effect is on the CNS, resulting in substantive alterations in function. Such effects of methylmercury are correlated with exposures during the fetal and neonatal period, when rapid proliferation of neurons occur (Weiss and Doherty, 1975; Harada, 1978).

All of these human terata resulting from xenobiotic exposures also have been identified in animal models, and have been investigated for mechanisms of action. Thus, not only must important surveillience techniques be applied to human study models, but effects and timing of exposures must be based upon a fundamental understanding of basic embryology in the human and animal systems investigated. For example, in the rodent, development of the vagina and cervix during the fetal/neonatal period demonstrates structural malformations and tumorigenesis but is different from exposures in humans.

Several considerations of the differences between fetal and adult physiology are relevant to a consideration of biologic markers, because the resulting internal dose or biologically effective doses might differ as a result of these properties. For instance, from a cardiopulmonary standpoint, the fetus maintains normal tissue oxygenation in the face of what in the adult would be considered pathologically low arterial O_2 tension, by maintaining a per-kilogram cardiac output of more than twice that of an adult. Therefore, the dissemination of blood constituents is much more rapid in the fetus than in the adult. Some specific differences are detailed in Table 17-1.

Much of the morphologic information

TABLE 17-1 Comparisons of Fetal and Adult Cardiovascular Functions

	Fetus	Adult
paO_2, torr	25	100
$paCO_2$, torr	48	40
pH	7.35	7.40
V•O_2, ml/min per kilogram	8	4
Hemoglobin concentration, g/dl	17.5	11.5
Hematocrit, %	55	35
O_2 content, ml/dl	16	15.4
O_2 content, mM	7	6.7
Blood volume, ml/kg	130	80
Descending aorta pressure, mm Hg	45	95
Pulmonary artery pressure, mm Hg	45	15
Cardiac output, ml/min per kilogram	200a	100
Systemic vascular resistance	Low	High
Vascular compliance	High	Low

[a]Calculated for right and left ventricles.

about the human fetus is based on the study of abortuses or inferences from studies of other organisms. Information about fetal organ function is even less direct, in that embryos aborted spontaneously usually have abnormal functioning of some organs, and fully developed organs are not comparable with fetal organs.

Target receptors, modulators, and regulators of steroids and protein hormones develop at different times during gestation. Furthermore, the developing fetus has a unique endocrine system, owing to the interdependence of the maternal-placental-fetal complex. The extraembryonic membranes contain key enzymes to metabolize steroids and prostaglandins that are absent or present in very low concentrations in the fetus; the fetal adrenal glands and liver contain key enzymes absent from the placenta. Dehydroepiandrosterone sulfate (DHEA-S) serves as a substrate for biosynthesis of placental estrone and estradiol—hormones that probably are important in mediating many maternal adaptations to pregnancy (Longo, 1983). Maternal urinary and serum estriol measurements have been important measures of fetal compromise; however, many xenobiotics, antibiotics, and glucocorticords are noted to alter estriol excretion. These interactions between normal physiology and drug therapy demonstrate how xenobiotics can

alter fetal and maternal function without compromising fetal survival.

The fetal adrenal gland also produces large amounts of cortisol, which is important in the maturation of the lung, pancreas, and other organs and which initiates hormonal events in extraembryonic tissues (including a decrease in progesterone and increases in estrogen production and prostaglandin synthesis). The stimuli for fetal adrenal hormone synthesis are unclear; fetal adrenocorticotropin (ACTH) plays a role, as do peptides derived from pro-opiomelanocortin (POMC) and several other growth factors.

Other endocrine systems peculiar to fetal life are the para-aortic chromaffin system active in catecholamine synthesis; the fetal intermediate pituitary, which secretes α-melanocyte-stimulating hormone and β-endorphin; and the posterior pituitary, which secretes arginine vasotocin, vasopressin, and pregnancy-specific proteins (Rosen, 1986), such as several pituitary-like hormones and neuropeptides.

The concentrations of many of these hormones and other polypeptides can serve as biologic markers. Various chemicals and toxicants can affect the fetal hypothalamus, and adrenals, the placenta, and other tissues, thereby inhibiting enzymes and metabolic pathways and altering the synthesis of and response to various hormones and growth factors. Because of the uniqueness of fetal hormones, the fetus may be susceptible to chemicals that the adult is not susceptible to. The extent to which this occurs remains to be demonstrated.

PLACENTAL INVOLVEMENT

The placenta provides biologic communication between mother and fetus while maintaining immunologic and genetic integrity of the two organisms. Placental tissues are embryonic in origin; however, the placenta functions autonomously during the first trimester. By the end of the first trimester, the fetal endocrine system develops sufficiently to influence placental function and provides hormone precursors to the placenta.

After implantation, the trophoblast invades the maternal endometrium. Two layers of the developing placenta are evident: the syncytiotrophoblast (adjacent to the endometrium) and the cytotrophoblast. The syncytiotrophoblast is derived from the precursor cytotrophoblast of the embryo, is the source of hormone production, and is in direct contact with the maternal blood supply. Hormonal products of the placenta include human chorionic gonadotropin (hCG); human chorionic somatomammotropin (hCS), also called human placental lactogen (hPL); and several other peptides that are not as well defined. The concentration of maternal serum hCG doubles every 2 days during the early weeks of pregnancy and peaks at about the tenth week of gestation. hCG is known to have several activities, but their significance is not entirely understood. For instance, hCG is luteotropic; it stimulates increased progesterone production by the corpus luteum cells. hCG also increases placental conversion of precursors to pregnenolone and progesterone and demonstrates thyroid-stimulating-hormone-like activities. The polypeptide hormone hPL, which contributes to increased glucose metabolism and mobilization of free fatty acids, is not detectable in the maternal blood until 4-5 weeks of gestation.

18

Molecular Biology:
Developing DNA Markers of Genotoxic Effects

This chapter briefly discusses the effect of molecular biology on prenatal diagnosis. The assays are discussed in detail in Chapters 9 and 12.

The ability to obtain DNA from the fetus has been possible for the past 10 years through amniocentesis after 16 weeks of gestation. However, techniques to sample chorion in the first trimester and to karyotype the sample directly without long-term tissue culture recently became possible. The chorionic biopsy consists of aspiration of chorionic villi through the cervical canal or transabdominally with ultrasound guidance. Results of chromosomal analysis are available within days, rather than the weeks required with conventional techniques. Without the need for cultured preparations, direct karyotyping could prove amenable to analysis of chromosomal aberrations possibly associated with early loss (1-6 weeks after implantation). Also, DNA is readily available from the collected tissue or from cultured cells derived from the sampling.

DETECTING HERITABLE GENETIC DAMAGE

One of the most interesting innovations in prenatal diagnosis is the use of DNA probes that reveal genetic markers near

specific genes (McDonough, 1985) (Table 18-1). DNA probes have been applied for prenatal diagnosis of cystic fibrosis, and predictive testing for the gene for Huntington's disease began. In addition, the gene for Duchenne muscular dystrophy has been sequenced and the so-called "recessive oncogene" responsible for familial predisposition to retinoblastoma was discovered. The number of probes available is increasing exponentially.

A genetic marker is a segment of DNA that lies near a unidentified gene that is involved in the disease etiology. With DNA probes and genetic markers, it might be possible to detect most of the more than 3,000 conditions caused by single-gene mutations.

Applications of molecular biology to clinical medicine will change the approach to diagnosis (McDonough, 1985). Molecular diagnosis—even during prenatal life—is possible with two related techniques (see Chapter 12 for details):

• **Restriction-fragment-length polymorphisms.** Highly specific restriction endonucleases cut DNA between particular base sequences. When altered by mutation, DNA is severed into fragments of a size different from normal. The homozygous and heterozygous states can be differen-

TABLE 18-1 Disorders Diagnosable by Analysis
of Cellular DNA

Validated Uses of DNA Analysis for Diagnosis
  Sickle-cell anemia
  β-thalessemia
  α-thalessemia
  Factor VIII deficiency
  Factor IX deficiency
  Phenylketonuria
  $α_1$-antitrypsin deficiency
  Huntington's disease
  Antithrombin III deficiency
  Ornithine transcarbamylase deficiency
  Duchenne muscular dystrophy
  Argininosuccinic acid dehydrogenase deficiency
  Osteogenesis imperfecta type II
  Congenital adrenal hyperplasia

Probable Uses of DNA Analysis for Diagnosis
  Fragile X syndrome
  Adult-onset polycystic kidney disease

Source: McDonough, 1985.

tiated by comparing abnormal fragment size with normal fragment size. Potentially, these can be used even if the gene leading to the disease state is unknown. Abnormalities of the genes for hemoglobin (whose deficiency results in sickle-cell anemia), growth hormone, and 21-hydroxylase (whose deficiency results in congenital adrenal hyperplasia) are conditions that can be diagnosed with this approach. Also, these have been used to diagnose several other conditions, including Huntington's disease, phenylketonuria, factor VIII and factor IX deficiencies, and $β$-thalassemia.

  • **Oligonucleotide probes.** When the precise DNA mutation is known, but the mutation cannot be discriminated with a restriction-enzyme cut, oligonucleotide probes representing the normal and abnormal sequences can be used to identify the genotypes. As increasing numbers of normal and mutated gene sequences become identified, the practicality of this technique will increase.

DNA technology is versatile. Every monogenic disorder potentially is diagnosable with DNA probes, as increasing numbers of probes and polymorphisms are recognized and restriction enzymes are developed.

The explication of the molecular map of the human genome will be accompanied by the development of functional correlations that might provide insights into the basic pathogenesis of most disorders, including those caused by chemical mutagens and physical factors (such as radiation). Within the next decade, many diseases and toxic conditions probably will be defined in molecular terms and become subject to diagnosis from a few microliters of blood (Ward et al., 1983). DNA probes might be used to detect chemical or food contaminants in the body. Tests based on such DNA probes could replace current assays, because of their greater sensitivity and speed.

Quality control may suffer as DNA probes are used more widely, particularly if commercial kits become available. Accuracy is essential, and reliability could be diminished because of such problems as incomplete DNA digestion, faulty hybridization, contamination, and mislabeling. Even in the best of hands, interpretation of these tests and associated family counseling require extensive experience and commitment. In addition, the validity of the tests in the absence of confirmatory assays is a problem. Although revolutionary developments in DNA probes have enormous potential as biologic markers in prenatal diagnosis, many challenges lie ahead.

MARKERS OF EXPOSURE

With the rapid development of monoclonal antibodies, radioimmunoassays, and molecular genetic technologies, new techniques have been developed to detect toxicants covalently bound to DNA to form adducts (Wogan and Gorelick, 1985; Perera, 1986; Wogan, 1988). Many chemicals that are active carcinogens or mutagens either are electrophilic or are converted to electrophilic metabolites. These may become bound to DNA, RNA, or proteins. The consequences of these adducts have not been clearly demonstrated, but they are thought to initiate carcinogenesis or mutagenesis (Wogan and Gorelick, 1985). DNA adducts have been measured in blood. Cord blood has also been used (Daffos et al., 1985; Reddy and Randerath, 1988). Theoretically, it should be possible to use amniotic cells or chorionic villus cells to determine the fetal exposure to genotoxic chemicals. The amount of tissue required by the assays is large for these sampling procedures. But improvements in the laboratory procedures might make the assays possible on smaller samples.

In humans, assessment of in utero exposure to DNA-damaging agents has been attempted by comparing SCE frequencies in blood from mothers and their offspring. A case report from Sweden described increased SCEs in four children of two laboratory technicians who worked during pregnancy (Funes-Cravioto et al., 1977). Ardito et al. (1980) compared SCE frequencies in smoking and nonsmoking mothers and their infants' cord blood and found that mean SCE frequency was slightly higher in mothers than in the newborns. No difference was found between frequencies in maternal or cord blood of smokers and nonsmokers. In a similar study of smoking mothers and alcoholic mothers (Seshadri et al., 1982), the SCE frequency only in drinking mothers was higher than that in controls (13.5 versus 10.95 SCE/cell), but the SCE rate in their infants was not significantly increased (9.71 versus 8.95 SCE/cell). In a separate analysis, neonates with neural tube defects were found to have higher rates of SCEs than normal babies (10.34 versus 8.95 SCE/cell) (Seshadri et al., 1982). A systematic study of infants with normal and reduced birthweights found no association of growth retardation with SCEs measured in cord and postnatal blood (Hatcher and Hook, 1981b).

19

Reproductive Immunology: Biologic Markers of Compromised Pregnancies

Women experiencing repeated pregnancy losses have been reported in all populations (Bloom, 1981). Many of these losses are probably not due to exposures to toxic chemicals. Therefore, it is important to be able to distinguish women experiencing recurrent pregnancy losses because of toxic effects from those experiencing losses for other reasons. This chapter discusses assessments that distinguish some mechanisms of spontaneous abortion. The chapter also discusses other immunologic assessments that might be used to characterize the changes that occur during normal pregnancies. These changes might be altered by toxicants, thereby causing a pregnancy loss.

MATERNAL IMMUNOLOGIC RECOGNITION AND REACTION DURING NORMAL PREGNANCY

Efforts to understand the immunology of human pregnancy have focused on extra-embryonic membranes, because the point of contact between maternal tissues and the conceptus is the trophoblast. Cells of the inner cell mass differentiate to become the embryo, and extraembryonic components form an interface with maternal blood and uterine cells (Faulk and McIntyre, 1983). This materno-trophoblastic interface exists at all sites of contact, including placenta, point of contact of the amnion and chorion (the amniochorion), spiral (uteroplacental) arteries, basal plate, and interstitial tissues.

The trophoblast is strategically important in potential maternal immune recognition and rejection. Its plasma membranes are unique, inasmuch as none expresses the polymorphic form of class I or class II transplantation histocompatibility antigens (human leukocyte antigens or hLAs); however, some cytotrophoblasts are reactive with monoclonal antibodies thought to recognize class I hLA. The general lack of transplantation antigens has led to speculation of trophoblastic immunologic neutrality; however, numerous investigators have shown that trophoblast membranes are not immunologically inert (Faulk and Hsi, 1983). The immunogens that signal and maintain maternal recognition are extraembryonic structures at the materno-trophoblastic interface, i.e., trophoblast antigens (Faulk and McIntyre, 1981).

The immune system in mammals consists of B lymphocytes, which are thought to arise from bone marrow and are responsible for antibody immunity, and T lymphocytes, which emerge from the thymus and are responsible for cell-mediated immunity.

Both lymphocyte populations are activated during normal human pregnancy.

B-lymphocyte activation by trophoblast antigens has been shown by elution of maternal antitrophoblast antibodies from homogenates of individual placentae (Faulk et al., 1974). T-lymphocyte activation by trophoblast antigens has been confirmed by demonstrating that trophoblast antigens cause maternal lymphocytes to release a lymphokine (migratory inhibition factor, or MIF), which is a quantitative measure of cell-mediated immunity (Rocklin et al., 1982).

An early step in the generation of cytotoxic reactions in cell-mediated immunity is allogeneic recognition of the target cell. Blocking allogeneic recognition by B-cell-produced antibodies to trophoblast can inhibit T-cell activation and the succeeding steps that result in cell death. Antibodies that impede such immune reactions are called blocking antibodies. Ample evidence from studies done in mice and human beings shows the presence of blocking antibodies in maternal blood and placental eluates during normal pregnancy (Faulk et al., 1974; Rocklin et al., 1982).

Some trophoblast antigens (McIntyre and Faulk, 1979a) and some antitrophoblast antibodies (McIntyre and Faulk, 1979b) can modulate allogeneic recognition reactions. Although antitrophoblast antibodies have been identified in some normal and abnormal pregnancies (McIntyre et al., 1984a), antitrophoblastic activity usually cannot be identified in serum taken during a normal pregnancy (Davies and Browne, 1985).

Antitrophoblast antibodies are not always found in serum taken during a pregnancy, for at least five reasons:

• Antibody-combining sites might be bound by trophoblast antigens in immune complexes.
• Trophoblast immunogens might stimulate blocking or incomplete antibody production.
• Autoanti-idiotypic antibodies might prevent antitrophoblast immunity.
• Inhibitors of antigen-antibody reactions or of their products (such as complement fixation) might mask the presence of antibodies.

• Antitrophoblast antibodies might not always be present in serum taken during a pregnancy.

A proposed model for immunologic response is presented in Figure 19-1. Studies of human serum taken during pregnancy have revealed the presence of circulating immune complexes (i.e., antigen in combination with its antibody) comprising five biochemically identifiable trophoblast and antitrophoblast components, two of which also can be identified at significantly lower concentrations in serum of nulliparous, nonpregnant women (Davies, 1985). Increased quantities of trophoblast antigens have been reported in maternal serum as pregnancy progresses (Faulk and McIntyre, 1983), and this has been cited as further support for immune complexes containing trophoblast antigen. Immune complexes might account for the difficulty in demonstrating antitrophoblast immunity in some maternal serum, because the antitrophoblastic components are captured in these complexes (Davies, 1985).

Blocking or incomplete antibodies have been suspected of being important manifestations of immunity in reproduction and cancer research (Gorer et al., 1959; Voisin et al., 1972). Whether the antigens responsible for host responses to transplants, cancers, tolerance induction, and pregnancy have chemical similarities is not known, but blocking or incomplete antibodies have been described in all these conditions.

In pregnancy research, human trophoblast antigens and antibodies to them have been shown to block mixed-lymphocyte-culture (MLC) reactions (McIntyre and Faulk, 1979a,b). Maternal antipaternal blocking immunity appears to be important in normal pregnancy, but is absent in some abnormal pregnancies (Rocklin et al., 1982; McIntyre et al., 1984a). Experiments in mice have shown that immunization with placental extracts and an additional, unrelated antigen promotes production of more blocking and less cytotoxic antibodies (Duc et al., 1985). Thus, blocking activity in maternal blood might explain the lack of antitrophoblast immunity in vitro.

CONTROL CIRCUIT IN HUMAN PREGNANCY

(TLX/Anti-TLX/Anti-anti-TLX)

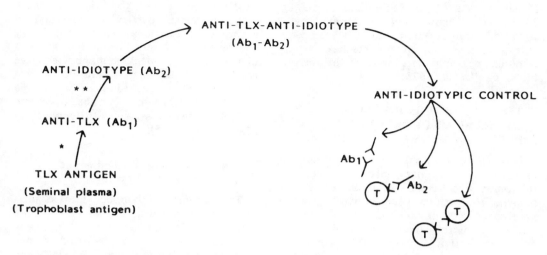

FIGURE 19-1 Proposed model of immunologic response. Source: Faulk et al., 1987.

Autoanti-idiotypic antibodies to anti-hLA antibodies in recipients of donor-specific transfusions have been found in serum of patients who lack serologic responses to hLA; that indicates that the presence or absence of detectable antibodies might reflect the amount of anti-idiotype (Reed et al., 1985). Serum from people alloimmunized through pregnancy, transfusion, or transplantation can react with autologous T lymphoblasts primed against the immunizing donor (Suciu-Foca et al., 1983).

Coupled with the finding that primed T cells display idiotypelike receptors for alloantigens, those observations have prompted the idea that the receptors induce formation of anti-idiotypic antibodies. Such autoanti-idiotypic antibodies to hLA can be found during and after pregnancy (Reed et al., 1983). Maternal anti-idiotypic antibodies to trophoblast antigens have not been studied in human pregnancies, but they should help to explain why anti-trophoblast antibodies cannot always be found in maternal serum.

Inhibitors of antigen-antibody reactions in complement-dependent assays (Torry et al., 1986) and complement-independent assays (McIntyre and Faulk, 1985) have been described. Inhibitors of inhibitors have also been noted (Faulk

et al., 1989a) that are heat-labile, sensitive to calcium concentrations, destroyed by Russell's viper venom, and absent from the plasma of patients with deficiency in clotting factor V. Some cytotrophoblasts in placental villi react with antibodies to factor V; factor V might play a role in modulating maternal antigen-antibody interactions within the placental bed. Knowledge of the inhibitors and inhibitors of inhibitors of antigen-antibody reactions is only beginning to emerge; these inhibitors might be important factors when maternal antitrophoblast antibodies are not detected, particularly if a sample is heated, an incorrect anticoagulant is used, or the calcium concentration is erroneous. There seems to be a novel and unexplored link between blood clotting and immunologic reactions.

Another possible explanation for the inability to detect maternal antitrophoblast antibodies in maternal serum taken during pregnancy is that the antibodies are not present. Antibodies were not thought to be a response to transplantation until Gorer et al. (1959) developed methods to reveal them; and antibodies were not thought to be an aspect of neonatal tolerance until Voisin et al. (1972) demonstrated their presence. In many cases, what was thought to be a lack of serologic reactivity to hLAs has been found to be an inhibition of reactivity by pregnancy-induced autoanti-idiotypic antibodies (Reed et al., 1983; Suciu-Foca et al., 1983). That matter will be clarified by time and more research, but it is reasonable to assume that all mothers have antibodies to trophoblast antigens in their blood (Faulk et al., 1978).

TESTS TO DETERMINE MARKERS OF MECHANISMS OF RECURRENT PREGNANCY LOSS

Infertility is classified as primary or secondary. Primary infertility occurs in women who have never conceived, and secondary infertility occurs in women who have conceived but have failed to conceive during 1 or more years of intercourse without contraception (Coulam, 1981). Recurrent spontaneous abortion comprises at least two groups of persons designated as primary and secondary spontaneous aborters (McIntyre et al., 1984a). Repeated pregnancy loss in primary aborters is thought to be associated with compatibility of trophoblast-lymphocyte-cross-reactive (TLX) antigens between mating partners, which results in lack of necessary protective or blocking maternal responses during pregnancy (McIntyre et al., 1986). The trophoblast antigens that elicited antibodies to cross-reacting antigens on lymphocytes were designated TLX antigens (Faulk and McIntyre, 1981).

Mixed-Lymphocyte-Culture Reactions

One of the best examples of cell-mediated immunity in pregnancy is the MLC reaction, in which lymphocytes from two persons are mixed and cultured under conditions that permit measurement of their DNA metabolism, which is an index of the intensity of reaction of one cell to the other. If the father's cells are sufficiently irradiated to prevent their immunologic response but their capacity to stimulate the mother's lymphocytes is retained, then the reaction is called a one-way MLC reaction. This test assesses the reaction of the mother's cells to the father's cells.

Perhaps the most convincing support for an important role of blocking factors in pregnancy comes from clinical investigations done in some conditions of abnormal pregnancies, particularly unexplained spontaneous abortions. Often, a primary spontaneous aborter does not produce a blood factor that blocks lymphocytes from reacting with her mate's cells in in vitro models of cell-mediated immunity. In some cases, that deficiency can be overcome by immunizing the woman with lymphocytes (Taylor et al., 1985).

Lymphocytotoxic Antibodies and Histocompatibility Typing

The presence of lymphocytotoxic antibodies in maternal serum never has been explained adequately, because these antibodies are not always identified in maternal serum (Kajino et al., 1988). When they are detected, it usually is in serum collected during a second or later pregnancy, although such antibodies have been reported in samples from first pregnancies (Davies and Brown, 1985); lymphocytotoxins probably are not of central importance in the immunobiology of human pregnancy. Lymphocytotoxins of pregnancy often are of broad reactivity and might react with antigens common to several class I hLAs (Konoeda et al., 1986).

Lymphocytotoxins in maternal serum taken during pregnancy traditionally have been characterized as anti-hLA. That clearly is not the case in secondary spontaneous abortion, inasmuch as cytotoxicity is removed from serum by absorption with hLA-negative trophoblast or platelets unrelated to hLAs (Faulk and McIntyre, 1986).

During the last several years, investigators have theorized that normal pregnancy requires maternal immunologic recognition of the TLX antigens inherited by the conceptus (Faulk and McIntyre, 1981) and that failure of recognition or inappropriate recognition results in faulty blastocyst implantation and ultimately spontaneous abortion. The nature of the immunogen is speculative, and results of animal-model studies have prompted some investigators to suggest that maternal recognition depends on incompatibility of major histocompatibility complex (MHC) antigens (Kiger et al., 1985). That suggestion is supported by the comparative success of outbred matings (as opposed to inbred matings), by the benefit of allogeneic, third-party leukocyte immunizations to primary aborters (McIntyre et al., 1986), and by the demonstration of a beneficial effect of mating with an MHC-incompatible male on pregnancy outcome in mice (Clark et al., 1986).

The role of MHC antigens in defining incompatibility has been controversial (Palm, 1974; Komlos et al., 1977; Gerencer et al., 1979; Schacter et al., 1979; Gill, 1983; Thomas et al., 1985). Reports have supported (Taylor and Faulk, 1981; Beer et al., 1981; McIntyre and Faulk, 1983; Unander and Olding, 1983) and refuted (Lauritsen et al., 1976; Caudle et al., 1983; Mowbray et al., 1983; MacQueen and Sanfilippo, 1984; Jeannet et al., 1985) an association between hLAs and reproductive performance. The variation in results can be explained by small sample sizes and lack of homogeneity of the populations investigated. Controversy involving association of hLAs and reproductive performance could be resolved in part by properly classifying recurrent spontaneous aborters and inexplicably infertile couples.

Immunologists often liken pregnancy to an allogeneic graft, inasmuch as the developing conceptus expresses paternal genes that are foreign to the mother. Men and women who share hLAs might be expected to be ideal reproducers, because chances for immunologic recognition and rejection by the mother would be low. However, genetic identity between mother and conceptus does not appear to be necessary for the pregnancy to continue; surrogate mothers and in vitro fertilization techniques have shown that a fertilized egg can develop successfully if it is genetically different from the recipient. Furthermore, couples that share hLAs often are not able to have successful pregnancies.

Women suffering from primary recurrent spontaneous abortions often have hLA (TLX) profiles more similar to those of their mates than is the case in normal childbearing couples (McIntyre et al., 1986). Those women do not manifest antipaternal humoral immunity, and their cell-mediated responses often are absent or suboptimal, as measured by mother-father MLC reactions. The cellular deficit appears to be intrinsic, in that it occurs regardless of the serum supplement used in the culture system. The deficit can be used to distinguish primary from secondary recurrent spontaneous abortion, because secondary aborters have extrinsic or acquired MLC-inhibitory activity (McIntyre and Faulk, 1983).

Autoimmunity Tests

Spontaneous abortion is common in patients with some autoimmune diseases, such as systemic lupus erythematosus (Derue et al., 1985), although how antibodies associated with these diseases interrupt pregnancies is unknown. The lupus anticoagulant is an antibody that reacts with the phospholipid component of prothrombinase (Hougie, 1985) and affects many in vitro blood coagulation tests (Shapiro and Thiagarajan, 1982). Preeclampsia and fetal growth retardation, which sometimes appear in pregnancies complicated by lupus anticoagulant, are associated with decreased prostacyclin production by maternal and fetal vascular tissues (Bussolino et al., 1980; Remuzzi et al., 1980). Identification of lupus anticoagulant in maternal blood taken during pregnancy should signal a high risk of pregnancy loss.

PROMISING MARKERS OF MATERNAL-FETAL INTERACTIONS

Trophoblast Antigens and Classes of Couples with Recurrent Spontaneous Abortions

Trophoblast antigen-1 (TA1) and trophoblast antigen-2 (TA2) were among the first human trophoblast antigenic groups to be identified with the use of polyclonal antisera (Faulk et al., 1978). Other polyclonal and monoclonal antibodies have been used to define several trophoblast antigens (Faulk and Hsi, 1983). The study of such antibodies is essential for building an understanding of trophoblast antigen functions in mammalian pregnancies, because trophoblast tissues account for the operational interfaces between maternal and extraembryonic cells in the allogeneic relationship of human pregnancy (Faulk, 1983).

Antisera to TA1 (anti-TA1) contain antibodies to syncytiotrophoblast and subpopulations of cytotrophoblast found in the amniochorion, basal plate (Wells et al., 1984a), spiral arteries (Wells et al., 1984b), and uterine interstitial tissues (Hsi et al., 1984a). In contrast with their lack of reactivity with normal somatic cells, anti-TA1 reacts with many human-transformed cell lines (Faulk et al., 1979). Normal, abnormal, and extraembryonic tissues from other species do not react with human anti-TA1.

Anti-TA1 inhibits MLC reactions without affecting nonspecific mitogenic stimulation of lymphocytes (McIntyre and Faulk, 1979b). The mechanism of MLC inhibition seems to be recognition and stimulation, rather than T-cell proliferation (McIntyre and Faulk, 1979c).

All normal human extraembryonic tissues have TA1 at the materno-trophoblastic interface, including the amniochorion (Hsi et al., 1982). Two immunohistologic observations of the amniochorion have been made: the cytotrophoblast forms a barrier having TA1 that is three to five cells thick (Faulk et al., 1982), and the amniotic epithelium does not have TA1 but expresses an antigen called amniotic antigen-3 (AA3) (Hsi et al., 1984b). In newborns with epidermolysis bullosa letalis (EBL; also called polydysplastic epidermolysis bullosa), the barrier of TA1-bearing cytotrophoblast is so thin that maternal uterine cells sometimes come into contact with amniotic epithelium-derived tissues (Faulk et al., 1988b). In EBL, amniotic epithelial cells react with TA1 and AA3 antibodies. The only other circumstance in which anti-TA1 reacts with amniotic epithelial cells is when they are transformed (Yeh et al., 1984). Amniotic epithelium is representative of somatic ectoderm (Faulk and McCrady, 1983). Babies with EBL have defective

ectodermal derivatives, such as junctional epidermolysis bullosa, and defective ectodermal thymus. The absence of AA3 from skin biopsies is a useful prenatal diagnostic criterion for EBL (Kennedy et al., 1985).

TA2 was established on the basis of three observations:

- Solubilized trophoblast membranes were partitioned into two peaks (TA1 and TA2) with chromatography.
- The antigens responsible for generating lymphocytotoxic antibodies were present in the second (TA2) peak of solubilized, chromatographed trophoblastic microvilli.
- Lymphocytotoxic antibodies were removed from antitrophoblast serum by absorption with lymphocytes (Faulk et al., 1978).

Absorption of anti-TLX antigens with trophoblast membranes from different placentas has indicated that these TLX antigens are allotypic (Faulk et al., 1980; McIntyre and Faulk, 1982). At least three groups of TLX antigens can be found with the use of rabbit antibodies (McIntyre et al., 1984b). One TLX antigen in the fertilized egg must be incompatible with the mother and be capable of signaling allogeneic recognition and immunologic protection through the generation of maternal blocking antibodies and suppressor cells (McIntyre and Faulk, 1985).

Maternal recognition of paternal TLX can be initiated by its immunogenicity in seminal plasma (Faulk and McIntyre, 1986). Failure of recognition or inappropriate recognition of TLX antigens could result in spontaneous abortion that occurs so early in the pregnancy that it goes undetected and the couple is thought to be infertile (Faulk and McIntyre, 1986).

Major Basic Protein

Major basic protein (MBP) is a protein whose blood concentrations rise by the sixth week of gestation and return to normal by 6 weeks after birth (Maddox et al., 1984). Pregnancy-associated MBP can be purified from human placentas and is biochemically indistinguishable from MBP found in eosinophil granules. MBP, which accounts for most of the granule protein (M.S. Peters et al., 1986), is toxic to mammalian cells in vitro (Gleich et al., 1979) and parasites in vivo (Kephart et al., 1984) and mediates inflammation in asthma (Frigas and Gleich, 1986). In human pregnancy, increases in MBP in peripheral blood are independent of either eosinophils or eosinophil proteins other than MBP (Maddox et al., 1983); immunohistologic techniques show increases to be localized in the extravillus trophoblast (Maddox et al., 1983).

Quantitative studies have indicated that MBP concentrations plateau by 20 weeks of gestation at more than 10 times the nonpregnant value, and they rise sharply in the third trimester in women who experience a spontaneous onset of labor (Wasmoen et al., 1987). The late increase, which accounts for 40% of the total increase in MBP during pregnancy, begins at least 3 weeks before onset of labor. Women who experience preterm labor have a similar increase; those with oxytocin-induced labor do not, nor do those with prolonged gestation (Coulam et al., 1987). These observations suggest that MBP concentration is a marker of the onset of labor.

Early Pregnancy Factor

Early pregnancy factor (EPF) is an immunosuppressive molecule that increases the ability of antilymphocyte antibodies to inhibit active, spontaneous rosette formation between lymphocytes and red cells (Rolfe et al., 1984). EPF is produced by the mother within 24 hours of fertilization, and it wanes in midpregnancy, at which time its function is assumed by a placental form of EPF. The functions of maternal EPF and embryonic EPF seem to be indistinguishable. In the mother, the molecule is assembled from an oviduct component (EPF-A) and an ovarian component (EPF-B) and is synthesized under the influence of prolactin and an ovum factor (Cavanagh et al., 1982).

Because EPF appears in maternal blood so soon after implantation, it provides a method to distinguish infertile couples from those who are becoming pregnant but aborting very early. It might also be useful in judging success of in vitro fertilization/embryo transfer programs.

During pregnancy, the molecule is present only when there is a viable embryo and could be used as an early marker of embryonic viability (Morton et al., 1982). Detection of EPF in a nonpregnant patient suggests a tumor of germ-cell origin (Rolfe et al., 1983).

PROMISING TECHNIQUES THAT MIGHT YIELD BIOLOGIC MARKERS

Fluorescence-Activated Cell Sorting

In the past several years, molecular biology, genetics, and immunology have converged and produced mutually advantageous techniques, such as molecular probes, gene cloning, and hybridoma formation. One of the most promising techniques for pregnancy immunology is fluorescence-activated cell sorting (FACS). FACS uses fluorochrome-labeled antibodies to identify and quantify membrane markers on cells in a heterogeneous mixture (Parks et al., 1979). For example, FACS has been used to measure the flux of fetal cells into maternal blood during pregnancy. The technique can be used to isolate immunologically marked cells from a complex mixture, such as blood, and it has been used to measure trophoblast membranes in the peripheral circulation of pregnant women (Kawata et al., 1984). This approach might make it possible to harvest fetal cells for cytogenetic investigations without resorting to the more invasive techniques of amniocentesis or chorionic villus biopsy.

Immunotherapy to Prevent Spontaneous Abortion

Immunotherapy to prevent primary spontaneous abortions was begun in 1979 (Taylor and Faulk, 1981). Primary aborting women were given leukocyte transfusions from nonpaternal blood donors (Taylor et al., 1985); the rationale was that nonpaternal leukocytes would express TLX allotypes foreign to the mother and cause her to mount a protective anti-TLX response to the infused cells that cross-reacted with TLX antigens on the blastocyst, thereby protecting the developing embryo from maternal immune rejection (Faulk and McIntyre, 1981, 1983; Beer et al., 1986). More than 45 couples receiving this immunotherapy had a rate of successful pregnancy comparable with that of normal childbearing women (Mowbray, 1987). No graft-versus-host reactions were observed in any of the offspring (Mowbray and Underwood, 1985; McIntyre et al., 1986), and no evidence of intrauterine growth retardation was found.

Immunopharmacology

Immunopharmacology brings a promising new approach to the study of biologic markers in pregnancy by joining two previously unconnected fields of medical investigation, particularly in relation to toxicology and environmental pollutants. Immunopathologic reactions can be grouped into four types of causal mechanisms that are central to understanding diagnostic, therapeutic, and prognostic variables of immunologic disease. Careful clinical identification of immunologic reactions and failed pregnancies in association with environmental exposures might help to clarify the action of toxic chemicals.

Type I lesions are mediated by immediate hypersensitivity reactions—such as anaphylaxis, urticaria, and angioedema—and usually can be attributed to a reaginic (IgE) antibody that fixes to receptors on tissue mast cells and blood basophils. When the reagin meets its antigen—which can be a drug, such as penicillin, or an environmental allergen, such as ragweed pollen—the mast cells or basophils degranulate and release mediators (such as histamine) that cause the signs and symptoms of immediate hypersensitivity reactions. Patients with preeclampsia have been found to have significant elevations of IgE in their blood (Alanen, 1984).

Type II immunopathologic reactions are caused by direct interaction of antibody and complement to cause cell lysis, usually with the collaboration of blood complement. An example of such antibodies in pregnancy is the maternal antipaternal lymphocytotoxin seen in secondary spontaneous aborters (McConnachie and McIntyre, 1984). Type II immune responses also can be caused by drugs, such as insulin, and they are represented by pathophysiologic conditions in maternal isoimmunization and erythroblastosis fetalis. Anti-D vaccine was developed for Rh-negative mothers with Rh-positive fetuses, to prevent rhesus isoimmunization, a type II reaction (Whitfield, 1976).

Type III immunopathologic reactions are caused by immune complexes, usually with the participation of complement. These reactions include urticarial skin eruptions, arthralgia or arthritis, lymphadenopathy, and fever. The reactions generally last 6-12 days and subside when the offending antigen is eliminated (Salmon, 1982). Environmental pollutants, such as mercury, can cause type III reactions (Roman-Franco et al., 1978) that are demonstrated by granular glomerular deposits of immunoglobulin and complement in renal biopsies. Similar immunohistologic findings have been reported in preeclampsia, but mercury intoxication was not evident in these cases (Matter and Faulk, 1980).

Type IV reactions are independent of antibody and complement and are mediated by lymphocytes. (That is commonly referred to as cell-mediated immunity.) Owing to the role of soluble lymphocyte products (e.g., lymphokines), there might be no such thing as a pure type IV reaction, but it is useful in thinking about diagnosis, treatment, and prognosis. Graft rejection and delayed-hypersensitivity reactions are generally thought to be type IV reactions (Turk, 1975), and these types of reactions are inhibited by TA1 and anti-TLX antibodies, as measured by MLC reaction (McIntyre and Faulk, 1979a). Type IV reactions are central in host defense reactions against some infectious diseases (Chandra and Newberne, 1977), and the lymphoid axis on which such immunity is based (i.e., T cells) is severely damaged by protein-calorie malnutrition. Protein-calorie malnutrition is a major factor in diminution of reproductive capacity as demonstrated by decreased ovulation. T-cell function, reproduction, and diet also are closely linked.

20

Cell Biology:
Identifying Biologic Markers
Expressed During Early Pregnancy

This chapter discusses the biologic processes that are important before and around the time of implantation. Many cellular and developmental stages at this early time of pregnancy are critical to the further development of the pregnancy. In the section of this report on female reproductive markers, one hormonal change (hCG) is discussed (see Chapter 15). In this section, many more potential markers are discussed, including cellular differentiation, diffusible cellular products, and additional hormonal concentration changes. The markers discussed here will lead to better understanding of the biologic processes and possible mechanisms of toxic action.

The greatest risk to successful gestation occurs around the time of implantation, when the maternal uterine environment and the embryo interact to establish pregnancy. The chance of a couple of proven fertility to conceive offspring in any menstrual cycle is about 25% (Vessey et al., 1976; Short, 1979), but it is difficult to determine the extent to which this low success rate is due to errors or dysfunctions in ovulation, fertilization, implantation, or later development. Livestock have a high incidence of early embryonic loss—up to 40% in pigs—and up to 50% of human conceptions are estimated to un-

dergo early embryonic termination (Leridon, 1977; Short, 1979).

Records of human IVF/ET programs indicate a 15-20% rate of completed pregnancy (Webb and Glasser, 1984); however, the fertilized embryos that are transferred are selected for apparent viability. IVF/ET failure is attributable to events related to implantation (Edwards et al., 1980; Webb and Glasser, 1984). Implantation errors constitute one of the largest causes of failure in reproductively competent persons in IVF programs (Fig. 20-1). The high risk of implantation failure might be compounded by xenobiotic agents introduced into the intrauterine environment.

In the United States and other countries, experimentation in humans is problematic because of ethical and legal restrictions (Andrews, 1984a,b). Knowledge of early human development necessarily depends more on comparative studies than does knowledge of other human biology. Although imperfect as analogies of human reproduction, examples of many mammalian reproductive systems are available for study, and comparative analysis has provided insight into identification of common mechanisms characteristic of the preimplantation period (Amoruso, 1981).

Animal experiments—particularly in

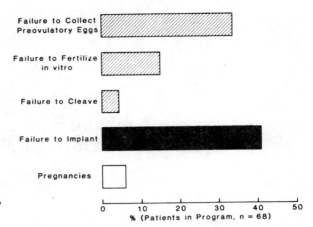

FIGURE 20-1 Relative contribution of different states of very early gestation to outcome of early IVF/ET. Summary of data on treatment group resulted in first successful IVF/ET pregnancy (Edwards et al., 1980). Of 79 women monitored during menstrual cycles, 68 underwent laparoscopy for attempted oocyte retrieval, which resulted in at-term birth of two normal infants. Source: Glasser et al., 1987a.

vitro laboratory models—are useful only to the extent that they mimic the specific processes of interest. The caution appropriate to the design and execution of laboratory models (Glasser, 1985) limits the opportunity to focus on a distinct target or toxicant-specific cellular or molecular events that alter reproductive or developmental outcome. The accuracy with which such events can be identified and analyzed is constrained in the experimental model by limitations in the understanding of the process under investigation and its putative role.

No specific or reliable markers of xenobiotic agents can be correlated with any cellular or molecular events of early mammalian development. Some data point to adverse influences of a variety of toxic agents during early development (Dixon, 1985); however, the data were derived from studies that were not stringently designed or executed and often were evaluated retrospectively. Their usefulness in identifying specific, sensitive markers is questionable. The few markers of early development that might prove useful (for instance, indexes of uterine epithelial and stromal cell cleavage rates, compaction, blastocoele formation, expression of embryonic mRNA, or expression and organization of trophectodermal cytokeratins) rarely have been used in studies of reproductive toxicology.

IMPLANTATION

Implantation of the mammalian embryo in the uterus of the maternal host is a unique interaction between two genetically dissimilar organisms. In most species, changes in the developing embryo and the uterus are coordinated closely, probably while the embryo is still in the oviduct. Disruption of this synchrony leads to implantation failure (Noyes et al., 1963).

The ideas that the uterus matures from a nonreceptive environment to one that is receptive to the blastocyst in response to changing concentrations of progesterone and estrogen and that implantation involves changes of the uterus, not of the blastocyst (Glasser, 1972; Psychoyos, 1973; Glasser and Clark, 1975), are confirmed by morphologic data (Nilsson, 1970; Schlafke and Enders, 1975), physiologic data (Glasser, 1972; Psychoyos, 1973; Glasser and Clark, 1975), and, to a lesser extent, biochemical correlates (Glasser, 1972; Glasser and Clark, 1975; Surani, 1975; Bell, 1979; Glasser and McCormack, 1982). Although estrogen and progesterone are essential to uterine development, their function in embryonic development is uncertain. With a single exception (Smith and Smith, 1971), in vitro blastocyst differentiation has been reported to proceed in the absence of steroid hormones (Marcal et al., 1975; Sherman and Wudl, 1976).

Conditions in the female reproductive tract necessary to maximize the opportunity for embryo implantation in the uterine environment are well understood (Psychoyos, 1973; Glasser and Clark, 1975; Glasser and McCormack, 1980, 1982). However, the understanding does not explain the specific effects of hormones, drugs, and toxic agents on pregnancy. More incisive methods of investigation—such as those developed for cell biology, immunology, and molecular biology—must be found to define the regulatory biology of blastocyst-endometrial interactions and to show how they can be interrupted by xenobiotic agents.

The implantation process is initiated when trophectoderm cells of the blastocyst come into intimate contact with the receptive uterine endometrium (Sherman and Wudl, 1976; Glasser and McCormack, 1980, 1982). Progressive phases of this process are controlled by molecules exchanged directly by cell-to-cell communication (Enders et al., 1981) and modulated by molecular signals from stromal-epithelial communication (Cunha et al., 1985). These molecules are expressed in response to the same steroids that synchronize the blastocyst and uterus from conception.

To interpret specific cell-to-cell interaction during implantation events, homogeneous populations of individual cell types involved directly in implantation—i.e., endometrial epithelial and stromal cells, blastocyst trophectoderm and ectoplacental cone cells, and trophoblast giant cells—are isolated. The cell populations can be cultured in vitro so that biochemical mechanisms that regulate their differentiation and interactions can be studied (McCormack and Glasser, 1980; Glasser and McCormack, 1981; Soares et al., 1985; Glasser and Julian, 1986; Glasser et al., 1987b). Recent developments for studying trophoblast interactions used three-dimensional culture systems, in which trophoblast cells are grown as free-floating spheroids (White et al., 1988a). Such trophoblast spheroids can be used to study interactions with explants or monolayer cultures of other tissues, e.g., endometrium (White et al., 1988b).

To apply the concept of biologic markers to the interaction of xenobiotic compounds with the early mammalian developmental processes, a research strategy that depends on animal experimental models can be formulated (Glasser, 1985). Cellular and biochemical methods can be used to identify and validate critical structural or functional markers of the regulatory processes involved in the differentiation of tissue during each step. Those markers provide a basis for selecting the most appropriate markers for use outside the laboratory.

ASSESSING ENDOMETRIAL SIGNALS

The role of the uterus is defined by the responses of its epithelial and stromal cells to a specific sequence of ovarian hormones (Psychoyos, 1973; Glasser and Clark, 1975; Glasser and McCormack, 1982). Whether the uterine epithelial cells respond directly to hormone instructions or indirectly to signals emanating from hormone-regulated uterine stromal cells is unknown (Cunha et al., 1985; Bigsby and Cunha, 1986). In uterine epithelial or stromal cells, xenobiotic agents might interfere with the binding of a steroid hormone to a target-cell receptor or might affect steps in the biochemical responses to hormonal regulation initiated by the binding of hormones. These effects might be independent or could be coupled, so a response could be additive or synergistic.

Use of receptor analysis to identify biologic markers of endometrial cell biology is limited by the difficulty of gaining access to markers and target cells.

Current data concerning effects of xenobiotic agents on the uterus describe responses of the whole uterus and do not reveal the extent to which each cell type contributes to the net uterine response or which cell type is at risk under different environmental conditions. If information of this nature were available, efforts could focus on risk reduction.

Homogeneous populations of endometrial cell types from the uterus can be isolated (McCormack and Glasser, 1980; Glasser and Julian, 1986) and their regulatory biology and differentiation studied in vitro.

Experiments with primary cultures of uterine epithelial cells must have confluent and polarized monolayers to obtain biologically relevant data identifying and assessing markers and their interaction with xenobiotic agents. This research design is useful, because the basal surface of the epithelial cells is accessible for experimentation and analyses, inasmuch as the cells are cultured on semipermeable, matrix-impregnated supports.

Tables 20-1 and 20-2 list candidate markers to assess the status of uterine epithelial or uterine stomal cells. These markers are not practical for medical monitoring because they are not readily accessible. Nevertheless, they are useful to identify critical targets, times, and processes.

Uterine Secretions

Early morphologic studies suggested that uterine secretions might be biochemical correlates of receptivity development. The quality and quantity of secretions that accumulate in the uterus during various phases of the reproductive cycle have been studied in several species, to identify phase-specific or hormone-specific secretory products that might be markers of particular facets of implantation. Few attempts have been made to examine critically which cell type is the source of each secretory product, whether proteins that are secreted are synthesized de novo, or whether the appearance of such possible markers represents selectively stimulated protein synthesis.

Analyses of human endometrial washings have not revealed large concentrations

TABLE 20-1 Putative Biologic Markers to Assess Status of Uterine Epithelial Cells

Cellular or Developmental Stage	Biologic Marker	Comments
Proliferation	Cell number; mitotic index; labeling index	These markers assess mitogenic response of uterine epithelial cells
	Short-term biosynthetic and metabolic index; profiles of apical versus basal cell surface and secretory proteins/glycoproteins	These markers evaluate physiochemical and biologic responses to regulatory factors (hormones, growth factors); can be used to evaluate analogues, congeners (phytoestrogens, catechol E); can use other cell and tissue models
Postmitosis	Long-term biosynthetic and metabolic index; differential response to steroid hormones; differential trafficking of apical versus basal cell surfaces and secretory proteins/glycoproteins	These markers assess hormonally regulated differentiation during transition of hostile to neutral uterus; changes blocked by castration; because of embryonic diapause in some animals, these markers can describe ability to reactivate blastocytes
Preimplantation	Timing of final stages of differentiation; protein/glycoprotein profiles of apical versus basal surface secretions; timing of rising titers of progesterone and estrogen; differential receptor response	These markers describe transition from neutral to sensitized or receptive uterus
Implantation	Biochemical index of terminal differentiation; different changes in steroid hormone receptors; specific early pregnancy factors	These markers describe uterine epithelial receptivity to blastocyst, attachment, uterine influence on the blastocyst, and initiation of stromal cell differentiation; growth factors not well studied

TABLE 20-2 Putative Biologic Markers to Assess Status of Uterine Stromal Cells

Cellular or Developmental Stage	Biologic Marker	Comments
Proliferation	Changes in number of cytoplasmic and nuclear endoplasmic reticulum	Not well studied
Postmitosis	Rate of cell division; continued changes in number of endoplasmic reticulum	These markers describe responsiveness of cells to steroids; not sensitive to deciduogenic stimuli; uterine stromal growth factors not well studied
Preimplantation	Number of endoplasmic reticulum and frequency of stromal mitosis	These markers assess uterine sensitivity
Implantation	Number of endoplasmic reticulum; maximal rate of stromal cell division	These markers describe stromal component of receptive uterus

of specific proteins. Most studies have found that uterine secretions consist mainly of common serum proteins (Wolf and Mastroianni, 1975; Roberts et al., 1976; Hirsh et al., 1977); however, transudation of serum proteins appears to be selective, but variable throughout the endometrial cycle (Beier and Beier-Hellwig, 1973). Electrophoretic analysis and more recent radiolabeling studies have revealed specific proteins not found in serum. These range from low-molecular-weight components—possibly glycoproteins—to proteins of 60-67 kilodaltons (Wolf and Mastroianni, 1975; Tzartos and Surani, 1979; Sylvan et al., 1981); several appear to be specific to the secretory phase of the endometrium during the menstrual cycle.

In uterine washings, various enzymes have been found at concentrations above those in serum; for example, glycosidase (Hansen et al., 1985), antitrypsin (Roberts et al., 1976; Casslen and Ohlsson, 1981), and fibrolytic activity (Werb et al., 1980) in human uterine fluid vary throughout the menstrual cycle. Those enzymes might be involved in the implantation process (Tzartos and Surani, 1979).

Maathuis and Aitken (1978) have shown that proteins are secreted throughout the proliferative and secretory phases of the cycle, and their concentrations are lower after ovulation. That is in accord with

the finding of lower fluid volume in the secretory phase (Clemetson et al., 1973). The finding does not preclude secretion of specific proteins into the uterus during this phase. Associated with lower fluid volume is increased potassium ion concentration, which is particularly high around the time of implantation. Other nonprotein components also change throughout the cycle. The concentration of fructose increases around the midsecretory phase, but glucose concentration changes little throughout the cycle (Douglas et al., 1970; Maathuis and Aitken, 1978).

In rats and mice, high estrogen concentrations elicit intrauterine secretions, particularly of proteins; high concentrations of progesterone reverse this effect (Armstrong, 1968; Surani, 1975; Aitken, 1977; Pratt, 1977; Fishel, 1979). Unique uterine secretory proteins have been reported—one protein appeared 18-20 hours after estrogen injection (Surani, 1975). That interval corresponds with the pulse of ovarian estrogen that is released by normal rats late on day 3 and during early phases of implantation on day 4. The appearance of unique proteins could be coincidental, and the relationship of induced proteins to a specific embryonic or endometrial function is unproved. Intrauterine proteins might be serum transudates, metabolic products, or degradation products that are unrelated to the specific process being studied.

Because they affect the uterine environment, uterine secretions probably help to regulate the blastocyst awaiting implantation. Such regulation might be:

- Direct, i.e., secretions might be information proteins that signal the blastocyst or adhesive proteins that increase cell-to-cell communication.
- Indirect, i.e., secretory proteins might serve as nutrients or as modulators of pH or isotonicity of the uterine environment.
- Passive, in that secreted proteins contribute to endometrial cell maintenance or the pharmacodynamics of the myometrium.

The opportunity for proteins to influence implantation success exists for approximately 72 hours in humans (Hertig and Rock, 1945; Hodgson and Pauerstein, 1976; Croxatto et al., 1978), but only 18-24 hours for species with short preimplantation periods (Glasser and McCormack, Webb and Glasser, 1984).

Studies of uterine secretions have been unrewarding in demonstrating a regulatory role for some proteins or in suggesting a cause-and-effect relationship that might make the marker useful to detect specific effects. In part, the difficulty arises from heterogeneity of the uterine secretions, which prevents us from distinguishing between the degrees to which oviductally and transepithelially transported stromal secretions contribute to the secretory profile.

Uterine Epithelial Cells

Hormone-regulated expression of specialized uterine epithelial cell functions—recognition, adhesion, and secretion—are related to differentiation of epithelial cell structure and functional polarity. Polarity depends on establishment of cross-linked interepithelial tight junctions and results in distinct apical and basal surface membrane domains. Development of cross-linked tight junctions coincides with active remodeling

of the apical surface. Both development and remodeling are stimulated by progesterone and occur in vivo immediately before implantation. Putative biologic markers of these processes are listed in Table 20-1.

Experimentally polarized uterine epithelial cells are necessary to analyze hormonal mechanisms that regulate specialized epithelial cell functions, and enough information has been collected to support an in vitro model of polarized epithelial cells. Proliferation, growth, and differentiation of polarity occur in epithelial cells isolated from immature rat uteri and cultured on matrix-impregnated filters in the presence of estrogen, progesterone, or both (Carson et al., 1988; Glasser et al., 1988; Glasser and Julian, 1989). The model approximates the in vivo situation by providing access to the epithelial cell through its basal surface. Supplemental regulatory factors stimulate polarized cells and validate the experimental model for use in studying expression of epithelial cell-specialized functions.

Stromal proteins, which can modify the hormonal response, have access to the cell through its basal surface. Profiles of glycoconjugates and proteins associated with epithelial cells in apical and basal surface secretions or in the apical surface membrane are analyzed during hormone-regulated proliferation, growth, and differentiation of filter-cultured epithelial cells. Differential changes in the apical surface membrane and its secretions are detected by differences in the protein/glycoprotein profiles and their distributions (Carson et al., 1988; Glasser et al., 1988; Glasser and Julian, 1989).

Uterine Stromal Cells

Studies of uterine stromal cells have produced a variety of biologic markers that can be used to monitor responses to regulatory agents and to identify processes that limit the responses. (Table 20-2).

It has been suggested that uterine epithelial cell response to estrogen (and perhaps progesterone) does not involve the steroid hormone receptors endogenous to those cells in fetal neonatal uteri (Cunha et al., 1985; Bigsby and Cunha, 1986). Rather, the response is indirect and is stimulated by interaction between steroids and hormone receptors in the underlying stromal cells. The extent to which these principles apply to cells in sexually mature animals remains to be investigated, but evidence points to modulation of uterine epithelial cells by stromal cells; furthermore, stromal cell decidualization might require signal transduction via epithelial cells (Lejeune and Leroy, 1980). Decidualization is a unique structural (and presumably functional) transformation of fibroblastlike cells of the uterine stroma to a distinctive tissue that is later discharged. The new tissue has giant, polygonal, multinucleate, endoreduplicative cells with abundant thin cytoplasmic filaments. These cells are rich in glycogen and lipids; specific alterations in nucleic acid (Glasser, 1975) and protein synthesis (Glasser, 1972; Glasser and Clark, 1975; Bell, 1979) are believed to be associated with the process. The decidual cell response can be induced in uteri, by blastocysts or by a variety of artificial stimuli in laboratory animals (Glasser, 1972). Decidualization is a progesterone-dependent process (Glasser and Clark, 1975), and its response to the blastocyst signals that the uterus has matured.

Research has provided many interesting clues and directions for potential markers, but has not yielded markers that are practical for medical monitoring. Further research should focus particularly on two processes sensitive to xenobiotic agents: stromal-epithelial cell communication and hormone-regulated differentiation of stromal cells and remodeling of their extracellular matrices.

Recombination of epithelial and mesenchymal cells from various tissues has been instrumental in showing that stromal cells give rise to directive and permissive factors that influence epithelial cell differentiation. Studying recombinations of uterine epithelial cells cultured on a matrix-impregnated filter with the stromal cells or their conditioned media ap-plied to the basal side of the filter promises to yield more detailed and specific data and permit analysis of functional differentiation. Accessibility to the basal secretory compartment also will permit research on the influence of xenobiotic agents on secretions from epithelial cells or on the effect of secretions on the morphologic and functional differentiation of uterine stromal cells. Studies probably will not yield useful markers in the immediate future, but they will determine whether cell-to-cell communication might be subject to toxic effects.

The role of decidual tissue remains to be defined, although several functions have been ascribed to it (Glasser and McCormack, 1980). Decidualization probably is a conservative mechanism in which trophoblast invasion of the endometrium is controlled and limited during establishment of the hemochorial placenta (Bryce and Teacher, 1908). Regardless of function, decidualization reflects a change in the synchronized endometrial substrate that follows blastocyst attachment, and it might be sensitive to xenobiotic agents.

Studies of in vivo and in vitro rat decidualization (Glasser and Julian, 1986; Glasser et al., 1987b) have demonstrated that non-coordinate expression the intermediate filament subunit, desmin, and vimentin is a correlate of hormone-regulated stromal cell differentiation. Desmin is marginally detectable in undifferentiated stroma and accumulates at a greater rate than cell protein. Vimentin expression is a marker of decidual cell growth; vimentin increases in proportion to decidual cell protein. Ninety-six hours after decidualization is initiated, the concentration of decidual cell desmin is equal to or greater than that of vimentin.

Inductive accumulation of the extracellular matrix proteins laminin and entactin, which are absent from undifferentiated stroma, is also a marker of decidualization (Wewer et al., 1985; Glasser et al., 1987b). Other indexes include decreased production and reorganization of fibronectin (Grinnell et al., 1982; Glasser et al., 1987b), expression of a decidual luteotropin (Markoff et al.,

1983; Maslar et al., 1986), and appearance of heparin sulfate proteoglycan and chondroitin sulfate proteoglycan (Wewer et al., 1985). These markers suggest that the surfaces and later the extracellular matrix of epithelial and stromal cells are remodeled in response to the hormonal interactions that regulate uterine receptivity to the blastocyst. The changes alter cell-to-cell interactions and accommodate the specialized attachment and invasive functions of the differentiating trophoblast. Interference with the remodeling of the stromal extracellular matrix into a basal, laminlike structure might have far-reaching consequences, not only for the programmed advance of trophoblast through stroma, but also for mobilized migration of B and T lymphocytes into the uterus as elements of implantation and establishment of the hemochorial placenta.

Those alterations are ordered by a precise program of hormone-specific synthesis of informational proteins (Glasser and Clark, 1975; Glasser and McCormack, 1980). One of the instructions might be provided by a luteotropinlike peptide hormone synthesized by decidual cells (Markoff et al., 1983; Maslar et al., 1986), thereby involving decidual cells in the regulation of endometrial response to the trophoblast. If that endocrine capability also occurs in humans, the preimplantation endometrial stroma might directly affect the intricate modulation of the preparatory processes required for implantation.

Evidence of an endocrine function of decidual tissue is the demonstration that human prolactin (hPRL) is a separate hormone discrete from growth hormone and that amniotic fluid contains extremely high concentrations of immunoreactive hPRL. Specific functions of hPRL in the female include regulation of postpartum lactation in mammary glands, reproductive cycle regulation, pregnancy maintenance, and embryonic growth and development. hPRL in amniotic fluid is involved in fetal osmoregulation. Suppression of maternal pituitary hPRL during pregnancy does not affect amniotic fluid concentrations. It has been determined that the source of amniotic fluid hPRL is decidualized endometrium of pregnancy.

Proliferative human endometrium cultured in the presence of progesterone—with or without estrogen—has been reported to produce immunoreactive hPRL (Daly et al., 1983a). Immunoreactive hPRL is an in vitro culture product of decidua from day 23 of the menstrual cycle through term (Daly et al., 1983b). The amount of hPRL produced is a function of the extent of decidualization and is progesterone-dependent. Production of hPRL by proliferative endometrium after 6 days in culture with progesterone in the absence of a blastocyst suggests that hPRL synthesis and secretion could be produced in vivo by early luteal endometrium, including predecidual cells. If hPRL synthesis and secretion could be produced in vivo, maturation of the uterus to a receptive environment might occur earlier in the human than anticipated and provide an endocrine basis for the success of early cleavage-stage embryos after IVF/ET. The specific decidual hormone in the circulation of women presumed to be pregnant then could be assayed, and a practical marker of endometrial differentiation around the time of implantation would be identified. Experimental data suggest that synthesis and secretion of this hormone occurs before the earliest time reported for hCG expression (Saxena et al., 1974).

Detection of markers of changes around the time of implantation may be possible, in light of recent studies of endometrial cell types and their interaction with the blastocyst (Copp, 1979). Interpretation of risk-assessment data developed in the laboratory must take into account functional polarity and epithelial-stromal cell communication, as well as possible toxic effects on matrix remodeling as an expression of hormonal regulation of these cells.

Investigations of Cervix and Vagina

Although studies of uterine epithelial and stromal cells have not yielded a practical biologic marker, investigations should be extended to the cervix and vagina. Vaginal and cervical changes reflect uterine changes, and the vagina and cervix are accessible; correlative studies of epithelial and stromal cells from these areas might identify markers useful for studies of human reproductive toxicology. Specific hormone-regulated changes in cervical glycoproteins (Chilton et al., 1981) and in the expression of cytokeratin patterns by differentiating vaginal epithelial cells (Kronenberg and Clark, 1985) offer strong support for such correlative studies.

TROPHOBLAST BIOLOGIC MARKERS

Prospective studies of human trophoblast development or studies seeking trophoblast signals of early human pregnancy (0 to 3 weeks) are so constrained by social and ethical restrictions (Andrews 1984a,b) that they are not practical. Use of nonhuman trophoblasts prevents the study of hCG as a peri-implantation marker, but laboratory and domestic animals offer many other advantages for critical experimental analysis. The following discussion outlines current understanding and suggests new approaches that might yield valid and accessible markers of trophoblast response to xenobiotic agents (Table 20-3).

Trophectoderm

The first cells to differentiate in the mammalian embryo are the trophectoderm cells. This differentiation occurs at embryo compaction during the cleavage stage, at which time the blastomeres assume an inside-outside orientation (Johnson et al., 1981). Although trophoblast cells do not contribute to embryo formation, they become an integral part of the placenta (Sherman and Wudl, 1976). Trophectoderm and its differentiated derivatives—the trophoblast giant cells (TGCs)—are involved intimately and structurally in most of the placental functions that are critical to viviparity (Billington, 1985). Thus, compaction and later the process of blastocoelation are critical points at which adverse effects of toxic agents could influence development.

The mammalian embryo comprises an inner cell mass—the presumptive embryo—and a blastocoele; both are surrounded by trophoblast cells. The trophectoderm cells covering the inner cell mass are termed "polar," and those surrounding the blastocoele are called "mural" (Fig. 20-2). Preimplantation and early postimplantation trophectoderm cells are proliferative and diploid.

When trophoblast cells lose contact with the inner cell mass or attach to the uterine epithelium, they lose their ability to divide. The trophoblast cells cease division and become giant cells, which have more DNA than other cells (Ilgren, 1983). In the human, the increase in DNA is accomplished mainly by cell fusion (the cytotrophoblast becomes the syncytiotrophoblast). In the mouse and the rat,

FIGURE 20-2 Schematic diagram of rat conception during midterm pregnancy. Dashed lines show regional limits of dissection to harvest individual groups of trophoblast giant cells. Diagram based on information presented by Davies and Glasser, 1968. Source: Glasser et al., 1987a.

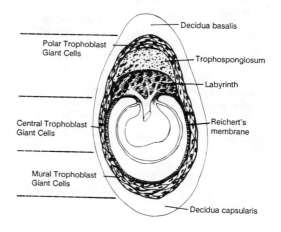

TABLE 20-3 Putative Biologic Markers to Assess Status of Trophoblast

Cellular or Developmental Stage	Biologic Marker	Comments
Syngamy	Number of cells with n + n versus 2n chromosomes	This marker might assess sensitivity of nuclear target 2n nucleus might be different target from an n + n nucleus; can evaluate nuclear determinant of sensitivity, such as maternal versus paternal genome; can use microinjection to introduce toxins to cytoplasm or nucleus
Totipotency	Number of viable embryos; 37-kilodalton one-cell embryo marker	These markers assess deletion of affected pronucleus, sensitivity of cleavage stages, expression of genome (paternal, maternal), and influence of xenobiotic on progression
Compaction	Allocation of cells inside versus outside; number of tight junctions and desmosomes; efficiency of ion channels	These markers assess effect on determination and differentiation of cell lineage between inner cell mass and trophectoderm
Blastocoelation	Macromolecular synthesis; expression of paternal genome markers; efficiency of proton pumps and ion channels	These markers assess transepithelial and paraepithelial transport; methods are improving rapidly
Early blastocyst	Histology of inner cell mass	Not well-studied markers of cell lineage; several models have been described (embryonic carcinoma cell model and embryonic stem cell model)
	Trophectoderm: concentrations of cytokeratins; changes in lectin-binding specificity; changes in profiles of surface and secretory proteins and glycoproteins; chorionic gonadotropin concentrations	These markers describe increasing complexity and cleavage stages of trophectoderm differentiation
	Concentration of early pregnancy factors	Role of these factors still unclear; existence is not confirmed in all species
Late blastocyst hatching	Loss of zona pellucida	This marker describes time of hatch and changes in biochemistry of zona pellucida
Recognition	Expression of trophectoderm recognition antigens	These markers describe immunologic response to embryo; can identify implantation defects
Attachment	Presence of specific lectin receptors and oligonucleotide acceptors; changes in surface and secretory proteins, and glycoconjugates; cytokeratin expression	These markers assess implantation and postrecognition attachment of trophectoderm and uterine epithelium; changes suggest reorganization of trophectoderm cell surface
Cessation of mitosis	Decrease in mitotic labeling index in mural trophectoderm followed by polar trophectoderm	This marker identifies effect of mitosis-inhibiting factors in initiating differentiation
Binucleation	Proportion of cells with 2n chromosomes; increase in nuclear and cytoplasmic areas	These markers assess endocycles and role of DNA in sensitivity

Cellular or Developmental Stage	Biologic Marker	Comments
Endoreduplication	Proportion of cells with > 4n chromosomes; shift in activity of DNA polymerases; patterns of DNA DNA fragments on southern blots or RNA fragments on northern blots	These markers assess sensitivity of trophoblast giant cells. (For instance, do changes in chromosome number alter sensitivity? Is entire genome being replicated? Are genes expressed differentially?)
Expression of specific proteins	Concentration of cytokeratins (40, 51, 55, 44 and 46 kilodaltons), actin, and tubulin	These are cell lineage markers for trophoblast giant cells; simple epithelial cell type not found in inner cell mass; markers also assess role of cytoskeleton in differentiation; changes in actin and tubulin not well studied, but do not appear to be specific responses
	Concentrations of early pregnancy factors	These markers assess establishment of trophoblast-uterine relationship; the role of these factors is not well defined
	Concentration of 37-kilodalton mitogen	This marker assesses fetal growth
	Concentrations of growth factors (insulin growth factor-I, -II, platelet-derived growth factor)	Role of these factors in organogenesis and fetal growth unknown
Secretion of tissue remodeling enzymes	Loss of adhesive properties, matrix proteins and receptors; migration and invasion of cell types; activity of specific enzymes, such as plasminogen activator	These markers assess sensitivity of differentiation to regulation by extracellular environment; specific enzymes not well studied; role of enzymes might be secondary
Synthesis and secretion of steroid hormones	Concentrations of progesterone, estrogen, and testosterone; density of steroid receptors	These markers can be used to study factors that initiate up and down regulation of steroids and explore possible autocrine and paracrine regulation; preimplantation synthesis and secretion have been validated only in pig, cow, and sheep; postimplantation validated in many species; role not well understood
Synthesis and polypeptide hormones	Humans: concentrations of hCG α and β subunits and hPL	These markers assess differentiation secretion of human cytotrophoblast to syncytiotrophoblast
	Concentrations of various pregnancy-associated factors	These measurements have been described, but roles are not adequately defined
	Rodents: concentrations of placental lactogens-PL-1 and PL-2	These markers assess differentiation of trophoblast giant cells; PL-1 and PL-2 are under different regulatory mechanisms

the DNA increase occurs in the nucleus via endomitotic and endoreduplicative mechanisms (Nagl, 1978; Ilgren, 1983).

Blastocyst attachment is an initial step in implantation, starting differentiation processes that manifest themselves in the establishment of a definitive placenta. Highly regulated structural and functional differentiation is found during the interval of blastocyst trophectoderm attachment to a receptive uterine epithelium and TGC apposition with elements of the maternal vascular system (Sherman and Wudl, 1976; Glasser and McCormack 1980, 1982). These modifications in trophoblast function support viability and the ordered patterns of embryonic growth and development. Progression of the trophoblast through the remodeled substrate of uterine decidual cells is ensured by increased secretion of progesterone by either the corpus luteum or the trophoblast cells.

Steroidogenesis

Studies of rat blastocysts and their trophoblastic outgrowths cultured in vitro (Beier and Beier-Hellwig, 1973) have yielded information on the patterns of secretion of various hormones. Rat blastocysts secrete progesterone at increased rates (0.1 to 0.5 pg/ml per blastocyst) during the initial phases of hatching—equivalent gestation day (EGD) 5—and outgrowth (EGD 6); the rate increases to 6 or 7 pg/ml per blastocyst on EGD 8-13 and then falls to a lower, but still high, rate of 4.5 pg/ml per blastocyst after EGD 14 (Fig. 20-3). Estradiol and testosterone patterns of the rat blastocyst and trophoblast outgrowths can be demonstrated, but they are so erratic as to be unreliable as markers of trophoblast differentiation.

Progesterone production by rat blastocyst or trophoblast outgrowths occurs during gestation (EGD 6 to 8) when maternal plasma progesterone already has increased to nearly 60 ng/ml (Glasser and McCormack, 1980). Maternal plasma progesterone plateaus at 90-120 ng/ml between EGD 10 and 12.

In the presence of this great pool of maternal plasma progesterone, no role has been identified for progesterone, estradiol, or testosterone contributed by the trophoblast cell. Trophoblast progesterone might have a paracrine role in ontogeny of decidual cell differentiation or an endocrine role in trophoblast regulation of its estradiol receptor (McCormack and Glasser, 1978) or in rat placental lactogen synthesis (Soares et al., 1985). If trophoblast steroidogenesis does have a regulatory role, how progesterone affects development, and the relative sensitivity of trophoblast steroidogenesis to toxic insult should be compared with the sensitivity of steroid hormone production by the corpus luteum.

Trophoblast Giant Cells and Placental Hormones

Differentiation of the fetal placenta is essential to embryonic development in mammals. In rodents, TGCs are an integral component of the fetal placenta. In vivo and in vitro primary and secondary TGCs are derived from mural and polar blastocyst trophectoderm, respectively. Cells of the ectoplacental cone are derived from polar trophectoderm (Ilgren, 1983) and

FIGURE 20-3 Steroid production by rat blastocyst outgrowths. Day 4 blastocysts (sp + = day 0) were recovered from uterus and cultured in groups of 10-15 in 3 ml of NCTC-135 plus 10% fetal calf serum in 35-mm plastic dishes. Medium was changed daily. Hormones assayed with specific radioimmunoassay of spent medium. Equivalent gestation day = age of blastocyst in culture equivalent to age that it would have been if left in utero. Source: Glasser et al., 1987a.

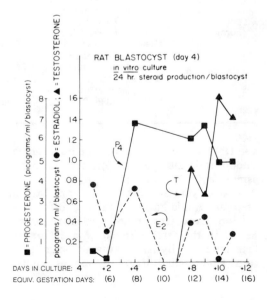

are precursors for additional secondary TGCs. TGCs cease to divide as a primary step in their differentiation from trophectoderm to ectoplacental cone cells; 15-25% of the trophectoderm cells become binucleate or multinucleate. Morphologically, TGCs are nondividing, polytene giant cells. The c number (number of copies of haploid DNA) in TGCs increases from about 2-4 to as much as 1,024, as a result of endoreduplication (Barlow and Sherman, 1972). In human trophoblast cells, no more than 40% become endoreduplicative (Friedman and Skehan, 1979). Functionally, TGCs express secretory proteolytic enzymes (proteases, collagenases, and elastases), steroid hormones (progesterone, testosterone, and estrogen), and, at midgestation, one or more peptide hormones (placental lactogens). Rodent trophoblast does not produce a hormone similar to hCG.

Placental peptide hormones display characteristics similar to hPRL and growth hormone (GH). For this reason, they have been termed placental lactogens (PLs) or chorionic somatomammotropins (Talamantes et al., 1980; Josimovich, 1983). PLs are the dominant trophic hormones affecting fetal development during the latter half of pregnancy (Brinsmead et al., 1981). They can regulate fetal tissues development directly. For example, ovine PL stimulates amino acid transport and ornithine decarboxylase activity in fetal tissues (Hurley et al., 1980; Freemark and Handwerger, 1982) and indirectly alters maternal protein, carbohydrate, and lipid metabolism (Kaplan and Grumback, 1981).

The biologic and biochemical characteristics of PLs vary among species, but all are secretory products of placental giant cells. In the human, PLs are produced by the syncytiotrophoblast (Watkins, 1978); in sheep, by the trophoblast binucleate cells (Martal et al., 1977; Watkins and Reddy, 1980; Wooding, 1981); and in rats, by the TGCs (McCormack and Glasser, 1980; Soares et al., 1985). These PL secretory cells differentiate from readily identifiable precursor cell populations—human cytotrophoblasts (Enders, 1965), sheep uninucleate cells (Wooding, 1981),

and mouse trophectoderm and ectoplacental cone cells (Rossant and Tamura-Lis, 1981; Ilgren, 1983). TGC differentiation has not been studied rigorously; the most thorough investigations have been done in rodents, primarily mice. The transformation of mouse ectoplacental cone cells to differentiated TGCs has been analyzed (Rossant and Ofer, 1977; Johnson and Rossant, 1981; Rossant and Tamura-Lis, 1981).

Rats and mice produce two types of PLs that can be distinguished biochemically, immunologically, and temporally by their appearance during pregnancy (Kelly et al., 1975; Robertson et al., 1982; Soares et al., 1985). The early form (PL-1) is present during midpregnancy (days 9-11 in the mouse; days 10-12 in the rat), and the late form (PL-2) predominates during the latter half of pregnancy. PL-1 has a higher molecular weight and is more acidic than PL-2 (Soares et al., 1985). Both are active in radioreceptor assays and bioassays for lactogenic hormones (Soares et al., 1985); however, neither is active in a growth hormone radioreceptor assay. Serum PL-2 has been measured in the mouse throughout pregnancy and around parturition (Soares et al., 1982; Soares and Talamantes, 1984). Ovaries have an inhibitory influence on serum PL-2 concentration; the fetus has a trophic influence (Soares and Talamantes, 1985). Distinct genetic differences in serum profiles of PL-2 have been reported (Soares et al., 1982). PLs are products of the trophoblast, rather than of other components of the placenta (Glasser and McCormack, 1981; Soares et al., 1985; Glasser and Julian, 1986), and the change from PL-1 to PL-2 that occurs in vivo has not been reproduced in vitro.

Differences in hormone production might be effects of local environment. For example, polar TGCs are in contact with maternal decidua basalis and the mesenchymally induced trophospongiosum of the chorioallantoic placenta; mural TGCs are between the decidua capsularis and the parietal endoderm (Davies and Glasser, 1968). Production might be mediated by signal differences from the different environments or by their interpretation by TGCs. That suggests that residual effects of xenobiotic agents in undifferentiated uterine stromal cells might influence trophoblast endocrine function when these cells decidualize. The absence of qualitative regional effects confirms that all TGCs are similar; functional differentiation of an individual TGC results in sequential expression of PL-1 and PL-2 by the same cell. Thus, the response of the single trophoblast target to toxic exposure is influenced by the developmental stage at which the trophoblast cell is at risk.

Expression of PL-1 during organogenesis might be coincidental but PL-1 and organogenesis could be more closely related—possibly through a reciprocal relationship with insulin growth factors (Adams et al., 1983). Thus, early exposure to a xenobiotic agent might produce more serious consequences than if exposure occurred as TGC differentiation and organogenesis were terminating.

Trophoblast Giant Cell Cytoskeleton

Morphologic and functional differentiation have been linked to alterations in the expression of cytoskeletal proteins; therefore, microtubules and intermediate filaments—both primary cytoskeletal proteins—in endocrine-competent TGCs (Glasser, 1986) have been analyzed to identify markers.

Tubulin and polymerized microtubules have been identified in TGCs, but the microtubule organizing center has not been described. TGCs from trophectoderm are unlike most other cells. A decentralized system for microtubule renucleation occurs at multiple sites throughout the cytoplasm, and microtubules may be sensitive and ubiquitous targets for xenobiotic agents. Disruption in microtubule assembly might have extensive effects on cell function.

In rodents, trophectoderm cells are the first embryonic cells to exhibit intermediate filament proteins (Glasser, 1986). Mouse preimplantation blastocyst trophectoderm cells displayed two intermediate filament proteins (54 and 46 kilodaltons), identified as cytokeratins. In contrast, midgestation TGCs displayed

not only major components of 54- and 46-kilodalton intermediate filament proteins, but 52-, 45-, 43-, and 40-kilodalton species. Of these additional keratins, the 52- and 40-kilodalton species were most prominent (Glasser and Julian, 1986).

No specific function has been assigned to any of the proteins of the various cytokeratin gene families, but these additional landmarks might be important in monitoring functional and morphologic differentiation of specialized cells (e.g., ectoplacental cone cells and other stem or precursor cell populations) (Venetianer et al., 1983).

In vitro analysis of morphologic and functional differentiation of blastocyst outgrowths and isolated TGCs have been effective in identifying valid biologic markers of processes essential to establishment of the fetal placenta (Copp, 1979). Other markers have not been discounted, but these experiments have determined PLs to be unique among the candidates. Identification of differentiated TGC as the only cellular source of these hormones is important. Further studies to assess the risk of a presumably pregnant female can now focus on the development of the trophoblast and its expressed secretions.

Ectoplacental Cone Cells

Ectoplacental cone cells arise from polar trophectoderm and are diploid and proliferative. Because of their numbers, endoreduplication, and differentiation in culture, they present a model that might resolve some questions posed by studies of other trophoblast cell populations. Mouse ectoplacental cone cells transplanted ectopically or cultured transform into giant cells (Rossant and Ofer, 1977; Johnson and Rossant, 1981; Rossant and Tamura-Lis, 1981). TGCs of rodents endoreduplicate their nuclear DNA; they are polyploid and synthesize proteins that are characteristic of TGCs, ectoplacental cone cells (Johnson and Rossant, 1981). The presence of the inner cell mass adjacent to ectoplacental cone cells is believed to maintain proliferation and inhibit transformation to TGCs (Rossant and

Ofer, 1977). As gestation progresses, some ectoplacental cells are pushed farther from the inner cell mass or its derivatives and transform into TGCs. Thus, ectoplacental cells provide a reservoir of precursor cells that contribute to trophoblast growth and expansion during the second half of pregnancy.

When ectoplacental cells lose contact with the inner cell mass and its derivatives (Rossant and Ofer, 1977; Wooding, 1982), they lose their ability to divide, and they become giant cells with nuclear DNA contents greater than 4 c. Very little information has been developed since the advent of recombinant DNA methods regarding the organization of the TGC genome or gene expression of genomic DNA, which can increase from 2 to 1,024 c. Extensive endoreduplication during the course of normal differentiation makes trophectoderm cells, ectoplacental cone cells, and TGCs unique mammalian cells—and thereby unique models to study gene expression. Studies of the locations of repetitive-DNA sequences and changes of protein expression during endoreduplication in TGCs offer opportunities to learn about the cellular and molecular base of differentiation.

Regulation of Trophoblast Hormone Synthesis

For the process of giant cell differentiation, the relative advantage of nuclear DNA endoreduplication compared with regulation of early postimplantation interactions is unknown. PL expression might correlate with replication of the entire trophoblast genome, rather than with amplification (Barlow and Sherman, 1972; Sherman et al., 1972). Signals that initiate PL-1 synthesis in the enlarging genome, signals that direct the sequence rearrangements obligatory for the transition of PL-1 to PL-2, and what initiates PL-2 synthesis are unclear. In analysis of human trophoblast, those questions have to do with regulatory foci coincident to cytotrophoblast differentiation. Xenobiotic agents might interrupt or redirect normal development of the trophoblast.

EXTRAPOLATION TO HUMAN TROPHOBLASTS

Experiments with in vitro models of trophoblast cells have identified the importance of their postmitotic differentiation and their derivatives in the control of normal mammalian development. Markers of structural and functional events critical to differentiation also have been reported.

Although rodent trophoblast experiments have enlarged understanding of some aspects of postimplantation biology, they have not yielded markers that reliably signal the status of the trophoblast during the high-risk periods before and around implantation in mice, rats, or humans. The residence time of the free blastocyst of these mammals in utero is rather short. In contrast, some livestock—such as sheep, pigs, and cows—have long uterine residence time of the free blastocyst (more than 12 days), and the blastocysts synthesize and secrete gonadal signals that mark the immediate preimplantation period (Heap et al., 1981). Generation of steroid implantation signals by blastocysts with short residence in utero (Dickmann and Dey, 1974) has not been confirmed (Heap et al., 1981).

Human Chorionic Gonadotropin

Human chorionic gonadotropin is a practical marker of human trophoblast cell development. The major recognized action of hCG is its role in regulation of steroidogenesis in the corpus luteum. Thus, hCG serves a pivotal function in pregnancy maintenance. hCG also has been implicated in regulation of steroidogenesis in the fetal testis and the fetal adrenal gland (Stock et al., 1971). Its role in the regulation of placental steroidogenesis is controversial. In situ hybridization has shown that the α-subunit mRNA of hCG is found in cytotrophoblasts, syncytiotrophoblasts, and the intermediate forms between those two definitive cell types (Pijnenborg et al., 1985). However, the β-subunit mRNA can be found only in the intermediate form and syncytiotrophoblast (Dreskin et al., 1970; de Ikonicoff and Cedard, 1973; Hoshina et al., 1985). Those findings argue strongly that trophoblast differentiation must be in progress before the β-subunit becomes available for dimerization with the α-subunit and before the rapid secretion of the intact hCG molecule.

The widespread use of the radioimmunoassay (RIA) for hCG derives from ready availability of specific antibodies at reasonable cost, ease of using the assay, extensive background on interpretation of plasma titers, and absence of more specific markers. (See Chapter 15 for discussion of the development of these assays and the problems with using them in epidemiologic studies.) Use of RIA for hCG has received general acceptance for determining whether pregnancy has begun and is based on the rationale that a rise in plasma hCG reflects the presence of a functional trophoblast and verifies the existence of pregnancy (Canfield et al., 1984).

Decrease in hCG titers signals interruption of pregnancy, but does not identify cause or tissue site of initial damage.

The earliest measurement of complete hCG depends on differentiation of enough intermediate and terminal syncytiotrophoblast cells to produce hCG that can be detected with the assay. This occurs 3-4 days after the blastocyst has become implanted in the uterine endometrium. Therefore, hCG is not an effective marker of events associated with transition from morula to blastocyst, of entry into the uterus, and of various events before and around implantation that occur during days 5-10 (Fig. 20-4). Saxena and colleagues (1974) reported that a luteotropic hCG-like factor is produced by the human blastocyst and is detectable immediately before implantation (day 6), but this has never been confirmed. Aggressive prospective research with animal or experimental models is required to identify trophoblast

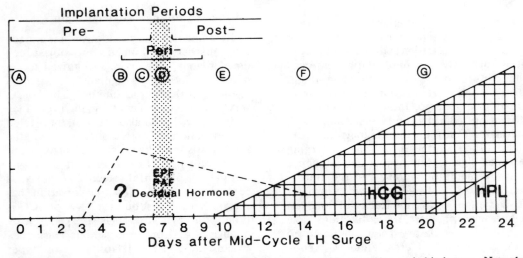

FIGURE 20-4 Some major events occurring in utero that define peri-implantation period in human. Note absence of confirmed markers available. Sensitive radioimmunoassays of hCG that use specific monoclonal antibodies do not detect hCG until day 10, at least 3 days after implantation and initiation of trophoblast differentiation. EPF and PAF are trophoblast products that might be used as markers of risk status of embryo. Decidual luteotropic hormone might serve as marker of endometrial condition. Source: Glasser et al., 1987a.

signals that might be expressed during the critical high-risk periods of days 5-7 and 5-10.

Placental Lactogens

Although PLs are not hormones of the period around implantation, they are important to fetal well-being. PLs are products of transcription and translation of the five-gene human placental lactogen (hPL) family. Translation and transcription take place in the syncytiotrophoblast, but not cytotrophoblast (McWilliams and Boime, 1980; Boime et al., 1982; Hoshina et al., 1982, 1985).

Biologic actions of hPL are interpreted in terms of its homology to growth hormone and prolactin. Effects of hPLs have been demonstrated in several nonprimates (Friesen, 1966), but its role in humans has not been resolved (Josimovich, 1966). Whether hPL directly affects fetal lipid and carbohydrate metabolism in response to transient fluctuations in nutrient availability or has a chronic role in modifying the set point and response time of various systems of intermediary metabolism is unclear.

PLs might have indirect effects and be mediated by reciprocal action with insulin growth factor (IGF-I) (Pijnenborg et al., 1985). Some data have demonstrated that hPL is not required to maintain human pregnancy (Parks et al., 1985).

An injury to trophoblast cells that prevents normal expression of hCG would interfere with early events involved in establishing the placenta and also interfere with hPL expression. However, the sensitivities of hPL and hCG genes and their mRNAs to particular xenobiotic agents could be different, and hPL and hCG gene products should be monitored. Effects on hPL would be detected only between weeks 2 and 3 of gestation, when hPL is scheduled to be secreted.

Pregnancy-Associated Factors

Owing to the lack of an hCG homologue in nonprimate mammals and the inability to detect very early status, further search is under way for macromolecules that are present in females during early gestation, but absent in the nonpregnant females. A directory of more than 20 factors specific or unique to pregnancy has been developed (Bohn, 1985). New proteins detected with immunologic methods in placental extracts or in sera of pregnant women continue to be reported (see Chapter 19 and Bell, 1983; Ellendorff and Koch, 1985); these proteins range in molecular weight from 25,000 to 2.1 million and are mainly glycosylated proteins. Detectable amounts of factors thought to be pregnancy-specific have been immunologically assayed in oviduct secretions and in seminal plasma.

The usefulness of these pregnancy-associated factors as markers is restricted by important problems (Chard, 1985). In many cases, a secretion has neither cell- nor tissue-specific origin. The presumed uniqueness of a putative marker might be related to design of an appropriate assay and to physiologic relevance of that assay. Much of the debate regarding specificity of the early pregnancy factors rests on the functional significance of rosette inhibition assay (Beverley, 1985). Analyzing and identifying the principal role of single-pregnancy factors has led to the suggestion that they are better understood if considered as categories of pregnancy-associated proteins, rather than as single, nominally specific factors (Table 20-4) (Chard, 1985).

Some proteins contribute to trophoblast maintenance and function (perhaps through their ability to bind steroids), some have protease inhibitory functions, and others have been implicated in the immunobiology of pregnancy. The time of synthesis onset and the identification or relevance of a suggested specific function are not clear. One protein, pregnancy protein 14 (PP-14, an α_1 microglobulin), has been shown to be in high concentrations in seminal plasma and increases rapidly in the plasma of pregnant females. For these reasons, it has been suggested that PP-14 functions at implantation and has some role in establishing the hemochorial placenta. No protein, even hCG, can be related unequivocally to specific events associated with fertilization, concep-

TABLE 20-4 Categorization of Pregnancy-Associated Factors

	Source	Functions	Control

Category 1
hCG, hPL,
  specific protein
  (SP-1),
  ß1-glyco-
  protein
	Trophoblast
	(syncytio-
	trophoblast)
		Regulation of
		  growth and
		  differentiation
		  (autocrine? para-
		  crine? endocrine?)
			Number of synthetic units;
			  changes in blood flow, export,
			  delivery; steroids; growth factors

Category 2
Pregancy protein 5,
  PAPP-A
	Trophoblast
	(syncytio-
	trophoblast)
		Local immune
		  and coagulation
		  reactions
			Not known

Category 3
Binding proteins,
  PZP
Pregnancy
protein 12
	Endometrium;
	  decidua; maternal
		Binds small molecules
		  molecules
			Estrogen and progesterone;
			  may represent maternal re-
			  sponse to pregnancy

Pregnancy
protein 14
	Liver		Does not include endometrial
			  surface and secretory proteins
			  and glycoconjugates

tion, or the period before or around implantation (Billington, 1985).

The questions of source and function of these proteins might be resolved in part through the study of differentiation of the human trophoblast cell in culture. A variety of methods, including recombinant-DNA technology, could be used to analyze the transition from cytotrophoblast to syncytiotrophoblast. The technology also would be excellent for investigating the consequences of introducing xenobiotic agents at different steps in trophoblast differentiation. Friedman and Skehan (1979) described a directory of morphologic and functional properties that characterized the transition of cytotrophoblastlike (CTL) cells of the BeWo choriocarcinoma cell line to syncytiotrophoblastlike (STL) cells. Cytologically, CTL and STL cells were identical with their counterparts in utero.

BeWo CTL cells constitute 96-99% of the cell types of the stored cell line. Cultured in the presence of subthreshold concentrations of methotrexate, the CTL cells differentiate. At the end of a 96-hour culture, more than 90% of the cells assume STL structure and function. When methotrexate is removed, the STL cells become CTL-like. Although methotrexate suppresses DNA synthesis, the DNA content per cell increases by 66% coincidentally with expression of increased hCG.

Very little work has been done with this model to study either cell invasiveness or hCG synthesis, nor have the intermediate forms been analyzed. A substantial data base has resulted from studies of molecular biology of gene regulation of hCG (Boime et al., 1982), the endocrine physiology of response to cyclic AMP and the molecular basis for second-messenger response (Hilf and Merz, 1985), and the cell biology of cytoskeletal changes (Friedman and Skehan, 1979; Glasser, 1986). Such data would make it possible to expand the utility of hCG as a marker of risk associated with exposure to xenobiotics in early pregnancy.

21

Physiologic Assessment of Fetal Compromise

This chapter discusses several clinical and laboratory procedures that have made the measurement of potential biologic markers of fetal development possible. Most of these assessments are performed during the organogenesis (beginning 14 days post-conception) and fetal growth periods.

Despite technologic and methodologic achievements, no specific biologic markers are available to indicate that exposure to a xenobiotic is associated with specific cellular, subcellular, or pharmacodynamic events. This chapter considers the general roles that various diagnostic methods and biologic markers could have in understanding regulation of fetal growth and differentiation and the way xenobiotic agents affect normal events.

Understanding the physiologic and endocrinologic bases of fetal development is a major goal of perinatal biology. During the last decade, many technologic developments have allowed precise evaluation of the fetus in utero and diagnosis of abnormalities. The following are examples of important contributions to biologic markers of pregnancy:

- Concentration of phospholipids, e.g., lecithin:sphingomyelin ratio and phosphatidylglycerol in amniotic fluid as related to fetal lung maturity.

- Genetic screening of amniotic cells.
- Concentration of estriol in maternal urine and serum.
- Direct measures of growth including femur length and biparietal diameters.

These clinical assessments have been most helpful in modifying clinical care and ascertaining appropriate in utero development. Maternal serum alpha-fetoprotein (AFP) screening is a test to assess fetal well-being. High concentrations of maternal AFP are found when fetuses have open neural tube defects (see Appendix). Increases in maternal serum AFP with no detectable amniotic fluid AFP and normal anatomic features as assessed with ultrasonography suggest that a pregnancy is at high risk for preterm labor, low birthweight, and fetal wastage.

Developing and using biologic markers of fetal growth and development are difficult. The fetus is not easily accessible, and obtaining fetal cells, tissue, or amniotic fluid for marker determination can be burdensome. Several fetal physiologic variables can be measured, but most are nonspecific, imprecise, and insensitive as markers of toxicity. Major hormonal and metabolic differences between the fetus and the adult hamper comparisons. Furthermore, fetal responses vary widely among species; a marker in one species might not be informative in another. The fetus consists of cells of numerous types, each with unique responses to various agonists; the reactions of cells and tissues vary with maturity and stage of development; and fetal maturity at birth differs among species. Many studies have been conducted on prenatal exposure to toxicants, such as methyl mercury (Koos and Longo, 1976; Clarkson, 1987), lead (Hoffer et al., 1984), alcohol (Jones et al., 1973), and tobacco (Longo, 1982). Despite knowledge of these and other chemical and toxicant effects, little is known of critical doses or exposures, critical periods during development, or how biologic variation produces response variation. Although one can observe and measure gross effects, few indexes detail more subtle cellular and subcellular changes.

Before 1960, most birth defects were considered to be genetic in origin. The fetus was believed to occupy a privileged site within the uterus, protected from the effects of environmental agents to which the mother might be exposed. Association between maternal rubella infection and abnormal fetal development was recognized in the early 1940s, but adverse effects of prenatal exposure to a drug were not reported until 20 years later, when thalidomide was determined to be the cause of serious developmental defects. A host of chemicals, drugs, and environmental agents have since been implicated as teratogens, and most drugs and chemicals are suspected of being able to cause a congenital malformation. Chemical teratogens have been classified by their relationship to fetal abnormalities (Simpson et al., 1982; Jones and Chernoff, 1984; Sever and Brent, 1986). Some drugs have potential adverse effects on the fetus (Table 21-1); others have a questionable relationship to fetal anomalies; still others, including several classed as "litogens"—drugs that have been mentioned in legal actions (Mills and Alexander, 1986)—seem to have no relationship to fetal abnormalities.

Most pregnant women are exposed to a variety of agents. The National Institutes of Health found that the average pregnant woman takes five drugs (including nutritional supplements) during the course of her pregnancy, and about half the total drug consumption occurs during organogenesis, i.e., during the first trimester (Sever and Brent, 1986; Shepard, 1986). Some 2-3% of developmental defects are known to be due to drugs and chemicals, and 65% are of unknown origin; the size of the

TABLE 21-1 Some Drugs with Potential Adverse Effects on the Neonate

Drug	Potential Adverse Effect
Azathioprine	Decreased immunologic competence (Lower et al., 1971; DeWitte et al., 1984)
Cannabis	Neurobehavioral abnormalities (Fried, 1980)
Chloramphenicol	Gray syndrome (Sutherland, 1959)
Cocaine	Neonate withdrawal; cerebral infarction (Chasnoff, 1985)
Diazepam	Floppy baby syndrome (Owen et al., 1972; Cree et al., 1973; Gillberg, 1977; Speight, 1977)
Ethanol	Cardiac malformations; mental retardation (Mulvihill, 1976)
Heroin/narcotics	Prenatal and postnatal growth retardation, neonate withdrawal; microcephaly (Stone et al., 1971; Zelson et al., 1971; Naeye et al., 1973; Rothstein and Gould, 1974; Fricker and Segal, 1978)
Lithium	Cardiac malformations (Schou et al., 1973)
Oxytocin	Hyperbilirubinemia (Sims and Neligan, 1975; Chew and Swan, 1977; Beazley, 1984)
Phenobarbital	Neonatal bleeding (Desmond et al., 1972)
Propranolol	Hypoglycemia and bradycardia (Cotrill et al., 1977)
Reserpine	Nasal congestion (Budnick et al., 1955)
Salicylates	Platelet dysfunction (Rumack et al., 1982)
Thiazides	Thrombocytopenia, electrolyte imbalance (Trimble et al., 1964; Alstatt, 1965; Anderson and Hanson, 1974)
Tobacco	Intrauterine growth retardation (Werler et al., 1985)
Warfarin	Bleeding disorders (Warkany, 1976; Hall et al., 1980)

latter category might reflect lack of knowledge and shows the need for markers to identify specific chemicals and their relation to defects. Table 21-1 lists potential adverse effects of several commonly used drugs.

Some environmental chemicals can harm the fetus and be particularly troublesome, because they can be pervasive and unsuspected, such as chlorbiphenyls, laboratory solvents, naphthalene, and organic mercury. The effects of exposure to polybrominated biphenyls are discussed in Chapter 24. This chapter focuses on assessment procedures. The need for biologic markers of exposure to and effects of these and other chemicals is enormous.

ULTRASONOGRAPHY

Of the tools used successfully for fetal diagnosis, few have attracted as much attention as ultrasonography (ultrasound). During the last decade, antenatal ultrasonographic resolution has increased dramatically, from detection of such gross defects as anencephaly to detection of subtle abnormalities of the brain, cardiovascular system, kidneys, and other organs (Table 21-2) (Callen, 1983). Although instrument resolution and operator skill have improved, few studies have rigorously examined diagnostic accuracy for specific anomalies. The accuracy of high-resolution ultrasonography is of critical importance. Anencephaly can be diagnosed ultrasonographically with 100% certainty (Chervenak et al., 1984). Anomalies manifested by two or more embryologically related markers, each detectable ultrasonographically (such as allobar holoprosencephaly) can be diagnosed reliably (Chervenak et al., 1985). Ultrasonography is less accurate for single abnormalities of dimension (such as microcephaly) or those in which the structure of interest cannot be clearly distinguished.

Use of ultrasonography has advanced rapidly for diagnosis of fetal size and, to a lesser extent, maturity. Some structural measurements are useful in these diagnoses, including biparietal diameter and head circumference, abdominal circum-

TABLE 21-2 Diagnostic Ultrasound and Biologic Indicators

MATURATIONAL OR MORPHOMETRIC BIOLOGICAL INDICATORS

Fetus
  Gross malformations
  Estimated fetal weight
Head: biparietal diameter circumference
  Abdomen: circumference
  Head/abdomen: circumference
  Femur: length
Heart: right/left ventricle malformations
Placenta
  Size
  Maturity

FUNCTIONAL OR PHYSIOLOGICAL BIOLOGICAL INDICATORS

Fetus
  Brain: cerebral blood flow velocity
Cardiovascular: heart rate; cardiac output; umbilical blood flow velocity; umbilical systolic/diastolic
Respiratory: breathing movements; somatic movements
Placenta
  Blood flow: uterine; umbilical

ference, and length of the femur and other long bones.

Ultrasonography also is valuable to diagnose fetal heart defects and cardiovascular function (Knochel et al., 1983). Combining real-time and static techniques has enabled practitioners to monitor cardiac structure, rhythm, chamber size, and pericardial effusion (Knochel et al., 1983). Measurement of vessel diameter using ultrasonography in conjunction with pulsed-doppler technology that measures blood-flow velocity through cardiac valves and in major vessels makes calculation of cardiac output and blood-flow rate possible. More than 100 major malfunctions and malformations, including those of the genitourinary tract, abdominal wall, and other systems, can be diagnosed with these techniques (Allan et al., 1984; Weaver, 1988). Doppler profiles of the arcuate artery at midgestation might identify fetuses at risk for death and intrauterine growth retardation and doppler studies during the third trimester could be used to assess the umbilical vessel to judge fetal well-being.

Ultrasonography increases the safety and effectiveness of other diagnostic methods, such as amniocentesis, chorionic villus sampling, fetoscopy, and fetal blood sampling (Ward et al., 1983; DeVore and Hobbins, 1984; Hobbins et al., 1985; Katayama and Roesler, 1986).

AMNIOCENTESIS

The ability to enter the amniotic cavity to sample amniotic fluid without appreciable risk to mother or fetus allows several diagnostic tests to be performed that are indicative of fetal well-being. Amniocentesis initially was used to estimate concentrations of bilirubin and related pigments in amniotic fluid and thereby identify hemolytic disease. Amniocentesis performed later in gestation is now used primarily to determine the relative concentration of surfactant-active phospholipids released from the fetal lung (i.e., measure the lecithin-to-sphingomyelin ratio), in an effort to assess the risk of respiratory distress syndrome in a premature infant. Amniocentesis is also used to measure individual phospholipids, such as phosphatidylglycerol; it is increasingly useful for identifying hereditary disorders (O'Brien, 1984; Johnson and Godmilow, 1988). Measurement of abnormal biochemical processes is useful in antenatal diagnosis of about 100 inborn errors of metabolism. All chromosomal anomalies—i.e., trisomy 21, 13, and 18 and triploidy—are potentially diagnosable from amniotic fluid samples (Roberts et al., 1983; McDonough, 1985). In addition, measurement of AFP in amniotic fluid can suggest the presence of neural tube defects, congenital nephrosis, or trisomy 21. Techniques of molecular biology can be applied to cell analyses for antenatal diagnosis of chromosomal abnormalities, genetic enzymatic defects, and DNA adducts.

Amniotic cells and fluids obtained through amniocentesis are among the best markers of development available and are now available at less than 14 weeks of gestation (Johnson and Godmilow, 1988). Combined ultrasound, amniocentesis, alpha-fetoprotein analysis, and other methods can detect prenatally more than 700 fetal

disorders (Weaver, 1988). The success of these genetic markers and markers of specific organ function demonstrates the potential for developing markers for toxicity screening. Fetal lung maturity markers can be modified with drug therapy; that demonstrates that lung maturation is sensitive to xenobiotic intervention. New applications or screens for embryonic tissues might provide similar specificity and sensitivity necessary to examine the ability of a conceptus to respond to selected xenobiotics. A similar clinical application likely to be more widely available is umbilical blood sampling with cordocentesis (see below) (Daffos et al., 1985). Blood samples would make it possible to diagnose fetal conditions, such as hypothyroidism, in utero and permit appropriate therapy to be initiated.

CHORIONIC VILLUS SAMPLING

Chorionic villus sampling (CVS) allows cells to be obtained that reflect the genetic constitution of the conceptus (Green et al., 1988). Used during weeks 10-12 of the first trimester, CVS provides diagnostic results within a few days to 2 weeks. A small sample of chorionic villi is aspirated through a flexible catheter that is passed through the cervix under sonographic guidance; alternatively, an aspiration needle can be passed transabdominally (Smidt-Jensen and Hahnemann, 1984). As with cells obtained through amniocentesis, heritable disorders can be diagnosed with enzyme assays, by chromosomal number or abnormality, and with DNA analysis (Jackson, 1985; Green et al., 1988; Rhoads et al., 1989). However, villi can be contaminated with maternal decidua, and chromosomal mosaicism (i.e., the incomplete expression of genes on all chromosomes) appears to be more common than in amniotic fluid cells.

Additional screening of genetic information, as well as enzymes and cDNA probes or DNA adducts, provides an important tool to assess risk associated with exposure to toxic substances. CVS is a valuable technique to assess the effects of xenobiotics during early organogenesis, when critical periods of structural organ development occur. As microanalytic methods are developed, xenobiotic concentrations in trophoblast tissue might also be determined.

FETOSCOPY

Fetoscopy to allow direct visualization of the fetus and placenta without disrupting pregnancy (Rodeck and Nicolaides, 1983a) would permit external fetal anomalies to be detected and allow fetal tissues and blood vessels to be sampled directly from the umbilical cord or placenta.

Fetal well-being in Rh isoimmunization can be assessed by measuring fetal and maternal antibody changes. Fetoscopy holds great promise for the detection of biologic markers; however, it is a research procedure carried out in few academic institutions. Until the risks of spontaneous abortion associated with fetoscopy are reduced, the general applicability of this technique for toxicity screening is limited.

FETAL BLOOD AND TISSUE SAMPLING

Fetal blood obtained in utero can be used for antenatal diagnosis of hemoglobinopathies, coagulation defects, metabolic and cytogenetic disorders, immunodeficiencies, and infections (Rodeck and Nicolaides, 1983b; Daffos et al., 1985). Originally, blood specimens were collected directly from umbilical or placental vessels with a fetoscope. More recently, percutaneous sampling of fetal cord blood with a needle and sonographic guidance (cordocentesis) has allowed fetal blood sampling with less risk to the fetus (Daffos et al., 1985). Analysis of fetal blood permits many conditions to be diagnosed and various biologic markers to be detected (Hobbins et al., 1985).

Fetal skin and liver have been sampled successfully and used for histologic and biochemical studies when amniocentesis or fetal blood sampling could not provide the necessary information. Biopsy instruments introduced under fetoscopic and ultrasonographic guidance have been used to collect samples for detection of genetic disorders, such as epidermolysis and ichthyosis, glycogen storage disease,

and ornithine transcarbamylase deficiency of the liver. The same tissues might be used to study more subtle markers, such as DNA probes.

MEASURING FETAL BODY AND BREATHING MOVEMENTS

In normal pregnancies, fetal movements increase from about 30 per 12 hours at 24 weeks of gestation to about 130 at 32 weeks and then decrease to about 100 at term; however, these values vary considerably (Ehrström, 1987). The fetus that is felt by the mother to move consistently is usually healthy. A sudden decrease in fetal activity suggests fetal compromise (Sadovsky and Polishuk, 1977).

Fetal breathing activity can be detected with ultrasonography and other techniques by 11 weeks. By 16 weeks, breathing movements are sufficiently intense to move small amounts of amniotic fluid in and out of the respiratory tract. During the third trimester, breathing movements average about 50 per minute and are episodic. In sheep, much of the activity occurs in association with periods of rapid eye movements and high-frequency, low-voltage, electrocortical activity (Koos, 1985). The frequency of fetal breathing movements decreases in association with hypoxemia, asphyxia, hypoglycemia, maternal smoking, ethanol ingestion, and other stresses. Although breathing and body movements are indexes of fetal well-being, changes in these movements are nonspecific and thus are crude biologic markers.

ELECTRONIC FETAL HEART-RATE MONITORING

Antepartum and intrapartum surveillance of fetal well-being with early detection of fetal distress is most commonly accomplished by monitoring the fetal heart rate. During uterine contractions, uteroplacental blood flow temporarily decreases. Antepartum electronic fetal heart rate monitoring is associated with a high specificity 99% but sensitivity of 40-50% (Lavery, 1982).

The normal fetal heart rate is 120-160 beats per minute. Rates higher than 160 beats per minute are recognized as tachycardia and lower than 120 beats per minute as bradycardia. Fetal heart rate variability is divided into short-term or beat-to-beat variability and long-term variability, which consists of regular crude sine waves with a cycle of approximately 6 beats per minute or greater.

Periodic changes in the fetal heart rate are evident. Early decelerations occur with uterine contraction and are thought to represent a vagal reflex due to mild transient hypoxia not associated with fetal compromise. Late decelerations are thought to be caused by myocardial hypoxia for the period of a contraction, when the deoxygenated bolus of blood from the placenta insufficient to support myocardial action. Variable decelerations differ in duration and timing as related to a uterine and are thought to represent umbilical cord compression. There are variable degrees of variable decelerations.

A reactive, reassuring nonstress test is one in which at least two fetal movements occur within 20 minutes with accelerations of fetal heart rate above baseline by 15 beats per minute, with long-term variability greater than 10 beats per minute and a baseline between 120 and 160 beats per minute (Keegan and Paul, 1980).

A contraction stress test can be performed by nipple simulation or intravenous infusion of oxytocin. An adequate test is achieved when three contractions are obtained within 10 minutes. A positive result demonstrates persistent late decelerations associated with more than 50% of the uterine contractions in 10 minutes. A negative result is a normal baseline fetal heart rate with no late decelerations. A negative result is associated with fetal survival for 1 week or more for more than 99% of patients (Collea and Holls, 1982).

Fetal heart-rate monitoring has been the subject of many reports. Unfortunately, empirical observations and associa-

tions have not been matched by deep understanding of the physiologic bases of the phenomena observed. Although of proven usefulness in the detection of fetal distress, changes in heart rate and in beat-to-beat variability are sufficiently nonspecific to restrict their role as markers.

BIOPHYSICAL PROFILE

Evaluation of several factors of fetal well-being have been combined into the "biophysical profile." Heart-rate reactivity is determined with electronic monitoring and diagnostic ultrasonography is used to assess gross body movements, muscle tone, breathing activity, and amniotic fluid volume. These factors are evaluated on an Apgar-like scale (Platt et al., 1985). Again, although apparently of empirical value, each factor is nonspecific and of only limited diagnostic value.

MAGNETIC RESONANCE IMAGING

Several reports of organ and tissue imaging with magnetic resonance imaging (MRI) during pregnancy have suggested the usefulness of MRI (F.W. Smith et al., 1983; I.R. Johnson et al., 1984; Kay and Mattison, 1986; McCarthy and Haseltine, 1987; Mattison and Angtuaco, 1988; Mattison et al., 1099); however, its safety during gestation has not been established. The high magnetic fields required might affect embryonic and fetal development, particularly in pregnant women who work with the instrument over a long term. In experimental animals, use of ^{31}P-MRI has great potential to determine functional metabolic correlates, temporal relationships, and intracellular actions of chemicals in the heart, liver, and placenta. Use of paramagnetic ions also might be applicable for specific visualization of the placenta and conceptus.

22

Biologic Markers of Exposure During Pregnancy: Pharmacokinetic Assessments

This chapter discusses pharmacokinetic considerations that are unique to pregnancy as well as standard considerations. These have important implications for the interpretation of concentrations of chemicals and their metabolites either from the maternal tissues or fluids or from the fetal tissues or fluids.

Issues related to pharmacokinetics have become central to toxicity risk assessments (Kuemmerle and Brendel, 1984; Fabro and Scialli, 1986). The new field of toxicokinetics and evaluation of biologic markers of exposure focuses on the presence and consequences of xenobiotic chemicals over specific intervals. Studies are being undertaken to assess the absorption, distribution, and excretion of a substance, as well as its metabolism to other substances and their distribution, metabolism, and elimination. With the rapid development of highly sophisticated and sensitive analytic techniques, exposures can be evaluated to determine an organism's risk of toxic effects. Metabolite profiles and measurements of tissue content, chemical half-life in various tissues and blood, and total excretion can be used to determine therapeutic efficacy or to indicate toxic response. At issue are what measurements should be made, what media or tissues should be used,

and what techniques can best determine the expression or potential expression of toxic action.

Many general pharmacokinetic considerations are relevant, including:

- Are the techniques for analyzing a chemical compound sufficiently sensitive to detect it at low, nontoxic concentrations?
- Do the techniques fulfill all aspects of quality-control standardization?
- Does the substance reach the organ or target site in a concentration sufficient to produce the effect noted?
- Does the effect stop when the organ concentration of the substance decreases to a particular point?
- What intervals are necessary for repeated administration of the substance to elicit or maintain the effect noted?
- What conditions in the organism will modify the effect noted, in light of an evaluation of the compound's structure-activity relationship?

In many instances, only small tissue samples can be obtained because of the inaccessibility of the conceptus, although new noninvasive procedures such as ultrasonography and magnetic resonance imaging are becoming available. The only tissues or fluids readily available from the conceptus are the placenta at delivery or after therapeutic interruption of pregnancy, placental tissue obtained by chorionic biopsy, amniotic fluid, on rare occasions fetal blood samples, and fetal tissue taken by biopsy before delivery (Table 22-1).

The following critical pharmacokinetic factors are peculiar to pregnancy:

• Dramatic and continuing physiologic and biochemical changes in mother and conceptus that can persist throughout gestation.
• Two separate and distinct genomes existing in the same organism (the mother).
• Two separate and distinct blood supplies with a unique interface at the trophoblast.
• Rapid and selective growth of specific cell types in the conceptus at particular stages of gestation.
• Direct and indirect interactions among mother, embryo or fetus, and placenta (Miller and Kellogg, 1985).

TABLE 22-1 Tissues and Fluids Available During Pregnancy for Laboratory Assessments

---

Fetal tissue analysis at delivery
Placenta
Cord blood
Fetal blood
Amniotic fluid
Skin
Hair
Adipose tissue
Urine
Feces

Maternal tissue and fluid analysis throughout gestation
Blood
Urine
Feces
Adipose tissue
Air
Hair
Milk
Endometrium

First-trimester fetal tissues and fluids
Chorionic villi (less than 10 weeks)
Amniotic fluid and cells

ASSESSMENTS FOR PHARMACOKINETIC ANALYSES

Initial considerations for pharmacokinetic analysis include the analytic technique involved, its sensitivity, and standardization of quality-assurance protocols associated with it. Quality assurance must be rigorous and described thoroughly in publications, particularly in cases where blood or tissue sample sizes are limited, as in samples from a conceptus. Problems in comparing results from different laboratories, for example, difficulties encountered in trace-metal analyses of human tissues (Friberg, 1983), can be alleviated by adopting universal standards for any analysis, regardless of technique. Necessary in quality assurance are control samples from established sources, preanalytic control of collection containers and patient information, statistical evaluation of all samples for maximal allowable deviations, and external monitoring and review programs (state, national, and international). Because the technique used for a chemical compound often is specific, this chapter does not review individual techniques, but suggests reviews for further consideration (Friberg, 1983; Kaul et al., 1983; Kay and Mattison, 1986; Miller et al., 1988).

Classic pharmacokinetic studies depend on single acute exposures to determine half-life of a chemical for a specific tissue or fluid compartment. More accurate pharmacokinetic analyses require repeated sampling of the same tissue or fluid compartment and constructing a curve from these data that relate tissue or fluid concentration of a chemical or its metabolites with time since exposure. Examination of the concentration-time curve makes it possible to establish the amount of chemical in the organ or fluid. Such analyses can be applied to the whole body by measuring the urinary, pulmonary, or fecal excretion of the compound.

In humans, blood concentrations of a chemical often are used to describe the characteristics of distribution; but the physicochemical characteristics of the chemical might be the primary factors determining absorption in and distribution to body compartments. Those characteristics include molecular weight, lipid solubility, ionization capability, protein-binding capability, and metabolism (Longo, 1972; Miller et al., 1976; Kuemmerle and Brendel, 1984; Mattison, 1986); and changed hormonal balance in pregnant women can affect all of them. Tissue and fluid compartments other than the ones available for testing can become "deep" compartments, in which the chemical appears to be irreversibly trapped; the placental interface might limit chemical transit to the conceptus. If the chemical is largely bound to plasma albumin, less of it will be found in fetal blood than in maternal blood, because less albumin is found in fetal circulation. However, the amount of free chemical—the critical factor—can be identical, and simply measuring one or two compartments might not accurately reflect a third compartment.

Metabolism of xenobiotics might differ substantially between pregnant and non-pregnant women. Biotransformation can reflect lower blood concentrations of a substance, as well as shorter half-lives. Such changes can be important to maintain therapeutically effective concentrations of a substance, e.g., phenytoin. Changes in half-life might reflect not only metabolism to inactive water-soluble metabolites, but also increased renal clearance of these metabolites. During pregnancy, renal plasma flow increases by 30%, while the glomerular filtration rate increases by 50% (Davison and Hytten, 1975).

Besides the physiologic changes in hepatic and renal function, body fat content usually increases, and mammary glands enlarge during pregnancy. Highly lipid-soluble compounds can be stored in those maternal depots. Compared with an adult, the fetus has little body fat; a lipid-soluble chemical is distributed to whatever lipid is available. Lipid-soluble chemicals usually concentrate in the fetal CNS because it composes much of the lipid in the conceptus, and a large percentage of the umbilical blood flow goes directly to the brain (Stave, 1978).

Accurate assessment of distribution and biotransformation of a chemical in the conceptus is hampered by inaccessibil-

ity. During the 1960s and early 1970s, clinical data became available from therapeutic interruptions of pregnancies during which drugs had been administered and products of conception obtained and analyzed for the parent chemical and its metabolites (Miller et al., 1976). Those investigations identified chemical distribution in single pregnancies at single points. Information also is needed from continuous sampling from individual patients over time. At term, continuous sampling can be achieved with fetal scalp blood sampling (Miller et al., 1976), in which multiple samples are obtained during the course of delivery.

Until recently, only direct fetal biopsy of tissues provided information on pharmacokinetic response; however, MRI has demonstrated movement of paramagnetic ions, such as manganese and gadolinium glutamic-pyruvic transaminase, between mother and fetus in primates (Kay and Mattison, 1986; Miller et al., 1987a,b, 1988; Mattison et al., 1988; Panigel et al., 1988). The potential to label chemicals with carbon[13] and determine their distribution is an example of noninvasive techniques that might resolve some problems; for example, identifying compartments that can be measured to give an accurate distribution index. Of the compounds evaluated most frequently, heavy metals and anticonvulsants have provided the most information on exposure.

Observations made during human pregnancy usually reflect chronic exposure throughout pregnancy, and tissue, maternal blood, and fluid concentrations reflect exposure, rather than response. The placenta has been used as an exposure index for many heavy metals, because they are concentrated in it (Miller and Shaikh, 1983). Maternal occupational exposure and lead content in human placentas at term have been measured, as have environmental exposure from bath and drinking water (Miller et al., 1987a). Other studies have correlated an increase in cadmium in the placenta with maternal cigarette-smoking (Miller et al., 1987a). The amounts and types of mercury compounds also have been confirmed in the placenta (Miller, 1983). CVS might provide an early screen for environmental exposures to heavy metals (Fabro and Scialli, 1986).

Early placental tissue and amniotic fluid cells might be useful to assess the presence of chemicals and their interactions with fetal tissues. Shum et al. (1979) demonstrated that birth defects due to benzoapyrene in the mouse were caused not only by the genetic makeup of the mother, but also depended on the fetal genome and the ability of the conceptus to metabolize benzoapyrene to putative reactive intermediates. That indicates that the conceptus might partially regulate the consequences of therapeutic or environmental exposures.

Phenytoin, an anticonvulsant, is also metabolized to putative reactive intermediates via mixed-function monoxygenases, whose expression can be controlled genetically (Martz et al., 1977). The incidence of fetal phenytoin syndrome might be low because the genetic makeup of many conceptuses contains the alleles of the less active monoxygenases. A case-reported woman who had two consecutive pregnancies during which she maintained phenytoin therapy, but bore only one child with phenytoin syndrome indicated that identical exposures can affect fetuses differently (Wong et al., 1985a). Those kinds of observations are not conclusive, but suggest that the fetus might respond uniquely to selected agents. If a specific embryonic cell population could be sampled and susceptibility of the conceptus to phenytoin determined, then therapy or environmental exposure could be modified. The trophoblast might be a tissue for such evaluations. Measurement of monoxygenases or other xenobiotic metabolizing enzymes could indicate the sensitivity of the conceptus.

The placenta selectively metabolizes xenobiotics and polycyclic aromatic hydrocarbons. Placentas of cigarette-smokers produce 8-10 times as much arylhydrocarbon hydroxylase (AHH) as placentas of nonsmokers (Welch et al., 1969), although not all placentas of smokers produce it. Cytochrome P_1-450 also has been identified in the human placenta (Song et al., 1985; Jaiswal et al., 1985a). Because benzoapyrene is a constituent of cigarette smoke

and is the substrate for AHH, the presence of benzoapyrene and its metabolites could be related to the effects of smoking on fetal development; but no dose-response relationship has been established for the effects of smoking and for benzoapyrene as a teratogen and perinatal carcinogen. AHH induction in the human placenta is related to the number of cigarettes smoked per day and the enzyme activity reaches its maximum at 20-25 cigarettes/day (Gurtoo et al., 1983). Demonstration of a dose-response relationship is difficult, because of individual variation in genetic composition and enzyme inducibility (Juchau, 1980; Gurtoo et al., 1983; Manchester et al., 1984).

The placenta might act as a filter to prevent passage of reactive substances into the conceptus. Fetal tissue—especially endothelium from the umbilical vein—did not produce AHH from pregnant women who smoked, but endothelial AHH activity was induced in primary cell cultures. The placenta might protect the fetus from exposure to low-concentration environmental pollutants (Manchester and Jacoby, 1984; Manchester et al., 1984). Amniotic fluid cells also could be used for these studies, but these cells usually are not available until the end of the first trimester.

Interactions with selected cellular constituents other than enzyme induction and inhibition can be assessed. Randerath and associates (Randerath et al., 1981, 1985; Lu et al., 1986) have suggested that examination of DNA adducts provides information about the nature of environmental chemical interaction with the genome. Selected reproductive tissues—especially the human placenta—have been used to determine whether exposure to cigarette smoke during pregnancy alters the DNA-adduct pattern (Everson et al., 1986, 1987).

CURRENT AND PROMISING MARKERS

Many markers of exposure to xenobiotics are specific to one chemical and its reactivity and distribution. For example, the pharmacokinetics of methylmercury can be monitored with blood or hair concentrations (Clarkson, 1987); but, if 2,3,7,8-tetrachlorodibenzo-p-dioxin is of concern, these concentrations are not as useful as milk or adipose tissue concentrations. Thus, the monitoring site is as important as the sensitivity of the analytic procedure.

Human hair reveals mercury exposure for many months, and a dose-response relationship can be established on the basis of hair and blood concentrations relative to toxic outcome (Clarkson, 1987). With the exception of methylmercury, no human dose-response relationships are available for prenatal exposures to xenobiotic compounds. The only similar examples are based on at-term placental tissue analyses for cadmium or lead and correlation of the results with environmental exposure, smoking, and pottery paint (Miller et al., 1988). Tissue analyses at delivery provide little useful information other than documentation of exposure. Monitoring maternal blood concentrations of drugs, such as phenytoin, has assisted in maintaining adequate control of seizures, but has shown no correlation with teratogenicity (Kuemmerle and Brendel, 1984).

Studies have demonstrated that blood from cigarette-smokers, cancer-chemotherapy patients, or patients on anticonvulsant therapy produces malformed embryos when added to a culture medium of embryos (Chatot et al., 1980; Klein et al., 1980, 1982; Carey et al., 1984). Such techniques have been proposed as a screen to identify populations at risk (Klein et al., 1982). However, epidemiologic investigations have not been undertaken rigorously.

Drosophila and bacteria have been used in bioassays to screen amniotic fluid (Bournias-Vardiabasis, 1985; Everson,

1987). These screening procedures could identify carcinogenic or teratogenic properties of amniotic fluid. Other programs have used cultured human cells to screen potential teratogens (Braun et al., 1979, 1982; Yoneda and Pratt, 1981; Pratt et al., 1982); however, human fluids have not been used. These procedures appear appropriate for human screening.

Recent studies measuring of DNA-adduct patterns in human placenta and animal tissue indicate that pattern of DNA adducts can be a basis for separating chemicals. Thus, the compounds or their reactive metabolites that directly alter cellular constituents might be identifiable (Everson et al., 1987). As noted earlier, some DNA adducts in the human placenta are correlated with cigarette-smoking (Everson et al., 1986). If very early tissue samples could be obtained from the conceptus and adducts could be measured, risk to the conceptus could be determined. CVS might prove useful in this regard.

Other promising techniques for investigating biologic markers of exposure are MRI and spectroscopy. Use of those techniques in pregnant women is substantially restricted; however, as knowledge concerning the safety of MRI in humans becomes more available, more extensive studies will be undertaken. MRI and spectroscopy are used primarily to examine the fetal structure for malformations and to detect placenta previa. However, the opportunity to follow the distribution of compounds selectively labeled with carbon[13] or paramagnetic ions is appealing. Studies in nonhuman primates have demonstrated the feasibility of this, as have studies in the perfused human placenta. Nonetheless, direct, noninvasive spectroscopy within specific embryonic and fetal tissues is in the future.

23

Conclusions and Recommendations

Maternal physiology is remolded during pregnancy to nurture a separate organism. The influence of xenobiotic agents must be considered for the conceptus and the mother; neither organism can be assessed separately, nor can knowledge of an agent in nonpregnant females be extrapolated to pregnant females.

During the past decade, several techniques to assess fetal risk have become available, and increasing numbers of chemicals have been recognized as having teratogenic and mutagenic potential. Nonetheless, no specific markers are available to indicate that exposure to a xenobiotic agent is directly associated with a cellular, subcellular, or pharmacodynamic event. Nor will such predictive markers be developed easily with existing methods. The greatest benefit of biologic markers is to establish risk of toxic response in mother or conceptus before the adverse response is expressed or becomes irreversible. No such biologic markers associated with pregnancy in animals or humans have been validated. Most descriptors of health status do not distinguish adverse health outcomes that resulted from environmental insult.

Evaluation of potential biologic markers is made difficult because of the following factors:

- Normative data are lacking for proposed markers.
- Multiple end points are possible for xenobiotic interaction with the mother or the conceptus.
- The effects of exposure are likely to depend on the stage of development of the fetus.
- Exacting validation is required for any putative markers of specific effects of xenobiotic exposure.

RESEARCH STRATEGIES

Developing biologic markers or new assays is complicated. Insufficient data are available to formulate a strategy that could be universally applied for studying populations exposed to specific toxicants or linking assessments of various health end points to specific exposures. Normal frequencies of various health end points are unknown, as are other risk factors for the various health end points. However, some areas for immediate research initiatives can be identified.

Recent research emphasizing cellular and molecular aspects of mammalian developmental and reproductive biology has identified an extensive directory of structural and functional markers. These could be used effectively to define a cause-and-effect relationship in which cell- or stage-specific events are pivotal to the differentiation of the embryo or particular endometrial cell type.

Studies designed to determine the success rate of human pregnancy are needed. The developmental and biochemical aspects of fertilization, implantation, and growth of embryonic and extraembryonic tissues also must be determined. Embryonic signals around the time of implantation might be essential moderators of growth and development in the singular allogeneic environment of mammalian pregnancy, and absence of signals or presence of abnormal signals might be responsible for failed or abnormal pregnancies.

Research has indicated that the trophoblast produces and orchestrates the signals that modulate maternal biochemical or immunologic acceptance or rejection of the blastocyst for implantation, but very little is known about this key trophoblastic function. The nature of the signals is not known, and this research should be expanded. New information in this basic aspect of cell-to-cell interaction should produce clinically relevant data to answer more general questions on the outcome of human pregnancy.

Other evaluations of maternofetal function can be obtained by pharmacokinetic studies, such as monitoring manganese in human placentas by magnetic resonance imaging. Those types of investigations offer an opportunity to study human placental function with noninvasive tests for biologic markers, and the approach should be encouraged.

Most toxicologic studies focus on populations or animals exposed to large doses. Knowledge of events at clinically relevant serum or tissue concentrations and at specific times in pregnancy is needed.

Chronic exposure conditions in toxicologic studies have different standards from acute exposure conditions during specific gestation intervals. Many markers selected for consideration are not indicators of specific responses to particular toxicants. The specificity of screening for a compound and its resulting insult has not been addressed as the capability to detect the compound and its association with such other constituents

as DNA adducts in selected tissues and fluids has been.

Biochemical, cellular, immunologic, and molecular assessments of tissues (e.g., chorionic villus sampling and maternal blood sampling) should be emphasized to develop assays for fetal assessments.

A substantial contribution could be made to environmental health research by designing and carrying out studies to determine the success rate of human pregnancy. The use of biologic markers of toxicity during pregnancy is intrinsic to such an investigation to establish quantitative and qualitative standards to be used in studies of pregnancy outcome. A few predictive tests of high-risk pregnancies have been developed, but very few tests measure preimplantation events. More information is needed about the fertilized egg before implantation.

At least three kinds of quantitative measurements allow determination of risk of pregnancy complications: measurements of alpha-fetoprotein, assessment of concentrations of environmental toxicants in tissues, and classical physiologic determinations. Diagnostic and prognostic limits for the first two categories are being studied, and decisions must be made as to what physiologic tests to use. At least two qualitative determinations could be used in assessing possible pregnancy complications: appearance of DNA adducts in extraembryonic fetal tissues and immunologic procedures that allow identification of high-risk couples, perhaps before they begin to produce fertilized eggs. Tests should be developed to classify high-risk couples according to the etiology of the problem; some spontaneously aborting women could then be treated according to the pathophysiologic characteristics of the underlying defect.

ACCOMPLISHING THE RESEARCH GOAL

The following are recommendations are related to funding and the research agenda:

• Encourage and support the continuing development of a data base identifying the factors that affect the various stages of pregnancy. This data base would identify stage- and cell-specific biologic markers that optimize detection of cause-and-effect relationships, as well as constitutive factors that predispose to the success or failure of any pregnancy in a neutral environment. No informative risk assessments can be made now, especially of the processes that characterize the pre-implantation and peri-implantation periods of mammalian development.

- Stimulate and support studies that focus on the response to exposures to xenobiotic agents during specific stages of gestation. The research would develop the data to characterize markers of exposure. Coupling these data with those developed according to the recommendation above would provide a basis for risk assessment.

- Encourage active and aggressive participation of established developmental and reproductive biologists in cellular and molecular toxicology.

- Establish postdoctoral fellowship programs that encourage new Ph.D.s and M.D.s in toxicology to train in genetic, immunologic, and biochemical aspects of cellular and molecular developmental and reproductive biology.

- Support a biennial forum to encourage interaction between active contributors to the various disciplines relevant to reproductive toxicology. Because of the breadth of biologic disciplines that could be brought together in such a forum, a revolving task force is recommended. With such a task force, a core group of scholars representing the various fields would be charged to make recommendations regarding research and to encourage research through position papers, forums, and demonstration projects funded by small amounts of research money. The core group should be augmented with a rotating group of scientists from the same and additional biologic disciplines.

BRINGING THE ASSAY OUT OF THE LABORATORY AND INTO THE PUBLIC HEALTH DOMAIN

In addition to the need for laboratory exploration of promising ideas, resources must be allocated to determine adequately the utility and validity of these ideas in human populations. Some markers will be useful only in carefully defined populations. Such limitations need to be defined.

Even when markers have been determined to be potentially sensitive and specific for use in the public health sector to establish exposure and disease potential, information collected from large, carefully documented studies is lacking, except for a few cases, such as alpha-fetoprotein. Because most exposed populations are likely to be small, a central clearinghouse for information from isolated studies of specific exposures will be needed to validate markers. The Food and Drug Administration and the Environmental Protection Agency have offices to collect case studies; however, an office is needed to compile and perhaps fund studies of exposed populations.

Measurement of indexes of exposure is complicated, not only because laboratory procedures are insufficiently refined, but also because obtaining samples is complex.

Consideration must be given to the fact that therapeutic options are available for adverse conditions confirmed in the fetal period. Options might include fetal surgery (e.g., bone-marrow transplantation and drainage of fluid in hydrocephalic fetuses) and choice of modes of delivery (e.g., cesarean section versus vaginal delivery for spina bifida). However, risk is associated with the use of any of these.

ASSESSMENTS OF THE STATUS OF SPECIFIC MARKERS RELATED TO PREGNANCY

The following recommendations concerning potentially useful biologic markers of exposure and effect during pregnancy represent current knowledge in clinical and basic sciences of reproduction and

TABLE 23-1 Biologic Markers Associated with Pregnancy and Possible Reproductive Hazards

Marker	Immediately Usable in Large-Scale Human Studies	Usable in Studies of Selected Individuals	Promising for Human Studies; Needs More Development	Animal Studies Needed
Exposure Markers—Concentrations of Toxicants or Metabolites in:				
Maternal				
Blood	+			
Urine	+			
Feces	+			
Hair	+			
Nails			+	
Tissue biopsy				
Endometrium			+	
Cervical mucus			+	
Uterine washings			+	
Fat		+		
Conception products		+		
Placenta				
Conception products		+		
Chorionic villus sample		+		
Placenta at delivery	+			
Amnion at delivery	+			
Chorion at delivery	+			
Umbilical blood	+			
Embryo-Fetus				
Conception products				
Tissue		+		
Blood		+		
Hair at delivery		+		
Nails at delivery			+	
Urine at delivery		+		
Feces			+	
Amniotic fluid		+		
Amniotic cells		+		
Foreskin	+			
Effect Markers				
Maternal				
Clinical history				
General health	+			
Reproductive history	+			
Menstrual history	+			
Steroid-LH-FSH concentrations		+		
Uterine prolactin			+	+
Immune status				
hLA types		+	+	
TLX antigens		+	+	
Histology				
Vaginal epithelium		+	+	
Endometrium		+	+	
Cervical mucus		+	+	
Globulin and other protein types		+		
Uterine blood flow		+	+	

Marker	Immediately Usable in Large-Scale Human Studies	Usable in Studies of Selected Individuals	Promising for Human Studies; Needs More Development	Animal Studies Needed
Placental				
Placental hormone concentrations:				
hCG	+			
hPL	+			
EPF		+	+	
cACTH			+	
cTSH			+	
PAPP-A			+	
Interferon			+	
Interleukin I			+	
Estradiol	+			
Estriol	+			
Progesterone	+			
Receptor number and affinity for:				
Beta-adrenergic			+	
Diazepam			+	
Glucocorticoids			+	
Epidermal growth factor			+	
Opiates			+	
Somatomedin			+	
Testosterone			+	
TCDD			+	
Binding site number for:				
IgG-Fc			+	
Low-density lipoproteins			+	
Retinol binding protein			+	
Transcobalamin II			+	
Transferrin			+	
Concentrations of associated proteins				
SP1			+	
Alpha-fetoprotein			+	
Alkaline phosphatase			+	
Transcobalamin I/II/III			+	
DNA adduct frequency			+	
Metabolic rates of:				
Nutrients			+	
Xenobiotics			+	
Transport efficiency of:				
Nutrient			+	
Xenobiotic			+	
Karyotyping	+			
Morphometry		+		
Histology of placenta tissues	+			
Embryo, Fetus, and Neonate				
Physical examination at delivery	+			
Apgar or Brazelton scales	+			
Presence of dysmorphology	+			

Marker	Immediately Usable in Large-Scale Human Studies	Usable in Studies of Selected Individuals	Promising for Human Studies; Needs More Development	Animal Studies Needed
Death	+			
Carcinogenesis	+			
Growth measures	+			
In utero				
Growth measures	+			
Death	+			
Presence of dysmorphology	+			
Plasma growth factors				
IGF			+	
EGF			+	
NGF			+	
Physiology				
Cardiac output (ultrasound or Doppler)		+		
Assessment of inborn metabolism errors or other genetic disorders		+		
Reproductive MIF		+		
Fetal breathing rate		+		
Techniques That Might Yield New Markers				
Analytic techniques for xenobiotics			+	
Amniocentesis		+		
Chorionic villus sampling	+			
Doppler or biophysical monitoring		+		
Karyotyping		+		
Fluorescent-activated cell sorting of fetal blood samples			+	
Fetoscopy			+	
Magnetic resonance imaging			+	
Conception products		+		
Measurements of tissue-specific antigens				
CEA		+		
AFP		+		
AChE		+		

development that can be applied to toxicologic studies during pregnancy. The techniques of basic research that might be useful to assess risk associated with toxic exposure in the conceptus and mother are considered, as well as the clinical application of the techniques.

Markers are divided into two major categories: markers of exposure and markers of effect (Table 23-1). Table 23-1 also lists laboratory techniques that might yield new markers. These markers have potential utility only during certain periods of the pregnancy. Subdivisions of particular periods during pregnancy reflect changing conditions, sensitivities, and responses. Three periods during which relevant exposures or events could occur are considered: before and around implantation, during organogenesis, and during fetal development until just after birth. Table 23-2 lists some biologic markers by the periods during pregnancy for which the marker is informative. The first period comprises any time before conception until the anticipated menstrual period in that cycle (14 days after conception). Organogenesis constitutes the interval from the anticipated menstrual period until 8 weeks of embryonic age. The fetal-peripartum period extends from 8 weeks of embryonic age until 24 hours after birth.

Markers of Effect Before and Around Implantation

Success of a pregnancy is established before and around implantation as a consequence of maternal immunologic status and synchronized interaction of the uterine endometrium and the developing embryo. The greatest risk to successful completion of pregnancy occurs during the interval encompassing conception, embryo attachment to the uterine epithelium, and embryo advancement into the differentiating stroma. Although failed pregnancies result from numerous mechanisms, including physical defects in the uterus or fallopian tube, endocrine abnormalities, and immunologic alterations, the risk before and around implantation might increase if xenobiotic agents are introduced into the intrauterine environment. Fragmentary data indicate adverse influences of toxic agents, but most investigations have not been designed stringently and their results have often been analyzed retrospectively. The lack of biologic markers might be attributed to the paucity of stage-specific indicators of developmental and reproductive events that could be used in prospective studies of xenobiotic exposure.

Epithelial cell markers are profiles of proteins and glycoconjugates that characterize the distinct apical and basolateral cell surfaces and their secretory compartments. Analyses of these profiles in animals provide probes to monitor progressive development from a neutral to an ova-receptive uterus, determine the role of the individual steroid hormones in regulating changes associated with epithelial cells, and monitor alterations in secretion rates. Also, biochemical and immunocytochemical matrix proteins—laminin, entactin, and fibronectin—and expression of decidual luteotropin enable the sensitivity of stromal cells to xenobiotic exposure to be determined.

Biologic indicators of endometrial status are more promising than practical; the endometrium is not readily accessible for monitoring. Extending the protocols developed to analyze endometrial cell differentiation to the cervix and vagina would identify biologic markers that are more readily accessible (and thus depend less on animal models) and more readily extrapolated to studies of health and environment.

Differentiation of rodent trophectoderm to trophoblast giant cells has proved to be a productive experimental model for identifying potential markers of the status of the developing embryo. Trophoblast antigens and placental hormones (hCG and hPL in humans and PL-1 and PL-2 in rodents) appear to be the most promising biologic markers.

Little useful information is available about the biochemical or physiologic regulation of placental lactogen synthesis and secretion by trophoblast giant cells. Recombinant-DNA technology, using specific molecular probes, is needed to identify the mechanism of progesterone regulation.

Around the time of implantation, the only proven indicator of implantation in humans is hCG. However, the presence of hCG reflects the presence of differentiated trophectoderm and the blastocyst, and hCG cannot be detected before blastocyst implantation. hCG is not a marker of a specific toxicant or response to a toxicant; therefore, it is not effective in identifying risk. However, hCG does reflect the status of successful or failed pregnancy. The effectiveness of hCG as a marker of early postimplantation development would be increased if the regulatory relationships between the undifferentiated (cytotrophoblast) and differentiated (syncytiotrophoblast) trophoblast cells and their intermediate stages were better understood.

Indicators of preimplantation status, such as early pregnancy factor (EPF) or pregnancy-associated proteins (some purported to originate in the trophoblast), have been proposed as biologic markers. Until these proposed indicators are established as markers and their usefulness is proved, there is no basis for attempting

TABLE 23-2 Biologic Markers Associated with Pregnancy and Possible Reproductive Hazards,
by Period of Gestation

Up to 14 Days After Conception
Useful:	hCG
	Clinical history
	Steroid concentrations
	Blood or urine concentrations
Promising:	EPF
	TLX
	HLA types
Potential:	Uterine prolactin concentration
	Uterine secretions of hormones
	Uterine secretions of xenobiotics

Organogenesis
Useful:	Sonography for size and location
	hCG
	Clinical history
	Hormone concentrations
	Sonographic measurements:
	Growth
	Size
	Movement
Promising:	Magnetic resonance imaging:
	Image dysmorphology
	Identify energy sources using P^{32}
	Assess xenobiotic or nutrient concentrations using C^{13}
Potential:	Chorionic villus sampling:
	Karyotyping
	Enzyme activity levels and isotypes
	Localization-metabolism of xenobiotics
	DNA-adducts
	Flow-activated sorting

During Fetal Growth Period Until Birth
Useful:	Sonography to assess:
	Growth
	Fetal breathing
	CNS function
	Dysmorphology
	Doppler blood flow measurement of:
	Uterine blood flow
	Umbilical blood flow
	Fetal heart rate
	Biophysical monitoring:
	Heart rate
	Amniocentesis to perform:
	Karyotyping
	Xenobiotic analysis
	Toxicity screen of embryo culture
	Alpha-fetoprotein
	Maternal serum-cord to measure concentrations of:
	CEA
	Alpha-fetoprotein
	HPL
	Xenobiotics
Promising:	Fetal cord sampling
	Fetoscopy
	Flow-activated sorting

to use them in assessing risk associated with toxic exposures during mammalian development.

The field of reproductive immunology has provided some potentially useful indicators for evaluating repeated pregnancy losses. This documentation and categorization can reveal cases in which therapeutic intervention, such as immunotherapy, would be useful.

Toxicology assesses the effects of toxicants and environmental pollutants on pregnancy outcome. However, some couples are at high risk for reproductive failure. Inclusion of high-risk couples in large epidemiologic studies of the effects of environmental factors on reproductive performance clouds the issue of pathophysiologic effects on normal patterns of reproduction, because such at-risk couples are not rare. At-risk couples should be identified for any population to be investigated.

The study of trophoblast antibodies is essential because trophoblast tissues form the operational interfaces between maternal and extraembryonic cells in allogeneic relationships of human pregnancy. None of the currently studied placental proteins has proved valuable in this regard, but not all placental proteins have been studied. AA3 should be studied, because it appears to be central to the allogeneic relationship of maternal endothelia with extraembryonic cytotrophoblasts.

Furthermore, inasmuch as placental perfusion with oxygen, nutrients, and antibodies depends on maternal blood flow through spiral arteries, it is important to develop techniques to measure spiral arterial function. Ultrasonography has provided a means to determine the effectiveness of spiral arterial function, but quantitative biochemical measures of placental bed perfusion would be more informative.

Organogenesis

Organogenesis is the period of greatest embryonic vulnerability to insult and permanent structural and functional alteration. Such alterations are difficult to assess, because the conceptus is inaccessible; no invasive procedures are performed earlier than week 8 of gestation, for fear of interrupting the pregnancy. Thus, morphologic and biochemical assessments of the embryo are nonexistent. However, advances in ultrasonography allow early visualization of the embryonic sac and the conceptus, and the placenta and its attachment to the uterus can be examined.

Biochemical assessments of trophoblast hormone production can be made by detection of hCG and hPL, associated proteins of undefined function (SP1, PAPP-A, and PP11/12), and steroids. Yet these only indicate general viability of the embryo and do not provide information concerning development. Markers of embryonic development, such as fetal red blood cells, can be assessed by applying fluorescent tagging techniques to maternal blood samples. Such techniques also might provide direct information concerning the development of specific organ systems (e.g., Müllerian inhibiting substance concentration) and the relative contributions of the embryonic genome and maternal genome in producing the effects of an insult. Immunologic tests for anti-RHO, lupus, and lymphocytotoxic antibody are useful in assessing viability of a pregnancy. Biophysical monitoring can be used to determine uterine vascular responses to maternal toxicity. Magnetic resonance imaging is being used increasingly to assess structure and to determine substance localization within fetal tissues. As molecular probes specific for events in organ development are developed, markers of toxicity of specific toxicants might become evident.

The Fetal and Neonatal Period

The fetal and neonatal periods of development are the most productive for assessing effects of xenobiotic agents and the pharmacokinetics of such agents, because the conceptus is sufficiently large and accessible during these periods. Tests performed during the fetal period might only document events that occurred before and around the time of implantation and during organogenesis, but increasingly sophisticated technology should make earlier assessments possible.

Biophysical monitoring provides information concerning heart rate and blood flow velocity in maternal, fetal, and placental vessels. Ultrasonography measures structure, fetal weight, growth (biparietal diameter, femoral length, cardiac function, placental size, and maturity), and function (chest wall movements and body movements). These assessments are used to develop normative data, but no markers have been associated with specific environmental exposures.

Fetal blood samples and tissue biopsies can be obtained with fetoscopy, and amniocentesis provides amniotic fluid and cells for evaluation. The combination of chorionic villus sampling and molecular probes for specific genetic disorders can be used to assess the capability of the conceptus to respond to environmental insult. Functional alterations can be demonstrated by inducing specific enzymes, such as AHH, or by interactions with specific cellular constituents, such as DNA adducts. Progress is expected to come from combining magnetic resonance imaging with molecular probes.

Major basic protein concentrations are markers of the onset of labor. Preterm labor—labor before 37 weeks of gestation—increases the risk of infant morbidity and mortality. The mechanisms causing preterm delivery and the effects of toxicants on the likelihood of preterm delivery need to be assessed.

MARKERS OF EXPOSURE

Biologic markers of exposure represent the detection and assessment of a specific chemical or its metabolites in an organism and demonstrate that exposure to a substance occurred. Unfortunately, this information, especially on a virtually inaccessible conceptus, usually is unavailable. Measuring the amount of a chemical in air, water, or food to which the pregnant female is exposed (environmental monitoring) is one means to determine exposure. Other methods include analyses of readily accessible maternal body products, such as blood, urine, feces, and hair.

DNA-adduct evaluation is used as an indicator of exposure to a compound that has a specific pattern or fingerprint on 2-D gel electrophoresis and has been dependent on monitoring specific fluids, such as blood and urine. However, with the advent of magnetic resonance imaging, specific substances can be monitored—especially paramagnetic ions—in specific organs without compromising pregnancy. Laboratory assessment of internal dose in fetal or embryonic compartments requires invasive techniques to obtain tissues or fluids.

If tests of human fluids could be refined so that the effects of nutrition could be distinguished from the effects of xenobiotics, in vitro tests could be used to screen populations for persons susceptible to adverse effects of exposure. For example, rat embryo culture and early mouse blastocysts have been used to screen serum from high-risk populations of women who repeatedly abort spontaneously. The rat embryo culture has been used to screen patients exposed to tobacco smoke, drugs, and anticonvulsants. Nonmammalian systems have been used to screen amniotic fluids. These in vitro systems have been proposed to test whether xenobiotics are producing toxic changes in the serum of patients exposed to them. Changes might be the production of reactive metabolites of the xenobiotic compound, nutritional alterations, or the products of maternal changes, such as changes in immunoglobulin G concentrations.

IV

Biologic Markers in
Neurodevelopmental Toxicology

24

Introduction

This section of the report examines many topics salient to the study of developmental markers in persons more than 24 hours old. Rather than discuss in detail the development of the many systems and processes of the human body, we focus the review on the development of the central nervous system. Then, using radiation as a paradigm, we examine the effects of timing and dose on central nervous system teratogenesis. The consequences of abnormal cell migration for host behavior are also examined. The critical topic of methodologic issues encountered in constructing inferences abut human effects from animal data is discussed next. Finally, neurodevelopmental outcomes that might be among the most sensitive markers of neurotoxicant-engendered dysfunction in humans are surveyed. Lead, the neurotoxicant with the largest neurotoxicologic data base, is used as a paradigm to examine some of the methodologic issues inevitably encountered in the pursuit of a valid collection of markers.

Many discussions of biologic markers and development focus on using markers to ascertain effects of xenobiotic substances on development. However, the influence of growth and development on xenobiotic disposition and effect is equally important. Biologic markers that might be useful at one age might not be useful at other ages.

Results of many assessments differ quantitatively and qualitatively from one age group to another. That is particularly true of concentrations of substances in blood and pharmacokinetic (absorption, distribution, metabolism, and elimination) characteristics in general. Therefore, interpretation of biologic markers of internal dose must take into account the developmental status of the subject.

The pharmacodynamics (biochemical and physiologic effects of drugs and their mechanisms of action) of target-organ response and short- and long-term consequences of exposures also vary with age. From birth until 2 or 3 years of age, major developmental changes occur in the CNS and major changes occur in the regulation of the endocrine system during puberty. Thus, the end-organ effects in these systems after exposure at different times of life are likely to be different, and the biologic markers used to assess the effects might be different.

In neurodevelopment, the demonstration of sensitive, specific markers that clearly link exposure and outcome has been rare; many variables intervene between toxicant exposure and demonstration of altered behavior. Demonstration of causal relationships between low-dose exposure and impairment has been particularly difficult. Establishing causal relationships often depends on inferences drawn from the combined results of animal experiments and epidemiologic studies; neither might be sufficient in isolation. Animal studies raise the question of species differences; some human teratogens or neurotoxicants might not cause measurable alterations in commonly used species, and some animal teratogens have not been shown to be human teratogens. It is hard for epidemiologic studies to establish causal links, owing to the complex exposure histories of the subjects to many compounds, inability to control exposure levels of the subjects, and the population-based statistical procedures.

Relationships between phenotypic effects of pollutants and behavioral alterations are increasingly apparent. The recent recognition that agents that affect physical development can be expressed at an early age in alterations of nervous system function—possibly as aberrant behavior—has given rise to a new discipline: behavioral teratology.

Before beginning the discussion of neurodevelopmental markers themselves, we digress briefly to consider issues of developing and using biologic markers to assess the status of children, using as a paradigm the exposure to and effects of halogenated aromatic hydrocarbons. Our examples illustrate the need for careful integration of chemical, pharmacodynamic and pharmacokinetic, analytic, and cellular methods in epidemiologically sound studies to assess the multifaceted nature of health effects.

BIOLOGIC MARKERS OF EXPOSURE: PHARMACOKINETIC CONSIDERATIONS

Therapeutic agents usually are administered in known amounts for defined periods, but accurate quantitative histories of exposure to environmental chemicals are rare. Because polychlorinated biphenyls (PCBs) are found in most adipose tissue samples from randomly selected persons (Mes et al., 1982; Kreiss, 1985), it is difficult to design experiments to decide what degree of contamination constitutes significant risk. These considerations emphasize the value of intensive study of populations with massive exposures, very high body burdens, or evidence of chemically induced disease in the mother during gestation. Infants exposed in utero under such circumstances probably would manifest more readily detectable physical and biochemical findings and serve as models for phenomena to be examined in offspring of less-exposed populations.

Many pharmacokinetic considerations are discussed in the previous part of the report. Here these are briefly discussed with regard to the specific situation of PCB exposures. Interpretation of concentrations of a toxicant, or drug, depends on knowledge of the amount and timing of exposure relative to sampling of serum and route of administration.

Highly sensitive and specific assays have been developed for various halogenated aromatic compounds (Albro et al., 1986). Isomer-specific detection and quantification in the parts-per-trillion range are possible. The ability to determine multiple compounds and their breakdown products is critical, because exposures rarely involve pure substances; the blood of people reportedly exposed in Japan and Taiwan to PCB-contaminated rice oil contained products of thermal degradation, such as dibenzofurans (Chen et al., 1981; Kashimoto et al., 1981). PCB isomers have different half-lives and different biologic effects with respect to enzyme induction (Safe et al., 1985), so interpretation of biologic markers of effects of these substances must account for chronic exposure and the presence of specific compounds at the time when the markers are examined.

Although quantification of environmental pollutants in biologic samples, including serum and tissues, is possible, pharmacokinetic interpretation of data remains difficult. Indeed, for such fat-soluble, extremely long-lived compounds, direct relationships among serum concentration (internal dose), concentrations at receptor (biologically effective dose), and biologic effect are difficult to establish (Evans et al., 1986). Furthermore, the distribution of lipophilic compounds can be altered by diet, disease states, and other drug or chemical exposure (Gibaldi and Perrier, 1982), as well as by developmental differences in metabolism, body size, and fat distribution from fetal life through adulthood.

Continuous changes in composition and amounts of fat stores and altered metabolism in pregnant women and children would be expected to have profound effects on the availability of compounds like halogenated aromatics to interact with receptors critical to their toxic effects and, consequently, affect the behavioral findings during childhood (Krauer et al., 1984; MacLeod and Radde, 1985). Pregnancy is associated with as much as a 25% increase in subcutaneous fat. Other pharmacokinetic changes during pregnancy include decreased and erratic absorption from the intestine, increased blood volume, increased total body water, decreased protein binding due to a decrease in serum albumin concentration and an increase in endogenous displacing substances, increased glomerular filtration rate, and generally increased renal and hepatic clearance of many xenobiotics. Basing interpretation of body burden of a given lipid-soluble, slowly cleared compound on an isolated blood or serum concentration is extremely difficult. Similarly, fat stores in the developing fetus change markedly with gestational age. Body fat increases from less than 1% of body weight at 28 weeks to 15% at term. Total body burden in the newborn depends not only on the total maternal exposure, but on when that exposure took place and on the gestational age at birth.

Mobilization of a compound from fat stores might increase its clearance via hepatic and renal mechanisms, but might also make more of it available at toxicologically important sites. Sampling of fat from biopsy samples could provide a better index of total body stores, but this still might not reflect toxicologically relevant concentrations. Mobilization might be particularly high during lactation, increasing the clearance of lipid-soluble compounds from the mother, but also increasing delivery to a nursing infant. Complex pharmacokinetic mathematical models could be helpful in assessing relevance of various tissues concentrations to body burden, quantification of exposure, and biologic effects. Such models will have to incorporate additional complexity based on multifactorial determination of outcomes. Establishing dose-response relationships in humans is further complicated by individual differences in biologic response to exposure (pharmacogenetics).

BIOLOGIC MARKERS OF EFFECT: PHARMACODYNAMIC CONSIDERATIONS

Toxic exposures that might affect the developing child can occur before or after birth. In assessing the relationship between exposure and effects, one must study the mother, the neonate, and the child.

Markers in the Mother

Most studies of pharmacologic effects of pregnancy focus on pharmacokinetics. However, pharmacodynamics and susceptibility to toxicity might also differ. For example, susceptibility to tetracycline-induced hepatic injury increases markedly during gestation, owing to unknown mechanisms.

Attempts to interpret data obtained from offspring of mothers exposed to polyhalogenated aromatics should include histories, physical examinations, and laboratory findings of the mothers. Again, people with massive exposure, such as those exposed to contaminated rice-cooking oil in Japan and Taiwan (Yusho and Yuchen) (Masuda et al., 1982; Lu and Wong, 1984), or histories of maternal illness, should be studied. In humans, dermatologic findings—especially chloracne—have been the most consistent clinical feature among people with massive exposure to polyhalogenated aromatics (Suskind, 1985). Histologic features of chloracne are not pathognomonic of exposure, but the combination of chloracne with measurement of exposure and of PCB concentrations in adipose tissue and plasma suggests chemical-induced disease. Other histologic findings in skin are not likely to be helpful in assessing the existence or extent of chemically mediated illness (Moses and Prioleau, 1985). Other physical features and clinical laboratory findings associated with massive exposure include weight loss, porphyria, hepatic dysfunction, and peripheral neuropathies (Mocarelli et al., 1986; Kimbrough, 1987). None is diagnostic of the PCB-induced disease.

Target organs and toxic manifestations vary widely among species (Safe, 1986). Recently, immunologic changes have been emphasized, particularly because thymic involution occurs in most animals susceptible to 2,3,7,8-tetrachlorodibenzo-*p*-dioxin (TCDD) toxicity. Subclinical changes in cell-mediated immunity have been reported in humans with possible TCDD exposure (Hoffman et al., 1986); none of the subjects examined had other manifestations of illness—including chloracne—and body burden of TCDD was not documented. Much work needs to be done in populations with clearly massive exposures, to verify whether any immunologic changes can serve as useful markers. No human data are available about effects of TCDD exposure during pregnancy.

Many effects of halogenated aromatics are mediated via a specific cytosolic receptor (Roberts et al., 1985; Denison et al., 1986). Induction of specific drug-metabolizing enzymes is the most widely recognized outcome of interaction of the compounds with receptors (Poland and Knutson, 1982; Denison and Wilkinson, 1985; Denomme et al., 1986; Okey et al., 1986). The toxicity of various halogenated aromatics in different species is correlated with the affinities of various analogues for the receptor and with characteristics of the receptor in specific species and strains. Results of animal studies show that genetically determined receptor differences are correlated with inducibility of specific cytochrome P-450 and with toxic outcomes of exposure to polycyclic aromatic hydrocarbons, including carcinogenesis and teratogenesis (Nebert

and Jensen, 1979). Results of studies of human peripheral blood lymphocytes suggest that inducibility of arylhydrocarbon hydroxylase (AHH) is correlated with susceptibility to lung cancer (Kellerman et al., 1973a,b; Kouri et al., 1984), but the mode of inheritance of inducibility in humans remains uncertain.

The above considerations suggest that a variety of markers might be used to assess biologic effects of halogenated aromatic exposure. If susceptibility to adverse effects depends in part on the nature of the aryl hydrocarbon receptor (possibly under genetic control), biochemical and molecular analyses of differences in human receptors might be helpful in correlating outcomes with exposures and in typing the population according to susceptibility. Such analyses could use human lymphocytes or skin cells. One strategy would be to include direct assessment of the presence and binding properties of the receptor, as well as molecular approaches to the receptor gene and the whole AHH gene complex (Whitlock, 1986).

The major effect of most halogenated aromatic compounds is to induce a pattern of enzyme activity similar to that produced by 3-methylcholanthrene. Some PCB and polybrominated biphenyl (PBB) isomers, however, also induce enzymes typically induced by phenobarbital (Safe et al., 1985). Enzyme induction in humans can be assessed with nontoxic, in vivo probes (e.g., caffeine for AHH), with examination of induction capacity in vitro in lymphocytes (Kellerman et al., 1973a,b; Kouri et al., 1984), with direct measurement of enzymes in tissues (liver or skin biopsies), and perhaps with molecular techniques, including quantification of mRNAs for specific forms of cytochrome P-450 (Jaiswal et al., 1985b). Enzyme activity can be influenced not only by genetic differences, but by a wide range of environmental exposures—such as cigarette smoke, diet (including consumption of charcoal-broiled foods or cruciferous vegetables), and medications (Okey et al., 1986). Pregnancy also affects activity of drug-metabolizing enzymes (Krauer et al., 1984).

Interpretation of data obtained in the diverse human population requires careful attention to experimental design—number of subjects and their clerical status, documentation of exposure, definition of sensitivity and specificity of methods, and selection of control populations. An ideal study would include a population with defined exposures (including large exposures), quantification of exposure in family members (especially fathers), determination of endogenous (perhaps genetic) and environmental variables that might alter outcome, detailed medical and biochemical analyses of family members, and thorough study of pregnancy outcomes. Drug-metabolizing enzymes should be analyzed (particularly during pregnancy); the metabolism of hormones critical to pregnancy also would be of interest, because steroid hormone metabolism might be altered by exposure to compounds such as TCDD (Okey et al., 1986).

Placental Markers

Placental markers were discussed in the previous part of the report. Here, the markers relevant to PCB exposure are discussed in the context of the need to study population longitudinally.

The placenta is obviously a critical organ in determining pregnancy outcomes; it is a potential target for environmental chemical toxicity and influences fetal exposure. No information is available from exposed people about placental tissue markers before delivery. Such techniques as chorionic villus biopsy (discussed previously) might prove useful in following pregnancies prospectively. The placenta early in gestation is different from the placenta at term; therefore, extrapolation of markers from different times in pregnancy should be undertaken cautiously. Similarly, vast differences among species in placental derivation and structure make cross-species extrapolation questionable.

At-term placenta markers include quantification of halogenated aromatic compounds, determination of histopathologic changes, determination of enzyme induction status, determination of presence and characteristics of aryl hydrocarbon receptor. Considerable data are available on at-term placentas of women exposed to contaminated rice oil in Taiwan; AHH activity and 7-ethoxycoumarin-*O*-deethylase (7-ECOD) activity were markedly increased in placentas from such pregnancies (Wong et al., 1985a), even though exposures occurred 3-4 years before pregnancy. The induction by other factors, such as cigarette-smoking (Fujino et al., 1984), had to be ruled out. In the rice-oil-exposed population, the AHH induction was greater than expected, as determined by the metabolic profile of benzo*a*pyrene in placental homogenates. Cotinine and thiocyanate were used as markers to ensure that the subjects were not smokers and were not exposed to substantial passive smoking. Some methodologic issues in studying placental tissue have been discussed (Wong et al., 1985a).

In more recent studies, specific cytochrome anti-P-450 isozyme 6 rabbit antibodies were used to detect an increase in one P-450 protein in microsomes from the placentas of women exposed to contaminated rice oil (T.K. Wong et al., 1986). Thus, direct measurement of P-450 activities, use of monoclonal antibodies, and perhaps direct molecular probes are promising markers of effects of exposure to enzyme inducers during late pregnancy.

Neither maternal blood PCB concentration nor clinical symptoms were correlated with AHH or 7-ECOD induction in the study (T.K. Wong et al., 1986). The placenta of one of the nine mothers with clear exposure and substantial clinical symptoms showed no induction. That is interesting, with respect to study of other environmental variables or individual differences in susceptibility related to endogenous and perhaps genetic mechanisms. Aryl hydrocarbon receptor has been measured in human placentas (Manchester et al., 1987), and correlations of induction with receptor characteristics will be of interest. Similarly, study of adducts of reactive metabolites—particularly of products of cigarette smoke—holds promise of quantifying exposures that lead to toxicologically relevant tissue interactions.

Placental enzyme induction by compounds that are metabolized to electrophilic toxicants (such as some components of cigarette smoke) might protect the fetus from exposure (Manchester and Jacoby, 1984; Manchester et al., 1984). High induction and placental metabolism might prevent compounds from crossing the placenta. Umbilical vein endothelial cells from placentas of women who smoke do not show enzyme induction in situ, but retain the capacity for in vitro induction (Manchester and Jacoby, 1984; Manchester et al., 1984). Similarly, decreased placental enzyme induction appears to be correlated with an increased risk of birth defects. However, it is uncertain whether that is true for nonmetabolized inducers, such as PCBs or TCDD, in which case induction of placental enzymes might not increase first-pass clearance of the compound. Studies have examined the roles of genetics and environmental variables on placental response to cigarette-smoke-mediated enzyme induction (Gottlieb and Manchester, 1986). Studies of placentas from pregnancies that produce identical and fraternal twins show that heredity has a role in determining outcome, and position of the placentas in utero might contribute greatly to the extent of induction.

Little information is available to assess and predict interactions of markers of placental exposure, receptors, enzyme induction, function (including steroid hormone metabolism), anatomy (including pathologic changes), and pregnancy outcomes. An ideal approach integrates assessment of each of these factors; it is unlikely that any marker will adequately predict outcomes that are multifactorial. Intensive study of placental tissue—including potentially early biopsies and at-term perfusion models—could lead to an integrated picture of critical determinants of outcome and their contributions.

Markers in the Newborn

The following discusses markers mentioned in the previous part of the report specifically as they relate to assessing effects of PCB exposure.

Newborns are unique pharmacokinetically and pharmacodynamically. Their fat content, total body and extracellular water, protein binding, and renal and hepatic clearance of compounds vary widely with gestational age and can be influenced by many prenatal factors (e.g., nutrition, disease, genetics, medication, and environmental exposures) (Boréus, 1973). Unique interaction of variables can lead to unusual susceptibility to toxic effects that do not occur in other age groups, such as kernicterus resulting from increased red-cell turnover (Karp et al., 1985), decreased protein binding of bilirubin (Karp et al., 1984), decreased hepatic capacity to metabolize bilirubin (Karp et al., 1984), and increased entry of bilirubin into the CNS because of immaturity of the blood-brain barrier (Karp et al., 1984). Newborns differ markedly from older subjects in organ sizes and physiologic roles (e.g., the fetal adrenal gland is a major drug-metabolizing organ). These considerations suggest that newborns differ from children and adults in responses to a variety of exposures and require different interpretations of biologic markers—or different markers—of internal dose and effect. Many data are available on kinetic differences, but more work is needed on receptor maturation, response to enzyme inducers, and pharmacodynamic differences, including unique CNS responses of the immature brain; e.g., phenobarbital and antihistamines produce excessive excitation in infants and children and sedation in adults at comparable blood concentrations (MacLeod and Radde, 1985).

Physical examination of newborns exposed to halogenated aromatic compounds should be comprehensive and quantitative (including weight, height, head circumference, and such other measurements as internal and external canthal distances), and should be performed by persons unaware of the nature of the exposure. Appropriate control subjects should be examined simultaneously.

Several features are possible sequelae of in utero exposure (Rogan, 1982), including low birthweight, conjunctivitis with enlarged sebaceous glands in the eyelids, natal teeth, pigmentary changes of gums and skin with deformed and pigmented nails, peculiar skin coloration, and chloracne. The findings of important dermatologic effects are interesting, given the similarity of sites of toxic effects in adults. Later neurologic development in children with these effects might also be abnormal (Harada, 1976), but such findings require further verification. The possible effects on immune function that appear in adults after TCDD exposure and effects in animals suggest developmental immunologic abnormalities (Lubet et al., 1984; Silkworth et al., 1984) and need to be evaluated in newborns and children.

Other than physical findings, few markers of exposure have been documented. Cord blood concentrations of environmental chemicals could be obtained and correlated with maternal and placental concentrations. A great deal is known about developmental patterns for a variety of drug-metabolizing enzymes in humans and the effects of inducing substances, such as the effects of phenobarbital on glucuronyl transferase (Catz and Yaffe, 1968). Probes, such as caffeine, would be worth using to assess the status of AHH-like cytochrome P-450 activity in infants with possible exposure to halogenated aromatics. Caffeine metabolism and clearance in the newborn are extremely limited and can be followed with a ^{13}C-labeled probe for breath testing (to estimate oxidative demethylation) or with analysis of specific urinary metabolites that reflect probable cytochrome P_1-450 activity (Lambert et al., 1986; Campbell et al., 1987). Determination of aryl hydrocarbon receptor and P_1-450 induction by molecular techniques in available cell types will add to a comprehensive picture of the effects of prenatal exposure on newborn drug-metabolizing capacity. Appropriate psychometric testing of newborns also should be performed for later comparison with results of followup studies.

Markers in Children

The extremely long half-lives of many compounds of interest warrant careful, long-term followup study of the consequences of exposure. Content of environmental chemicals in breast milk should be documented if an infant is breast-fed. Concentrations might vary during a given breast-feeding and over an extended period as the maternal stores of lipid-soluble compounds are redistributed into breast milk.

Accurate clinical descriptions of the skin with photographic records and histopathologic examination are necessary. Routine growth characteristics and appropriate assessments of CNS development should be documented. Given the current focus on immunologic effects of halogenated aromatic compounds, knowledge of responses to routine childhood immunizations and documentation of histories of infectious disease, allergy, and autoimmune disease will be helpful. The roles of environmental exposures and altered immunologic status (Marshall, 1986) are controversial, and verifiable markers and clinically accurate descriptions of outcomes are badly needed.

The ultimate goal of studies of the long-term consequences of in utero exposure to environmental toxicants is to assess the effects of growth and development on handling of and response to the toxicants and the effects of the toxicants on the very processes of ontogeny. Standard estimates of half-lives of various halogenated aromatic compounds are typically based on sparse data on adults. The developing child changes dramatically in body composition, and the change likely has major effects on distribution and clearance of lipid-soluble compounds. Renal and hepatic clearance rates of most compounds generally are greater in children (until puberty) than in adults (MacLeod and Radde, 1985).

Kinetics of environmental contaminants depend on what compounds a subject is exposed to over a long period. For example, enzyme induction could result from persistence of the inducing substance or from an effect manifested long after exposure. The latter situation is exemplified by changes in steroid-hormone metabolism at puberty in rats that received one dose of phenobarbital during gestation (Gupta and Yaffe, 1981). It is uncertain whether a similar phenomenon occurs in humans, but long-term effects caused by short-exposures are possible; any effects noted in children who were exposed prenatally to halogenated aromatic compounds will have to be separated into effects of chronic exposures to compounds with very long half-lives and persistent effects of previous acute exposures.

A safe, noninvasive approach to assessing in vivo phenotype with respect to cytochrome P_1-450 induction would be extremely useful as a tool for comparing development in exposed and control populations. Oxidative demethylation of caffeine is mediated by cytochrome P_1-450 and can be monitored with breath tests that use carbon-13 and measurements of urinary-metabolite ratios (Lambert et al., 1986; Campbell et al., 1987). Such a procedure can be repeated on the same subjects. For example, metabolism is high in children after the neonatal period and later declines to values characteristic of adults; patterns of metabolism at the time of puberty differ between males and females (Lambert et al., 1986). The relatively rapid P-450-mediated clearance of most compounds in prepubertal children might make it more difficult to assess the effects of enzyme inducers. Many children exhibit caffeine clearance as great as that in adults in whom it is maximal because it is induced by cigarette-smoking (Campbell et al., 1987). Some of the most dramatic effects of exposure to environmental inducers might be noted at puberty, when drug-metabolism rates normally decline to adult values.

The simultaneous observation of symptoms, monitoring of persistent specific chemicals, assessment of inducibility of AHH in lymphocytes, and continual monitoring of P_1-450 metabolism status will allow determination of the health implications of exposures and induction. Growing understanding of individual differences in inducibility, perhaps related to differences in aryl hydrocarbon receptor characteristics, might make it possible to divide the population with respect to these variables and health outcomes, including long-range susceptibility to cancer.

25

Developmental Neurobiology
of the Central Nervous System

The developing nervous system has a unique sensitivity to interference by exogenous agents, including environmental agents, specific cytotoxins, and ionizing radiation. Several structural abnormalities that occur in the prenatal CNS (e.g., anencephaly, spina bifida, hydrocephalus, and anophthalmia) can be produced in laboratory animals with the appropriate choice of species, test agent, and stage of intrauterine development.

This chapter provides background and a conceptual base for demonstrating that the vertebrate nervous systems develops through distinct processes and the establishment of neurochemical systems. The morphogenetic processes include:

- Cytogenesis.
- Transformation of neuronal precursors (neuroblasts) from a mitotic population into a population of irreplaceable, non-mitotic neurons.
- Morphogenetic migration of this neuron population to its appropriate position in the neuronal architecture.
- Death of selected members of the primordial neuron population ("morphogenetic cell death," according to Saunders, 1966) that contributes to the final makeup of neuronal groups.
- Overt cytodifferentiation in the

cytoplasm and on the cell surface, leading to the formation of specialized cells (glia and neurons) and processes (axons, dendrites, and synapses).

The chapter also reviews the complex array of endogenous neurochemicals that lead electric impulses across the synaptic junctions between neurons. Neuronal communication has several important features, including:

- The complex cellular architecture of the neuronal system, which involves multiple connections, redundancies, and positive and negative feedback loops.
- The synthesis, storage, release, and takeup of multiple neurochemicals in many neurons.
- Almost complete dependence of neurochemical synthesis on peripheral availability of amino acids.

BASIC MORPHOGENESIS

The purpose of this selective review is to demonstrate that dysgenesis of the CNS can be understood in terms of two developmental events: neuron death and neuron migration. The two events and their biologic consequences can be considered as biologic markers of neuronal development.

If known, these events in the development of the CNS could serve as an effective model for the analysis of normal and abnormal development in other organ systems. Cell death is important in the development of the limbs, oral cavity, and secondary palate; cell migration is essential for normal development of the gonads, hematopoietic system, and immune system.

Several other components of normal differentiation could also serve as useful and effective biologic markers. These include:

- Ontogeny of the neural cell adhesion molecule.
- Patterns of axoplasmic flow and axon growth.
- Ontogeny of dendritic patterns.
- Expression of neuron transmitters.

However, cell death and cell migration are concentrated on here, because they represent basic developmental events that are readily monitored and are known to be associated with normal morphogenesis. Neuronal death and migration can be manipulated to cause abnormal development within the CNS, and experimental production of cellular derangements can cause behavioral alterations in animals exposed to toxicants at specific periods of CNS cytomorphogenesis.

The initial event of vertebrate CNS development is an alteration in the embryonic surface ectoderm by the chorda-mesoderm or its structural analogue. The alteration, referred to as the primary inductive stimulus, is apparently chemical. The region of the ectoderm that receives the stimulus becomes committed to the expression of the neuronal phenotype. It is called the neural plate, and it is formed on the nineteenth day of intrauterine life—embryonic day 19 (ED 19)—in the human, on ED 7 in the mouse, and on ED 9.5 in the rat (Hoar and Monie, 1981). Coincidentally with formation of the neural plate, the neural crest is recognized as a distinct group of cells at the junction of the neural plate and the remainder of the surface ectoderm.

The neural crest is the primary source of a wide array of neurons and mesodermal cells. A series of complex, well-organized alterations in the cytoskeleton of the cells in the neural plate, now properly called the neural epithelium, lead to an elevation of the plate that results in the formation of the neural groove. The raised sides of the neural groove fuse in the apical midline on the dorsal surface of the embryo to form the neural tube. At first, the interior of the neural tube is in direct communication with the fluid-filled amniotic cavity. Separation of the neural tube from the amniotic cavity occurs with the closing of the anterior and posterior neuropores of the tube. The closing occurs rapidly in mammals, being completed on ED 25-27 in the human, ED 9.0-9.5 in the mouse, and ED 10.5-11.0 in the rat (Hoar and Monie, 1981).

BASIC CYTOGENESIS

The closing of the neural tube starts a period of rapid proliferation followed by discrete waves of migration and cytodifferentiation. Capacity of the cells of the neural epithelium to proliferate occurs in a time-dependent, orderly sequence that results in the presence of a mitotic gradient from the cephalic to the caudal end of the embryo. However, specific regions of the CNS (e.g., the cerebellum and the cerebral cortex) give evidence of prolonged proliferative capacity. The capacity for cell division is retained well beyond birth in some parts of the brain in humans.

Almost all neurons originate in the descendant cells of the neural plate that form the primitive neural tube. The cells line the central canal of the neural tube and give evidence of a characteristic proliferative pattern. Nuclei near the central canal can be observed in the various phases of mitosis. At this point, the mitotic cells are connected by tight junctions and form an internal limiting membrane. These cells form a similar attachment, the external limiting membrane, on the lateral surface of the neural tube (Kaufman, 1966).

A precise pattern of DNA synthesis, interkinetic nuclear migration, mitosis, and postmitotic nuclear migration occurs

(Sidman et al., 1959; Fujita et al., 1964; Kauffman, 1966; Langman et al., 1966). The neural epithelium is now referred to as the primitive ependymal zone (Sidman et al., 1959) or the matrix cell layer (Fujita et al., 1964). Most cells enter the S phase of DNA synthesis when their nuclei are in the periphery. On completion of the S phase, the cells round up with their surface membranes still held at the internal limiting membrane. The nuclei (now at the 4c stage of mitosis) migrate within the cytoplasm to the medial surface of the layer of cells, where they complete mitosis. The posttelophase nuclei migrate within the cytoplasm of the new daughter cells and again arrive at the periphery.

This in-and-out nuclear (interkinetic) migration, with an interspersed S phase and a period of mitosis, contributes to a large increase in the size of the neural tube. The neuroblasts that are programed to differentiate lose contact with the central canal and, presumably as postmitotic cells, migrate from the neural epithelium and come to populate the mantle layer. On completing their migration, these cells begin processes that allow them to complete their differentiation.

Development of Major Subdivisions

The anatomic disposition of the CNS is the result of a series of developmental events within the neural tube. The basic morphology of the brain and, in particular, the formation of its major subdivisions—telencephalon, diencephalon, mesencephalon, metencephalon, and myelencephalon—arise from differential mitosis and selective cell death (Bergquist and Kallen, 1954; Bergquist, 1964). The basic pattern also occurs in the developing spinal cord, where neuroblasts in the anterior (basal plate) region have a higher initial mitotic index than those in the dorsal (alar plate) region (Corliss and Robertson, 1963). Migrating neuroblasts in the anterior region therefore become postmitotic neurons earlier (Langman and Haden, 1970). In general, neuroblasts formed in the cerebral cortex conform to the basic pattern, albeit with some exceptions—for example, the Cajal-Retzius cells in layer I. How-

ever, in the cortex, cells closest to the central canal migrate, as neuroblasts, out of the primitive ependymal zone earlier than more peripheral cells. This "inside-out" pattern occurs as a result of specific spatial and temporal gradients that cause large neurons to be produced before small ones (Hicks et al., 1961; Langman and Welch, 1967; Jacobson, 1978).

Cell death is another part of normal development of the CNS. It occurs throughout neurogenesis (Kallen, 1965) and plays a necessary role in the formation of several regions of the CNS, two of which deserve special mention. In the limbs and axial musculature, the two components of the complex of motor nerves and striated muscle develop independently, and lack of effective contact between the two cellular elements leads to the degeneration and death of both (Jacobson, 1978; Vrbova et al., 1978). In addition, the nuclei of motor neurons are characterized by overproduction of cells; cells whose peripheral processes fail to make contact with developing myotubes undergo a normal sequence of degeneration and death. In the developing eye, cell death also plays a prominent role. During development, the neural retina exists as typical neural epithelium. However, unlike the cerebral cortex, this structure does not develop "inside out." The largest and most peripheral neurons (the ganglion cell layer) are formed first on ED 11 in mice, whereas bipolar and photoreceptor cells are formed on ED 13 (Sidman, 1961). Retinal neuron formation continues postnatally, ceasing on postnatal day (PN) 6 (Sidman, 1961; Young, 1985). After formation of central connections by the optic nerve, selective postnatal death of neurons formed earliest in development, the ganglion cells, occurs throughout the retina (Sengelaub et al., 1986).

Specific Development of Neuronal Type

Altman (1986) classified neurons into three principal types on the basis of their developmental origin: macroneurons, mesoneurons, and microneurons. Typically, macroneurons are large cells with long axons; the motor neurons of the spinal cord are macroneurons. Mesoneurons function primarily as relay cells, such as the relay neurons of the dorsal column nuclei. Microneurons are local elements that contribute to the fine circuitry of a given brain region; examples are the granule cells of the olfactory bulb, the cerebellar cortex, and the hippocampal dentate gyrus.

The three kinds of neurons differ in ontogeny. Macroneurons tend to form and differentiate early, during the embryonic period. Microneurons, at least in the rat, are formed and complete their development during the postnatal period (Altman, 1966; Pellegrino and Altman, 1979). Mesoneurons are intermediate in this context.

Neuronal ontogeny is best documented in the development of the cerebellum. The macroneuron component, the Purkinje cell, forms first and during embryonic development—ED 11 in the mouse (Uzman, 1960) and ED 14-15 in the rat (Altman and Bayer, 1978). The Golgi cells form next. The cerebellar microneurons form from a proliferative population on the surface of the cerebellar cortex, i.e., the external granule cell layer. The external granule cells are intensely mitotic in the first 7-10 days after birth in rats. The postmitotic neuroblasts formed in this region then migrate centrally through the molecular layer and come to reside as the internal granule cell layer, or granule cells, beneath the layer of Purkinje cells. This pattern of proliferation and migration occurs sequentially, so subpopulations of the granule cells are formed between PN 10 and PN 25 in rats (Pellegrino and Altman, 1979).

The pattern is similar in the ontogeny of the granule cells of the dentate gyrus, which make up the microneuronal compartment of the hippocampus. These cells also originate in the primitive ependymal zone surrounding the lateral ventricle. They migrate and continue their mitotic activity over the first 2 weeks after birth in rats and take up their position as postmitotic neurons in a specific pattern. The oldest cells are deposited as the top row of granule cells in contact with the superficial plexiform layer, and the younger cells end up in the basal layer (Altman and Das, 1966; Jacobson, 1978; Cowan et al., 1980).

NEUROCHEMISTRY OF NEURONAL COMMUNICATION

Biochemically, the nervous system functions as sets of connecting pathways of cells that send and receive information by releasing specific chemicals that translate changes in the electric properties of cell membranes into intracellular activity of enzymes. These chemical events usually occur within very small spaces that separate most nerve cells and their effectors, i.e., the synaptic clefts. Many cell-cell connections are short and involve only cells near each other. However, others are very long and connect the releasing cells with distant organs or targets by long cellular processes or indirectly through the circulation. The cellular architecture of the nervous system includes multiple connections, redundancies, recurrent pathways, negative and positive feedback loops, autoreceptors, densely and often highly arborized projections, and a variety of structures and cell types. In only a few instances have the connections between regions of the CNS been comprehensively mapped; in most cases, the efferent and afferent networks in even a fairly well-defined region (such as the locus ceruleus) are known only very incompletely.

The term "neurotransmitters" is used here to include all chemical substances that carry signals between cells, including neuromodulators and other cell-cell signaling chemicals. Only a small fraction of the very large number of cells in the human brain release the few well-characterized neurotransmitters. That is, the neurotransmitters released by most neurons are either unknown or poorly characterized. Moreover, many neurons are now known to contain and release more than one neurotransmitter; that greatly increases the complexity of information processing between cells. The old concept of the brain as a computing machine, in which cells or nodes in the system were only "on" or "off," has been replaced by an awareness of gradations in neural states. In addition, events can persist and influence the outcome of later events.

Communication between cells is affected by linked biochemical events involving cascades of second and third messengers, such as the cyclases, phosphodiesterases, calcium-binding proteins, and other enzymes and proteins (Snyder, 1984). RNA-directed protein synthesis can also be part of the messenger sequence of neurotransmission (Kandel and Schwartz, 1981; Gusella et al., 1984). These events take place within cells, so their biochemical products might never be released into compartments other than the immediate synaptic milieu. Other neurons regulate more distant events, such as the release of trophic hormones from the pituitary, uptake of nutrients from the gut, and function of smooth muscle in the peripheral vasculature. These products of neurotransmission can be measured physiologically as changes in circulating hormones, intestinal absorption, or venous blood pressure, for instance.

In addition to mediating intercellular communication, endogenous neuroactive substances play an important role in development. In the early stages of brain development, some of the classic neurotransmitters—such as gamma-aminobutyric acid (GABA) and norepinephrine—play a trophic role, guiding the formation and fixation of axonal projections and synaptic connections (Black et al., 1984). The latter neurochemical-dependent processes of enervation have been elegantly demonstrated in the model of neuronal development of Hoffer et al. (1987), in which fetal brain is transplanted into the chamber of the eye in rodents and its development is monitored chemically, morphologically, and electrophysiologically.

As shown in Figure 25-1, the fundamental neurochemical cycle of neurons and glia in the nervous system comprises the processes of uptake, transport, synthesis, storage, and release. Neuronal uptake, which is kinetically highly efficient and saturable at low concentrations, can serve several functions, including termination of cell stimulation by removing the neuroactive compound from the receptor, resupply of intracellular pools for later release, and provision of precursors for the synthesis of neurotransmitters, such as choline for acetylcholine, tyrosine for dopamine and norepinephrine, and tryptophan for serotonin. Transport is a critical process in neurons, particularly those with long axonal projections and extensive dendritic arborizations, in which enzymes and other materials synthesized in the cell body must be moved to the terminals. Synthesis of neurotransmitters involves highly regulated pathways. Some of the pathways can be used in other metabolic processes, in which case the synthetic pathway in neurons is usually distinguished by kinetic properties, rate-limiting cofactors, or compartmentation. Other synthetic pathways involve RNA-directed synthesis and enzymatic cleavage of large polypeptide precursors for the formation of neuroactive peptides, such as the enkephalins and so-called gut peptides. Storage in neurons involves compartmentation and intracellular transport of precursors and products by mechanisms that protect these substances from enzymatic degradation or hydrolysis. Storage can also provide a dosimetric func-

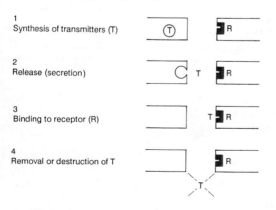

1
Synthesis of transmitters (T)

2
Release (secretion)

3
Binding to receptor (R)

4
Removal or destruction of T

FIGURE 25-1 Four biochemical steps in synaptic transmission: synthesis of neurotransmitter (T), release of transmitter synaptic cleft, binding of transmitter to postsynaptic receptor, and removal or destruction of transmitter substance.

tion by packaging neurotransmitters in minimal releasable amounts (or quanta). The dosimetric function can be important for maintaining trophic relations between cells, as has been demonstrated for cholinergic pathways in the peripheral nervous system. Release is the exocytotic process in which neurotransmitters are secreted by cells into the extracellular space. The release is usually ion-dependent and hinges ultimately on changes in intracellular free-calcium concentrations, which appear to control the fusion processes necessary for exocytosis.

After release, neurotransmitters act by binding to receptors on cells. Most receptors in the nervous system are membrane-bound and react to substances that reach the outer membrane of the receiving cell. Neuronal receptors can be grouped in complexes, such as the interrelated set of benzodiazepine, GABA, and chloride ionophore receptors in the GABA-ergic pathway. Activation of neuronal receptors translates into biochemical events in the receptive cell (such as activity of adenyl cyclases) that are then linked in a functional cascade of phosphorylation reactions that can stimulate or inhibit other enzyme activities, alter the nature and permeability of the cell membrane, and manifest other functions (Kandel and Schwartz, 1981).

The problems of obtaining access to the essential biochemical processes so as to use them as markers of neurobiologic function are exemplified in amino acid neurochemistry. Some amino acids—such as glutamate, aspartate, and glycine—are neurotransmitters in brain and spinal cord. However, the largest quantities of these amino acids in the body are involved in general intermediary metabolism; only a small fraction is reserved for the specific role of cell-cell communication.

The brain does not synthesize amino acids for neurotransmission, nor are they derived from catabolism within the brain. Neurons obtain amino acids for neurotransmission by removing them from the circulation through previously described high-affinity uptake processes. Building-block amino acids, such as tyrosine and tryptophan, are required for synthesis of other neurotransmitters; the brain must obtain these amino acids from the circulation. Because of that absolute dependence, changes in the peripheral availability of amino acids might be expected to alter the concentrations of neurotransmitters and, consequently, affect the function of some neural pathways in the brain. Conversely, high amounts of some amino acids in the diet, such as the excitatory neurotransmitters glutamate and aspartate, might be expected to be neurotoxic. The potential neurotoxicity of dietary amino acids, particularly during development, has received some attention recently, because of the increasing use of aspartame, a simple derivative of aspartate, as a sweetening agent (Sved, 1983). A body of evidence from neuropharmacologic research (Sved, 1983) indicates that alterations in circulating

TABLE 25-1 Biochemical Markers of Development and Cell Injury in the Nervous System

Biochemical Marker	Indicator For
Central Spinal Fluid Marker	
Protein I	Status of synaptic membranes of CNS neurons
D2 (neural tube)	Status of synaptic membranes of CNS neurons
B50	Status of synaptic membranes of CNS neurons
P5D 95	Postsynaptic receptors
Myelin basic protein (MS)	Status of oligodendroglia and myelin sheath
Myelin-associated glycoprotein	Oligodendroglia
GFAP	Astrocytes (gliomas)
Brain Tissue Markers	
Protein III	Cell loss (nerve terminals)
Synapsin I	Cell loss (nerve terminals)

concentrations of some amino acids (such as tryptophan) can affect CNS neurochemistry. However, the implications of the biochemical modifications for functional changes, such as neuronal activity in serotoninergic pathways, are not yet clear.

Reliable inferences regarding the status of function in pathways that use amino acids cannot be drawn from measurements of peripheral amino acid metabolism. Some research attempted to develop an index of CNS cholinergic function with arteriovenous difference in blood choline concentration as a marker (E. Silbergeld, Environmental Defense Fund, personal communication, 1987). No consistent differences that could be correlated with major changes in cholinergic function were ever found.

Thus, chemical indicators or biologic markers of neuronal function (see Table 25-1 for a partial list) are difficult to obtain, particularly outside the nervous system itself. However, other cell processes might be investigated, such as cell death, turnover of membranes, and cellular differentiation (O'Callaghan and Miller, 1983; Bondy, 1985). The utility of chemical indicators was demonstrated in studies of neurotoxicity in animals exposed to the neurotoxicant trimethyltin (O'Callaghan and Miller, 1984; Harry et al., 1985).

26

Morphologic, Neurochemical, and Behavioral Responses to Toxic Agents

This chapter discusses the importance of timing and dose for effects in developing organisms. Then, specific biologic markers of neurodevelopment are discussed, ranging from the association of minor physical anomalies with behavioral effects, to measurement of neurochemical concentrations, to behavioral assessment of complex processes.

EFFECTS OF TIME OF EXPOSURE AND DOSE: IRRADIATION AS A PARADIGM

The extensive use of ionizing radiation as an embryotoxic agent has been of paramount importance in the delineation of several important concepts that have particular relevance to current efforts in developmental toxicology. Studies in radiation teratology are unique, in that the physics of ionizing radiation has allowed scientists to produce effects in the embryo directly, with no concern for the moderating influence of the so-called placental barrier.

Work from four laboratories established two basic concepts: stage specificity and the relation of dosage to response (Hicks, 1953; Rugh, 1953; Russell and Russell, 1954; Wilson, 1954). All the early studies provided convincing evidence that the production of specific congenital malfor-mations depended on the stage of development at which radiation was administered and that the severity of the effect produced was a function of the dosage. The preimplantation stages of development were usually shown to be relatively radio-resistant when neonatal death was assessed, although Rugh (1959) reported that low doses of radiation during this period produced a low incidence of exencephaly in survivors.

Wilson (1954) showed the influence of dose, especially on the production of eye defects in rat fetuses. Irradiation with 25 rads on ED 9 produced eye defects in 6% of the fetuses. The magnitude of response increased with increasing dose; exposure to 200 rads produced eye defects in 72% of the fetuses. The embryo, under the experimental conditions, became radioresistant with time. Irradiation on ED 11 with either 25, 50, or 100 rads produced no eye defects in survivors, although all survivors were affected when the dose was 200 rads. Hicks (1953) also showed time-dependent responses to ionizing radiation in studies of specific CNS defects. Hicks noted that irradiation of the embryo early in development produced forebrain defects, whereas irradiation during fetal and neonatal life resulted primarily in cerebellar malformations.

As summarized by Rugh (1953), "the inert

primordium or the totally differentiated cell will be relatively resistant in terms of morphologic change. The actively differentiating intermediate stage or stages will be highly radiosensitive since they are in the process of transformation (differentiation)." Hicks developed a mechanistic framework to validate the concept (Hicks, 1959; Hicks and D'Amato, 1966). He hypothesized that the radiosensitive cells (neuroblasts) were localized to the region of the neural epithelium that was active in DNA synthesis and that the immediate consequence of ionizing radiation was massive and extensive death of cells of that type. Cell death was a transient phenomenon, and the embryo was capable of repairing the damage quickly. He postulated that the malformations observed at term were the consequence of a balance between the initial damage and the degree of regulation and regeneration inherent in that region of the embryo at the time of injury.

Kallen (1965) documented that the developing nervous system is capable of stage-dependent regeneration. The idea could be validated under conditions in which regenerative capacity was absent and radiation would produce specific cell death and specific neuronal deficits. As documented below, the microneurons of the cerebellum and the dentate gyrus, which develop postnatally, satisfy the necessary conditions, and selective radiation of these regions at appropriate periods of development produces specific structural and functional deficits.

MICRONEURONAL RADIATION

Cerebellum

Hicks et al. (1962) reported that irradiation of the cerebellum of the 6-day-old rat with 200 R produced extensive damage to the external granule cell layer. The layer retained its regenerative capacity and formed an ectopic granule cell layer within 4 days. However, the initial damage also interfered with Purkinje cells: their form was altered, and the association of Purkinje cells and granule cells formed later was abnormal. Altman and coworkers

(Altman et al., 1969; Pellegrino and Altman, 1979) confirmed that observation in a series of experiments that effectively verified the idea that cell killing without substantial repair led to severe cellular deficits within the cerebellum. They used the knowledge that different microneuronal populations that arise in the external granule layer do so at specific times after birth; the basket cells were formed on PN 6-7, the stellate cells on PN 8-11, and the granule cells on PN 8-21. They used sequential irradiation to monitor morphogenesis and later performance in a variety of tasks. In the first group, focal irradiation of the cerebellar cortex with 200 R on PN 4 and 5 produced cerebellar disorganization similar to that observed by Hicks et al. (1962). In a second group of animals, the focal irradiation consisted of 200 R on PN 4-5 (as above) followed by 150 R on PN 7, 9, 11, 13, and 15. With the fractionated regimen, all derivatives of the external granule cell layer failed to form—an effect that resulted in severe motor deficits. Postponing the initial irradiation to PN 8 and 12 produced selective neuronal deficits (stellate and late-forming granule cells in one group, late-forming granule cells in another group) with corresponding selective behavioral effects.

Initial observations failed to show any postural or motor deficits in the second group of rats that were irradiated many times postnatally (Pellegrino and Altman, 1979). In fact, in experiments with a motor-driven rotating rod, the irradiated animals performed better than controls; that is, they fell off the rod less frequently. However, in open-field tests, the irradiated rats were observed to be significantly more active than control animals. Ambulation in the open field is affected by agitation. It was concluded that microneuronal hypoplasia in the cerebellum that does not produce demonstrable locomotor deficits can nevertheless lead to hyperactivity at an age when the animals tend to be the most active (2 months). In adult animals, the difference disappeared.

The experiments documented a strong correlation between the developmental history of a neuronal population and its contribution to the behavioral hierarchies within the animal. The studies are a logical extension of the classical studies reviewed earlier and confirm the observation that the primary effect of irradiation of sensitive cell populations is cell death. It is clear, however, that fractionated irradiation of developing microneurons is not accompanied by extensive regeneration. Hence, highly specific and highly reproducible cell deficits can be produced, and their behavioral consequences can be monitored.

Hippocampus

The same approach has been used to evaluate the effects of microneuronal hypoplasia on the hippocampus (Altman, 1986). Focal x irradiation of the hippocampus, begun immediately after birth, prevents the formation of nearly 85% of the granule cells of the dentate gyrus.

The rats were then tested in the same protocols as were the animals mentioned above with cerebellar microneuronal hypoplasia. They with hippocampal microneuronal hypoplasia were found to be extremely hyperactive when tested in the open field (Bayer et al., 1973). They were also extremely active, compared with control rats, in the running wheels (Peters and Brunner, 1976). The irradiated rats displayed other behavioral changes usually associated with hippocampal damage, including disappearance of spontaneous alternation in a T maze and deficits in passive avoidance learning (Bayer et al., 1973). The rats showed deficits at all ages; the deficits differed in severity with age and between tests. Further studies indicated that, as long as the learning tasks ranged from very easy to moderately difficult for normal rats, the irradiated animals performed as well as normal rats (Altman, 1987). However, when the tasks were more difficult, the irradiated animals were significantly impaired in tactile and visual discrimination, in acquisition learning, and in reversal learning.

Thus, selective microneuronal hypoplasia in the hippocampus leads to selective behavioral effects. Altman (1986) has postulated that these, and other, selective effects on microneuronal populations can provide the anatomic basis for minimal brain dysfunction under the influence of a broad spectrum of environmental factors, such as alcohol, lead, and glucocorticoids.

APPLICATION TO OTHER TOXIC SUBSTANCES

Several kinds of toxic agents can generate structural and behavioral alterations in animals and humans that result from fetal and perinatal exposure. Experimental studies of time and dose effects should be done with these agents, as has been done for irradiation.

Hicks et al. (1961), Hicks and D'Amato (1963), and Berry and Eayrs (1966) showed that one effect of irradiation during fetal life was the alteration of migration patterns in the layers of the cerebral cortex. A similar effect was observed in the mouse fetus as a result of subjecting the dams to hypervitaminosis A (Langman and Welch, 1967).

Miller (1986) recently showed that exposure of pregnant rats to ethanol from ED 6 to ED 23 produced effects similar to those seen after irradiation, i.e., a deficit in cortical neurons and an alteration in their migration patterns. Although not documented, the cytotoxicity of ingested ethanol on cortical neuroblasts is a possible underlying mechanism of this observation.

Similarly, the administration of lead to rats immediately after birth results in altered hippocampal cytodifferentiation, including the presence of smaller numbers of granule cells (Petit et al., 1983; Kawamoto et al., 1984) and behavioral changes. Observation of those effects is confounded by the severe effects of lead on brain endothelial cells (Winder et al., 1983), so a specific effect on the granule cell population cannot be ruled out.

RELATIONSHIP BETWEEN MINOR PHYSICAL ANOMALIES AND BEHAVIORAL EFFECTS

Agents that influence physical development are likely to alter behavior. The relationship between phenotypic and behavioral responses to pollutants is clearly exemplified in the study of minor physical abnormalities (MPAs) and behavioral pathology. These observations also suggest a potentially useful set of markers for CNS dysfunction.

The relationship between MPAs and behavioral aberrance was first observed in schizophrenics (Waldrop and Halverson, 1972; Goldfarb and Botstein, in press). The following were observed in increased proportion in schizophrenics: excessively fine hair that stands on end, multiple hair whorls, excessively large or small head circumference, epicanthal folds, hypertelorism, low-set ears, adherent earlobes, high arched palate, curved fifth finger, simian palmar crease, spaced toes, and partial syndactyly. That most of those changes are primarily ectodermal in origin suggests that the timing and pathogenesis of the events were shared by alterations in the CNS.

The study of MPAs was extended to other behavioral aberrations. In a study of normal 2.5-year-olds, Waldrop et al. (1968) found that, because the number of MPAs was significantly correlated with restless, aggressive impulsive behavior, the MPAs might have been indicators of hyperactivity. Behavior was stable when the subjects were followed up to the age of 5 years. The number of MPAs was found to be negatively correlated with verbal IQ (Rosenberg and Weller, 1973), with full-scale IQ (Waldrop and Halverson, 1972; Firestone and Prabhv, 1983), and with academic achievement (Halvorsen and Victor, 1976). An apparent sexual dimorphism in the relationship between MPAs and behavior has been observed. Boys with high MPA scores tend to be hyperactive, and girls with high MPA scores seem to display more inhibited, intractable behavior (Waldrop et al., 1976; O'Donnell and Van Tuinan, 1979). Quinn et al. (1977) classified infants according to MPA number into low, middle, and high groups. At 2 years of age, high-MPA boys were more irritable and had a higher incidence of night awakening, and high-MPA girls were less active and more withdrawn.

MPAs are increased in autistic children (Steg and Rapoport, 1974). Quinn and Rapoport (1974) found an association between increased MPAs and aggression and hyperactivity, but not anxiety, in boys. In the same sample, dopamine β-hydroxylase activity in blood was correlated with MPA score. MPA scores were higher in hyperactive and retarded boys (Rapoport et al., 1974) and in siblings who were considered mentally normal.

Offspring with high MPA scores are more likely to have been the products of complicated pregnancies (e.g., with toxemia or prematurity) than of uncomplicated pregnancies (Simonds and Aston, 1981). They are also more likely to have siblings and parents with high MPA scores; that suggests that both genetic and nongenetic mechanisms were involved in the pathogenesis (Smalley et al., 1988).

Those observations link a class of relatively easily identified and measurable changes in physical structure with abnormal CNS development with behavioral deficits. Both the physical structures and the CNS are ectodermal in origin. Thus, the two classes of tissue might have responded in an analogous way to a given noxious agent. The inference to be drawn is that each class of observations (physical and behavioral) could serve as a marker of the other. MPAs in newborns have also been found in dose-dependent relationship with umbilical cord blood lead concentrations (Needleman et al., 1984).

NEUROCHEMICAL EFFECTS

For the assessment of nervous system status with chemical methods, samples of various bodily fluids are taken and constituent materials are analyzed (see Table 25-1). The methods have required the development of high-performance assays, because the amounts of sample usually available and the concentrations present in the relevant compartments are small. The fluids are blood, urine, and cerebrospinal fluid (CSF). Blood and urine are of less utility, because of their remoteness from the nervous system and the contribution of nonneuronal sources to the amounts of most substances in these compartments. CSF is not routinely available, inasmuch as its sampling requires medical oversight and involves danger. Advanced techniques of imaging have recently enlarged the ways in which biochemical reactions and events in the brain can be measured; these are discussed in the Chapter 30, because they present the most important new opportunities for obtaining markers of neuropsychiatric function.

Neurochemical characteristics have been studied in only two major intoxication states: lead poisoning and brain damage induced by N-methyl-4-phenyltetrahydropyridine (MPTP). The neurochemistry of lead poisoning has been extensively studied in animals (Silbergeld and Hruska, 1980; Winder et al., 1983). Only recently have attempts been made to extrapolate from the results of those studies to the development of biologic markers in humans. Silbergeld and Chisolm (1976) studied monoamine metabolites in urine of lead-exposed children. As shown in Figure 26-1, there is a correlation between blood lead content and 24-hour urinary excretion of the dopamine metabolite homovanillic acid (HVA) in those children. HVA was measured before initiation of chelation therapy, within a week after the children were removed from lead-contaminated environments (in all cases, lead paint). Over the long term, urinary HVA content was reduced,

as was blood lead content, although both blood lead and urinary HVA remained higher in treated lead-exposed children than in age-matched controls. Other neurochemicals reported to be altered by lead in animal models—such as GABA and enkephalins— are less available to clinical measurement, because they require CSF, and have not been investigated in humans.

MPTP is a contaminant in some "designer drugs" or meperidine derivatives with opiatelike characteristics. After the remarkable finding that some addicts had acute-onset neurologic disorders that were indistinguishable from so-called idiopathic Parkinson's disease (Langston et al., 1983), attention focused on MPTP as the active pathologic agent. MPTP was found to be a specific basal ganglia toxin that damages the same nigrostriatal dopaminergic pathways that are affected in parkinsonism (Kopin and Markey, 1988). The mechanism of action of MPTP involves uptake into dopaminergic neurons, interaction with oxidases within the neurons, and selective cell killing, possibly by the generation of free-radical oxygen or hydroxyl radicals. As a dopaminergic toxin, MPTP could be expected to reduce output of dopamine metabolites from brain into CSF; this has been demonstrated in primate models of intoxication (Kopin and Markey, 1988).

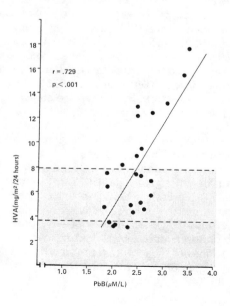

FIGURE 26-1 Correlation between blood lead and 24-hour urinary excretion in children. Source: Silbergeld and Chisolm, 1976.

BEHAVIORAL EFFECTS

Behavior is the observable response of an organism to changes in the external or internal environment. That definition includes actions ranging from reflex responses to the solving of complex problems and performance on psychometric tests and in social situations. Behavioral analysis can proceed at any of those levels.

The investigator of neurobehavioral function is confronted with a choice between attempting to characterize a possible deficit by measuring a single, isolated function and scaling more complex integrative behaviors. The choice might present a trade-off between the precision of a test that measures a single function (or as close to a single function as possible) and the relative imprecision of a test instrument that measures the subject's integrative capacity in a more complex single-demand task or in a number of tasks. The use of measures of single functions can be more effective in detecting neurotoxic deficits than gross-performance tests, because they are more focused and demanding (Smith, 1985). Tests of integrative capacity might yield increased sensitivity, but lead to less precision as to the location and degree of a deficit.

A number of outcomes are presented below, roughly in increasing order of functional complexity. In each instance, their utility when testing for lead exposures is noted, not because that represents their sole utility, but because the effects of lead exposure is the most widely studied toxicity in humans.

• **Nerve conduction time.** Measurements of nerve conduction time have been useful in assessing neurotoxicants that act on myelin development or on Schwann cells. Occupational exposures to lead at doses that did not produce symptoms were associated with impairments in nerve conduction (Seppalainen and Hernberg, 1980). Children exposed to lead from industrial sources were found to have dose-dependent nerve-conduction slowing (Landrigan et al., 1976). Because normal variation is large, assessment of nerve conduction time is not well suited as a screening device in neurotoxicant exposure.

• **Sensory psychophysical functions and scotopic vision.** Measures of visual acuity in bright light were not related to lead exposure in primates. Visual acuity in the dark was diminished in primates exposed to lead (Bushnell et al., 1977). The results suggest specific impairment of rod function.

• **Evoked potentials.** Because evoked potentials recorded from the scalp or spinal cord reflect activity in multisynaptic pathways, evoked potentials can provide useful information in assessing sensory transmission from the periphery to the cerebral cortex. Three types of evoked potentials have been recorded: auditory evoked potentials for which the stimuli are generally pure tones or clicks, but can be phonemes or words; visual evoked potentials for which the stimuli are stroboscopic light flashes or checkerboard patterns on computer monitors; and somatosensory evoked potentials for which the stimuli are brief electric impulses delivered to the skin. Auditory evoked potentials have been found to be altered in exposure to lead (Otto et al., 1981), carbon monoxide (Groll-Knoff et al., 1978), and trichloroethylene (Winneke et al., 1978). Visual evoked potentials have been found to be affected in exposure to xylene (Seppalainen et al., 1981), methyl mercury (Iwata, 1980), and n-hexane (Seppalainen et al., 1979). Somatosensory evoked potentials have been reported to be altered in exposure to lead (Seppalainen, 1978), but not in exposure to n-hexane severe enough to be symptomatic (Zappoli et al., 1978).

• **Auditory discrimination.** Auditory discrimination can be tested against various masking backgrounds, such as taped sounds containing a signal against background noises of increasing loudness (e.g., electric fan sounds or noise from a cafe) (Goldman et al., 1970). In studies of asymptomatic children, those with the higher concentrations of lead in their teeth had lower scores on this measure (Needleman et al., 1979).

• **Vibration sense.** Vibration sense can be measured with a number of methods; the most simple (and least sensitive) uses a tuning fork. Computerized methods with increased sensitivity and precision are available (Maurissen and Weiss, 1980). Decreased vibratory sensitivity is observed in many conditions that involve the central and peripheral nervous systems. Among them are systemic disease, such as diabetes, chronic liver failure, pernicious anemia, peripheral neuropathies, syphilis, spinal-cord lesions, and uremia; exposure to pharmaceuticals, such as isoniazid, phenytoin, vincristine, and glutethimide; and exposure to chemicals, such as acrylamide, arsenic, n-hexane, mercury, and methylbutylketone (Maurissen, 1985).

• **Motor function.** Quantitative measures of tremor have been evaluated in methyl mercury exposure. Quantitatively increased tremors were shown when clinical studies were noninformative. Studies of motor patterns of children classified as to lead exposure are under way. One study, which measured on-task behavior of children differentially exposed to lead, found that high-lead children spent more time in off-task behavior in the classroom—e.g., out-of-chair activity, staring out of the window, and talking to classmates (Needleman and Bellinger, 1981).

• **Attention.** Attention is a complex function of arousal, vigilance, and resistance to distraction. A number of measures of it are available. In the Continuous Performance Test, a measure of vigilance (Rosvold et al., 1956), the subject is presented with a set of stimuli (letters of the alphabet) on a screen for short intervals in rapid succession. The critical stimulus (the letter X) is presented at a predetermined probability. The demand task is to press the response key when the critical stimulus appears. The number of errors of omission, commission, and latency to response can be determined. Reaction time with various intervals of delay has been used to discriminate among children who differed as to lead exposure

(Needleman et al., 1979; Yule et al., 1981; Hunter et al., 1985).

• **Visual-motor integration.** Measures of visual-motor integration, such as the Bender-Gestalt test (Trillingsgaard et al., 1985), have found wide application in the study of brain damage in children. A number of skills are called on: spatial visualization, eye-hand coordination, and visual-spatial memory.

• **Speech and language function.** Speech and language function is the sum of many competences and can be perturbed at many levels. At the perceptual level, auditory acuity for pure tones and the ability to screen out background distractions can be measured (Goldman et al., 1970). Short-term memory and the ability to discriminate patterns are testable (Seashore Rhythm Test). The ability to comprehend language is testable with many instruments, such as the Token Test, the verbal subtests of the WISC-R, and the Illinois Test of Psycholinguistic Abilities.

• **Psychometric intelligence.** Studies of psychometric intelligence in children have been widely used in recent years to study lead toxicity, exposure to polybrominated biphenyls (PBBs), and fetal alcohol exposure. Three tests have been used generally: the Bayley Scales of Infant Development for children between 6 months and 3 years old, the McCarthy Scales for children more than 3 years old but less than 5 years old, and the Wechsler Intelligence Scales revised for children over 5 years old. The most sensitive subscales of these instruments appear to be the verbal and general cognitive index. Numerous studies of effects of lead exposure at low dose in children have shown IQ deficits after control of relevant covariates (U.S. EPA, 1986).

• **Social behavior.** Attention to cognitive, perceptual, and motor competences should not direct attention away from social behavior of children. Three groups of investigators have rated classroom behavior in lead-exposed children with structured questionnaires (Needleman et al., 1979; Yule et al., 1981; Hatzakis et al., 1987). Teachers blind to a subject's lead exposure reported a dose-dependent increase in nonadaptive classroom behavior, such as distractibility, inability to work independently, disorganization, hyperactivity, impulsivity, and inability to follow directions (Needleman et al., 1979).

27

Methodologic Issues of Extrapolation from Animal Studies to Human Toxicant Exposure

This chapter briefly reviews some of the major approaches to relating animal findings in functional teratology to the assessment of potential human health hazards. The approaches include:

- Investigation of underlying mechanisms of functional alterations observed in animals.
- Investigation in animals of normal and abnormal development of functional end points that are comparable in humans.
- Direct comparison of functional effects seen in animals and humans when data are available on both.

Intrauterine exposures to some teratogenic agents have been linked to gross physical malformations in both humans and animals. Structural abnormalities are often profiled as syndromes, e.g., the fetal alcohol syndrome (Clarren and Smith, 1978) and the fetal hydantoin syndrome (Hanson et al., 1976). Interest in people with subtle functional effects after low-dose exposures and people without overt anomalies has increased. Since Wilson (1973) included functional alterations in a list of possible effects of exposure to developmental toxicants, research in the subject has expanded greatly. All functional systems are theoretically at risk at some point in their development

and maturation. Only a few functional systems have been studied, and that situation is changing (Kimmel and Buelke-Sam, 1981; Kavlock and Grabowski, 1983; Riley and Vorhees, 1986). Unlike studies that evaluate multiple structural changes after exposures, studies of postnatal function typically evaluate effects in a single organ system or on a single end point, e.g., the CNS or immune deficiency.

Additional complications are encountered in cross-species comparisons of postnatal functional alterations, because species often vary both in their responsiveness or susceptibility to toxic insult and in the manner in which they manifest toxicity. Examples of research aimed at overcoming those problems with each approach are discussed.

INVESTIGATION OF UNDERLYING MECHANISMS

One way of relating animal findings and human hazard is to evaluate underlying structural, biochemical, and physiologic correlates of overt functional changes seen in animals. The rationale is that the determination of the target and degree of toxicity produced by developmental exposure will yield information relevant to the human situation.

Mirmiran and colleagues (1985) recently

reviewed the relationships between behavioral alterations observed in humans and animals and the underlying neurochemical and electrophysiologic disturbances observed in rats that were exposed to pharmaceutical agents during development. Table 27-1 presents some of the findings.

The immaturity of the blood-brain barrier and greater accumulation of many of these compounds in the developing brain make the fetal brain a major target of its mother's medication. Mirmiran et al. (1985) have shown that neonatal exposure of rats to clonidine, an antihypertensive agent, and clomipramine, an antidepressant that acts on norepinephrine and serotonin neurotransmission, suppresses rapid-eye-movement (REM) sleep in the developing rats. In adulthood, the offspring rats showed hyperactivity, hyperanxiety, reduced sexual behavior, disturbed sleep patterns, and smaller cerebral cortex.

STUDY OF COMPARABLE FUNCTIONAL END

A second approach to determining the relationship of animal findings in postnatal functional studies to the human situation is to select a functional response that is comparable across species. A variety of potentially relevant end points are available, e.g., sleep patterns, neonatal vocalizations, and suckling patterns. The development of these end points and their sensitivity to toxic insult could be compared directly across species.

The startle reflex is valuable in such an effort for several reasons:

- Startle can be elicited in all mammals, including humans.
- The startle reflex is mediated via simple neuronal circuits.

TABLE 27-1 Sequelae of Developmental Exposure to Drugs in Humans and Animals[a]

Drug	Effects in Humans	Effects in Animals	REM[b] Sleep Deprivation Effects	Relevant Transmitter System
Clonidine	Smaller head circumference, questionable neurologic status, increased myoclonic jerks during sleep	Hyperactivity, delayed motor development	+ + +	Norepinephrine
Diazepam	Low Apgar score, reluctance to eat	Hyperactivity, decreased male sexual behavior, decreased startle reflex	+	Gamma-amino-isobutyric acid
Imipramine-like agents	Poor sucking, irritability	Hyperactivity, decreased male sexual behavior, smaller brain	+ + +	Norepinephrine, acetylcholine, serotonin
Reserpine	Anorexia, lethargy	Smaller brain, altered activity, altered startle reflex	+	Norepinephrine, dopamine, serotonin

[a]Data from Mirmiran et al., 1985.
[b]REM = rapid eye movement.

• The startle reflex is modulated via several neurotransmitter systems.

• The startle reflex can be measured at early ages in many species.

• The startle reflex is quantifiable.

• Inhibitory or excitatory effects can be determined.

• Startle displays different types of plasticity.

Davis (1984) has reviewed aspects of the mammalian startle reflex. It consists of a characteristic, very rapid sequence of muscular responses elicited by a sudden, intense stimulus. Under comparable circumstances, the more intense the stimulus, the greater the response. The graded amplitude of the mammalian startle response can be detected in direct muscle recordings (e.g., electromyographic recordings from a limb or muscles involved in blinks) or in the output from transducers that measure cage movements when whole-body startle is measured. A standard feature of this reflex is its very short latency; the response occurs only milliseconds after the onset of the eliciting stimulus.

Although the neural circuitry that mediates startle is at lower levels of the CNS, higher neural networks can modulate it. Nearly all defined neurotransmitter systems interact to modulate the startle response (Fechter, 1974; Davis and Aghajanian, 1976; Davis and Sheard, 1976; Handley and Thomas, 1979; Davis and Astrachan, 1981; Gallager et al., 1983; Holson et al., 1985). In the spinal cord and facial motor nucleus, serotonin and norepinephrine increase auditory startle and glycine tonically inhibits it; it appears that GABA can also inhibit the response at this level. Supraspinally, dopamine and perhaps GABA receptor stimulation increases startle, and serotonin activation depresses it. Startle is also modulated in several brain regions distant from the primary startle pathway itself. Therefore, the reflex can provide a sensitive indicator of function after toxicant exposure. Developmental insults that result in changes in a neurotransmitter system might be expressed as changes in the latency, amplitude, or modification of the response. The type of change observed can suggest which systems have been affected by exposure.

Auditory startle has been used often in studies of animal developmental toxicology. Recently, automated procedures have been applied in such studies, thus allowing more specific characterization of changes in this reflex. Automated procedures for stimulus presentation and data collection have yielded useful information for evaluating sensitization, habituation, prepulse inhibition, and reflex modification by prior associative learning after toxicant exposure (Hoffman, 1984). Startle thus represents a potentially powerful tool in developmental toxicology for investigating sensorimotor reactivity. The simplicity of the response and the plasticity displayed within it across animal species, including humans, suggest that specific efforts to investigate the comparability of startle alterations in animals and humans after developmental insult are warranted.

DIRECT COMPARISONS BETWEEN ANIMALS AND HUMANS

A third approach to determining the relationships among human and animal developmental toxicity is to compare observed effects when data are available for several species. Few human behavioral-teratology studies have been reported, and most experimental behavioral-teratology studies have used rodents. The comparisons outlined here reflect that situation. In addition, similarities and differences in design and conduct between experimental and clinical research must be considered in any comparison of results. The similarities between the two include the following:

• Physical growth and development are the most commonly measured end points.

• Several behavioral subsystems are assessed with a battery of functional tests.

• Experimental and control subjects are matched for maternal and environmental characteristics.

• The majority of studies are designed to provide descriptive information.

In human studies, weight and motor development usually are measured for 1-2 years after birth. In rodent studies, weight is monitored repeatedly, most often throughout the duration of the study, and assessments of preweaning reflex development are often carried out as well.

A battery of functional tests usually are used for neurobehavioral evaluation in both human and animal studies. The use of a single assessment technique that incorporates multiple evaluations is most common in human research. The Apgar test (Apgar, 1953) is used routinely 1 and 5 minutes after birth; it consists of a 10-point scale based on five components: appearance (skin color), heart rate, latency of the cry reflex, muscle tone, and respiration. The Bayley scale of infant development (Bayley, 1969) is used commonly for evaluation of older infants; it contains sensory, motor, verbal, and cognitive items, and results are summarized in motor and mental development scores. Neurobehavioral function in rodents is evaluated with a test battery that often includes assessment of reflex and sensorimotor development, activity level, and some evaluation of learning ability. Each category of function is evaluated with separate tests.

Another similarity in study designs is the use of experimental and control subjects matched for maternal and environmental characteristics. In human studies, mothers are matched as closely as possible for age, parity, and nutritional and socioeconomic status. In animal studies, maternal weight, parity, diet, and housing conditions routinely are controlled across groups.

Both clinical and experimental studies designed to evaluate neurobehavioral outcomes after prenatal drug or chemical exposures provide primarily descriptive information. The methods permit a description of functional deficits after insult, but not of underlying physiologic or neurochemical mechanisms responsible for the observed behavioral alterations. As noted above, that situation is changing in animal studies. Multidisciplinary efforts can provide information on the mechanisms involved and thus might suggest types of intervention that could be effective in alleviating or improving clinical outcomes.

Several basic differences in design and conduct between human and animal studies are common. They include differences in the relative age range, in timing of administration of tests, in attempts at standardization, and in methods of reporting results (Adams, 1986).

The relative age range studied is broader in much of the animal research than in human studies. Practically, it is very difficult to follow a prenatally exposed person for more than a year or two after birth. In the best of circumstances, clinical investigators can extend evaluations to 6 or 8 years of age. Funding, time requirements, and population mobility and attrition all contribute to the difficulty. In contrast, rodent studies often include behavioral evaluations into early adulthood. Such a longitudinal approach can be even more valuable if testing is identical across the age span studied.

In human research, several functions usually are evaluated at a single age. Neurobehavioral function in rodents most often is evaluated through separate testing at different ages. Furthermore, unlike the results of multifunctional evaluations in humans, rodent performance across tests is not integrated into a single value or score.

A greater effort is made in human than in animal studies to perform testing on infants in a comparable behavioral state, e.g., alert, drowsy, or asleep. Such control contributes to both absolute response levels and decreased variability in behavioral data collected in infants and children (Clifton and Nelson, 1976). Animal researchers at best attempt to control such factors by balancing time of day during testing across experimental groups.

Developmental delays in physical, motor, and cognitive end points are considered more important in human than in animal studies. Such delays can be assessed only during particular periods of development. Their biologic meaning in rodents after prenatal exposures is not clear, and they have been viewed as problematic (Tilson and Wright, 1985). One contributing factor might be the time disparity in postnatal developmental schedules between humans and rodents, i.e., months and years versus days.

Finally, a characteristic difference between human and animal studies involves the method of reporting results. Clinical studies typically identify the incidence of behavioral dysfunction in individual control versus individual exposed subjects. Animal data usually are presented in terms of the presence or absence of group mean differences. Thus, the incidence of affected (and nonaffected) rodent offspring in the exposed group is not available.

In the light of these differences in design, conduct, and reporting between human and animal behavioral-teratology studies, findings are discussed below if appropriate data were available for comparison. The number of reported human studies was the limiting factor in the following brief overview; once those were identified, the animal literature was evaluated. In all cases, if human subsystem dysfunction was reported, corroborative evidence was found in the animal data *if* a comparable end point had been evaluated. In that manner, effects observed after developmental exposures to lead, mercury, PCBs, phenytoin, ethanol, and methadone are compared. Articles reviewing the spectrum of effects observed in humans and animals are cited in the following discussion.

Table 27-2 summarizes the comparability of neuromotor effects after exposure to particular toxicants. Delayed motor development has been reported in both humans and rodents exposed to lead (Rutter, 1980; Reiter, 1982), mercury (Reuhl and Chang, 1979), alcohol (Abel, 1980), and phenytoin (Hanson et al., 1976; Vorhees, 1983). Cerebral palsy and seizure disorders have been reported in humans developmentally exposed to mercury, and motor dysfunction and increased susceptibility to seizure induction have been reported in rats and mice. Developmental exposures to PCBs have resulted in motor dysfunction in both humans (Jacobson et al., 1984) and mice (Tilson et al., 1979). A prolonged neonatal abstinence syndrome with neuromotor sequelae has been identified in human infants and rodents prenatally exposed to methadone (Hutchings, 1983). The specific motor alterations observed in humans and rodents were not always identical, but the normal behavioral repertoires of the two also are different. The data do indicate that the motor systems of humans and rodents are susceptible to disruption after developmental exposures to the agents in question.

The clinical relevance of experimental data on cognitive functions is more difficult to evaluate. Tests of human and animal cognitive abilities might evaluate very different functions; i.e., a rodent

TABLE 27-2 Examples of Motor Dysfunction After Behavioral-Teratogen Exposures[a]

Agent	Effects in Humans	Effects in Rodents
Lead	Delayed growth and motor development, motor incoordination, deficits in fine motor control	Delayed growth and motor development
Mercury	Delayed growth and motor development, cerebral palsy, seizure disorders	Delayed growth and motor development, ataxia, seizure susceptibility
PCBs	Depressed reflexes, delayed motor development	Neuromotor weakness, poor balance
Phenytoin	Delayed growth and motor development	Delayed growth and motor development
Ethanol	Delayed growth and motor development	Delayed growth and motor development
Methadone	Neonatal abstinence syndrome: tremors, sleep disturbances, hyporeflexia, irritability	Neonatal abstinence syndrome: hyperactivity, lability of state, sleep disturbances

[a]Data from Adams, 1986.

brain is not capable of the many complex functions evaluated in human assessments. Techniques used to assess cognitive function measure responses that are modified by sensory and motivational processes, as well as motor capabilities. However, performance on such tests can provide useful information concerning the postexposure integrity of underlying systems that contribute to an animal's or child's ability to learn, process, store, and retrieve relevant information. Table 27-3 summarizes some of the cognitive deficits noted after developmental insult. Reduced general intelligence, as measured on standardized tests, has been found in some children who were exposed prenatally to lead (Needleman et al., 1979), mercury (Harada, 1976), ethanol (Streissguth et al., 1984), and phenytoin (Hanson et al., 1976). The IQs of those children are often less than 70. Mental retardation is one of the most serious results of exposure to the agents in question. Attentional deficits have been reported in some children prenatally exposed to lead, mercury, and ethanol. Experimental studies have indicated impairments in visual recognition memory in infants exposed to PCBs (Jacobson et al., 1985) and increased reaction times during a vigilance task in ethanol-exposed children.

The animal literature indicates impaired learning and memory abilities in rodents after developmental exposures to the agents. Performance deficits on avoidance tasks have been reported after exposures to lead (Kimmel et al., 1978), mercury (Spyker et al., 1972), PCBs (Tilson et al., 1979), ethanol (Abel, 1980), and phenytoin (Vorhees, 1983). Results of water-maze tasks have indicated impaired function in rodents exposed to mercury (Spyker et al., 1972), ethanol (Abel, 1980), and phenytoin (Vorhees, 1983). Hughes and Sparber (1979) found that prenatal mercury exposure disrupted operant performance.

It is interesting that prenatal methadone exposure does not appear to alter cognitive performance in either humans or animals (Hutchings, 1983). The fact that both humans and animals showed no effects on cognitive performance supports the utility of experimental-teratology data in assessing toxic effects.

Sensory/perceptual processes have not been carefully evaluated in most behavioral-teratology studies (Adams and Buelke-Sam, 1981; Ison, 1984). As shown in Table 27-4, clinical case reports have suggested that some persons exposed in utero to ethanol (Clarren and Smith, 1978) or to phenytoin (Hill et al., 1974) have unspecified hearing defects. Visual impairments have been reported in some children exposed in utero to phenytoin (Wilson et al., 1978). However, specific sensory functions have not been evaluated in animals that have been exposed prenatally to alcohol or phenytoin. Vorhees (1983) reported delayed development of auditory responsiveness in rats treated with phenytoin prenatally.

TABLE 27-3 Examples of Cognitive Dysfunction After Behavioral-Teratogen Exposures[a]

Agent	Effects in Humans	Effects in Rodents
Lead	Decreased general intelligence, decreased attention span, impaired verbal ability	Impaired learning ability on passive avoidance and T-maze tasks
Mercury	Decreased general intelligence, decreased attention span	Learning deficits on many tasks
PCBs	Impaired visual recognition memory	Impaired learning on avoidance tasks
Phenytoin	Decreased general intelligence	Impaired spatial learning on water-maze tasks
Ethanol	Decreased general intelligence, decreased attention span, delayed reaction time	Learning deficits on many tasks
Methadone	None reported	None reported

[a]Data from Adams, 1986.

TABLE 27-4 Examples of Sensory/Perceptual Processing Dysfunction After Behavioral-Teratogen Exposures[a]

Agent	Effects in Humans	Effects in Rodents
Lead	Decreased visual acuity, altered brain electrophysiologic response to visual stimulation	Decreased visual acuity, altered brain electrophysiologic response to visual stimulation
Mercury	Unspecified hearing defects, altered reactivity to visual and auditory stimuli	Increased reactivity to auditory startle stimuli
PCBs	None specifically evaluated	None specifically evaluated
Phenytoin	Unspecified hearing defects and visual impairments in some cases	Delayed development of auditory startle response
Ethanol	Unspecified hearing defects in some cases	None specifically evaluated
Methadone	Deficits in visual, auditory, and tactile perception, but no specific sensory deficits	Hyperreactivity to aversive stimuli

[a]Data from Adams, 1986.

Decreases in visual acuity have been reported to occur in children (Rummo et al., 1979), rats (Fox et al., 1977, 1979; Fox and Wright, 1982), and monkeys (Bushnell et al., 1977) after exposure to inorganic lead. Altered electrophysiologic brain activity in response to visual stimulation has also been found in lead-exposed children (Otto et al., 1981, 1982, 1985; Otto and Reiter, 1983) and rats (Fox et al., 1979).

Fetal exposure to methylmercury has been reported to produce postnatal alterations in reactivity to visual and auditory stimulation in humans (Harada, 1976, 1977). Studies done on prenatally exposed rats have shown increased reactivity to auditory stimuli (Buelke-Sam et al., 1985).

Wilson et al. (1979) reported that children prenatally exposed to methadone had deficits in visual, auditory, and tactile perception, but these were interpreted as resulting from poor attentional and strategic processing abilities, rather than from specific sensory deficits. Lodge (1976) reported alterations in brain electrophysiologic responses to visual stimuli and hypersensitivity to auditory stimulation in children exposed to methadone before birth. The integrity of sensory functioning has not been specifically evaluated in studies carried out in rodents, but hypersensitivity to aversive stimulation has been reported by Hutchings (1983).

In nearly all the cases outlined above, the clinical problem was identified before the development of animal models to explore such toxicity. The literature of animal behavioral teratology has expanded greatly in recent years and suggests that a number of additional drugs and chemicals warrant clinical investigation. However, the potential value of animal data in predicting human hazard cannot be determined fully until clinical studies to look for behavioral dysfunctions identified in experimental animals are designed and conducted.

DATA INTERPRETATION

We have discussed many problems that contribute to the difficulty of evaluating the relevance of animal data in identifying potential health hazards in developing humans. The first three issues that follow bear on the interpretation of both human and animal data; the last two are related to problems in cross-species extrapolation of results.

The first problem centers on determining whether behavioral-teratology findings are a result of primary developmental toxicity or are secondary to maternal toxicity or to primary toxicity produced in other organ systems, e.g., liver or kidney. The heart of the issue is the relative susceptibility of the developing organism (whether human or animal) to toxic insult. If postnatal dysfunction only accompanies maternal toxicity, such information might be of value in alerting clinicians to monitor such pregnancies more closely. If postnatal functional deficits are obtained in the absence of overt toxic signs, the potential for selective developmental toxicity must be considered in human risk assessment.

Genetics might play a large role in susceptibility or expression of postnatal dysfunction. Genetic makeup could predispose the parents or offspring to greater sensitivity to a toxicant. Toxicants could produce postnatal dysfunction via genetic mechanisms, or such dysfunction could be transmitted to later generations (cf. Fujii et al., 1987; Stoetzer et al., 1987). Thus, whether developmental-toxicology studies are performed with inbred strains of mice or in the highly diverse human and whether one or both parents have been exposed to a toxicant are important considerations in evaluating and interpreting postnatal functional data.

The postnatal environment can have an impact on developmental toxicity, maximizing or minimizing the expression of damage. In human studies, the role of maternal socioeconomic status is great in that regard, as well as contributing to overall prenatal and perinatal status. In animal research, controlling litter size is one means of standardizing this factor. In both types of research, the degree of environmental experience and enrichment might contribute to the manifestations of toxicity.

Once those issues are considered, two additional aspects must be dealt with in determining the relevance of animal data to the human situation. The first concerns the disparity in timing of organ-system development across species. The rat gestation period covers approximately 3 weeks, and CNS development, including cell differentiation and migration, continues into the immediate neonatal period. More CNS development occurs in humans during the 9-month gestation period, although completion of histogenesis does not occur until well after birth. Agent exposure is timed specifically in animal studies, and care is taken to standardize doses within treatment groups. Such control usually is not available in human studies, and retrospective investigations often rely on maternal reporting of exposure to drugs and when it occurred.

Postnatal development schedules also differ between humans and animals. Postnatal development through puberty in a rat requires approximately 6 weeks. It takes years to reach that stage in humans. Thus, prenatal and postnatal disparities in timing must be accounted for, as well as the variations in agent exposures during the comparative process.

28

Lead as a Paradigm for the Study of Neurodevelopmental Toxicity

Because lead is the best-studied neurotoxin, it is a useful paradigm through which to understand the problems encountered in the search for valid and effective markers of exposure, effect, and vulnerability.

A variety of markers of exposure to lead have been used over the last 3 decades. The utility, sensitivity, and specificity of these markers vary widely. Newer analytic techniques that measure lead in smaller samples and at lower concentrations have permitted numerous epidemiologic studies of effects to be conducted in many countries (Needleman, 1988). This has resulted in sharper hypothesis formation. Also, the tissue concentration of lead considered to be toxic has been lowered.

MARKERS OF EXPOSURE TO LEAD

There have been many improvements in the biologic markers of exposure to lead. The following discussion includes the relatively imprecise use of surrogate markers and the more precise physiologic assessments. However, each marker has characteristics that limit its use and interpretation.

Surrogate Markers

When direct sampling of body tissues is not possible, investigators use surrogates to measure burden, such as employment history (e.g., length of employment in the lead industry and type of job) and residence history (distance from a lead source—a smelter or a heavily traveled roadway). These surrogates for actual tissue concentrations are subject to large measurement errors, which reduce the specificity (power) of a study.

Blood Lead

The most common marker of exposure is lead concentration in the blood. It has two weaknesses:

- Blood might not accurately reflect concentration in the tissue of interest, e.g., the brain or the testis.
- Blood lead content reflects only recent exposure (Rabinowitz et al., 1976).

Therefore, measurement of current concentrations of blood lead could result in misclassification of subjects whose exposure ended before the sampling. By blurring the distinction between exposed and unexposed groups or amplifying the

297

error in measuring critical tissue concentrations, these weaknesses in the biologic marker can increase the false-negative rate of a study.

Hair Lead

Hair seems to be an ideal tissue by which to index body burden. Not only is keratin an avid sink for lead, but the tissue is accessible, and its collection is painless. It is difficult, however, to determine whether the lead measured in an assay of hair is externally deposited or represents internal dose. The concentration of lead in hair is also sensitive to the type and frequency of shampooing (Limic and Valkovic, 1986).

Erythrocyte Protoporphyrin

Lead interferes with heme production at several sites. It competes with iron in the presence of ferrochelatase and blocks the entry of iron into the heme pocket (Needleman, 1988). Instead, zinc enters the heme pocket, and the entry results in an abnormal heme compound, zinc protoporphyrin (ZPP). ZPP is easily measured and is correlated with internal lead burden (Piomelli et al., 1973). It is used widely as a screening tool, because it is inexpensive and insensitive to contamination with lead on the skin. The assay does have the disadvantage of missing early lead exposures, and it might be insensitive to lead at internal doses below 30 μg/dl. Because ZPP or its extracted compound—free erythrocyte protoporphyrin—is also increased in iron deficiency, this assay for lead exposure can be confounded by the iron status of the subject; this lowers the specificity of the assay. ZPP measurement is one example of a biologic marker that can be considered a marker of either exposure or effect.

Chelatable Lead

The most critical measure of exposure is the amount of lead available to general circulation—also known as the chelatable-lead pool. The amount of chelatable lead in children is determined by administering a calculated dose of edathamil calcium disodium ($CaNa_2EDTA$), 500 mg/m^2 of body surface area, and measuring the resulting excretion of lead in the urine over a stipulated period. Ratios of urinary lead to the EDTA dose are calculated; ratios greater than 0.6 μg of lead to 1 mg of EDTA over 8 hours or 1.0 μg of lead to 1 mg of EDTA over 24 hours are considered evidence of excess lead storage and indications for therapeutic chelation. Although the chelatable-lead pool is the best index of metabolically available lead, its measurement requires close supervision, often in an inpatient setting.

Bone Lead

Lead is taken up by bone and is associated with the apatite crystal, much as calcium is (Schwartz et al., 1988; Silbergeld et al., 1988). In children with long, intense exposure, x-ray pictures of the long bones can show increased densities at the metaphyses. That is thought to be due not to the presence of lead, but to increased calcification secondary to the effect of lead on osteoblast activity (Silbergeld et al., 1988). Its specificity as a marker is good, because few conditions yield similar x-ray findings. However, the sensitivity of the assessment is low, and rather intense exposures to x rays are required to show the change (Silbergeld et al., 1988).

Tooth Lead

Bone biopsy is an impractical measure, but lead in dentin of deciduous teeth is a good marker for integrated exposure over time (Altshuller et al., 1962; Needleman et al., 1972, 1974; Steenhout and Pourtois, 1981). This tissue is correlated with blood lead content obtained 3 years before shedding. Subjects with increased dentin lead were found to have lower IQ scores, shorter attention spans, and greater speech and language impairment than controls, after adjustment for relevant socioeconomic and biologic covariates (Needleman et al., 1979).

X-ray Fluorescence of Bone

In vivo x-ray fluorescence of bones has been used successfully to classify exposure in industrial settings (Ahlgren et al., 1980). Also, x-ray fluorescence of in situ deciduous teeth has been reported to be a useful marker (Bloch et al., 1977). The exposure time in the trials was long, so the technique is difficult to use in screening studies of children.

MARKERS OF EFFECT

The definition of "adverse health effect" is controversial. When a single molecule enters a cell, it binds to a ligand and thus alters the state of the cell. That might be considered an adverse health effect. Considerable debate has surrounded the question of when a biochemical change becomes a health effect.

Hernberg (1972) described a model that clarified the issue considerably. If lead dose were plotted on the abscissa and relative frequency of given effect were plotted on the ordinate, measurement of a large number of outcomes would be expected to produce a display like Figure 28-1. Figure 28-2 displays the resulting relationship between intensity of effect and definition of toxicity. The dose-effect relationship is assumed to be monotonic, but could be linear, exponential, or logarithmic.

The display in Figure 28-2 takes into consideration the effect of individual judgment on definition of toxicity. For noncritical, non-rate-limiting systems, there would be some effect with small intensity that all observers would agree is not an adverse health effect; that is, the effect bears no measurable relationship to longevity, reproductive efficiency, or vigor of the host. For some effect with large intensity, agreement could be obtained that the health effect was adverse. Between these two boundaries lies the domain of controversy. Efforts to rationalize the decision-making process and achieve consensus have used probability estimation techniques (U.S. EPA, 1989). Such efforts have the virtue of measuring the range of uncertainty and making the degree of agreement explicit.

Table 28-1 lists various responses to lead, threshold blood lead concentrations for alterations (measures of sensitivity), and estimates of specificity. The thresholds are based on a consensus of investigators doing research on responses to lead exposure.

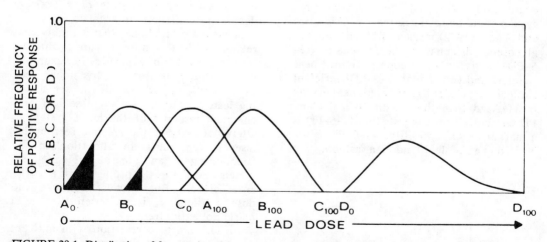

FIGURE 28-1 Distribution of frequencies of four putative responses to internal lead dose. A represents effect beginning at very low doses; B, C, and D are effects at higher lead doses. D might represent death. A_0, B_0, C_0, and D_0 represent doses above which first response would occur. A_{100}, B_{100}, C_{100}, and D_{100} represent doses above which all subjects show effects. Source: Needleman, 1987.

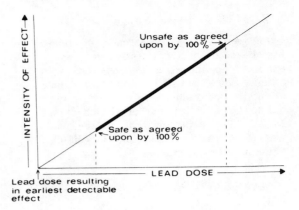

FIGURE 28-2 Intensity of given effect as function of dose. Thick part of line represents areas of disagreement as to whether health effect has occurred. Source: Needleman, 1987.

MARKERS OF SUSCEPTIBILITY

A given internal dose of lead does not produce an effect of the same intensity in all people. Young children are more vulnerable to effects, perhaps because of increased metabolic rate, differences in blood-brain barrier permeability, or stage of neurogenesis or synaptogenesis. Women appear more susceptible to effects of lead than men. Therefore, sex and age may be considered markers of vulnerability to lead.

Rogan and colleagues (1986) studied black children and found that children with high free erythrocyte protoporphyrin had low delta-aminolevulinic acid dehydrase and suggested that the enzyme system could be used as a marker. McIntire and Angle (1972) measured blood lead and glucose-6-phosphate dehydrogenase (G-6-PD) activity in students living near smelters and found that G-6-PD-deficient subjects had significantly higher erythrocyte lead than those with normal G-6-PD content. That suggests that G-6-PD is a marker of susceptibility. G-6-PD is inherited in a simple, Mendelian fashion.

METHODOLOGIC CONSIDERATIONS IN THE ESTABLISHMENT AND EVALUATION OF MARKERS OF DEVELOPMENT

Differentiating Markers of Exposure and Effect

At first glance, the difference between markers of exposure and markers of effect seems self-evident. However, the concentration of an internal dose of a toxicant, which seems clearly a marker of exposure, involves the organism's inherent responsiveness to the stimulus (i.e., external dose of the agent). That becomes obvious when one considers that subjects exposed to the same external dose of a toxicant demonstrate different internal doses.

A biochemical change that is highly correlated with the dose of an agent can serve as a marker of exposure. For example, free erythrocyte protoporphyrin has been used in this regard by clinicians evaluating lead exposure. An increase is a biochemical response of the heme system to altered activity of the enzyme ferrochelatase; it represents an alteration in the functioning of an integrated biochemical system and shares the properties of a marker of effect. But, because it tracks lead dose predictably, it has been used as a marker of exposure.

In addition to correlation with toxicant dose, two other attributes of a candidate exposure marker—sensitivity and specificity—influence its utility as an exposure

TABLE 28-1 Responses to Lead

Biologic Effect	Threshold, $\mu g/dl$[a]	Specificity
Heme pathway effects		
5'Pyri	10	Moderate
Decreased δ-aminolevulinic acid dehydrase activity	10	Moderate
Increased free erythrocyte protoporphyrin concentration	15	Moderate
Decreased heme production	40	Low
Basophilic stippling	50	High
Frank anemia	60	Low
Central nervous system effects		
Altered electrophysiologic responses		
Quantitative EEG abnormalities appear	35	Low
Evoked potential abnormalities appear	10	Low
Psychologic deficits		
IQ deficits increased	15	Low
Attention deficits appear	10	Low
Speech and language deficits appear	35	Low
Other effects		
1,25-vitamin D hydroxylase activity altered	12	High
Na-K ATP-ase activity altered	20	Low

[a]Threshold blood lead concentration for indicating that an effect has occurred.

marker. These concepts are discussed in detail in the appendix. Sensitivity is the probability that an effect will be detected by the marker. Specificity is the probability that the lack of effect will be correctly identified, i.e., that the marker does not falsely indicate an effect. Together, these two characteristics and the prevalence of the effect in the population determine the predictive value of the marker.

The higher the specificity of a given marker (i.e., the lower its probability of association with other diseases or exposures), the greater its utility in diagnosing disease or exposure. The specificity of a marker of effect is related to its proximity in the causal chain to the health effect in question. That in turn depends on the definition of adverse health effect and on the process of causal inference.

ADVERSE HEALTH EFFECTS

Causal Inferences in Toxicant Exposure

Establishment of a causal nexus between the exposure variable or marker and a dis-

ease is essential in defining the relationships between the variables. In nonexperimental studies, investigators must rely on experimental design and model-building to determine the most reasonable causal model. Scientists must balance type I errors (accepting spurious causal hypotheses as true) and type II errors (rejecting true causal hypotheses as spurious). Avoiding type I errors has been emphasized; less attention has been given to the risk of type II errors. In making causal judgments, the following errors are often encountered:

- **Overvaluing the importance of the criterion $p = 0.05$.** Many investigators dismiss findings in which the p is greater than 0.05. The use of this fixed boundary is not logical, but is a long-held preference that has become conventional.

- **Improper causal specification or modeling.** Many studies attempting to reduce confounding control for many variables that might be markers of effect. That results in reducing the variance properly attributable to the causal agent under study.

• **Arguing for lack of effect from studies with inadequate power to find an effect.** Power measures the probability that a study can find an effect if it is present. It is a function of three variables: the number of subjects studied, the effect size under scrutiny, and the significance criterion selected by the examiner. Many studies that have argued for a lack of causal effect have done so on the basis of sample sizes with less than a 50% chance of finding an effect at a statistical significance of 0.05.

• **Assigning shared variance to the confounder.** If the main effect (e.g., exposure variable) is highly associated with another variable, the effect might not be detected in regression analyses. This problem is referred to as collinearity of effect and confounders. The order in which the variables are considered in the regression equation will affect the conclusions of the analyses.

• **Evaluating studies in isolation.** Many review articles discuss studies in isolation, tabulate their advantages and disadvantages, and then count the studies that favor and do not favor a causal effect. That type of evaluation seriously degrades the data. A relatively new form of review, meta-analysis, or quantitative summaries of many studies, promises to improve on the evaluation of multiple studies.

29

Conclusions and Recommendations

The preceding chapters have discussed a wide range of biologic markers of neurodevelopmental effect. Having discussed potential biologic markers of effect (summarized in Table 29-1), we should mention biologic markers of exposure to various toxicants known or thought to have neurodevelopmental effects (Table 29-2).

Experience with the study of populations exposed to lead indicates that it is important to approach the study of neurodevelopmental toxicity with batteries of biologic markers of effect, and careful assessment of characteristics of the subjects and other potential exposures. The complexity of the process being assessed is important. Assessments of simple functions can focus the site of neurotoxic deficits, but may not be as sensitive as integrated tasks. The determination of the appropriate set of biologic markers is an iterative process, involving assessments and reassessments in potentially exposed populations.

There are several aspects of developing organisms that make the assessment of possible effects of toxic exposure difficult. Differentiation is not a continuous process; therefore, the timing of the exposure as well as dose received, determine the nature and extent of effects.

Consequently, epidemiologic studies cannot simply combine groups of individuals without knowledge of the developmental status of those individuals at the time of exposure.

Another aspect of toxicity in developing systems is that the effects may come and go. That is, subjects need to be followed longitudinally to fully characterize the neurotoxic deficits. Hence, knowledge of the developmental status of the subjects at the time of testing is also important.

In the following discussions of areas of research that promise to yield useful biologic markers of neurodevelopment.

MODELS OF NEURODEVELOPMENT

Attempts to develop mechanistic models of postnatal behavior require understanding of the development of the central nervous system and of how its components respond to a toxic insult. Radiation is useful for assessing perinatal toxicity. It interferes with normal development in two basic ways: by killing specific cell populations and by altering migration patterns of surviving cells. Those morphogenetic alterations have severe consequences, especially in microneuronal populations. Whether this model has universal applicability remains to be deter-

TABLE 29-1 Summary of Some Markers of Central Nervous System Development

Marker	Usable in a Screening Study[a]	Usable in Population Subgroups as Secondary Assessment[b]	Usable Only in Studies of Special Populations[c]	Remarks
Growth variables: length, head size, etc.	+			Low specificity
Developmental landmarks	+			Low specificity, low sensitivity
Minor physical anomalies	+			Low specificity
Nerve conduction time		+		—
Psychophysical measures		+		—
Evoked potentials			+	—
Pure-tone hearing		+		—
Auditory discrimination		+		—
Speech and language competence		+		—
Vibratory sense		+		—
Motor function	+			—
Attention		+		High sensitivity, low specificity
Visual-motor perception		+		—
Psychometric intelligence		+		—
Social behavior			+	—
Positron emission tomographic scan			+	—
Magnetic resonance imaging scan			+	—

[a] + = sufficiently validated and safe for application in field studies, although might warrant further refinement.
[b] + = sufficiently demanding in terms of subject and interviewer efforts that should be used as secondary assessment in multistage assessment battery.
[c] + = too invasive or too demanding for use on broad scale.

mined. However, results with other agents and the unique biology of microneurons suggest that the model has much promise.

NEUROENDOCRINE AND NEUROIMMUNOLOGIC MARKERS

The nervous system has major interactions with other systems in the body, such as the endocrine, reproductive, and immune systems. Through complex networks of control and feedback, the nervous system produces changes in endocrine and immune functions that are important, particularly during development. Maturation of the reproductive system, for instance, is under neural control through the hypothalamic-pituitary axis. From the perspective of research on biologic markers, those interactions provide peripheral signals of neuron action that might be directly measurable when the primary neurochemical signals are not detectable.

Nervous System-Endocrine Interactions

Neural signals processed through the hypothalamus control the release of gonadotropins that regulate pituitary release of luteinizing hormone (LH) and follicle-stimulating hormone (FSH) (see discussions in parts of report on female or male

TABLE 29-2 Status of Some Markers of Exposure

Agent	Marker	Advantages	Disadvantages
Lead	Blood lead concentrations	Easily obtained	Measures only recent exposures
	Hair lead concentrations	Easily obtained	Subject to contamination
	Free erythrocyte protoporphyrin concentrations	Easily obtained, not subject to contamination	Low sensitivity at lead concentrations $<30\ \mu g/dl$, low specificity
	Tooth lead concentrations	Integrative marker	Hard to obtain
	X-ray fluorescence of bone	In vivo integrative	Sensitivity uncertain
	Provocative chelation response	Sensitive to tissue burden	Requires 8 hours of observation
Cadmium	Blood cadmium concentrations	Easily obtained	Blood not critical site
	Urinary cadmium concentrations	Measures excess saturation	Depends on metallothioneine
Mercury	Blood mercury concentrations	Easily obtained blood not critical target	Measures only recent exposure;
	Urinary mercury concentrations	Easily obtained	Measures only recent exposure
	Hair mercury concentration	Easily obtained	Indirect measure

reproductive system). Chemical signals can be directly measured as concentrations of gonadotropins and gonadal hormones in blood and indirectly measured by secondary physiologic events (for instance, ovulation, luteinizing-hormone releasing hormone surge, and lactation). Serum prolactin has been measured and correlated with hypothalamic dopaminergic function, because of dopamine's role as a prolactin-inhibiting factor (Memo et al., 1986). Hyperprolactinemic states, sometimes with galactorrhea, are associated with deficiencies in hypothalamic dopaminergic neurotransmission (Ferrari and Crosignani, 1986); similarly, acromegaly, a syndrome of disordered growth-hormone release, is associated with decreased hypothalamic dopamine release (Hanew et al., 1987). Treatment of those two conditions involves administration of dopamine receptor agonists (Memo et al., 1986).

Little clinical use has been made of endocrine factors as biologic markers of neurochemical function, except in disorders of hypothalamic-pituitary function in which response to infused dopamine has been monitored by measuring LH and FSH (Nicoletti et al., 1986). There should be an increased use of markers of endocrine status to make inferences of neurochemical function.

Nervous System-Immune System Interactions

As noted by Pert and colleagues (1985), the central nervous system and the immune system have in common many specific cell-surface recognition sites or receptors for peptides. Human peripheral monocytes might also have receptors for amino acid neurotransmitters (Malone et al., 1986). Cells of the immune system—T cells, monocytes, B cells, and alveolar macrophages-have been found to contain and respond to specific neuroactive peptides (Pert et al., 1985; Zhu et al., 1985). Monocytes, a heterogeneous population of cells in blood that undergo differentiation into macrophages in the presence of particular stimulation, demonstrate chemotaxis as an important part of their function in inflammation and repair processes. In addition to the classic chemotactic stimuli, such as bacterial material and complement activation, the neuropeptides have been recently shown to elicit monocyte chemotaxis, among them opiates, substance P, bombesin, and cholecystokinin (Pert et al., 1985). Elastin peptides also modulate monocyte ion fluxes (Jacob et al., 1987). Those findings indicate strong interactions and communication between the brain and cells of the immune system.

It is now well known that the T_4 antigen, a membrane receptor for the acquired immune deficiency syndrome (AIDS) virus, is present on some cells in the human brain (Pert et al., 1986). The involvement of the brain in the late stages of AIDS is now a well-characterized part of the disease.

Neuroleukins are a new class of growth factors—present in muscle, brain, and other organs—that promote growth and survival of spinal and sensory neurons in culture, as well as affecting B cell maturation (Gurney et al., 1986a,b). Neuroleukins are secreted by T cells in response to stimulation by such lectins as concanavalin A. Measurement of neuroleukins in bone marrow or in T cell secretions might yield a useful index of neurologic function, particularly during development of the nervous and immune systems.

Markers of immunologic function are accessible, and sophisticated methods for their measurement have been developed in the last decade. The possibility that monitoring some aspects of immune function could provide markers of neuroimmunologic interaction has not been explored.

NEUROCHEMICAL MARKERS

Recent Advances in Neurochemical Methods

Application of the biologic-markers paradigm to studies of neurodevelopmental toxicology is restricted by the complexity and inaccessibility of many functional parts of the nervous system. Some of these problems might be overcome by examining systems that are substantially controlled by neuronal processes, such as some aspects of endocrine and immune function. In addition, major technologic advances have been made in the methods available for studying the nervous system noninvasively in vivo.

Computed axial tomography (CAT) has provided vast improvements in visualizing structures of organs in the body. CAT scanning has been heavily used in neuropsychiatric disease, and its ability to reveal structural abnormalities in the brains of schizophrenics is among the many accomplishments of these new techniques (Zec

and Weinberger, 1986). Even more exciting, however, has been the recent development of positron emission tomography (PET) and magnetic resonance imaging (MRI) technologies that allow visualization of physiologic and biochemical processes as they occur in the brain (Battistin and Gerstenbrand, 1986).

PET couples the fine visualization of CAT scanning with the ability to detect positrons emitted from unstable isotopes. Fluorine-18 and carbon-11 are often-used positron emitters; they can be used to tag chemicals of neurologic interest, such as drugs that bind to specific neuronal receptors or metabolic precursors of neurotransmitters (cf. Battistin and Gerstenbrand, 1986). In addition, fluorine-18-tagged 2-deoxyglucose can be used to reveal the degree of cell metabolic activity in brain regions (Alavi et al., 1986). Because neurally active cells—cells that receive neural stimuli or process signals—are metabolically active, they take up more 2-deoxyglucose and can be identified by increased density of fluorine-18 in PET scanning with fluorinated 2-deoxyglucose derivatives.

PET scans of patients have demonstrated that schizophrenics have a higher density of dopamine receptors in basal ganglia (D.F. Wong et al., 1986) and that people with parkinsonism have a lower uptake of fluorine-18-tagged dopa (the precursor of dopamine) than control subjects (Leenders et al., 1986). Interestingly, the parkinsonism patients did not show increased uptake of the fluorine-18-tagged postsynaptic-receptor ligand spiperone—a finding that suggests that denervation supersensitivity did not occur in these patients (Leenders et al., 1986).

The theoretical basis of quantitative interpretation of PET densitometry is controversial; however, remarkable images of regional changes in neurochemistry can be obtained with such techniques, thus overcoming (at least qualitatively) the barriers to obtaining samples of brain tissue. PET scanning appears also to have some use in diagnosis of preclinical states, such as MPTP-induced brain damage (Calne et al., 1985).

MRI detects the spin resonance of some

atoms and thus yields information on energy state. That is useful for studying biochemical reactions involving energy transfer from phosphate groups, such as adenosine triphosphate (ATP). Because many events in neurotransmission involve phosphorylation-the transfer of high-energy phosphates from ATP to proteins—this detection method is particularly attractive for the in vivo study of neurochemistry. Some types of neuronal structures, such as myelin, can be selectively imaged with MRI. MRI is applied with increasing frequency; it provides considerable advantages in spatial delineation and will greatly add to information on central nervous system neurochemistry involving changes in energy state.

Surrogate Cell Systems

The inaccessibility of neurons has been approached ingeniously with the study of surrogate cell systems. Platelets and red cells contain some of the same biochemical apparatus as neurons, including receptors, high-affinity uptake, enzymes, and storage and releasing processes (Pletscher, 1968; Murphy, 1976), and thus serve indirectly as media to assess biologic markers of events in neurons, particularly drug response. However, because of the blood-brain barrier, receptors on central nervous system neurons and peripheral neurons may be different (e.g., for serotonin and benzodiazepine (Snyder, 1984)). Surrogate cells could not be used to characterize the possible differences.

Platelets have binding sites for the neurotoxin MPTP (del Zompo et al., 1986). Red cells also absorb choline by processes somewhat similar to those in neurons (Houck et al., 1988). Red cell choline uptake has been studied in patients with disorders thought to involve deficits in central nervous system cholinergic function, such as Alzheimer's disease. The results of studies done so far are of interest, but of unknown clinical utility. The clearest example of surrogate monitoring has been the measurement of peripheral esterases in workers exposed to organophosphates (Levine et al., 1986).

Neuronal esterase, the site of action of these neurotoxins, can be studied in blood. As demonstrated by Levine and co-workers (1986), esterase activity in circulating monocytes and red cell cholinesterase activity are decreased after toxic exposures. Monocyte esterase might be an even more sensitive marker of organophosphate exposure than the more commonly used red cell cholinesterase (Levine et al., 1986). Other accessible cell systems need to be evaluated for their use as sensitive surrogate indicators of neuronal response.

Appendix

Appendix

Assessing the Validity of Biologic Markers:
Alpha-Fetoprotein

The previous chapters associated with pregnancy issues have discussed potential biologic markers for use in toxicity evaluations during pregnancy; however, only alpha-fetoprotein has been evaluated in sufficient depth to allow for a rigorous evaluation of fetal and embryonic abnormalities. This detailed analysis is included to review and establish criteria for evaluating any proposed biologic marker of toxicity.

The identification of biologic markers that indicate exposure, effect, or susceptibility is a complicated process involving studies animals, refinements in laboratory assays, and studies in special human populations. Moreover, even when a marker has been validated in such studies, its use in larger populations is not straightforward. This chapter discusses experience in the implementation of large populations of alpha-fetoprotein (AFP) in maternal serum and amniotic fluid to screen for neural tube defects (NTDs) in fetuses.

The NTDs anencephaly and spina bifida are among the most common and serious of congenital malformations (Warkany, 1971). Both arise in the first month of pregnancy. They are sometimes fatal and sometimes result in lifelong handicap. During the past decade, it has become possible to detect the presence of neural-tube defects during fetal life by using concentrations of AFP as a biologic marker (Gastel et al., 1980; Mizejewski and Porter, 1985). Reviewing the use of AFP as a marker of fetal maldevelopment provides a useful framework for considering some practical aspects of the use of biologic markers of effect. Biologic markers sometimes are used primarily to assess the connection between a particular exposure and a particular disease, and they are sometimes used primarily to help in making personal or public health decisions. AFP is discussed here in the latter context, but the remarks are equally applicable to the former context.

Anencephaly is absence of the cranial vault with a degenerated or rudimentary brain. The defect is incompatible with extrauterine life; many affected infants are spontaneously aborted, and the remainder are stillborn or die soon after birth. Spina bifida is characterized by defective closure of the spinal cord. If the spinal cord tissue and meninges protrude through the vertebrae, the condition is called spina bifida cystica. The defect can occur anywhere on the spinal column, but is most common in the lumbar region. The protruding nervous tissue is sometimes covered with skin. Many affected fetuses die, but many are born alive and, depending on the

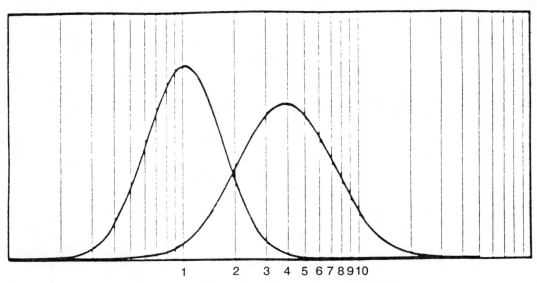

MSAFP (Multiples of the Unaffected Median)

FIGURE A-1 Distribution of concentration of maternal serum alpha-fetoprotein (MSAFP) in a population of unaffected pregnancies (curve on left) and in a population of pregnancies in which the fetus has spina bifida cystica (curve on right). The abscissa represents the gestational age-corrected concentration of MSAFP (using the date of the last menstrual period to date specimen) plotted in terms of multiples of the (unaffected) median (MOMs). The ordinate represents the frequency of MOM values with in each of the two populations. Source: Reprinted with permission from Adams et al., 1984. Copyright 1984 by C.V. Mosby Co.

location and seriousness of the lesion, can lead relatively normal lives, in spite of physical handicap. Recent advances in medical and surgical therapy have made it possible for many severely affected children to survive (Bamforth and Baird, 1989).

AFP was discovered about 30 years ago. It can be detected in the maternal serum by 29 days after conception (Bergstrand and Czar, 1957). It is synthesized early by the yolk sac and liver and later almost exclusively by the liver. In fetal serum, the concentration reaches a peak at about 13 weeks of gestation and decreases thereafter. AFP is excreted in the urine and therefore is found in the amniotic fluid (Crandall, 1981). AFP concentration is increased in maternal serum during pregnancy. Some fetal conditions, including anencephaly and spina bifida cystica, raise the amniotic fluid and maternal serum concentrations above normal (Bergstrand, 1986). The finding of increased AFP in maternal serum/amniotic fluid, or both

thus constitutes an indication of open NTDs. (High concentrations are also found when the fetus has a ventral wall defect or when there are twins.) Figure A-1 shows the distribution of maternal serum AFP concentrations in normal pregnancies (curve on left) and pregnancies with spina bifida cystica in the fetus (curve on right).

The screening process for NTD begins with the measurement of maternal serum AFP during the second trimester of pregnancy, preferably in week 16 of gestation. That measurement is the start of a multistage process, which is outlined in Figure A-2. The following discusses the various aspects of marker validity and concentrate on this first stage, but the principles apply to all stages in the screening; indeed, they apply to most medical decision-making processes and in particular to the use of any marker of disease or exposure (Galen and Gambino, 1975).

Experience with the use of AFP demonstrates the importance of determining the

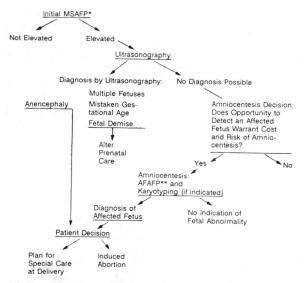

Initial MSAFP*

Not Elevated   Elevated

Ultrasonography

Diagnosis by Ultrasonography: No Diagnosis Possible

Multiple Fetuses
Anencephaly   Mistaken Ges-   Amniocentesis Decision:
               tational Age   Does Opportunity to
               Fetal Demise   Detect an Affected
                                       Fetus Warrant Cost
                                       and Risk of Amnio-
                      Alter   centesis?
                      Prenatal
                      Care   Yes   No

               Amniocentesis:
               AFAFP** and
               Karyotyping (if indicated)

          Diagnosis of   No Indication of
          Affected Fetus   Fetal Abnormality

     Patient Decision

Plan for   Induced
Special Care   Abortion
at Delivery

*MSAFP = maternal serum alpha-fetoprotein
**AFAFP = amniotic fluid alpha-fetoprotein

FIGURE A-2 Multistage prenatal detection of neural tube defects. Source: Reprinted with permission from Adams et al., 1984. Copyright 1984 by C.V. Mosby Co.

predictive value of a test in the general population and developing a multistep, multimarker screening protocol.

ASSESSING THE VALIDITY OF BIOLOGIC MARKERS

As noted earlier, an ideal biologic marker should be a sensitive and specific indicator of disease or exposure and (in the context of a specific use) should be suitably predictive of the disease or exposure. The terms "sensitive," "specific," and "predictive" are defined below, followed by a discussion of how maternal serum AFP qualifies as a marker.

"Sensitivity" refers to the ability of a marker to indicate the presence of disease or exposure when disease or exposure is present. Sensitivity is therefore usually measured as a conditional probability, that is, the probability that the marker will indicate disease, given the presence of disease. In the notation of Table A-1, the sensitivity of a marker is $P(M+|D+)$. Sensitivity is therefore the complement of the false-negative probability of the marker, $P(M-|D+)$. The sensitivity of markers is rarely, if ever, perfect (1.0)—there are always false-negatives—but a good marker will have a sensitivity close to 1.0.

"Specificity" refers to the ability of a marker to indicate the absence of disease or exposure when disease or exposure is absent. In the notation of Table A-1, the specificity of a marker is $P(M-|D-)$. Specificity is the complement of the false-positive probability, $P(M+|D-)$. A good marker will have a specificity close to 1.0.

TABLE A-1 Probabilities of Marker Presence and Absence, Conditional on Disease Presence and Absence

Marker	Disease[a]	
	Present	Absent
Positive	$P(M+|D+)$	$P(M+|D-)$
Negative	$P(M-|D+)$	$P(M-|D-)$
Total	$P(D+)$	$P(D-)$

[a] P, probability; M, marker; D, disease.

Suppose that an investigator is working with a new marker of exposure to a toxic chemical and determines its sensitivity and specificity from test data as presented in Table A-2. The marker has been measured in 1,000 persons with the disease (exposure) and 1,000 persons without the disease. The results are encouraging: the sensitivity is 0.95 (950/1,000), and the specificity is 0.99 (990/1,000).

There are only 5% false-negatives and 1% false-positives. Because the numbers of exposed and unexposed persons in the evaluation are equal, the a priori probability of exposure in the total sample is 0.5.

TABLE A-2 Cross-Classification of Marker and Disease: Hypothetical Data from a Case-Control Study

Marker	Disease	
	Present	Absent
Positive	950	10
Negative	50	990
Total	1,000	1,000

Another quantity important in evaluating the validity of a marker is its predictive value. Table A-3 shows the probabilities of disease presence and absence, conditional on the presence or absence of the marker.

TABLE A-3 Probabilities of Disease Presence and Absence, Conditional on Marker Presence and Absence

Marker	Disease[a]		
	Present	Absent	Total
Positive	$P(D+\mid M+)$	$P(D+\mid M)$	$P(M+)$
Negative	$P(D-\mid M+)$	$P(D-\mid M-)$	$P(M-)$

[a]P, probability; M, marker; D, disease.

The focus here is on the rows, rather than on the columns as in Table A-1. The predictive value of a positive (PVP) test for the marker is $P(D+\mid M+)$, that is, the probability of the disease given a positive test. That probability is the complement of the false-positive probability, $P(D-\mid M+)$, the probability that there is no disease when there is a positive test for the marker. The predictive value of a negative (PVN) test is $P(D-\mid M-)$, which is the complement of the false-negative probability, $P(D+\mid M-)$. In applying a test for a disease or exposure, it is important to consider the data both as they are presented in Table A-3 and as they are presented in Table A-1. Consider the data in Table A-2 focusing on the marker (rows), rather than on the exposure (columns). Here the false-positive rate, the probability that the marker indicates exposure when there is none, is about 0.01 (10/960). The PVP is about 0.99 (950/960), and the PVN is about 0.95 (990/1,040).

Note that the data in Table A-2 have been gathered in such a way that the number exposed is equal to the number not exposed. That is, the a priori probability of disease in the sample is 0.5—a typical case-control study design. Suppose that the a priori probability of disease in question in the population at large is low, about 1%. Table A-4 shows how the marker would work in practice in this population. The data in Table A-4 exhibit the same sensitivity and specificity (0.95 and 0.99, respectively) as the data in Table A-2, but very different PVP. Here the PVP is about 0.49, that is, only half the people with positive test results will have the disease. The sole cause of the differing PVPs is the difference in the a priori probability of the disease in the two settings. One setting is that of a study of the marker where the disease is common (50%) by design, and the other setting is that of the use of the marker in practice where the disease is relatively infrequent. In the example, because the disease is infrequent, the PVN is high (more than 0.99).

TABLE A-4 Cross-Classification of Markers and Disease: Hypothetical Data from a Population Study

Marker	Disease		
	Present	Absent	Total
Positive	95	100	195
Negative	5	9,900	9,905
Total	100	10,000	10,100

The example reflects a common outcome of the transition from laboratory bench to community or clinical practice: very good tests can perform poorly. In the clinical or community setting, it is most important to know how likely it is that a positive test indicates disease truly and how likely it is that a negative test indicates the absence of disease truly. The predictive values of tests can be increased in two ways—by increasing the sensitivity and/or the specificity or by choosing the individuals or populations for testing so that the a priori probability of the disease or exposure is high.

In many situations, it is not possible to change the sensitivity without changing the specificity, and vice versa. The situation with maternal serum AFP is a case in point. Figure A-1 shows that the AFP distributions in normal and affected pregnancies overlap. If a particular concentration of AFP, the "cut-point," is chosen as an indication of the presence of an abnormal fetus, there will be false-positives and false-negatives, because some normal pregnancies are associated with higher AFP concentrations than are some affected pregnancies. If the cut-point is chosen so that the test is very sensitive—so that nearly all affected pregnancies fall above the cut-point—there will be more false-positives, and the specificity will be low. But, if the cut-point is chosen so that nearly all normal pregnancies fall below, there will be more false-negatives, and the sensitivity will be low.

Regarding the possibilities for increasing the PVP of a test by testing only people with a relatively high a priori probability of exposure or disease, recall the marked contrast in the PVP values calculated from the data in Tables A-2 and A-4. In those two examples, there was no difference in the sensitivity or specificity of the test, but only differences in the a priori probabilities of exposure. A major reason for disappointment in the practical application of a test is that it is indiscriminately applied in populations where the a priori probability of exposure or disease is low, so the false-positives greatly outnumber the true-positives.

In many situations, a test with a low PVP is applied as a screening test. Persons with a positive result can be followed by more definitive tests. The definitive tests are usually not used as a first step, because they are expensive and invasive. The use of a screening test also allows the use of the definitive test on persons who have a relatively high a priori probability of having the disease or exposure of interest, which increases the operational PVP of the definitive test (even "definitive" tests are rarely perfect). The currently recommended process for screening for NTDs through the use of maternal serum AFP testing is an example of this approach.

VALIDITY OF MATERNAL SERUM AFP MEASUREMENT AS A BIOLOGIC MARKER OF NTDs

As indicated earlier, AFP is normally present in amniotic fluid and maternal serum, and it is present in increased concentrations in the presence of anencephaly and spina bifida cystica. Measuring the second-trimester concentration of maternal serum AFP can be used to identify the likelihood that a pregnant woman is carrying a fetus with anencephaly or spina bifida cystica, a ventral wall defect, or twins; other defects can also be predicted. These likelihoods can be used by the woman and her physician to decide whether to bear the risks and costs of further diagnostic procedures. Ultrasonography can identify multiple fetuses and fetuses with anencephaly. The most common cause of increased maternal serum AFP is mistaken gestational age, which can be checked by ultrasonography. When an increase in maternal serum AFP cannot be explained by one of the factors assessed with ultrasonography, the likelihood of a fetus with spina bifida cystica or ventral wall defect can be weighed against the cost and risks of amniocentesis to obtain fluid for assay for amniotic fluid AFP (see Fig. A-2). That assay can be considered as the definitive test for spina bifida cystica, but also is invasive and is associated with an increased risk of abortion.

The results of maternal serum AFP tests are usually presented in terms of multiples of the median (MOMs) for normal pregnancies. From the specific MOM and some characteristics of the mother (such as race, geographic area, and weight), one can estimate the odds of having an affected fetus. Rather than make a presentation in terms of odds, for the purpose of this discussion we consider a maternal serum AFP concentration of at least 2.5 MOM as "abnormal" and give an illustration of sensitivity, specificity, and PVP. (As noted above, the choice of a different MOM as the "abnormal" cut-point will change the sensitivity, specificity, and predictive values of the procedure.)

The data are derived from a synthesis of the UK Collaborative Study results (Wald and Cuckle, 1980) and are adjusted so that the base is a hypothetical cohort of 100,000 screened pregnant women (data on ventral wall defects and other defects are not included). Table A-5 puts the data on maternal serum AFP in the format of the other tables presented here; the 1,000 multiple pregnancies that would be expected among 100,000 pregnancies have been omitted from the table. The sensitivity of this test is 0.84, meaning that 16% of babies with anencephaly or spina bifida cystica would be missed.

TABLE A-5 AFP by Presence or Absence of Neural Tube Defect[a]

Marker	NTD Present	Absent	Total
Positive	334	3,254	3,588
Negative	66	95,346	95,412
Total	400	98,600	99,000

[a]Data derived from UK Collaborative Study data.
Source: Wald and Cuckle, 1980.

The specificity is 0.97, indicating that there are 3% false-positives. NTDs are rare complications of pregnancy; the false-positives greatly outnumber the true positives, and the PVP is about 0.09. The PVN is quite good at this stage of the screening process, more than 0.99, and a woman with a negative test has only a 7/10,000 risk of having a fetus with anencephaly or spina bifida cystica, only about

one-sixth of her a priori probability, about 40/10,000 in the UK at the time these data were obtained.

After staging through ultrasound and amniocentesis with evaluation of amniotic fluid AFP (see Fig. A-2), the UK data look like those presented in Table A-6. The sensitivity of the total screening package for open NTD is 0.81, the specificity is very nearly 1.0, and the PVP is almost 0.99. In all, the process is able to detect 81% of the fetuses affected by open NTDs at the risk of a fairly small number of false-positives. If one includes the risk of fetal death due to amniocentesis (estimated to be about 0.5-1.5% of the 3,588 amniocentesis performed), the net benefit of the program is 324 fetuses with open NTDs detected balanced against the 4 false-positives and 10-30 fetal deaths due to complications of amniocentesis.

TABLE A-6 AFP by Presence or Absence of Neural Tube Defect[a]

Total Screen	NTD Present	Absent	Total
Positive	324	4	328
Negative	76	98,596	98,672
Total	400	98,600	99,000

[a]Data derived from UK Collaborative Study data.
Source: Wald and Cuckle, 1980.

It is obvious that AFP screening could be a monumental failure if stopped at the maternal serum stage. There would be enormous errors, the normal fetuses with positive test results greatly outnumbering the affected fetuses. Therapeutic action taken on the basis of maternal serum AFP tests would be wrong most of the time. However, when properly used as the first stage of a screening process, maternal serum AFP evaluation is useful.

References

References

Abel, E. L. 1980. Fetal alcohol syndrome: Behavioral teratology. Psychol. Bull. 87:29-50.

Abraham, G. E., G. B. Maroulis, and J. R. Marshall. 1974. Evaluation of ovulation and corpus luteum function using measurements of plasma progesterone. Obstet. Gynecol. 44:522-525.

Abyholm, T., J. Kofstad, K. Molne, and S. Stray-Pedersen. 1981. Seminal plasma fructose, zinc, magnesium and acid phosphatase in cases of male infertility. Int. J. Androl. 4:75-81.

Acott, T. S., D. J. Johnson, H. Brandt, and D. D. Hoskins. 1979. Sperm forward motility protein: Tissue distribution and species cross reactivity. Biol. Reprod. 20:247-252.

ACS (American Chemical Society), Committee on Environmental Improvement and Subcommittee on Environmental Analytical Chemistry. 1980. Guidelines for data acquisition and data quality evaluation in environmental chemistry. Anal. Chem. 52:2242-2249.

Adams, J. 1986. Clinical relevance of experimental behavioral teratology. NeuroToxicology 7:19-34.

Adams, J., and J. Buelke-Sam. 1981. Behavioral assessment of the postnatal animal: Testing and methods development, pp. 233-258. In C. A. Kimmel and J. Buelke-Sam, Eds. Developmental Toxicology. New York: Raven Press.

Adams, M. J., Jr., G. C. Windham, L. M. James, F. Greenberg, J. A. Clayton-Hopkins, C. B. Reimer, and G. P. Oakley, Jr. 1984. Clinical interpretation of maternal serum alpha-fetoprotein concentrations. Am. J. Obstet. Gynecol. 148:241-254.

Adams, N. R. 1976. Pathological changes in the tissue of infertile ewes with clover disease. J. Comp. Pathol. 86:29-35.

Adams, N. R. 1977. Morphological changes in the organs of ewes grazing oestrogenic subterranean clover. Res. Vet. Sci. 22:216-221.

Adams, N. R. 1983. Sexual behavior of ewes with clover disease treated repeatedly with oestradiol benzoate or testosterone propionate after ovariectomy. J. Reprod. Fertil. 68:113-117.

Adams, N. R., and G. B. Martin. 1983. Effects of oestradiol on plasma concentrations of luteinizing hormone in ovariectomized ewes with clover disease. Aust. J. Biol. Sci. 36:295-303.

Adams, N. R., H. Hearnshaw, and C. M. Oldham. 1981. Abnormal function of the corpus luteum in some ewes with phyto-oestrogenic infertility. Aust. J. Biol. Sci. 34:61-65.

Adams, S. O., S. P. Nissley, S. Handwerger and M. M. Rechler. 1983. Developmental patterns of insulin-like growth factor-I and -II synthesis and regulation in rat fibroblasts. Nature 302:150-153.

Aedo, A.-R., P. H. Pedersen, S. C. Pedersen, and E. Diczfalusy. 1980. Ovarian steroid secretion in normally menstruating women. II. The contribution of the corpus luteum. Acta Endocrinol. 95:222-231.

Ahlgren, L., B. Haeger-Aronsen, S. Mattsson, and A. Schutz. 1980. In-vivo determination of lead in the skeleton after occupational exposure to lead. Br. J. Ind. Med. 37:109-113.

Aitken, R. J. 1977. Changes in the protein content of mouse uterine flushings during normal pregnancy and delayed implantation, and after ovariectomy and oestradiol administration. J. Reprod. Fertil. 50:29-36.

Alanen, A. 1984. Serum IgE and smooth muscle antibodies in pre-eclampsia. Acta Obstet. Gynecol. Scand. 63:581-582.

Alavi, A., R. Dann, J. Chawluk, J. Alavi, M. Kushner, and M. Reivich. 1986. Positron emission tomography imaging of regional cerebral glucose metabolism. Semin. Nucl. Med. 16:2-34.

Albertini, R. J. 1985. Somatic gene mutations in vivo

as indicated by the 6-thioguanine resistant T-lympho-cytes in human blood. Mutat. Res. 150:418-422.

Albertini, R.J., K.L. Castle, and W.R. Borcherding. 1982. T-cell cloning to detect the mutant 6-thiogua-nine-resistant lymphocytes present in human peripheral blood. Proc. Natl. Acad. Sci. USA 79:6617-6621.

Albertini, R.J., L.M. Sullivan, J.K. Berman, C.J. Greene, J.A. Stewart, J.M. Silveira, and J.P. O'Neill. 1988. Mutagenicity monitoring in humans by autoradio-graphic assay for mutant T lymphocytes. Mutat. Res. 204:481-492.

Albro, P.W., W.B. Crummet, A.E. Dupuy, Jr., M.L. Gross, M. Hanson, R.L. Harless, R.D. Hileman, D. Hilker, C. Jason, and J. L. Johnson. 1986. Methods for the quantitative determination of multiple, specific polychlorinated dibenzo-p-dioxin and dibenzofuran isomers in human adipose tissue in the parts-per-trillion range. An interlaboratory study. Anal. Chem. 57:2717-2725.

Alexander, N.J. 1973. Ultrastructural changes in rat epididymis after vasectomy. Z. Zellforsch. Mikrosk. Anat. 136:177-182.

Alexander, N. J. 1982. Male evaluation and semen analysis. Clin. Obstet. Gynaecol. 25:463-482.

Allan, L.D., D.C. Crawford, R.H. Anderson, and M.J. Tynan. 1984. Echocardiographic and anatomical correlations in fetal congenital heart disease. Br. Heart J. 52:542-548.

Allen, K.S., H.Y. Kressel, P.H. Arger, and H.M. Pol-lack. 1989. Age-related changes of the prostate: Evaluation by MR imaging. Am. J. Roentgenol. 152:77-81.

Altland, K., M. Kaempher, M. Forssbohm, and W. Wer-ner. 1982. Monitoring for changing mutation rates using blood samples submitted for PKU screening. Prog. Clin. Biol. Res. 103(Pt. A):277-287.

Altman, J. 1966. Autoradiographic and histological studies of postnatal neurogenesis. II. A longitudinal investigation of the kinetics, migration and transforma-tion of cells incorporating tritiated thymidine in infant rats, with special reference to postnatal neurogenesis in some brain regions. J. Comp. Neurol. 128:431-474.

Altman, J. 1986. An animal model of minimum brain dysfunction, pp. 241-304. In M. Lewis, Ed. Learning Disabilities and Prenatal Risk. Urbana, Ill.: University of Illinois Press.

Altman, J. 1987. Morphological and behavioral markers of environmentally induced retardation of brain de-velopment: An animal model. Environ. Health Per-spect. 74:153-168.

Altman, J., and S. A. Bayer. 1978. Prenatal development of the cerebellar system in the rat. I. Cytogenesis and histogenesis of the deep nuclei and the cortex of the cerebellum. J. Comp. Neurol. 179:23-48.

Altman, J., and G. P. Das. 1966. Autoradiographic and histological studies of postnatal neurogenesis. I. A longitudinal investigation of the kinetics, migration and transformation of cells incorporating tritiated

thymidine in neonate rats, with special reference to postnatal neurogenesis in some brain regions. J. Comp. Neurol. 126:337-389.

Altman, J., W. J. Anderson, and K. A. Wright. 1969. Reconstitution of the external granule layer of the cerebellar cortex in infant rats after low-level x-irradia-tion. Anat. Rec. 163:453-472.

Altshuller, L. F., D. B. Halak, B. H. Landing, and R. A. Kehoe. 1962. Decidous teeth as an index of body burden of lead. J. Pediatr. 60:224-29.

Altstatt, L. B. 1965. Transplacental hyponatremia in the newborn infant: Report of 4 cases. J. Pediatr. 66:985-988.

Amann, R. P. 1970a. Sperm production rates, pp. 433-482. In A. D. Johnson, W. R. Gomes, and N. L. VanDemark, Eds. The Testis, Vol. 1. Development, Anatomy, and Physiology. New York: Academic Press.

Amann, R. P. 1970b. The male rabbit. IV. Quantitative testicular histology and comparisons between daily sperm production as determined histologically and daily sperm output. Fertil. Steril. 21:662-672.

Amann, R. P. 1981. A critical review of methods for evaluation of spermatogenesis from seminal characteris-tics. J. Androl. 2:37-58.

Amann, R. P. 1986. Detection of alterations in testicular and epididymal function in laboratory animals. En-viron. Health Perspect. 70:149-158.

Amann, R. P. 1987. Function of the epididymis in bulls and rams. J. Reprod. Fertil. Suppl. 34:115-131.

Amann, R. P., and J. O. Almquist. 1962. Reproductive capacity of dairy bulls. VI. Effect of unilateral vasec-tomy and ejaculation frequency on sperm reserves: Aspects of epididymal physiology. J. Reprod. Fertil. 3:260-268.

Amann, R. P., and W. E. Berndtson. 1986. Assessment of procedures for screening agents for effects on male reproduction: Effects of dibromochloropropane (DBCP) on the rat. Fundam. Appl. Toxicol. 7:244-255.

Amann, R. P., and S. S. Howards. 1980. Daily sper-matozoal production and epididymal spermatozoal reserves of the human male. J. Urol. 124:211-215.

Amann, R. P., and J. T. Lambiase, Jr. 1969. The male rabbit. 3. Determination of daily sperm production by means of testicular homogenates. J. Anim. Sci. 28:369-374.

Amann, R. P., G. J. Killian, and A. W. Benton. 1973. Differences in the electrophoretic characteristics of bovine rete testis fluid and plasma from the cauda epididymidis. J. Reprod. Fertil. 35:321-330.

Amann, R. P., S. R. Hay, and R. H. Hammerstedt. 1982. Yield, characteristics, motility and cAMP content of sperm isolated from seven regions of ram epididymis. Biol. Reprod. 27:723-733.

Amelar, R. D., and L. Dubin. 1977. Semen analysis, pp. 105-140. In R. D. Amelar, L. Dubin, and P. C. Walsh. Male Infertility. Philadelphia: W. B. Saunders.

Ames, B. M., J. Mccann, and E. Yamasaki. 1975. Methods for detecting carcinogens and mutagens with the Salmonella/mammalian-microsome mutagenicity test. Mutat. Res. 31:347-364.

Ammer, H., A. Henschen, and C. H. Lee. 1986. Isolation and amino-acid sequence analysis of human sperm protamines P1 and P2. Occurrence of two forms of protamine P2. Biol. Chem. Hoppe Seyler 367:515-522.

Amoroso, E. C. 1981. Viviparity, pp. 3-25. In S. R. Glasser and D. W. Bullock, Eds. Cellular and Molecular Aspects of Implantation. New York: Plenum Press.

Anderson, D. C. 1974. Sex-hormone-binding globulin. Clin. Endocrinol. 3:69-96.

Anderson, G. G., and T. M. Hanson. 1974. Chronic fetal bradycardia: Possible association with hypokalemia. Obstet. Gynecol. 44:896-898.

Andrews, L. B. 1984a. Legal issues raised by in vitro fertilization and embryo transfer, pp. 11-36. In D. P. Wolf and M. M. Quigley, Eds. Human *In Vitro* Fertilization and Embryo Transfer. New York: Plenum Press.

Andrews, L. B. 1984b. Ethical considerations in *in vitro* fertilization and embryo transfer, pp. 403-423. In D. P. Wolf and M. M. Quigley, Eds. Human *In Vitro* Fertilization and Embryo Transfer. New York: Plenum Press.

Angell, R. R., R. J. Aitken, P. F. A. van Look, M. A. Lumsden, and A. A. Templeton. 1983. Chromosome abnormalities in human embryos after in vitro fertilization. Nature 303:336-338.

Annest, J. L., J. L. Pirkle, D. Makuc, J. W. Neese, D. D. Bayse, and M. G. Kovar. 1983. Chronological trend in blood lead levels between 1976 and 1980. N. Engl. J. Med. 308:1373-1377.

Anthony, C. T., B. J. Danzo, and M.C. Orgebin-Crist. 1984. Investigations on the relationship between sperm fertilizing ability and androgen-binding protein in the restricted rat. Endocrinology 114:1413-1418.

Apgar, V. 1953. A proposal for a new method of evaluation of the newborn infant. Curr. Res. Anesth. Analg. 32:260-267.

Aragona, C., and H. G. Friesen. 1975. Specific prolactin binding sites in the prostate and testis of rats. Endocrinology 97:677-684.

Archer, P. G., M. Bender, A. D. Bloom, J. G. Brewen, A. V. Carrano, and R. J. Preston. 1981. Guidelines for cytogenetic studies in mutagen-exposed human populations, pp. 1-35. In A. D. Bloom, Ed. Guidelines for Studies of Human Populations Exposed to Mutagenic and Reproductive Hazards. New York: March of Dimes Birth Defects Foundation.

Ardito, G., L. Lamberti, E. Ansaldi, and P. Ponzetto. 1980. Sister-chromatid exchanges in cigarette-smoking human females and their newborns. Mutat. Res. 78:209-212.

Armstrong, D. T. 1968. Hormonal control of human uterine fluid retention in the rat. Am. J. Physiol. 214:764-771.

Armstrong, E. G., P. H. Ehrlich, S. Birken, J. P. Schlatterer, E. Siris, W. C. Hembree, and R. E. Canfield. 1984. Use of a highly sensitive and specific immunoradiometric assay for detection of human chorionic gonadotropin in urine of normal, nonpregnant, and pregnant individuals. J. Clin. Endocrinol. Metab. 59:867-874.

Arora, R., Dinakar, N., and M. R. N. Prasad. 1975. Biochemical changes in the spermatozoa and luminal contents of different regions of the epididymis of the rhesus monkey, Macaca mulatta. Contraception 11:689-700.

Aschheim, S., and B. Zondek. 1928. Die schwangerschaftsdiagnose aus dem harn durch nachweis des hypophysenvorderlappenhormons. II. Pracktische und theoretische ergebnisse aus den harnuntersuchungen. Klin. Wschr. 7:1453-1457.

Ashford, N. A. 1986. Policy considerations for human monitoring in the workplace. J. Occup. Med. 28:563-568.

Ashford, N. A., C. J. Spadafor, and C. C. Caldart. 1984. Human monitoring: Scientific, legal and ethical considerations. Harvard Environ. Law Rev. 8:263-363.

Ashley-Montague, M. F. 1957. The Reproductive Development of the Female with Special Reference to the Period of Adolescent Sterility. New York: Julian Press. 234 pp.

Atland, K., M. Kaempfer, M. Forszbohm et al. 1982. Monitoring for changing mutation rates using blood samples submitted for PKU screening, pp. 277-287. In Human Genetics, Part A: The Unfolding Genome. New York: Alan R. Liss.

Attramadal, A., C. W. Bardin, G. L. Gunsalus, N. A. Musto, and V. Hansson. 1981. Immunocytochemical localization of androgen binding protein in rat Sertoli and epididymal cells. Biol. Reprod. 25:983-988.

Auroux, M., C. Collin, and M. L. Couvillers. 1985. Do nonspermatozoal cells mainly stem from spermiogenesis? Study of 106 fertile and 102 subfertile men. Arch. Androl. 14:73-80.

Austin, C. R., and R. V. Short, Eds. 1982. Reproduction in Mammals, Vol. 1: Germ Cells and Fertilization, 2nd ed. New York: Cambridge University Press.

Awa, A. A., S. Honda, S. Neriishi, H. Shimba, T. Amano, and H. B. Hamilton. 1981. An interim report of the cytogenetic study of the offspring of atomic bomb survivors in Hiroshima and Nagasaki. Abstract for the 6th International Congress of Human Genetics, held September 13-18, 1981 in Jerusalem, Israel.

Axelrod, L., R. M. Neer, and B. Kliman. 1979. Hypogonadism in a male with immunologically active, biologically inactive luteinizing hormone: An exception of a venerable rule. J. Clin. Endocrinol. Metab. 48:279-287.

Ayala, A. R., B. C. Nisula, H. C. Chen, G. D. Hodgen, and G. T. Ross. 1978. Highly sensitive radioimmunoassay for chorionic gonadotropin in human urine. J. Clin. Endocrinol. Metab. 47:767-773.

Ayd, F. J., Jr. 1961. A survey of drug-induced extrapyramidal reactions. J. Am. Med. Assoc. 175:1054-1060.

Babich, H., D. L. Davis, and G. Stotzky. 1981. Dibromo-chloropropane (DBCP): A review. Sci. Total Environ. 17:207-221.

Bailey, R. E. 1950. Inhibition with prolactin of light-induced gonad increase in white-crowned sparrows. Condor 52:247-251.

Baird, D. D., and A. J. Wilcox. 1985. Cigarette smoking associated with delayed conception. J. Am. Med. Assoc. 253:2979-2983.

Baird, D. D., A. J. Wilcox, and C. R. Weinberg. 1986. Use of time to pregnancy to study environmental exposures. Am. J. Epidemiol. 124:470-480.

Baker, H. W., H. G. Burger, D. M. de Kretser, D. W. Lording, P. McGowan, and G. C. Rennie. 1981. Factors affecting the variability of semen analysis results in infertile men. Int. J. Androl. 4:609-622.

Baker, T. G. 1971. Radiosensitivity of mammalian oocytes with particular reference to the human female. Am. J. Obstet. Gynecol. 110:746-761.

Baker, T. S., K. M. Jennison, and A. E. Kellie. 1979. The direct radioimmunoassay of oestrogen glucuronides in human female urine. Biochem. J. 177:729-738.

Bamforth, S. J., and P. A. Baird. 1989. Spina bifida and hydrocephalus: A population study over a 35-year period. Am. J. Hum. Genet. 44:225-232.

Barbieri, R. L., J. A. Canick, A. Makris, R. B. Todd, I. J. Davies, and K. J. Ryan. 1977. Danazol inhibits steroidogenesis. Fertil. Steril. 28:809-813.

Barfield, A., J. Melo, E. Coutinho, F. Alvarez-Sanchez, A. Faundes, V. Brache, P. Leon, J. Frick, G. Bartsch, W. H. Weiske, P. Brenner, D. Mishell, Jr., G. Bernstein, and A. Ortiz. 1979. Pregnancies associated with sperm concentrations below 10 million/ml in clinical studies of a potential male contraceptive method, monthly depot medroxyprogesterone acetate and testosterone esters. Contraception 20:121-127.

Barlow, P. W., and M. I. Sherman. 1972. The biochemistry of differentiation of mouse trophoblast: Studies on polyploidy. J. Embryol. Exp. Morphol. 24:447-465.

Barlow, P., and C. G. Vosa. 1970. The Y chromosome in human spermatozoa. Nature 226:961-962.

Barnes, A. B. 1979. Menstrual history of young women exposed in utero to diethylstilbestrol. Fertil. Steril. 32:148-153.

Barros, C., J. Gonzelez, E. Herrera, and E. Bustos-Obregon. 1978. Fertilizing capacity of human spermatozoa evaluated by actual penetration of foreign eggs. Contraception 17:87-92.

Bartsch, G., and H. P. Rohr. 1977. Ultrastructural sterology. A new approach to the study of prostatic function. Invest. Urol. 14:301-306.

Bartsch, G., H. R. Muller, M. Oberholzer, and H. P. Rohr. 1979. Light microscopic sterological analysis of the normal human prostate of benign prostatic hyperplasia. J. Urol. 122:487-491.

Bartsch, G., G. Egender, H. Hubscher, and H. Rohr. 1982. Somometrics of the prostate. J. Urol. 127:1119-

1121.

Battistin, L., and F. Gerstenbrand, Eds. 1986. Pet and NMR: New Perspectives in Neuroimaging and in Clinical Neurochemistry. New York: Liss. 518 pp.

Bayer, S. A., R. L. Brunner, R. Hine, and J. Altman. 1973. Behavioural effects of interference with the postnatal acquisition of hippocampal granule cells. Nature 242:222-224.

Bayley, N. 1969. Manual for the Bayley Scales of Infant Development. New York: Psychological Corp. 178 pp.

Baukloh, V., H. G. Bohnet, M. Trapp, W. Heeschen, W. Feichtinger, and P. Kemeter. 1985. Biocides in human follicular fluid. Ann. N.Y. Acad. Sci. 442:240-250.

Beazley, J. M., and B. Alderman. 1984. Neonatal hyperbilirubinaemia following the use of oxytocin in labour. Br. J. Obstet. Gynecol. 82:265-271.

Bedford, J. M. 1966. Development of the fertilizing ability of spermatozoa in the epididymis of the rabbit. J. Exp. Zool. 163:319-329.

Bedford, J. M. 1967. Effects of duct ligation on the fertilizing ability of spermatozoa from different regions of the rabbit epididymis. J. Exp. Zool. 166:271-281.

Bedford, J. M. 1975. Maturation, transport, and fate of spermatozoa in the epididymis, pp. 303-317. In D. W. Hamilton and R. O. Greep, Eds. Handbook of Physiology, Section 7: Endocrinology. Volume V. Male Reproductive System. Washington, D.C.: American Physiological Society.

Bedford, J. M. 1976. Adaptations of the male reproductive tract and the fate of spermatozoa following vasectomy in the rabbit, rhesus monkey, hamster and rat. Biol. Reprod. 14:118-142.

Bedford, J. M. 1977. Evolution of the scrotum: The epididymis as the prime mover, pp. 171-182. In J. H. Calaby and C. H. Tyndale Biscoe, Eds. Reproduction and Evolution. Canberra City: Australian Academy of Science.

Bedford, J. M. 1978a. Anatomical evidence for the epididymis as the prime mover in the evolution of the scrotum. Am. J. Anat. 152:483-507.

Bedford, J. M. 1978b. Influence of abdominal temperature on epididymal function in rat and rabbit. Am J. Anat. 152:509-521.

Bedford, J. M. 1979. Evolution of the sperm maturation and sperm storage functions of the epididymis, pp. 7-21. In D. W. Fawcett and J. M. Bedford, Eds. The Spermatozoon. Maturation, Motility, Surface Properties and Comparative Aspects. Baltimore, Md.: Urban & Schwarzenberg.

Bedford, J. M., and G. W. Cooper. 1978. Membrane fusion events in the fertilization of vertebrate eggs, pp. 65-125. In G. Poste and G. L. Nicolson, Eds. Membrane Fusion. Cell Surface Reviews, Vol. 5. Amsterdam: Elsevier North-Holland.

Bedford, J. M., and R. P. Millar. 1978. The character of sperm maturation in the epididymis of the ascrotal hyrax, *Procarvia capensis* and armadillo, *Dasypus novemcinctus.* Biol. Reprod. 19:396-406.

Bedford, J. M., and L. Nicander. 1971. Ultrastructural changes in the acrosome and sperm membranes during maturation of spermatozoa in the testis and epididymis of the rabbit and monkey. J. Anat. 108:527-543.

Bedford, J. M., H. Calvin, and G. W. Cooper. 1973. The maturation of spermatozoa in the human epididymis. J. Reprod. Fertil. (Suppl. 18):199-213.

Beer, A. E., J. F. Quebbeman, J. W. Ayers, and R. F. Haines. 1981. Major histocompatibility complex antigens, maternal and paternal immune responses and chronic habitual abortions in humans. Am. J. Obstet. Gynecol. 141:987-999.

Beer, A. E., J. F. Quebbeman, and X. Zhu. 1986. Nonpaternal leucocyte immunization in women previously immunized with paternal leucocytes: Immune responses and subsequent pregnancy outcome, pp. 261-268. In D. A. Clark and B. A. Croy, Eds. Reproductive Immunology 1986: Proceedings of the 3rd International Congress of Reproductive Immunology Held in Toronto, Canada, 1-5 July 1986. Amsterdam: Elsevier Science.

Beier, H. M., and K. Beier-Hellwig. 1973. Specific secretory proteins of the female genital tract. Acta. Endocrinol., 180(Suppl.):404-425.

Bell, J. L. 1969. Comparative study of immunological tests for pregnancy diagnosis. J. Clin. Pathol. 22:79-83.

Bell, S. C. 1979. Protein synthesis during deciduoma morphogenesis in the rat. Biol. Reprod. 20:811-821.

Bellinger, D. C., H. L. Needleman, A. Leviton, C. Waternaux, M. B. Rabinowitz, and M. L. Nichols. 1984. Early sensory-motor development and prenatal exposure to lead. Neurobehav. Toxicol. Teratol. 6:387-402.

Bellisario, R., R. B. Carlsen, and O. P. Bahl. 1973. Human chorionic gonadotropin: Linear amino acid sequence of the alpha subunit. J. Biol. Chem. 248:6796-6809.

Belsey, M. A., R. Eliasson, A. H. Gallegos, K. S. Moghissi, C. A. Paulsen, and M. R. N. Prasad. 1980. Laboratory Manual for the Examination of Human Semen and Semen-Cervical Mucus Interaction. Singapore: Press Concern.

Benoit, J. 1926. Recherches anatomiques, cytologiques et histophysiologiques sur les voies excrétrices du testicule chez les mammifères. Arch. Anat. Histol. Embryol. 5:175-412.

Benziger, D. P., and J. Edelson. 1983. Absorption from the vagina. Drug Metabol. Rev. 14:137-168.

Bergman, A., A. Amit, M. P. David, Z. T. Homonnai, and G. F. Paz. 1981. Penetration of human ejaculated spermatozoa into human and bovine cervical mucus. I. Correlation between penetration values. Fertil. Steril. 36:363-367.

Bergquist, H. 1964. The formation of the front part of the neural tube. Experientia 20:92-93.

Bergquist, H., and B. Källén. 1954. Notes on the early histogenesis and morphogenesis of the central nervous system in vertebrates. J. Comp. Neurol. 100:627-659.

Bergstrand, C. G. 1986. Alphafetoprotein in paediatrics. Acta Paediatr. Scand. 75:1-9.

Bergstrand, C. G., and B. Czar. 1957. Paper electrophoretic study of human fetal serum proteins with demonstration of a new protein fraction. Scand. J. Clin. Lab. Invest. 9:277-286.

Berndtson, W. E. 1977. Methods for quantifying mammalian spermatogenesis: A review. J. Anim. Sci. 44:818-833.

Berry, M., and J. T. Eayrs. 1966. The effects of x-irradiation on the development of the cerebral cortex. J. Anat. 100:707-722.

Berry, S. J., R. Sterner, D. S. Coffey, and L. L. Ewing. 1985. Methods for monitoring canine prostate size: Internal and external caliper measurements. Prostate 6:303-314.

Beverley, P. C. L. 1985. The functional significant of the human E rosette receptor (CD2), pp. 93-104. In F. Ellendorff and E. Koch, Eds. Early Pregnancy Factors. Ithaca, N.Y.: Perinatology Press.

Beyler, S. A., and L. J. Zaneveld. 1979. The role of acrosin in sperm penetration through human cervical mucus. Fertil. Steril. 32:671-675.

Bibbo, M., W. B. Gill, F. Azizi, R. Blough, V. S. Fang, R. L. Rosenfield, G. F. B. Schumacher, K. Sleeper, M. G. Sonek, and G. L. Wied. 1977. Follow-up study of male and female offspring of DES-exposed mothers. Obstet. Gynecol. 49:1-8.

Bigbee, W. L., R. G. Langlois, M. Swift, and R. H. Jensen. 1989. Evidence for an elevated frequency of in vivo somatic cell mutations in ataxia telangiectasia. Am. J. Hum. Genet. 44:402-408.

Bigbee, W. L., A. J. Wyrobek, R. G. Langlois, R. H. Jensen, and R. B. Everson. In press. The effects of chemotherapy on in vivo frequency of glycophorin a "null" variant erythrocytes. Mutat. Res.

Bigsby, R. M., and G. R. Cunha. 1986. Estrogen stimulation of deoxyribonucleic acid synthesis in uterine epithelial cells which lack estrogen receptors. Endocrinology 119:390-396.

Billington, W. D. 1985. Pregnancy proteins, early pregnancy factor and maternal immune responses. A summary and an assessment of their relationships, pp. 267-272. In F. Ellendorff and E. Koch, Eds. Early Pregnancy Factors. Ithaca, N.Y.: Perinatology Press.

Birgersson, L., and E. D. B. Johansson. 1983. Inhibition of progesterone synthesis by Win 32,729 in women. Acta Endocrinologia. 103(Suppl. 256):145.

Birken, S., and R. E. Canfield. 1977. Isolation and amino acid sequence of COOH-terminal fragments from the beta subunit of human choriogonadotropin. J. Biol. Chem. 252:5386-5392.

Birken, S., R. Canfield, G. Agosto, and J. Lewis. 1982. Preparation and characterization of an improved beta-COOH-terminal immunogen for generation of specific and sensitive antisera to human chorionic gonadotropin. Endocrinology 110:1555-1563.

Birken, S., E. G. Armstrong, M. A. G. Kolks, L. A. Cole, G. M. Agosto, A. Krichevsky, J. L. Vaitukaitis, and R. E. Canfield. 1988. Structure of the human chorionic gonadotropin beta-subunit fragment from pregnancy urine. Endocrinology 123:572-583.

Birnberg, C. H., R. Kurzrok, and A. Laufer. 1958. Simple test for determining ovulation time. J. Am. Med. Assoc. 166:1174-1175.

Bitman, J., H. C. Cecil, S. J. Harris, and G. F. Fries. 1968. Estrogenic activity of o,p'-DDT in the mammalian uterus and avian oviduct. Science 162:371-372.

Bjersing, L., and H. Carstensen. 1967. Biosynthesis of steroids by granulosa cells of the porcine ovary in vitro. J. Reprod. Fertil. 14:101-111.

Black, I. B., J. E. Adler, C. F. Dreyfus, G. M. Jonakait, D. M. Katz, E. F. La Gamma, and K. M. Markey. 1984. Neurotransmitter plasticity at the molecular level. Science 225:1266-1270.

Blandau, R. J., and R. E. Rumery. 1964. The relationship of swimming movements of epididymal spermatozoa to their fertilizing capacity. Fertil. Steril. 15:571-579.

Blaquier, J. A. 1971. Selective uptake and metabolism of androgens by rat epididymis. The presence of a cytoplasmic receptor. Biochem. Biophhys. Res. Commun. 45:1076-1082.

Blaquier, J. A., M. S. Cameo, D. Stephany, A. Piazza, J. Tezon, and R. J. Sherins. 1987. Abnormal distribution of epididymal antigens on spermatozoa from infertile men. Fertil. Steril. 47:302-309.

Blatt, J., R. J. Sherins, and D. G. Poplack. 1981. Evidence of normal testicular function in boys following chemotherapy for acute lymphoblastic leukemia. Presented at Seventy-second Annual Meeting of the American Association for Cancer Research/Seventeenth Annual Meeting of the American Society of Clinical Oncology, 27 April-2 May, 1981, in Washington, D.C. Abstracts (Eng.) 22(Mar.): Paper No. C-303. (Abstract.)

Bloch, P., G. Garavaglia, G. Mitchell, and I. M. Shapiro. 1977. Measurement of lead content of children's teeth in situ by x-ray fluorescence. Phys. Med. Biol. 20:56-63.

Block, E. 1952. Quantitative morphological investigations of the follicular system in women. Variations at different ages. Acta Anat. 14:108-123.

Bloom, A. D., Ed. 1981. Guidelines for Studies of Human Populations Exposed to Mutagenic and Reproductive Hazards: Proceedings of a Conference on the Evaluation of Human Populations Exposed to Potential Mutagenic and Reproductive Hazards, Washington, D.C., Jan. 26-27, 1981. White Plains, N.Y.: March of Dimes Birth Defects Foundation. 163 pp.

Blum, M. D., R. R. Bahnson, C. Lee, T. W. Deschler, and J. T. Grayhack. 1985. Estimation of canine prostatic size by in vivo ultrasound and volumetric measurement. J. Urol. 133:1082-1086.

Bochkov, N. P., N. P. Kuleshov, A. N. Chebotarev, V. I. Alekhin, and S. A. Midian. 1974. Population cytogenetic invvestigation of newborns in Moscow. Humangenetik 22:139-152.

Böhmer, T., S. C. Weddington, and V. Hansson. 1977. Effect of testosterone propionate on levels of carnitine and testicular androgen binding protein (ABP) in rat epididymis. Endocrinology 100:835-838.

Bohn, H. 1985. Biochemistry of pregnancy proteins: An overview, pp. 127-139. In F. Ellendorff and E. Koch, Eds. Early Pregnancy Factors. Ithaca, N.Y.: Perinatology Press.

Boime, L., M. Boothby, M. Hoshima, S. Daniels-McQueen, and R. Darnell. 1982. Expression and structure of human placental hormone genes as a function of placental development. Biol. Reprod. 26:73-91.

Boisselle, A., and R. R. Tremblay. 1979. New therapeutic approach to the hirsute patient. Fertil. Steril. 32:276-279.

Boitani, C., C. L. C. Chen, A. N. Margioris, I. Gerendai, P. L. Morris, and C. W. Bardin. 1986. Pro-opiomelanocortin-derived peptides in the testis: Evidence for a possible role in Leydig and Sertoli cell function. Med. Biol. 63:251-258.

Bondareff, W. 1973. Age changes in the neuronal environment, pp. 1-18. In M. Rockstein and M. L. Sussman, Eds. Development and Aging in the Nervous System. New York: Academic Press.

Bondy, S. 1985. Especial considerations for neurotoxicological research. CRC Crit. Rev. Toxicol. 14:381-402.

Bordy, M. J., J. H. Shaper, and L. L. Ewing. 1984. Trophic influences of luteinizing hormone on steroidogenesis by Percoll-separated rat Leydig cells in culture. Ann. N.Y. Acad. Sci. 438:329-345.

Bostofte, E., J. Serup, and H. Rebbe. 1982a. Relation between sperm count and semen volume, and pregnancies obtained during a twenty-year follow-up period. Int. J. Androl. 5:267-275.

Bostofte, E., J. Serup, and H. Rebbe. 1982b. Relation between morphologically abnormal spermatozoa and pregnancies obtained during a twenty-year follow-up period. Int. J. Androl. 5:379-386.

Botero-Ruiz, W., N. Laufer, A. H. DeCherney, M. L. Polan, F. P. Haseltine, and H. R. Behrman. 1984. The relationship between follicular fluid steroid concentration and successful fertilization of human oocytes in vitro. Fertil. Steril. 41:820-826.

Bouckaert, P. X., J. L. H. Evers, W. H. Doesburg, L. A. Schellekens, and R. Rolland. 1986. Patterns of changes in glycoproteins, polypeptides, and steroids in the peritoneal fluid of women during the periovulatory phase of the menstrual cycle. J. Clin. Endocrinol. Metab. 62:293-299.

Boué, A., J. Boué, and A. Gropp. 1985. Cytogenetics of pregnancy wastage. Adv. Hum. Genet. 14:1-57.

Bournias-Vardiabasis, N. 1985. Use of amniotic fluid cells for testing teratogens, pp. 315-331. In A. M.

Goldberg, Ed. *In Vitro* Toxicology: A Progress Report from the Johns Hopkins Center for Alternatives to Animal Testing. Alternative Methods in Toxicity, Vol. 3: In Vitro Toxicology. New York: Liebert.

Boveris, A., E. Cadenas, R. Reiter, M. Filipkowski, Y. Nakase, and B. Chance. 1980. Organ chemiluminesence: Noninvasive assay for oxidative radical reactions. Proc. Natl. Acad. Sci. USA 77:347-351.

Bower, P. A., P. C. Yelick, and N. B. Hecht. In press. Both protamine 1 and 2 are expressed in mouse, hamster, and rat. Biol. Reprod.

Boyar, R. M., J. W. Finkelstein, H. Roffwarg, S. Kapen, D. Weitzman, and L. Hellman. 1973. Twenty-four-hour luteinizing hormone and follicle-stimulating hormone secretory patterns in gonadal dysgenesis. J. Clin. Endocrinol. Metab. 37:521-525.

Bradley, M. O., and G. Dysart. 1985. DNA single-strand breaks, double-strand breaks, and crosslinks in rat testicular germ cells: Measurements of their formation and repair by alkaline and neutral filter elution. Cell Biol. Toxicol. 1:181-195.

Brandriff, B., L. Gordon, L. Ashworth, G. Watchmaker, A. Carrano, and A. Wyrobek. 1984. Chromosomal abnormalities in human sperm: Comparisons among four healthy men. Hum. Genet. 66:193-201.

Brandriff, B. F., L. A. Gordon, D. Moore II, and A. V. Carrano. 1988a. An analysis of structural aberrations in human sperm chromosomes. Cytogenet. Cell. Genet. 47:29-36.

Brandriff, B. F., L. A. Gordon, L. K. Ashworth, and A. V. Carrano. 1988b. Chromosomal aberrations induced by in vitro irradiation: Comparisons between human sperm and lymphocytes. Environ. Mol. Mutagen. 12:167-177.

Brandt, H., T. S. Acott, D. J. Johnson, and D. D. Hoskins. 1978. Evidence for epididymal origin of bovine sperm forward motility protein. Biol. Reprod. 19:830-835.

Branton, C., and G. W. Salisbury. 1947. Morphology of spermatozoa from different levels of the reproductive tract of the bull. J. Anim. Sci. 6:154-160.

Braun, A. G., D. J. Emerson, and B. B. Nichinson. 1979. Teratogenic drugs inhibit tumour cell attachment to lectin-coated surfaces. Nature 282:507-509.

Braun, A. G., C. A. Buckner, D. J. Emerson, and B. B. Nichinson. 1982. Quantitative correspondence between the in vivo and in vitro activity of teratogenic agents. Proc. Natl. Acad. Sci. USA 79:2056-2060.

Brawer, J. R., and F. Naftolin. 1979. The effects of oestrogen exposure on hypothalamic tissue, pp. 19-33. In Sex, Hormones and Behaviour. CIBA Foundation Symposium 62 (New Series). Amsterdam: Excerpta Medica.

Brawer, J. R., H. Schipper, and B. Robaire. 1983. Effects of long term androgen and estradiol exposure on the hypothalamus. Endocrinology 112:194-199.

Brewen, J. G., R. J. Preston, and N. Gengozian. 1975. Analysis of x-ray-induced chromosomal translocations in human and marmoset spermatogonial stem cells. Nature 253:468-470.

Brinsmead, M. W., B. J. Bancroft, G. D. Thorburn, and M. J. Waters. 1981. Fetal and maternal ovine placental lactogen during hyperglycaemia, hypoglycaemia and fasting. J. Endocrinol. 90:337-343.

Brooks, D. E. 1979a. Influence of androgens on the weights of the male accessory reproductive organs and on the activities of mitochondrial enzymes in the epididymis of the rat. J. Endocrinol. 82:293-303.

Brooks, D. E. 1979b. Carbohydrate metabolism in the rat epididymis: Evidence that glucose is taken up by tissue slices and isolated cells by a process of facilitated transport. Biol. Reprod. 21:19-26.

Brooks, D. E. 1980. Carnitine in the male reproductive tract and its relation to the metabolism of the epididymis and spermatozoa, pp. 219-235. In R. A. Frankel and J. D. McGarry, Eds. Carnitine Biosynthesis, Metabolism, and Functions. New York: Academic Press.

Brooks, D. E. 1981. Metabolic activity in the epididymis and its regulation by androgens. Physiol. Rev. 61:516-555.

Brooks, D. E. 1982. Epididymal functions and their hormonal regulation. Aust. J. Biol. Sci. 36:205-221.

Brooks, D. E. 1983. Effect of androgens on protein synthesis and secretion in various regions of the rat epididymis, as analysed by two-dimensional gel electrophoresis. Mol. Cell. Endocrinol. 29:255-270.

Brooks, D. E., and K. Tiver. 1984. Analysis of surface proteins of rat spermatozoa during epididymal transit and identification of antigens common to spermatozoa, rete testis fluid and cauda epididymal plasma. J. Reprod. Fertil. 71:249-257.

Brown, C. R., K. I. von Glos, and R. Jones. 1983. Changes in plasma membrane glycoproteins of rat spermatozoa during maturation in the epididymis. J. Cell Biol. 96:256-264.

Bruce, W. R., R. Furrer, and A. J. Wyrobek. 1974. Abnormalities in the shape of murine sperm after acute testicular x-irradiation. Mutat. Res. 23:381-386.

Brusick, D., F. J. de Serres, R. B. Everson, M. L. Mendelsohn, J. V. Neel, M. D. Shelby, and M. D. Waters. 1981. Monitoring the human population for mutagenic effects: Detection of gene mutations and dna damage. In A. D. Bloom, Ed. Guidelines for Studies of Human Populations Exposed to Mutagenic and Reproductive Hazards. New York: March of Dimes Birth Defects Foundation,

Bryce, T. H., and J. H. Teacher. 1908. Contributions to the Study of Early Development and Imbedding of the Human Ovum. Glasgow: MacLehose. 93 pp.

Budnick, I. S., S. Leikin, and L. E. Hoeck. 1955. Effect in the newborn infant of reserpine administered antepartum. A.M.A. Am. J. Dis. Child. 90:286-289.

Buelke-Sam, J., C. A. Kimmel, J. Adams, C. J. Nelson, C. V. Vorhees, D. C. Wright, V. St. Omer, B. A. Korol, R. E. Butcher, M. A. Geyer, J. F. Holson, C. Kutscher, and M. J. Wayner. 1985. Collaborative behavioral teratology study: Results. Neurobehav. Toxicol. Teratol. 7:591-624.

Bulbrook, R. D., J. L. Hayward, and C. C. Spicer. 1971. Relation between urinary androgen and corticoid excretion and subsequent breast cancer. Lancet 2:395-398.

Burkhart, J. G., J. Benzinger, K. Svensson, and H. V. Malling. 1985. An evaluation of heterologous antibodies to lactate dehydrogenase-C in the detection of mutations. Mutat. Res. 148:135-149.

Burmer, G., and L. Loeb. In press. Mutations in C-Ki-ras- oncogene during progressive stages of human colon carcinoma. Proc. Natl. Acad. Sci. USA.

Burroughs, C. D., H. A. Bern, and E. L. R. Stokstad. 1985. Prolonged vaginal cornification and other changes in mice treated neonatally with coumestrol, a plant estrogen. J. Toxicol. Environ. Health 15:51-61.

Bush, T. L., and E. Barrett-Connor. 1985. Noncontraceptive estrogen use and cardiovascular disease. Epidemiol. Rev. 7:89-104.

Bushnell P. J., R. E. Bowman, J. R. Allen, and R. J. Marlar. 1977. Scotopoic vision deficits in young monkeys exposed to lead. Science 196:333-335.

Bussolino, F., C. Benedetto, M. Massobrio, and G. Camussi. 1980. Maternal vascular prostacyclin activity in pre-eclampsia (letter). Lancet 2:702.

Bygdeman, M., and R. Eliasson. 1969. Distribution of prostaglandins, fructose and acid phosphatase in human seminal plasma. Andrologie 1:5-10.

Callen, P. W. 1983. Ultrasonography in Obstetrics and Gynecology. Philadelphia: W. B. Saunders Co. 346 pp.

Calne, D. B., J. W. Langston, W. R. Martin, A. J. Stoessl, T. J. Ruth, M. J. Adam, B. D. Pate, and M. Schulzer. 1985. Positron emission tomography after MPTP: Observations relating to the cause of Parkinson's disease. Nature 317:246-248.

Calne, D. B., A. Eisen, E. McGeer, and P. Spencer. 1986. Alzheimer's disease, Parkinson's disease and motoneurone disease: A biotropic interaction between ageing and environment? Lancet 2:1067-1072.

Calvin, H. I., and J. M. Bedford. 1971. Formation of disulphide bonds in the nucleus and accessory structures of mammalian spermatozoa during maturation in the epididymis. J. Reprod. Fertil. (Suppl. 13):65-75.

Campbell, K. L. 1985. Methods of monitoring ovarian function and predicting ovulation: Summary of a meeting. Res. Frontiers Fertil. Regulation 3:1-16.

Campbell, M. E., S. P. Spielberg, and W. Kalow. 1987. A urinary metabolite ratio that reflects systemic caffeine clearance. Clin. Pharmacol. Ther. 42:157-165.

Canfield, R. E., S. Birken, P. Ehrlich, and G. Armstrong. 1984. Immunochemistry of human chorionic gonadotropin. Adv. Exp. Med. Biol. 176:199-215.

Canfield, R. E., J. F. O'Connor, S. Birken, A. Krichevsky, and A. J. Wilcox. 1987. Development of an assay for a biomarker of pregnancy and early fetal loss.

Environ. Health Perspect. 74:57-66.

Carey, S. W., N. W. Klein, W. T. Frederickson, G. Seakett, R. M. Greenstein, P. Sehgal, and M. Elliott. 1984. Analysis of sera from monkeys with histories of fetal wastage and the identification of teratogenicity in sera from human chronic spontaneous aborters using rat embryo cultures. Trophoblast Res. 1:347-360.

Carr, D. W., and T. S. Acott. 1984. Inhibition of bovine spermatozoa by caudal epididymidal fluid: I. Studies of a sperm motility quiescence factor. Biol. Reprod. 30:913-925.

Carrano, A. V., L. H. Thompson, P. A. Lindl, and J. L. Minkler. 1978. Sister chromatid exchange as an indicator of mutagenesis. Nature 271:551-553.

Carrano, A. V., J. L. Minkler, D. G. Stetka, and D. H. Moore, II. 1980. Variation in the baseline sister chromatid exchange frequency in human lymphocytes. Environ. Mutag. 2:325-337.

Carreau, S., M. A. Drosdowsky, C. Pisselet, and M. Courot. 1980. Hormonal regulation of androgen-binding protein in lamb testes. J. Endocrinol. 85:443-448.

Carreau, S., M. A. Drosdowsky, and M. Courot. 1984. Androgen-binding proteins in sheep epididymis: Characterization of a cytoplasmic androgen receptor in the ram epididymis. J. Endocrinol. 103:273-279.

Carson, D. D., J.-Y. Tang, J. A. Julian, and S. R. Glasser. 1988. Vectorial secretion of proteoglycans by polarized rat uterine epithelial cells. J. Cell Biol. 107:2425-2435.

Casillas, E. R., and S. Chaipayungpan. 1979. The distribution of carnitine and acetylcarnitine in the rabbit epididymis and the carnitine content of rabbit spermatozoa during maturation. J. Reprod. Fertil. 56:439-444.

Casslen, B., and K. Ohlsson. 1981. Alpha 1-antitrypsin—complexation and inactivation in the uterine fluid of IUD-users. Acta Obstet. Gynecol. Scand. 60:103-107.

Cate, R.L., R.J. Mattaliano, C. Hession, R. Tizard, N. M. Farber, A. Cheung, E.G. Ninfa, A.Z. Frey, D. J. Gash, E.P. Chow, R.A. Fisher, J.M. Bertonis, G. Torres, B.P. Wallner, K.L. Ramachandran, R.C. Ragin, T.F. Mangonaro, D.T. MacLaughlin, and P.K. Donahoe. 1986. Isolation of the bovine and human genes for Mullerian inhibiting substance and expression of the human gene in animal cells. Cell 45:685-698.

Catt, K. J., M. L. Dufau, and J. L. Vaitukaitis. 1975. Appearance of hCG in pregnancy plasma following the initiation of implantation of the blastocyst. J. Clin. Endocrinol. Metab. 40:537-540.

Catz, C., and S. J. Yaffe. 1968. Barbiturate enhancement of bilirubin conjugation and excretion in young and adult animals. Pediat. Res. 2:361-370.

Caudle, M. R., N. S. Rote, J. R. Scott, C. DeWitt, and M. F. Barney. 1983. Histocompatibility in couples with recurrent spontaneous abortion and normal fertility. Fertil. Steril. 39:793-798.

Cavanagh, A. C., H. Morton, B. E. Rolfe, and A. A.

Gidley-Baird. 1982. Ovum factor: A first signal of pregnancy? Am. J. Reprod. Immunol. 2:97-101.

Cavicchia, J. C. 1979. Fine structure of the monkey epididymis: A correlated thin-section and freeze-cleave study. Cell Tissue Res. 201:451-458.

Chandley, A. C. 1984. Infertility and chromosomal abnormality. Oxford Rev. Reprod. Biol. 6:1-46.

Chandley, A. C., P. Edmond, S. Christie, L. Gowans, J. Fletcher, A. Franckiewicz, and M. Newton. 1975. Cytogenetics and infertility in man. I. Karyote and seminal analysis: Results of a five-year survey of men attending a subfertility clinic. Ann. Hum. Genet. 39:231-254.

Chandra, R. K, and P. M. Newberne. 1977. Nutrition, Immunity and Infection. New York: Plenum Press. 246 pp.

Chang, T. S. K., and P. C. Albertsen. 1984. Interpretation and utilization of the zona pellucida-free hamster egg penetration test in the evaluation of the infertile male. Sem. Urol. 2(2):124-130.

Chang, T. S. K., and B. R. Zirkin. 1978. Distribution of sulfhydryl oxidase activity in the rat and hamster male reproductive tract. Biol. Reprod. 18:745-748.

Channing, C. P., and S. P. Coudert. 1976. Contribution of granulosa cells and follicular fluid to ovarian estrogen secretion in rhesus monkey in vivo. Endocrinology 98:590-597.

Chapman, R. M. 1983. Gonadal injury resulting from chemotherapy. Am. J. Ind. Med. 4:149-161.

Chard, T. 1985. Biological and clinical significance of pregnancy proteins, pp. 29-41. In F. Ellendorff and E. Koch, Eds. Early Pregnancy Factors. Ithaca, N.Y.: Perinatology Press.

Chasnoff, I. J., W. J. Burns, S. H. Schnoll, and K. A. Burns. 1985. Cocaine use in pregnancy. N. Engl. J. Med. 313:666-669.

Chatot, C. L., N. W. Klein, J. Piatek, and L. J. Pierro. 1980. Successful culture of rat embryos on human serum: Use in the detection of teratogens. Science 207:1471-1473.

Chellman, G. J., J. S. Bus, and P. K. Working. 1986. Role of epididymal inflammation in the induction of dominant lethal mutations in Fischer 344 rat sperm by methyl chloride. Proc. Natl. Acad. Sci. USA 83:8087-8091.

Chen, P. H., K. T. Chang, and Y. D. Lu. 1981. Polychlorinated biphenyls and polychlorinated dibenzofurans in the toxic rice-bran oil that caused PCB poisoning in Taichung. Bull. Environ. Contamin. Toxicol. 26:489-495.

Chervenak, F. A., M. A. Farley, L. Walters, J. C. Hobbins, and M. J.Mahoney. 1984. When is termination of pregnancy during the third trimester morally justifiable? N. Eng. J. Med. 310:501-504.

Chervenak, F. A., G. Isaacson, J. C. Hobbins, U. Chitkara, M. Tortora, and R. L. Berkowitz. 1985. Diagnosis and management of fetal holoprosencephaly. Obstet. Gynecol. 66:322-326.

Chew, W. C., and I. L. Swan. 1977. Influence of simul-taneous low amniotomy and oxytocin infusion and other maternal factors on neonatal jaundice: A prospective study. Br. Med. J. 1:72-73.

Chilton, B. S., S. V. Nicosia, and D. P. Wolf. 1981. Separation and characterization of endocervical cells, pp. 464-466. In S. R. Glasser and D. W. Bullock, Eds. Cellular and Molecular Aspects of Implantation. New York: Pergamon Press.

Chisolm, J. J., and E. K. Silbergeld. 1981. Increased excretion of homovanillic acid (HVA) in urine by young children with increased lead absorption, pp. 565-568. In Proceedings of International Conference on Heavy Metals in the Environment. Edinburgh: CEP Consultants.

Christensen, A. K., and K. C. Peacock. 1980. Increase in Leydig cell number in testes of adult rats treated chronically with an excess of human chorionic gonadotropin. Biol. Reprod. 22:383-391.

Chubb, C., and C. Desjardins. 1982. Vasculature of the mouse, rat, and rabbit testis-epididymis. Am. J. Anat. 165:357-372.

Chubb, C., and C. Nolan. 1985. Animal models of male infertility: Mice bearing single-gene mutations that induce infertility. Endocrinology 117:338-346.

Chulavatnatol, M., S. Panyim, and D. Wititsuwannakul. 1982. Comparison of phosphorylated proteins in intact rat spermatozoa from caput and cauda epididymidis. Biol. Reprod. 26:197-207.

Church, G. M, and W. Gilbert. 1984. Genomic sequencing. Proc. Natl. Acad. Sci. USA 81:1991-1995.

Cicero, T. J., R. D. Bell, W. G. Wiest, J. H. Allison, K. Polakoski, and E. Robins. 1975. Function of the male sex organs in heroin and methadone users. N. Engl. J. Med. 292:882-887.

Clark, D. A., A. Chaput, and D. Tutton. 1986. Active suppression of host-vs-graft reaction in pregnant mice. VII. Spontaneous abortion of allogeneic CBA/J x DBA/2 fetuses in the uterus of CBA/J mice correlates with deficient non-T suppressor cell activity. J. Immunol. 136:1668-1675.

Clark, R. V. 1988. Male hypogonadism, pp. 514-516. In J. W. Hurst, Ed. Medicine for the Practicing Physician, 2nd ed. Boston: Butterworths.

Clark, R. V., and R. J. Sherins. 1986. Use of semen analysis in the evaluation of the infertile couple, pp. 253-266. In R. J. Santen and R. S. Swerdloff, Eds. Male Reproductive Dysfunction: Diagnosis and Management of Hypogonadism, Infertility, and Impotence. New York: Marcel Dekker.

Clarkson, T. W. 1987. The role of biomarkers in reproductive and developmental toxicology. Environ. Health Perspect. 74:103-107.

Clarren, S. K., and D. W. Smith. 1978. The fetal alcohol syndrome. N. Engl. J. Med. 298:1063-1067

Clemetson, C. A. B., J. D. Kim, T. P. S. De Jesus, V. R. Mallikarjuneswara, and J. H. Wilds. 1973. Human uterine fluid potassium and the menstrual cycle. J. Obstet. Gynecol. Br. Comm. 80:553-561.

Clermont, Y. 1972. Kinetics of spermatogenesis in mammals: Seminiferous epithelium cycle and spermatogonial renewal. Physiol. Rev. 52:198-236.

Clift, A. F. 1945. Observations on certain rheological properties of human cervical secretion. Proc. R. Soc. Med. 39:1-9.

Clifton, D. K., and W. J. Bremner. 1983. The effect of testicular x-irradiation on spermatogenesis in man. J. Androl. 4:387-392.

Clifton, R. K., and M. N. Nelson. 1976. Developmental study of habituation in infants: The importance of paradigm, response system, and state, pp. 159-205. In T. J. Tighe and R. N. Leaton, Eds. Habituation: Perspectives from Child Development, Animal Behavior, and Neurophysiology. Hillsdale, N.J.: Lawrence Erlbaum.

Clifton, D. K., and R. A. Steiner. 1983. Cycle detection: A technique for estimating the frequency and amplitude of episodic fluctuations in blood hormone and substrate concentrations. Endocrinology 112:1057-1064.

Coffey, D. S. 1986. The biochemistry and physiology of the prostate and seminal vesicles, pp. 233-274. In P. C. Walsh, R. F. Gittes, A. D. Perlmutter, and T. A. Stamey, Eds. Campbell's Urology, 5th edition. Vol. 1. Philadelphia: W. B. Saunders.

Coffey, D. S., and J. T. Isaacs. 1981. Control of prostate growth. Urology 17(Suppl. 3):17-24.

Cohen, J., M. P. Ooms, and J. T. M. Vreeburg. 1981. Reduction of fertilizing capacity of epididymal spermatozoa by 5alpha-steroid reductase inhibitors. Experientia 37:1031-1032.

Cohen, J. S., R. H. Knop, G. Navon, and D. Foxall. 1983. Nuclear magnetic resonance in biology and medicine. Life Chem. Repts. 1:281-457.

Cohn, D. F., Z. T. Homonnai, and G. P. Paz. 1982. Diphenylhydantoin excretion in the semen of treated epileptics. Isr. J. Med. Sci. 8:509-511.

Cole, K. D., J. C. Kandala, and W. S. Kistler. 1986. Isolation of the gene for the testis-specific H1 histone variant H1t. J. Biol. Chem. 261:7178-7183.

Cole, R. J., J. Cole, L. Henderson, N. A. Taylor, C. F. Arlett, and T. Regan. 1983. Short-term tests for transplacentally active carcinogens. A comparison of sister-chromatid exchange and the micronucleus test in mouse foetal liver erythroblasts. Mutat. Res. 113:61-75.

Collea, J. V., and W. M. Holls. 1982. The contraction stress test. Clin. Obstet. Gynecol. 25:707-717.

Collins, J. A., W. Wrixon, L. B. Janes, and E. H. Wilson. 1983. Treatment-independent pregnancy among infertile couples. N. Engl. J. Med. 309:1201-1206.

Collins, W. P., P. O. Collins, M. J. Kilpatrick, P. A. Manning, J. M. Pike, and J. P. P. Tyler. 1979. The concentrations of urinary oestrone-3-glucuronide, LH and pregnanediol-3alpha-glucuronide as indices of ovarian function. Acta Endocrinol. 90:336-348.

Committee on Biological Markers of the National Research Council. 1987. Biological markers in environmental health research. Environ. Health Perspect.

74:3-9.

Connell, C. J., and A. M. Donjacour. 1985. A morphological study of the epididymides of control and estradiol-treated prepubertal dogs. Biol. Reprod. 33:951-969.

Cooke, B. A., and M. H. F. Sullivan. 1985. Review the mechanism of LHRH agonist action in gonadal tissues. Mol. Cell. Endocrinol. 41:115-122.

Cooper, T. G. 1986. The Epididymis, Sperm Maturation, and Fertilization. Berlin: Sprenger Verlag. 281 pp.

Cooper, T. G., and D. E. Brooks. 1981. Entry of glycerol into the rat epididymis and its utilization by epididymal spermatozoa. J. Reprod. Fertil. 61:163-169.

Cooper, T. G., and D. W. Hamilton. 1977. Phagocytosis of spermatozoa in the terminal region and gland of the vas deferens of the rat. Am. J. Anat. 150:247-267.

Cooper, T. G., C. H. Yeung, D. Nashan, and E. Nieschlag. 1988. Epididymal markers in human infertility. J. Androl. 9:91-101.

Copp, A. J. 1979. Interaction between inner cell mass and trophectoderm of the mouse blastocyst. II. The fate of the polar trophectoderm. J. Embryol. Exp. Morphol. 51:109-120.

Corliss, C. E., and G. G. Robertson. 1963. The pattern of mitotic density in the early chick neural epithelium. J. Exp. Zool. 153:125-140.

Corvol, P., A. Michaud, J. Menard, M. Freifeld, and J. Mahoudeau. 1975. Antiandrogenic effect of spirolactones: Mechanism of action. Endocrinology 97:52-58.

Cosentino, M. J., H. Takihara, J. W. Burhop, and A. T. K. Cockett. 1984. Regulation of rat caput epididymidis contractility by prostaglandins. J. Androl. 5:216-222.

Cottrill, C. M., R. G. McAllister, Jr., L. Gettes, and J. A. Noonan. 1977. Propranolol therapy during pregnancy, labor and delivery: Evidence for transplacental drug transfer and impaired neonatal drug disposition. J. Pediatr. 91:812-814.

Coulam, C. B. 1981. The diagnosis and treatment of infertility, Chapter 50. In J. J. Sciarra, Ed. Gynecology and Obstetrics. Hagerstown, Md.: Harper & Row.

Coulam, C. B., T. L Wasmoen, R. Creasy, P. Siiteri, and G. Gleich. 1987. Major basic protein as a predictor of preterm labor: A preliminary report. Am. J. Obstet. Gynecol. 156:790-796.

Courot, M. 1964. Some results obtained in the irradiation with x-rays of testes of lambs, pp. 279-286. In W. Carlson and F. Gassner, Eds. Effects of Ionizing Radiation on the Reproductive System. New York: Pergamon Press.

Courot, M. 1981. Transport and maturation of spermatozoa in the epididymis of mammals. Prog. Reprod. Biol. 8:67-79.

Cowan, W. M., B. B. Stanfield, and K. Kishi. 1980. The development of the dentate gyrus. Curr. Top. Dev. Biol. 15:103-157.

Crabo, B. 1965. Studies on the composition of epididymal content in bulls and boars. Acta Vet. Scand. 6(Suppl. 5):1-94.

Crabo, B., and B. Gustafsson. 1964. Distribution of sodium and potassium and its relation to sperm concentration in the epididymal plasma of the bull. J. Reprod. Fertil. 7:337-345.

Crain, J. L., and A. A. Luciano. 1983. Peritoneal fluid evaluation in infertility. Obstet. Gynecol. 61:159-164.

Crandall, B. F. 1981. Alpha-fetoprotein: A review. CRC Crit. Rev. Clin. Lab. Sci. 15:127-185.

Cree, J. E., J. Meyer, and D. M. Hailey. 1973. Diazepam in labour: Its metabolism and effect on the clinical condition and thermogenesis of the newborn. Br. Med. J. 4:251-255.

Croxatto, H. B., M. E. Ortiz, S. Diaz, R. Hess, J. Balmaceda, and H. D. Croxatto. 1978. Studies on the duration of egg transport by the human oviduct. II. Ovum location at various intervals following luteinizing hormone peak. Am. J. Obstet. Gynecol. 132:629-634.

Csapo, A. I., M. O. Pulkkinen, B. Ruttner, J. P. Sauvage, and W. G. Wiest. 1972. The significance of the human corpus luteum in pregnancy maintenance. I. Preliminary studies. Am. J. Obstet. Gynecol. 112:1061-1067.

Cumming, D. C., J. C. Yang, R. W. Rebar, and S. S. C. Yen. 1982. Treatment of hirsutism with spironolactone. J. Am. Med. Assoc. 247:1295-1298.

Cummins, J. M. 1976. Effects of epididymal occlusion on sperm maturation in the hamster. J. Exp. Zool. 197:183-190.

Cunha, G. R., L. W. K. Chung, J. M. Shannon, and B. A. Reese. 1980. Stromal-epithelial interactions in sex differentiation. Biol. Reprod. 22:19-42.

Cunha, G. R., R. M. Bigsby, P. S. Cooke, and Y. Sugimura. 1985. Stromal-epithelial interactions in adult organs. Cell Different. 17:137-148.

Curtis, G., and M. Fogel. 1970. Creatinine excretion: Diurnal variation and variability of whole and part-day measures. A methodologic issue in psychoendocrine research. Psychosom. Med. 32:337-350.

Cutler, G. B., Jr., and D. L. Loriaux. 1980. Adrenarche and its relationship to the onset of puberty. Fed. Proc. 39:2384-2390.

Czeizel, A. 1982. Epidemiological follow-up study on mutagenic effects in self-poisoning persons, pp. 639-646. In T. Sugimura, S. Kondo, and H. Takebe, Eds. Environmental Mutagens and Carcinogens. New York and Tokyo: Liss and University of Tokyo Press.

Czekala, N. M., J. K. Hodges, and B. L. Lasley. 1981. Pregnancy monitoring in diverse species by estrogen and bioactive luteinizing hormone determinations in small volumes of urine. J. Med. Primatol. 10:1-15.

Dacheux, J. L. 1977. Reinvestigation of the variation in total phospholipid content of the spermatozoa of the rat and the ram during epididymal transport. IRCS Medical Science 5:18.

Dacheux, J. L., and J. K. Voglmayr. 1983. Sequence of sperm cell surface differentiation and its relationship to exogenous fluid proteins in the ram epididymis. Biol. Reprod. 29:1033-1046.

Dacheux, J. L., T. O'Shea, and M. Paquignon. 1979. Effects of osmolality, bicarbonate and buffer on the metabolism and motility of testicular, epididymal and ejaculated spermatozoa of boars. J. Reprod. Fertil. 55:287-296.

da Cunha, M. F., M. L. Meistrich, M. M. Haq, L. A. Gordon, and A. J. Wyrobek. 1982. Temporary effects of AMSA (4'-(9-acridinylamino)methane sulfon-m-anisidide) chemotherapy on spermatogenesis. Cancer 49:2459-2462.

Daffos, F., M. Capella-Pavlovsky, and F. Forestier. 1985. Fetal blood sampling during pregnancy with use of a needle guided by ultrasound: A study of 606 consecutive cases. Am. J. Obstet. Gynecol. 153:655-660.

Dahl, K. D., N. M. Czekala, P. Lim, and A. J. Hsueh. 1987. Monitoring the menstrual cycle of humans and lowland gorillas based on urinary profiles of bioactive follicle-stimulating hormone and steroid metabolites. J. Clin. Endocrinol. Metab. 64:486-493.

Daly, D. C., I. A. Maslar, and D. H. Riddick. 1983a. Prolactin production during in vitro decidualization of proliferative endometrium. Am. J. Obstet. Gynecol. 145:672-678.

Daly, D. C., I. A. Maslar, and D. H. Riddick. 1983b. Term decidua response to estradiol and progesterone. Am. J. Obstet. Gynecol. 145:679-683.

Danzo, B. J., and B. C. Eller. 1975. Androgen binding to cytosol prepared from epididymides of sexually mature castrated rabbits: Evidence for a cytoplasmic receptor. Steroids 25:507-524.

Danzo, B. J., and B. C. Eller. 1979. The presence of a cytoplasmic estrogen receptor in sexually mature rabbit epididymides: Comparison with the estrogen receptor in immature rabbit epididymal cytosol. Endocrinology 105:1128-1134.

Danzo, B. J., M.-C. Orgebin-Crist, and D. O. Toft. 1973. Characterization of a cytoplasmic receptor for 5α-dihydrotestosterone in the caput epididymidis of intact rabbits. Endocrinology 92:310-317.

Danzo, B. J., T. G. Cooper, and M.-C. Orgebin-Crist. 1977. Androgen binding protein (ABP) in fluids collected from the rete testis and cauda epididymidis of sexually mature and immature rabbits and observations on morphological changes in the epididymis following ligation of the ductuli efferentes. Biol. Reprod. 17:64-77.

David, G., J. P. Bisson, F. Czyglik, P. Jouannet, and C. Gernigon. 1975. Anomalies morphologiques du spermatozoïde humain. 1. Propositions pour un système de classification. J. Gynecol. Obstet. Biol. Reprod. 4:17-36.

Davies, J., and S. R. Glasser. 1968. Histological and fine structural observations on the placenta of the rat. Acta Anat. (Basel) 69:542-608.

Davies, M. 1985. Antigenic analysis of immune complexes formed in normal human pregnancy. Clin. Exp. Immunol. 61:406–415.

Davies, M., and C. M. Browne. 1985. Anti-trophoblast antibody responses during normal human pregnancy. J. Reprod. Immunol. 7:285-297.

Davis, M. 1984. The mammalian startle response, pp. 287-351. In R. C. Eaton. Ed. Neural Mechanisms of Startle Behavior. New York: Plenum Press.

Davis, M., and G. K. Aghajanian. 1976. Effects of apomorphine and haloperidol on the acoustic startle response. Psychopharmacology 47:217-223.

Davis, M., and D. I. Astrachan. 1981. Spinal modulation of acoustic startle: Opposite effects of clonidine and d-amphetamine. Psychopharmacology 75:219-225.

Davis, M., and M. H. Sheard. 1976. p-Chloroamphetamine (PCA): Acute and chronic effects on habituation and sensitization of the acoustic startle response in the rat. Eur. J. Pharmacol. 35:261-273.

Davison, J. M., and F. E. Hytten. 1974. Glomerular filtration during and after pregnancy. J. Obstet. Gynaecol. Br. Commonw. 81:588-595.

de Ikonicoff, L. K. and L. Cedard. 1973. Localization of human chorionic gonadotropic and somatomammotropic hormones by the peroxidase immunohistoenzymologic method in villi and amniotic epithelium of human placentas (from six weeks to term). Am. J. Obstet. Gynecol. 116:1124-1132.

de Jong, F. H. 1979. Inhibin-fact or artifact. Mol. Cell. Endocrinol. 13:1-10.

de Jong, F. H. 1987. Inhibin—its nature, site of production and function. Oxford Rev. Reprod. Biol. 9:1-53.

DeKlerk, D. P., D. S. Coffey, L. L. Ewing, I. R. McDermott, W. G. Reiner, C. H. Robinson, W. W. Scott, J. D. Strandberg, P. Talalay, P. C. Walsh, L. G. Wheaton, and B. R. Zirkin. 1979. Comparison of spontaneous and experimentally induced canine prostatic hyperplasia. J. Clin. Invest. 64:842-849.

DeLap, L. W., S. S. Tate, and A. Meister. 1977. Gammaglutamyl transpeptidase and related enzyme activities in the reproductive system of the male rat. Life Sci. 20:673-680.

Delehanty, J., R. L. White and M. L. Mendelsohn. 1986. Approaches to determining mutation rates in human DNA. Mutat. Res. 167:215-232.

Delongeas, J. L., J. L. Gelly, R. Hatier, and G. Grignon. 1984. Postnatal evolution of oxidative metabolism of the epididymis in rats. C.R. Acad. Sci. [III] 298:19-22. (In French)

Del Rio, A. G., and R. Raisman. 1978. cAMP in spermatozoa taken from different segments of rat epididymis. Experientia 34:670-671.

De Luca, H. F. 1978. The hormonal nature of vitamin D function, pp. 249-270. In J. Dumont and J. Nunez, Eds. Hormones and Cell Regulation, Volume 2. Amsterdam: Elsevier/North-Holland.

Del Zompo, M., F. Bernardi, U. Bonuccelli, R. Maggio, M. Bajorek, M. Arnone, and G. U. Corsini. 1986.

Properties of [^{3}H]-MPTD binding sites in human blood platelets. Life Sci. 39:1855-1891.

Dempsey, J. L., R. S. Seshadri, and A. A. Morley. 1985. Increased mutation frequency following treatment with cancer chemotherapy. Cancer Res. 45:2873-2877.

Denison, M. S., and C. F. Wilkinson. 1985. Identification of the Ah receptor in selected mammalian species and induction of aryl hydrocarbon hydroxylase. Eur. J. Biochem. 147:429-435.

Denison, M. S., L. M. Vella, and A. B. Okey. 1986. Structure and function of the Ah receptor for 2,3,7,8-tetrachlorodibenzo-p-dioxin. Species difference in molecular properties of the receptors from mouse and rat hepatic cytosols. J. Biol. Chem. 261:3987-3995.

Denomme, M. A., K. Homonko, T. Fujita, T. Sawyer, and S. Safe. 1986. Substituted polychlorinated dibenzufuran receptor binding affinities and aryl hydrocarbon hydroxylase induction potencies—a QSAR analysis. Chem. Biol. Interact. 57:175-187.

De Rosis, F., S. P. Anastasio, L. Selvaggi, A. Beltrame, and G. Morianio. 1985. Female reproductive health in two lamp factories: Effects of exposure to inorganic mercury vapour and stress factors. Br. J. Ind. Med. 42:488-494.

Derue, G. J., H. G. Englert, E. N. Harris, A. E. Gharavi, S. H. Morgan, and R. G. Hull. 1985. Fetal loss in systemic lupus: Association with anticardiolipin antibodies. J. Obstet. Gynaecol. 5:207-209.

Desmond, M. M., R. P. Schwanecke, G. S. Wilson, S. Yasunaga, and I. Burgdorff. 1972. Maternal barbiturate utilization and neonatal withdrawal symptomatology. J. Pediatr. 80:190-197.

DeVore, G. R., and J. C. Hobbins. 1984. Antenatal diagnosis of congenital structural anomalies with ultrasound, pp. 1-55. In R. W. Beard and P. W. Nathanielsz, Eds. Fetal Physiology and Medicine, 2nd ed. New York: Marcel Dekker.

DeWitte, D. B., M. K. Buick, S. E. Cyran, and M. J. Maisels. 1984. Neonatal pancytopenia and severe combined immunodeficiency associated with antenatal administration of azathioprine and prednisone. J. Pediatr. 105:625-628.

Dickmann, Z., and S. K. Dey. 1974. Steroidogenesis in preimplantation rat embryo and its possible influence on morula-blastocyst transformation and implantation. J. Reprod. Fertil. 37:91-93.

Dixon, R. L. 1985. Reproductive Toxicology. New York: Raven Press. 350 pp.

diZerega, G. S., and G. D. Hodgen. 1980. The primate ovarian cycle: Suppression of human menopausal gonadotropin-induced follicular growth in the presence of the dominant follicle. J. Clin. Endocrinol. Metab. 50:819-825.

Dobson, R. L., and J. S. Felton. 1983. Female germ cell loss from radiation and chemical exposures. Am. J. Ind. Med. 4:175-190.

Donnez, J., S. Langerock, and K. Thomas. 1982. Peritoneal fluid volume and 17 beta-estradiol and progesterone concentrations in ovulatory, anovulatory, and postmenopausal women. Obstet. Gynecol. 59:687-692.

Dorfman, R. I., and R. A. Shipley. 1956. Androgens: Biochemistry, Physiology and Clinical significance. New York: John Wiley & Sons. 590 pp.

Dorrington, J. H., Y. S. Moon, and D. T. Armstrong. 1975. Estradiol-17 beta biosynthesis in cultured granulosa cells from hypophysectomized immature rats: Stimulation by follicle-stimulating hormone. Endocrinology 97:1328-1331.

Doshi, M. L. 1986. Accuracy of consumer performed in-home tests for early pregnancy detection. Am. J. Public Health 76:512-514.

Dott, H. M., and J. T. Dingle. 1968. Distribution of lysosomal enzymes in the spermatozoa and cytoplasmic droplets of bull and ram. Exp. Cell Res. 52:523-540.

Dougherty, R. C., M. J. Whitaker, S.-Y. Tang, R. Bottcher, M. Keller, and D. W. Kuehl. 1981. Sperm density and toxic substances: A potential key to environmental health hazards, pp. 263-278. In J. D. McKinney, Ed. Environmental Health Chemistry: The Chemistry of Environmental Agents as Potential Human Hazards. Ann Arbor, Mich.: Ann Arbor Science.

Douglas, C. P., J. S. Garrow, and E. W. Pugh. 1970. Investigation into the sugar content of endometrial secretion. J. Obstet. Gynaecol. Br. Commonw. 77:891-894.

Doull, J. 1980. Factors influencing toxicology, pp. 70-83. In J. Doull, C. D. Klaassen, and M. O. Amdur, Eds. Casarett and Doull's Toxicology: The Basic Science of Poisons, 2nd ed. New York: Macmillan.

Dreskin, R. B., S. S. Spicer, and W. B. Greene. 1970. Ultrastructural localization of chorionic gonadotropin in human term placenta. J. Histochem. Cytochem. 18:862-874.

Duc, H. T., A. Masse, P. Bobe, R. G. Kinsky, and G. A. Voisin. 1985. Deviation of humoral and cellular alloimmune reactions by placental extracts. J. Reprod. Immunol. 7:27-39.

Dudley, K., J. Potter, M. F. Lyon, and K. R. Willison. 1984. Analysis of male sterile mutations in the mouse using haploid stage expressed cDNA probes. Nucleic Acids Res. 12:4281-4293.

Dym, M., H. G. M. Raj, and H. E. Chemes. 1977. Response of the testis to selective withdrawal of LH or FSH using antigonadotropic sera, pp. 97-124. In P. Troen and H. R. Nankin, Eds. The Testis in Normal and Infertile Men. New York: Raven Press.

Dyson, A. L., and M.-C. Orgebin-Crist. 1973. Effects of hypophysectomy, castration and androgen replacement upon the fertilizing ability of rat epididymal spermatozoa. Endocrinology 93:391-402.

Eddy, E. M., R. B. Vernon, C. H. Muller, A. C. Hahnel, and B. A. Fenderson. 1985. Immunodissection of sperm surface modifications during epididymal maturation. Am. J. Anat. 174:225-237.

Eddy, E. M. 1987. The spermatozoon. In E. Knobil

and J. D. Neill. The Physiology of Reproduction. New York: Raven Press.

Edmonds, D. K., K. S. Lindsay, J. F. Miller, E. Williamson, and P. J. Wood. 1982. Early embryonic mortality in women. Fertil. Steril. 38:447-453.

Edwards, O. M., R. I. S. Bayliss, S. Millen. 1969. Urinary creatinine excretion as an index of the completeness of 24-hour urine collections. Lancet 2:1165-1166.

Edwards, R. G., P. C. Steptoe, and J. M. Purdy. 1980. Establishing full-term human pregnancies using cleaving embryos grow in vitro. Br. J. Obstet. Gynaecol. 87:737-756.

Egozcue, J., C. Templado, F. Vidal, J. Navarro, F. Morer-Fargas, and S. Marina. 1983. Meiotic studies in a series of 1100 infertile and sterile males. Hum. Genet. 65:185-188.

Ehrenberg, L. 1988. Dose monitoring and cancer risk, pp. 23-31. In H. Bartsch, K. Hemminki, and I. K. O'Neill, Eds. Methods for Detecting DNA Damaging Agents in Humans: Applications in Cancer Epidemiology and Prevention. IARC Scientific Publications, No. 89. Lyon: International Agency for Research on Cancer.

Ehrenberg, L., and S. Osterman-Golkar. 1980. Alkylation of macromolecules for detecting mutagenic agents. Teratogenesis Carcinog. Mutagen. 1:105-127.

Ehrhardt, A. A., and H. F. L. Meyer-Bahlburg. 1979. Psychosexual development: An examination of the role of prenatal hormones, pp. 41-50. In Sex, Hormones and Behaviour. CIBA Foundation Symposium 62 (New Series). Amsterdam: Excerpta Medica.

Ehrlich, P. H., W. R. Moyle, Z. A. Moustafa, and R. E. Canfield. 1982. Mixing two monoclonal antibodies yields enhanced affinity for antigen. J. Immunol. 128:2709-2713.

Ehrlich, P. H., Z. A. Moustafa, A. Krichevsky, S. Birken, E. G. Armstrong, and R. E. Canfield. 1985. Characterization and relative orientation of epitopes for monoclonal antibodies and antisera to human chorionic gonadotropin. Am. J. Reprod. Immunol. Microbiol. 8:48-54.

Ehrström, C. 1987. Fetal movement monitoring in nomal and high-risk pregnancy. Acta. Obstet. Gynecol. 80(Suppl.):1-32.

Elce, J. S., E. J. Graham, G. Zboril, L. Leyton, E. Perez, H. B. Croxatto, and A. de Ioannes. 1986. Monoclonal antibodies to bovine and human acrosin. Biochem. Cell Biol. 64:1242-1248.

El Etreby, M. F. 1980. The role of contraceptive steroids in the pathogenesis of pituitary tumours in various experimental animals and in man, pp. 211-2212. In G. Faglia, M. A. Giovanelli, and R. M. MacLeod, Eds. Pituitary Microadenomas. Proceedings of the Serono Symposia, Vol. 29. New York: Academic Press.

Eliasson, R. 1975. Analysis of semen, pp. 691-713. In S. J. Behrman and R. W. Kistner, Eds. Progress in Infertility, 2nd ed. Boston: Little, Brown & Company.

Eliasson, R. 1981. Analysis of semen, pp. 381-399. In

H. Burger and D. de Kretser, Eds. The Testis. New York: Raven Press.

Eliasson, R. 1982. Biochemical analysis of human semen. Int. J. Androl. (Suppl. 5):109-119.

Eliasson, R. 1983. Morphological and chemical methods of semen analysis for quantitating damage to male reproductive function in man, pp. 263-275. In V. B. Vouk and P. J. Sheehan, Eds. Methods for Assessing the Effects of Chemicals on Reproductive Function. SCOPE 20. New York: John Wiley & Sons.

Eliasson, R., and N. Virji. 1985. LDH-C4 in human seminal plasma and its relationships to testicular function. II. Clinical aspects. Int. J. Androl. 8:201-214.

Elinder, C.-G., G. Oberdörster, and L. Gerhardsson. In press. In T. W. Clarkson, L. Frieberg, G. Nordberg, and P. R. Sager, Eds. Biological Monitoring of Metals. New York: Plenum Press.

Ellis, G. B., and C. Desjardins. 1984. Mapping episodic fluctuations in plasma LH in orchidectomized rats. Am. J. Physiol. 247:E130-E135.

Enders, A. C. 1965. Formation of syncytium from cytotrophoblast in the human placenta. Obstet. Gynecol. 25:378-386.

Enders, A. C., D. J. Chavez, and S. Schlafke. 1981. Comparison of implantation in utero and in vitro, pp. 365-382. In S. R. Glasser and D. W. Bullock, Eds. Cellular and Molecular Aspects of Implantation. New York: Plenum Press.

Eng, L. A., and G. Oliphant. 1978. Rabbit sperm reversible decapacitation by membrane stabilization with a highly purified glycoprotein form seminal plasma. Biol. Reprod. 19:1083-1094.

Evans, H. J. 1982. Sister chromatid exchanges and disease states in man, pp. 183-228. In S. Wolff, Ed. Sister Chromatid Exchange. New York: John Wiley & Sons.

Evans, R. W., and B. P. Setchell. 1979. Lipid changes during epididymal maturation in ram spermatozoa collected at different times of the year. J. Reprod. Fertil. 57:197-203.

Evans, W. E., J. J. Schentag, and W. J. Jusko. 1986. Applied Pharmacokinetics, Principles of Therapeutic Drug Monitoring. 2nd ed. Spokane, Wash.: Applied Therapeutics. 1272 pp.

Evenson, D. P., Z. Darzynkiewicz, and M. R. Melamed. 1980. Relation of mammalian sperm chromatin heterogeneity to fertility. Science 210(4474):1131-1133.

Everson, R. B. 1987. A review of approaches to the detection of genetic damage in the human fetus. Environ. Health Perspect. 74:109-117.

Everson, R. B., E. Randerath, R. M. Santella, R. C. Cefalo, T. A. Avitts, K. Randerath. 1986. Detection of smoking-related covalent DNA adducts in human placenta. Science 231:54-57.

Everson, R. B., E. Randerath, T. A. Avitts, H. A. J. Schut, and K. Randerath. 1987. Preliminary investigations of tissue specificity, species specificity, and strategies for identifying chemicals causing DNA adducts in human placenta. Prog. Exp. Tumor Res. 31:86-103.

Everson, R. B., E. Randerath, R. M. Santella, T. A. Avitts, I. B. Weinstein, and K. Randerath. 1988. Quantitative associations between DNA damage in human placenta and maternal smoking and birth weight. J. Natl. Cancer Inst. 80:567-576.

Ewing, L. L., and B. R. Zirkin. 1983. Leydig cell structure and steroidogenic function. Recent Prog. Horm. Res. 39:599-635.

Ewing, L. L., C. Desjardins, D. C. Irby, and B. Robaire. 1977. Synergistic interaction of testosterone and oestradiol inhibits spermatogenesis in rats. Nature 269:409-411.

Ewing, L. L., J. C. Davis, and B. R. Zirkin. 1980. Regulation of testicular function: A spatial and temporal view. Int. Rev. Physiol. 22:41-115.

Ewing, L. L., B. R. Zirkin, and C. Chubb. 1981. Assessment of testicular testosterone production and Leydig cell structure. Environ. Health Perspect. 38:19-27.

Fabre, F., S. Carreau, and M. A. Drosdowsky. 1979. Androgen binding protein (ABP) was demonstrated in testicular and epididymal cytosols of adult guinea pig. Ann. Endocrinol. 40:15-16.

Fabro, S., and A. R. Scialli. 1986. Drug and Chemical Action in Pregnancy: Pharmacologic and Toxicologic Principles. New York: Marcel Dekker. 524 pp.

Fain-Maurel, M. A., J. P. Dadoune, and J. F. Reger. 1983. Surface changes in monkey spermatozoa during epididymal maturation and after ejaculation, pp. 159-162. In J. André, Ed. The Sperm Cell: Fertilizing Power, Surface Properties, Motility, Nucleus and Acrosome, Evolutionary Aspects. The Hague: Nijhoff.

Falck, B. 1959. Site of production of oestrogen in rat ovary as studied in micro-transplants. Acta Physiol. Scand. 47(Suppl. 163):1-101.

Faulk, W. P. 1983. Idiopathic spontaneous abortion [editorial]. Am. J. Reprod. Immunol. 3:48-49.

Faulk, W. P., and B.-L. Hsi. 1983. Immunobiology of human trophoblast membrane antigens, pp. 535-570. In Y. W. Loke and A. Whyte, Eds. Biology of Trophoblast. Amsterdam: Elsevier.

Faulk, W. P., and E. McCrady. 1983. Role of the extraembryonic membranes in transplantation analogies of human pregnancy, pp. 215-227. In R. Duncan and M. Weston Smith, Eds. The Encyclopedia of Ignorance, Vol. 3. Oxford: Pergamon Press.

Faulk, W. P., and J. A. McIntyre. 1981. Trophoblast survival. Transplantation 32:1-5.

Faulk, W. P., and J. A. McIntyre. 1983. Immunological studies of human trophoblast: Markers, subsets and functions. Immunol. Rev. 75:139-175.

Faulk, W. P., and J. A. McIntyre. 1986. Role of anti-TXL antibody in human pregnancy, pp. 106-114. In D. A. Clark and B. A. Croy, Eds. Reproductive Immunology. Proceedings of the 3rd International Congress of Reproductive Immunology held in Toronto, Canada, 1-5 July 1986. Amsterdam: Elsevier Science.

Faulk, W. P., M. Jeannet, W. D. Creighton, and A. Carbonara. 1974. Immunological studies of human placenta. Characterization of immunoglobulins on trophoblastic basement membranes. J. Clin. Invest. 54:1011-1019.

Faulk, W. P., A. Temple, R. E. Lovins, and N. Smith. 1978. Antigens of human trophoblast: A working hypothesis for their role in normal and abnormal pregnancies. Proc. Natl. Acad. Sci. USA 75:1947-1951.

Faulk, W. P., C. Yeager, J. A. McIntyre, and M. Ueda. 1979. Oncofetal antigens of human trophoblast. Proc. R. Soc. Lond. (B) 206:163-182.

Faulk, W. P., J. A. McIntyre, and B. L. B. Hsi. 1980. Transplantation analogies of the materno-fetal relationship in human pregnancy, pp. 143-150. In J. L. Touraine, J. Traeger, H. Betuel, J. Brochier, J. M. Dubernard, J. P. Revillard, and R. Triau, Eds. Transplantation and Clinical Immunology, Vol. 11. Amsterdam: Excerpta Medica.

Faulk, W. P., B. L. Hsi, J. A. McIntyre, C. J. G. Yeh, and A. Mucchielli. 1982. Antigens of human extraembryonic membranes. J. Reprod. Fertil. (Suppl. 31):181-199.

Faulk, W. P., C. B. Coulam, and J. A. McIntyre. 1987. Reproductive Immunology: Biomarkers of compromised pregnancies. Environ. Health Perspect. 74:119-127.

Faulk, W. P., D. S. Torry, and J. A. McIntyre. 1988a. Effects of serum versus plasma agglutination of antibody-coated indicator cells by human rheumatoid factors. Clin. Immunol. Immunopathol. 46:169-176.

Faulk, W. P., B. L. Hsi, C. J. G. Yeh, J. A. McIntyre, and P. J. Stevens. 1988b. Epidermolysis bullosa letalis: An immunogenetic disease of extraembryonic ectoderm? Am. J. Obstet. Gynecol. 158:150-157.

Fawcett, D. W. 1975. Ultrastructure and function of the Sertoli cell, pp. 21-55. In D. W. Hamilton and R. O. Greep, Eds. Handbook of Physiology, Section 7: Endocrinology. Volume V. Male Reproductive System. Washington, D.C.: American Physiological Society.

Fawcett, D. W., and R. D. Hollenberg. 1963. Changes in the acrosome of guinea pig spermatozoa during passage through the epididymis. Z. Zellforsch. Mikrosk. 60:276-292.

Fawcett, D. W., and D. M. Phillips. 1969. Observations on the release of spermatozoa and on changes in the head during passage through the epididymis. J. Reprod. Fertil. (Suppl. 6):405-418.

Fechter, L. D. 1974. Central serotonin involvement in the elaboration of the startle reaction in rats. Pharmacol. Biochem. Behav. 2:161-172.

Fédération CECOS (Fédération des Centres d'Etude et de Conservation de Sperme Humain), D. Schwartz, and M. J. Mayaux. 1982. Female fecundity as a function of age. Results of artificial insemination in 2193 nulliparous women with azoospermic husbands. N. Engl. J. Med. 306:404-406.

Felicio, L. S., J. F. Nelson, and C. E. Finch. 1986. Prolongation and cessation of estrous cycles in aging C57BL/6J mice are differentially regulated events. Biol. Reprod. 34:849-858.

Fenech, M., and A. A. Morley. 1986. Cytokinesis-block micronucleus method in human lymphocytes: Effect of in vivo ageing and low dose X-irradiation. Mutat. Res. 161:193-198.

Ferguson-Smith, M. A. 1983. Prenatal chromosome analysis and its impact on the birth incidence of chromosome disorders. Br. Med. Bull. 39:355-364.

Ferrari, C., and P. G. Crosignani. 1986. Medical treatment of hyperprolactinemic disorders. Hum. Reprod. 1:507-514.

Feuers, R. J., and J. B. Bishop. 1986. The response of kinetic parameters from selected enzymes to variant loci in congenic mice and implications for in vivo biochemical mutation tests, pp. 375-382. In C. Ramel, B. Lambert, and J. Magnusson, Eds. Genetic Toxicology of Environmental Chemicals, Part B: Genetic Effects and Applied Mutagenesis. Proceedings of the Fourth International Conference on Environmental Mutagens held in Stockholm, Sweden, June 24-28, 1985. New York: Liss.

Fiddes, J. C., and H. M. Goodman. 1981. The gene encoding the common alpha subunit of the four human glycoprotein hormones. J. Mol. Appl. Genet. 1:3-18.

Finch, C. E. 1976. The regulation of physiological changes during mammalian aging. Q. Rev. Biol. 51:49-83.

Finch, C. E. 1980. The relationships of aging changes in the basal ganglia to manifestations of Huntington's chorea. Ann. Neurol. 7:406-411.

Finch, C. E. 1981. Neural and endocrine mechanisms in aging: A synopsis, pp. 537-557. In R. T. Schmike, Ed. Biological Mechanisms in Aging. NIH Publ. No. 81-2194. Washington, D.C.: U.S. National Institutes of Health.

Finch, C. E., and R. G. Gosden. 1986. Animal models for the human menopause, pp. 3-34. In L. Mastroianni, Jr., and C. A. Paulsen, Eds. Aging, Reproduction, and the Climacteric. New York: Plenum Press.

Finch, C. E., L. S. Felicio, C. V. Mobbs, and J. F. Nelson. 1984. Ovarian and steroidal influences on neuroendocrine aging processes in female rodents. Endrocrinol. Rev. 5:467-497.

Fink, E., W. B. Schill, F. Fiedler, F. Krassnigg, R. Geiger, and K. Shimamoto. 1985. Tissue kallikrein of human seminal plasma is secreted by the prostate gland. Biol. Chem. Hoppe Seyler 366:917-924.

Firestone, P., and A. N. Prabhu. 1983. Minor physical anomalies and obstetrical complications: Their relationship to hyperactive, psychoneurotic and normal children and their families. J. Abnorm. Child Psychol. 11:207-216.

Fischer, S. G., and L. S. Lerman. 1983. DNA fragments differing by single base-pair substitutions are separated in denaturing gradient gels: Correspondence with melting theory. Proc. Natl. Acad. Sci. USA 80:1579-

1583.

Fishel, S. B. 1979. Analysis of mouse uterine proteins at pro-oestrus during early pregnancy and after administration of exogenous steroids. J. Reprod. Fertil. 55:91-100.

Flickinger, C. J. 1972. Alterations in the fine structure of the rat epididymis after vasectomy. Anat. Rec. 173:277-299.

Flickinger, C. J. 1982. The fate of sperm after vasectomy in the hamster. Anat. Rec. 202:231-239.

Foote, R. H. 1969. Research techniques to study reproductive physiology in the male, pp. 81-110. In Techniques and Procedures in Animal Science Research. Albany, N.Y.: American Society of Animal Science.

Forest, M. G., and J. Bertrand. 1972. Studies of the protein binding of dihydrotestosterone (17-beta-hydroxy-5 alpha-androstan-3 one) in human plasma in different physiological conditions and effect of medroxyprogesterone (17-hydroxy-6 alpha-methyl-4 pregnene-3,20 dione 17-acetate). Steroids 19:197-214.

Fournier-Delpech, S., G. Colas, M. Courot, R. Ortavant, and G. Brice. 1979. Epididymal sperm maturation in the ram: Motility, fertilizing ability and embryonic survival after uterine artificial insemination in the ewe. Ann. Biol. Anim. Biochim. Biophys. 19:597-605.

Fournier-Delpech, S., G. Colas, and M. Courot. 1981. The first cleavage of tubal sheep eggs after fertilization with epididymal or ejaculated spermatozoa. C.R. Acad. Sci. III 292:515-517. (In French)

Fowle, J. R., III. 1984. Workshop Proceedings: Approaches to Improving the Assessment of Human Genetic Risk—Human Biomonitoring. Report No. EPA-600/9-84-016. Washington, D.C.: Office of Health and Environmental Assessment, U.S. Environmental Protection Agency. 40 pp.

Fowler, R. E., R. G. Edwards, D. E. Walters, S. T. H. Chan, and P. R. Steptoe. 1978. Steroidogenesis in preovulatory follicles of patients given human menopausal and chorionic gonadotropins as judged by the radioimmunoassay of steroids in follicular fluid. J. Endocrinol. 77:161-169.

Fox, D. A., and A. Wright. 1982. Evidence that low level developmental lead exposure produces toxic amblyopia. Soc. Neurosci. Abs. 8:81.

Fox, D. A., J. P. Lewkowski, and G. P. Cooper. 1977. Acute and chronic effects of neonatal lead exposure on development of the visual evoked responses in rats. Toxicol. Appl. Pharmacol. 40:449-461.

Fox, D. A., J. P. Lewkowski, and G. P. Cooper. 1979. Persistent visual cortex excitability alterations produced by neonatal lead exposure. Neurobehav. Toxicol. 1:101-106.

Franchimont, P., U. Gaspard, A. Reuter, and G. Heynen. 1972. Polymorphism of protein and polypeptide hormones. Clin. Endocrinol. (Oxford) 1:315-336.

Frankel, A. I., and K. B. Eik-Nes. 1970. The metabolism of steroids in the rabbit epididymis. Endocrinology 87:646-652.

Franks, P. A., N. K. Laughlin, D. J. Dierschke, R. E. Bowman, and P. A. Meller. 1986. Effects of lead on luteal function in rhesus monkeys. Biol. Reprod. 34(Suppl. 1):186.

Fredricsson, B. 1979. Morphologic evaluation of spermatozoa in different laboratories. Andrologia 11:57-61.

Freeman, J. M., Ed. 1985. Prenatal and Perinatal Factors Associated with Brain Disorders. NIH Pub. 85-1149. [Bethesda, Md.]: U.S. Department of Health and Human Services, National Institutes of Health.

Freemark, M., and S. Handwerger. 1982. Ovine placental lactogen stimulates amino acid transport in rat diaphragm. Endocrinology 110:2201-2203.

Frenkel, G. P., R. Kaplan, G. Yedwab, Z. T. Homonnai, and P. F. Kraicer. 1978. The effect of caffeine on rat epididymal spermatozoa: Motility, metabolism and fertilizing capacity. Int. J. Androl. 1:145-152.

Freund, M. 1966. Standards for the rating of human sperm morphology. A cooperative study. Int. J. Fertil. 11:97-180.

Freund, M. 1968. Semen analysis, pp. 593-627. In S. J. Behrman and R. W. Kistner, Eds. Progress in Infertility. Boston: Little, Brown and Company.

Friberg, L. 1983. Quality control in laboratories testing for environmental pollution, pp.811-829. In T. W. Clarkson, G. F. Nordberg, and P. R. Sager, Eds. Reproductive and Developmental Toxicity of Metals. New York: Plenum Press.

Fricker, H. S., and S. Segal. 1978. Narcotic addiction, pregnancy, and the newborn. Am. J. Dis. Child 132:360-366.

Fried, P. A. 1980. Marihuana use by pregnant women: Neurobehavioral effects in neonates. Drug Alcohol Depend. 6:415-424.

Friedman S. J., and P. Skehan. 1979. Morphological differentiation of human choriocarcinoma cells induced by methotrexate. Cancer Res. 39:1960-1967.

Friend, D. S., and N. B. Gilula. 1972. Variations in tight and gap junctions in mammalian tissues. J. Cell Biol. 53:758-776.

Fries, H., S. J. Nillius, and F. Pettersson. 1974. Epidemiology of secondary amenorrhea. II. A retrospective evaluation of etiology with special regard to psychogenic factors and weight loss. Am. J. Obstet. Gynecol. 118:473-479.

Friesen, H. G. 1966. Lactation induced by human placental lactogen and cortisone acetate in rabbits. Endocrinology 79:212-215.

Frigas, E., and G. J. Gleich. 1986. The eoesinophil and the pathophysiology of asthma. J. Allergy Clin. Immunol. 77:527-537.

Frisch, R. E. 1980. Pubertal adipose tissue: Is it necessary for normal sexual maturation? Evidence from the rat and human female. Fed. Proc. 39:2395-2400.

Frisch, R. E. 1985. Fatness, menarche, and female fertility. Perspect. Biol. Med. 28:611-63.

Frommer, D. J. 1964. Changing aged of the menopause. Br. Med. J. 5405:349-351.

Frost, J. K. 1974. Gynecologic and obstetric cytopathology, pp. 634-728. In E. R. Novak and J. D. Woodruff, Eds. Novak's Gynecologic and Obstetric Pathology, 7th ed. Philadephia: W. B. Saunders.

Fuccillo, D. A. 1985. Application of the Avidin-Biotin Technique in microbiology. Biotechniques 3:494-501.

Fujii, T., H. Nakanishi, S. Morimoto, and N. Hara. 1987. Pharmacological assessment of the functional effects of maternal exposure to drugs: Transmission of the effects to the offspring of subsequent generations, pp. 159-174. In T. Fujii and P. M. Adams, Eds. Functional Teratogenesis: Functional Effects on the Offspring After Parental Drug Exposure. Tokyo: Teikyo University Press.

Fujimoto, H., R. P. Erickson, M. Quinto, and M. P. Rosenberg. 1984. Post-meiotic transcription in mouse testes detected with spermatid cDNA clones. Biosci. Rep. 4:1037-1044.

Fujino, T., K. Gottlieb, D. K. Manchester, S. S. Park, D. West, H. L. Gurtoo, R. E. Tarone, and H. V. Gelboin. 1984. Monoclonal antibody phenotyping of interindividual differences in cytochrome P-450-dependent reactions of single and twin human placenta. Cancer Res. 44:3916-3923.

Fujita, S., M. Horii, T. Tanimura, and H. Nishimura. 1964. H^3-thymidine autoradiographic studies on cytokinetic responses to x-ray irradiation and to thio-TEPA in the neural tube of mouse embryos. Anat. Rec. 149:37-48.

Funes-Cravioto, F., C. Zapata-Gayon, B. Kolmodin-Hedman, B. Lambert, J. Lindsten, E. Norberg, M. Nordenskjold, R. Olin, and A. Swesson. 1977. Chromosome aberrations and sister-chromatid exchange in workers in chemical laboratories and a rotoprinting factory and in children of women laboratory workers. Lancet 2:322-325.

Gabel, C. A., E. M. Eddy, and B. M. Shapiro. 1979. After fertilization, sperm surface components remain as a patch in sea urchin and mouse embryos. Cell 18:207-215.

Gagnon, C., R. J. Sherins, D. M. Phillips, and C. W. Bardin. 1982. Deficiency of protein-carboxyl methylase in immotile spermatozoa of infertile men. N. Engl. J. Med. 306:821-825.

Gagnon, C., D. Harbour, E. De Lamirande, C. W. Bardin, and J.-L. Dacheux. 1984. Sensitive assay detects protein methylesterase in spermatozoa: Decrease in enzyme activity during epididymal maturation. Biol. Reprod. 30:953-958.

Gagnon, C., E. de Lamirande, and R. J. Sherins. 1986. Positive correlation between the level of protein-carboxyl methylase in spermatozoa and sperm motility. Fertil. Steril. 45:847-853.

Galen, R. S., and S. R. Gambino. 1975. Beyond Normality: The Predictive Value and Efficiency of Medical Diagnoses. New York: John Wiley & Sons. 237 pp.

Gallager, D. W., J. H. Kehne, E. A. Wakeman, and M.

Davis. 1983. Development changes in pharmacological responsivity of the acoustic startle reflex: Effects of picrotoxin. Psychopharmacology 79:87-93.

Ganjam, V. K., and R. P. Amann. 1976. Steroids in fluids and sperm entering and leaving the bovine epididymis, epididymidal tissue, and accessory sex gland secretions. Endocrinology 99:1618-1630.

Gardiner, M. R., and M. E. Narin. 1969. Studies on the effect of cobalt and selenium in clover disease of ewes. Aust. Vet. J. 45:215-222.

Gastel, B., J. E. Haddow, J. C. Fletcher, and A. Neale, Eds. 1980. Maternal Serum Alpha-Fetoprotein: Issues in the Prenatal Screening and Diagnosis of Neural Tube Defects. Rockville, Md.: U.S. Public Health Service, Office of Health Research, Statistics, and Technology. 201 pp.

Gaunt, S. J. 1983. Spreading of a sperm surface antigen within the plasma membrane of the egg after fertilization in the rat. J. Embryol. Exp. Morphol. 75:259-270.

Gay, V. L., and N. W. Dever. 1971. Effects of testosterone propionate and estradiol benzoate—alone or in combination—on serum LH and FSH in orchidectomized rats. Endocrinology 89:161-168.

Generoso, W. M., K. T. Cain, M. Krishna, and S. W. Huff. 1979. Genetic lesions induced by chemicals in spermatozoa and spermatids of mice are repaired in the egg. Proc. Natl. Acad. Sci. USA 76:435-437.

Generoso, W. M., J. C. Rutledge, K. T. Cain, L. A. Hughes, and P. W. Braden. 1987. Exposure of female mice to ethylene oxide within hours after mating leads to fetal malformation and death. Mutat. Res. 176:269-274.

Gerencer, M., A. Drazancic, I. Kuvacic, Z. Tomaskovic, and A. Kastelan. 1979. HLA antigen studies in women with recurrent gestational disorders. Fertil. Steril. 31:401-404.

Gibaldi, M., and D. Perrier. 1982. Pharmacokinetics, 2nd ed. New York: Marcel Dekker. 494 pp.

Gill, T. G., 3d. 1983. Immunogenetics of spontaneous abortions in humans. Transplantation 35:1-6.

Gillberg, C. 1977. Floppy Infant Syndrome and maternal diazepam [letter]. Lancet 2:244.

Ginsburg, K. A., K. S. Moghissi, and E. L. Abel. 1988. Computer-assisted human semen analysis. Sampling errors and reproducibility. J. Androl. 9:82-90.

Gizang-Ginsberg, E., and D. J. Wolgemuth. 1985. Localization of mRNAs in mouse testes by in situ hybridization: Distribution of α-tubulin and developmental stage specificity of pro-opiomelanocortin transcripts. Dev. Biol. 111:293-305.

Glasser, S. R. 1972. The uterine environment in implantation and decidualization, pp. 776-833. In H. A. Balin and S. R. Glasser, Eds. Reproductive Biology. Amsterdam: Excerpta Medica.

Glasser, S. R. 1975. A molecular bioassay for progesterone and related compounds, pp. 456-465. In B. W. O'Malley and J. Hardman, Eds. Methods in Enzymology, Vol. 36. Hormone Action, Part A, Steroid Hormones. New York: Academic Press.

Glasser, S. R. 1985. Laboratory models of implantation, pp. 219-238. In R. L. Dixon, Ed. Reproductive Toxicology. New York: Raven Press.

Glasser, S. R. 1986. Current concepts of implantation and decidualization, pp. 127-154. In G. Huszar, Ed. The Physiology and Biochemistry of the Uterus in Pregnancy and Labor. Boca Raton, Fla.: CRC Press.

Glasser, S. R., and J. H. Clark. 1975. A determinant role for progesterone in the development of uterine sensitivity to decidualizaton and ovo-implantation, pp. 311-345. In C. L. Markert and J. Papaconstantinou, Eds. The Developmental Biology of Reproduction. New York: Academic Press.

Glasser, S. R., and J. A. Julian. 1986. Intermediate filament of the midgestation rat trophoblast giant cell. Dev. Biol. 113:356-363.

Glasser, S. R., and J. A. Julian. In press. UCLA Symposium on Early Embryogenesis and Paracrine Signals. New York: Liss.

Glasser, S. R., and S. A. McCormack. 1979. Estrogen-modulated uterine gene transcription in relation to decidualization. Endocrinology 104:1112-1121.

Glasser, S. R., and S. A. McCormack. 1980. Functional development of rat trophoblast and decidual cells during establishment of the hemochorial placenta. Adv. Biosci. 25:165-197.

Glasser, S. R., and S. A. McCormack. 1981. Separated cell types as analytical tools in the study of decidualization and implantation, pp. 217-239. In S. R. Glasser and D. W. Bullock, Eds. Cellular and Molecular Aspects of Implantation. New York: Plenum Press.

Glasser, S. R., and S. A. McCormack. 1982. Cellular and molecular aspects of decidualization and implantation, pp. 245-310. In H. M. Beier and P. Karlson, Eds. Proteins and Steroids in Early Pregnancy. New York: Springer-Verlag.

Glasser, S. R., J. A. Julian, M. I. Munir, and M. J. Soares. 1987a. Biological markers during early pregnancy: Trophoblastic signals of the peri-implantation period. Environ. Health Perspect. 74:129-147.

Glasser, S. R., S. Lampelo, M. I. Munir, and J. Julian. 1987b. Expression of desmin, laminin and fibronectin during in situ differentiation (decidualization) of rat uterine stromal cells. Differentiation 35:132-142.

Glasser, S. R., J. A. Julian, G. L. Decker, J.-Y. Tang, and D. D. Carson. 1988. Development of morphological and functional polarity in primary cultures of immaturae rat uterine epithelial cells. J. Cell Biol. 107:2409-2423.

Gleich, G. J., E. Frigas, D. A. Loegering, D. L. Wassom, and D. Steinmuller. 1979. Cytotoxic properties of the eopsinophil major basic protein. J. Immunol. 123:2925-2927.

Glover, T. D. 1969. Some aspects of function in the epididymis. Experimental occlusion of the epididymis in the rabbit. Int. J. Fertil. 14:215-221.

Glover, T. D. 1982. The epididymis, pp. 544-555. In G. D. Chisholm and D. I. Williams, Eds. Scientific Foundations of Urology, 2nd ed. London: Heinemann Medical Books Ltd.

Gloyna, R. E., and J. D. Wilson. 1969. A comparative study of the conversion of testosterone to 17beta-hydroxy-5alpha-androstan-3-one (dihydrotestosterone) by prostate and epididymis. J. Clin. Endocrinol. Metab. 29:970-977.

Gold, B., H. Fujimoto, J. M. Kramer, R. P. Erickson, and N. B. Hecht. 1983. Haploid accumulation and translational control of phosphoglycerate kinase-2 messenger RNA during mouse spermatogenesis. Dev. Bio. 98:392-399.

Goldberg, E. 1977. Isozymes in testes and spermatozoa, pp. 79-124. In M. C. Rattazzi, J. C. Scandalios, and G. S. Whitt, Eds. Isozymes. Current Topics in Biological and Medical Research, Vol. 1. New York: Liss.

Goldberg, E., D. Sberna, T. E. Wheat, G. J. Urbanski, and E. Margoliash. 1977. Cytochrome c: Immunofluorescent localization of the testis-specific form. Science 196:1010-1012.

Golditch, I. M. 1972. Post contraceptive amenorrhea. Obstet. Gynecol. 39:903-908.

Goldman, R., R. Woodcock, and M. Fristoe. 1970. Test of Auditory Discrimination. Circle Pines, Minn.: American Guidance Service Corporation.

Gonder, J. C., R. A. Proctor, and J. A. Will. 1985. Genetic differences in oxygen toxicity are correlated with cytochrome P-450 inducibility. Proc. Natl. Acad. Sci. USA 82:6315-6319.

Gonzalez Echeverria, F., P. S. Cuasnicu, A. Piazza, L. Pineiro, and J. A. Blaquier. 1984. Addition of an androgen-free epididymal protein extract increases the ability of immature hamster spermatozoa to fertilize in vivo and in vitro. J. Reprod. Fertil. 71:433-437.

Goodman, A. L., and G. D. Hodgen. 1977. Systemic versus intraovarian progesterone replacement after luteectomy in rhesus monkeys: Differential patterns of gonadotropins and follicle growth. J. Clin. Endocrinol. Metab. 45:837-840.

Goodman, A. L., W. E. Nixon, D. K. Johnson, and G. D. Hodgen. 1977. Regulation of folliculogenesis in the cycling rhesus monkey: Selection of the dominant follicle. Endocrinology 100:155-161.

Gordon, D. L., D. J. Moore, T. Thorslund, and C. A. Paulsen. 1965. The determination of size and concentration of human sperm with an electronic particle counter. J. Lab. Clin. Med. 65:506-512.

Gordon, D. L., J. E. Herrigel, D. J. Moore, and C. A. Paulsen. 1967. Efficacy of Couler Counter in determining low sperm concentrations. Am. J. Clin. Pathol. 47:226-228.

Gorer, P. A., Z. B. Mikulska, and P. O'Gorman. 1959. The time of appearance of isoantibodies during the homograft response to mouse tumours. Immunology 2:211-218,

Gorski, R. A. 1968. Influence of age on the response to paranatal administration of a low dose of androgen. Endocrinology 82:1001-1004.

Gorski, R. A. 1971. Gonadal hormones and the perinatal development of neuroendocrine function, pp. 237-290. In L. Martini and W. F. Ganong, Eds. Frontiers in Neuroendocrinology, 1971. New York: Oxford University Press.

Gosden, R. G. 1985. The Biology of Menopause: The Causes and Consequences of Ovarian Aging. Orlando, Fla.: Academic Press. 188 pp.

Gosden, R. G., S. C. Laing, L. S. Felicio, J. F. Nelson, and C. E. Finch. 1983. Imminent oocyte exhaustion and reduced follicular recruitment mark the transition to acyclicity in aging C57BL/6J mice. Biol. Reprod. 28:255-260.

Gottlieb, K. A., and D. K. Manchester. 1986. Twin study methodology and variability in xenobiotic placental metabolism. Teratogenesis Carcinog. Mutagen. 6:253-263.

Gougeon, A. 1982. Rate of follicular growth in the human ovary, pp. 155-163. In R. Rolland, E. V. Van Hall, S. G. Hillier, K. P. McNatty, and J. Schoemaker, Eds. Follicular Maturation and Ovulation: Proceedings of the IVth Reinier de Graaf Symposium, Nijmegen, Aug. 20-22, 1981. International Congress Series, 560. Amsterdam: Excerpta Medica.

Gravanis, A., J. R. Zorn, G. Tanguy, C. Nessmann, L. Cedard, and P. Robel. 1984. The "dysharmonic luteal phase" syndrome: Endometrial progesterone receptor and estradiol dehydrogenase. Fertil. Steril. 42:730-736.

Green, B. B., J. R. Daling, N. S. Weiss, J. M. Liff, and T. Koepsell. 1986. Exercise as a risk factor for infertility with ovulatory dysfunction. Am. J. Public Health 76:1432-1436.

Green, J. E., A. Dorfmann, S. Jones, S. Bender, L. Patton, and J. D. Schulman. 1988. Chorionic villus sampling: Experience with an initial 940 cases. Obstet. Gynecol. 71:208-212.

Greenberg, J., and W.-G. Forssmann. 1983. Studies of the guinea pig epididymis. II. Intercellular junctions of principal cells. Anat. Embryol. 168:195-209.

Griffin, J. E., III, and J. D. Wilson. 1987. Disorders of the testis, pp. 1907-1818. In Harrison's Principles of Internal Medicine, 11th ed. New York: McGraw Hill.

Grinnell, F., J. R. Head, and J. Hoffpauir. 1982. Fibronectin and cell shape in vivo: studies on the endometrium during pregnancy. J. Cell Biol. 94:597-606.

Griswold, M. D. 1988. Protein secretion of sertoli cells. Int. Rev. Cytol. 110:133-156.

Groll-Knoff, E., M. Haider, H. Hoeller, H. Jenkner, and H. G. Stidl. 1978. Neuro- and psychophysiological effects of moderate carbon monoxide exposure, pp. 424-430. In D. A. Otto, Ed. Multidisciplinary Perspectives in Event Related Brain Potential Research. EPA 600/9-77-043. Washington, D.C.: U.S. Environmental Protection Agency.

Grollman, S. 1967. The Human Body. Its Structure and Physiology. New York: Macmillan. 543 pp.

Grumbach, M. M., J. C. Roth, S. L. Kaplan, and R. P. Kelch. 1974. Hypothalamic-pituitary regulation of puberty in man: Evidence and concepts derived from clinical research, pp. 115-166. In M. M. Grumbach, G. D. Grave, and F. E. Mayer, Eds. Control of the Onset of Puberty. New York: John Wiley & Sons.

Guerrero, R., T. Aso, P. F. Brenner, Z. Cekan, B. M. Landgren, K. Hagenfeldt, and E. Diczfalusy. 1976. Studies in the pattern of circulating steroids in the normal menstrual cycle. I. Simultaneous assays of progesterone, pregnenolone, dehydroepiandrosterone, testosterone, dihydrotestosterone, androstenedione, oestradiol and oestrone. Acta Endocrinol. 81:133-149.

Guesdon, J. L., C. Jouanne, and S. Avrameas. 1983. Use of antibody conjugates in enzyme and erythro-immunoassay, pp. 197-205. In S. Avrameas, P. Druet, R. Masseyeff, and G. Feldmann, Eds. Immunoenzymatic Techniques. Proceedings of the Second International Symposium on Immunoenzymatic Techniques held in Cannes, France, 16-18 March 1983. Amsterdam: Elsevier.

Guillebaud, J., I. S. Fraser, G. D. Thorburn, and G. Jenken. 1977. Endocrine effects of danazol in menstruating women. J. Int. Med. Res. 5(Suppl. 3):57-66.

Gunsalus, G. L., N. A. Musto, and C. W. Bardin. 1980. Bidirectional release of a Sertoli cell product, androgen binding protein, into the blood and seminiferous tubule, pp. 291-297. In A. Steinberger and E. Steinberger, Eds. Testicular Development, Structure, and Function. New York: Raven Press.

Gupta, C., and S. J. Yaffe. 1981. Reproductive dysfunction in female offspring after prenatal exposure to phenobarbital: Critical period of action. Pediatr. Res. 15:1488-1491.

Gupta, R. C., M. V. Reddy, and K. Randerath. 1982. ^{32}P-postlabeling analysis of non-radioactive aromatic carcinogen-DNA adducts. Carcinogenesis 3:1081-1092.

Gurney, M. E., B. R. Apatoff, G. T. Spear, M. J. Baumel, J. P. Antel, M. B. Bania, and A. T. Reder. 1986a. Neuroleukin: A lymphokine product of lectin-stimulated T cells. Science 234:574-581.

Gurney, M. E., S. P. Heinrich, M. R. Lee, and H. S. Yin. 1986b. Molecular cloning and expression of neuroleukin, a neurotrophic factor for spinal and sensory neurons. Science 234:566-574.

Gurtoo, H., C. J. Williams, K. Gottlieb, A. I. Mulhern, L. Caballes, J. B. Vaught, A. J. Marinello, and S. K. Bansal. 1983. Population distribution of placental benzo(a)pyrene metabolism in smokers. Int. J. Cancer 31:29-37.

Gusella, J. F., R. E. Tanzi, M. A. Anderson, W. Hobbs, K. Gibbons, R. Raschtchian, T. C. Gilliam, M. R. Wallace, N. S. Wexler, and P. M. Conneally. 1984. DNA markers for nervous system diseases. Science 225:1320-1326.

Guyda, H. J., and H. G. Friesen. 1973. Serum prolactin levels in humans from birth to adult life. Pediatr.

Res. 7:534-540.

Haas, G. G., Jr., D. B. Cines, and A. D. Schreiber. 1980. Immunologic infertility: Identification of patients with antisperm antibody. N. Engl. J. Med. 303:722-727.

Haddock, L., G. Lebron, R. Martinex, J. F. Cordero, L. W. Freni-Titulaer, F. Carrion, C. Cintron, and L. Gonzalez. 1985. Premature sexual development in Puerto Rico: Background and current status, pp. 358-379. In J. A. McLachlan, Ed. Estrogens in the Environment II: Influences on Development. New York: Elsevier.

Hafez, E. S. E., and M. R. N. Prasad. 1976. Functional aspects of the epididymis, pp. 31-43. In E. S. E. Hafez, Ed. Human Semen and Fertility Regulation in Men. St. Louis: C. V. Mosby.

Hahn, J., R. H. Foote, and E. T. Cranch. 1969. Tonometer for measuring testicular consistency of bulls to predict semen quality. J. Anim. Sci. 29:483-489.

Hales, B. F., C. Hachey, and B. Robaire. 1980. The presence and longitudinal distribution of the glutathione S-transferases in rat epididymis and vas deferens. Biochem. J. 189:135-142.

Hales, B. F., S. Smith, and B. Robaire. 1986. Cyclophosphamide in the seminal fluid of treated males: Transmission to females by mating and effect on pregnancy outcome. Toxicol. Appl. Pharmacol. 84:423-430.

Hall, J. L. 1981. Relationship between semen quality and human sperm penetration of zona-free hamster ova. Fertil. Steril. 35:457-463.

Halperin, D. S., and P. C. Sizonenko. 1983. Prepubertal gynecomastia following topical inunction of estrogen-containing ointment. Helv. Paediatr. Acta 38:361-366.

Halpern, B. N., T. Ky, and B. Robert. 1967. Clinical and immunological study of an exceptional case of reaginic type sensitization to human seminal fluid. Immunology 12:247-258.

Halvorson, C. F., Jr., and J. B. Victor. 1976. Minor physical anomalies and problem behavior in elementary school children. Child Dev. 47:281-285.

Hamilton, D. W. 1972. The mammalian epididymis, pp. 268-337. In H. Balin and S. Glasser, Eds. Reproductive Biology. Amsterdam: Excerpta Medica.

Hamilton, D. W. 1975. Structure and function of the epithelium lining the ductuli efferentes, ductus epididymidis, and ductus deferens in the rat, pp. 259-301. In D. W. Hamilton and R. O. Greep, Eds. Handbook of Physiology, Section 7: Endocrinology. Volume V. Male Reproductive System. Washington, D.C.: American Physiological Society.

Hamilton, D. W. 1981. Evidence for alpha-lactalbumin-like activity in reproductive tract fluids of the male rat. Biol. Reprod. 25:385-392.

Hamilton, D. W., and D. W. Fawcett. 1970. In vitro synthesis of cholesterol and testosterone from acetate by rat epididymis and vas deferens. Proc. Soc. Exp. Biol. Med. 133:693-695.

Handelsman, D. J., and S. Staraj. 1985. Testicular size: The effects of aging, malnutrition and illness. J. Androl. 6:144-151.

Handley, S. L., and K. V. Thomas. 1979. Potentiation of startle response by d- and l-amphetamine: The possible involvement of pre- and postsynaptic alpha-adrenoreceptors and other transmitter systems in the modulation of a tactile startle response. Psychopharmacology 64:105-111.

Hanew, K., A. Sasaki, S. Sato, M. Goh, and K. Yoshinaga. 1987. Growth hormone inhibitory and stimulatory action of L-dopa in patients with acromegaly. J. Clin. Endocrinol. Metab. 64:255-260.

Haney, A. F., W. S. Maxon, and D. W. Schomberg. 1986. Compartmental ovarian steroidogenesis in polycystic ovary syndrome. Obstet. Gynecol. 68:638-644.

Hansen, P. J., F. W. Bazer, and R. M. Roberts. 1985. Appearance of ß-hexosaminidase and other lysosomal-like enzymes in the uterine lumen of gilts, ewes and mares in response to progesterone and oestrogens. J. Reprod. Fertil. 73:411-424.

Hanson, J. W., H. J. Hoffman, and G. T. Ross. 1975. Monthly gonadotropin cycles in premenarcheal girls. Science 190:161-163.

Hanson, J. W., N. C. Myrianthopoulos, M. A. Harvey, and D. W. Smith. 1976. Risks to the offspring of women treated with hydantoin anticonvulsants, with emphasis on the fetal hydantoin syndrome. J. Pediatr. 89:662-668.

Hansson, V., and O. Djoseland. 1972. Preliminary characterization of the 5α-dihydrotestosterone binding protein in the epididymal cytosol fraction. In vivo studies. Acta Endocrinol. 71:614-624.

Hansson, V., E. M. Ritzén, F. S. French, and S. N. Nayfeh. 1975. Androgen transport and receptor mechanisms in testis and epididymis, pp. 173-201. In D. W. Hamilton and R. O. Greep, Eds. Handbook of Physiology, Section 7: Endocrinology. Volume V. Male Reproductive System. Washington, D.C.: American Physiological Society.

Harada, M. 1978. Congenital Minamata disease: Intrauterine methylmercury poisoning. Teratology 18:285-288.

Harada, Y. 1976. Intrauterine poisoning. Clinical and epidemiological studies and significance of the problem. Bull. Inst. Constit. Med., Kumamoto Univ. (Suppl. 25):1-60.

Harada, Y. 1977. Congenital minamata disease, pp. 209-239. In T. Tsubaki and K. Irukayama, Eds. Minamata Disease: Methylmercury Poisoning in Minamata and Niigata Japan. Tokyo: Kodansha.

Hargreave, T. B., and R. A. Elton. 1983. Is conventional sperm analysis of any use? Br. J. Urol. 55:775-779.

Hargreave, T. B., and S. Nilson. 1983. Seminology, pp. 56-74. In T. B. Hargreave, Ed. Male Infertility. New York: Springer-Verlag.

Harrington, J. M., G. F. Stein, R. O. Rivera, and A. V. de Morales. 1978. The occupational hazards of formulating oral contraceptives: A survey of plant employees. Arch. Environ. Health 33:12-15.

Harris, H., D. A. Hopkinson, and E. B. Robson. 1974. The incidence of rare alleles determining electrophoretic variants: Data on 43 enzyme loci in man. Ann. Hum. Genet. 37:237-253.

Harry, G. J., J. F. Goodrum, M. R. Krigman, and P. Morell. 1985. The use of synapsin I as a biochemical marker for neuronal damage by trimethyltin. Brain Res. 326:9-18.

Hassold, T. J. 1985. The origin of aneuploidy in humans, pp. 103-115. In V. Dellarco, P. Voytek, and A. Hollaender, Eds. Aneuploidy: Etiology and mechanisms. New York: Plenum.

Hatcher, N. H., and E. B. Hook. 1981a. Somatic chromosome breakage in low birth weight newborns. Mutat. Res. 83:291-299.

Hatcher, N. H., and E. B. Hook. 1981b. Sister chromatid exchange in newborns. Hum. Genet. 59:389-391.

Hatzakis, A., A. Kokkevi, K. Katsouyanni, K. Maravelias, J. F. Salaminios, A. Kalandidi, and A. Koutselinis. 1987. Psychometric intelligence and attentional performance deficits in lead-exposed children, pp. 204-209. In S. E. Lindberg and T. C. Hutchinson, Eds. Heavy Metals in the Environment, Vol.1. New Orleans/Edinburgh: CEP Consultants.

Healy, D. L., G. P. Chrousos, H. M. Schulte, R. F. Williams, P. W. Gold, E. E. Baulieu, and G. D. Hodgen. 1983. Pituitary and adrenal responses to the antiprogesterone and anti-glucocorticoid steroid RU 486 in primates. J. Clin. Endocrinol. Metab. 57:863-865.

Heap, R. B., A. P. F. Flint, and J. E. Gadsby. 1981. Embryonic signals and maternal recognition, pp. 311-326. In S. R. Glasser and D. W. Bullock, Eds. Cellular and Molecular Aspects of Implantation. New York: Plenum Press.

Hecht, N. B. 1987a. Gene expression during spermatogenesis, pp. 90-101. In M. C. Orgebin-Crist and B. J. Danzo, Ed. Cell Biology of the Testis and Epididymis. Annals of the New York Academy of Sciences, Vol. 513. New York: New York Academy of Sciences.

Hecht, N. B. 1987b. Detecting the effects of toxic agents on spermatogenesis using DNA probes. Environ. Health Perspect. 74:31-40.

Hecht, N. B. 1987c. Regulation of gene expression during mammalian spermatogenesis, pp. 151-193. In J. Rossant and R. Pedersen, Eds. Experimental Approaches to Mammalian Embryonic Development. New York: Cambridge University Press.

Hecht, N. B. In press. Mammalian protamines and their expression, pp. 347-373. In G. Stein, J. Stein, and L. Hnilica, Eds. Histones and Other Basic Nuclear Proteins. Boca Raton, Fla.: CRC Press.

Hecht, N. B., K. C. Kleene, R. J. Distel, and L. M. Silver. 1984. The differential expression of the actins and tubulins during spermatogenesis in the mouse. Exp. Cell Res. 153:275-280.

Hecht, N. B., K. C. Kleene, P. C. Yelick, P. A. Johnson, D. D. Pravtcheva, and F. H. Ruddle. 1986a. Mapping of haploid expressed genes: Genes for both mouse

protamines are located on chromosome 16. Somatic Cell Mol. Genet. 12:203-208.

Hecht, N. B., P. A. Bower, S. H. Waters, P. C. Yelick, and R. J. Distel. 1986b. Evidence for haploid expression of mouse testicular genes. Exp. Cell Res. 164:183-190.

Heinrichs, W. L., R. J. Gellert, J. L. Bakke, and N. L. Lawrence. 1971. DDT administered to neonatal rats induces persistent estrus syndrome. Science 173:642-643.

Heller, C. G., and Y. Clermont. 1964. Kinetics of the germinal epithelium in man. Rec. Prog. Horm. Res. 20:545-575.

Hemeida, N. A., W. O. Sack, and K. McEntee. 1978. Ductuli efferentes in the epididymis of boar, goat, ram, bull, and stallion. Am. J. Vet. Res. 39:1892-1900.

Henneberry, M., M. F. Carter, and H. L. Neiman. 1979. Estimation of prostatic size by suprapubic ultrasonography. J. Urol. 121:615-616.

Hennig, B. 1975. Change of cytochrome c structure during development of the mouse. Eur. J. Biochem. 55:167-183.

Henry, J. P. 1980. Present concept of stress theory, p. 557-571. In E. Usdin, R. Kvetnansky, and I. J. Kopin, Eds. Catecholamines and Stress: Recent Advances. Developments in Neuroscience, Vol. 8. New York: Elsevier-North Holland.

Henzl, M. R. 1986. Contraceptive hormones and their clinical use, pp. 643-682. In S. S. C. Yen and R. B. Jaffe, Eds. Reproductive Endocrinology: Physiology, Pathophysiology and Clinical Management. 2nd ed. Philadelphia: W. B. Saunders Company.

Herbst, A. L., and H. A. Bern. 1981. Developmental Effects of Diethylstilbestrol (DES) in Pregnancy. New York: Thieme-Stratton. 203 pp.

Herbst, A. L., S. J. Robboy, R. E. Scully, and D. C. Poskanzer. 1974. Clear-cell adenocarcinoma of the vagina and cervix in girls: Analysis of 170 registry cases. Am. J. Obstet. Gynecol. 119:713-724.

Herbst, A. L., M. M. Hubby, R. R. Blough, and F. Azizi. 1980. A comparison of pregnancy experience in DES-exposed and DES-unexposed daughters. J. Reprod. Med. 24:62-69.

Hernberg, S. 1972. Biological effects of low lead doses, pp. 617-629. In International Symposium: Environmental Health Aspects of Lead. Luxembourg: Commission of the European Communities, Directorate General for Dissemination of Knowledge, Center for Information and Documentation.

Hernberg, S. 1980. Biochemical and clinical effects and responses as indicated by blood concentration, pp. 367-399. In R. L. Singhal and J. A. Thomas, Eds. Lead Toxicity. Baltimore: Urban & Schwarzenberg.

Hertig, A. T., and J. Rock. 1945. Two human ova of the pre-villous stage having developmental age of about seven and nine days respectively. Contrib. Embryol. 31:65-84.

Hicks, S. P. 1953. Developmental malformations produced by radiation. A timetable of their development. Am. J. Roentgenol. 69:272-293.

Hicks, S. P. 1959. Radiation as an experimental tool in mammalian developmental neurology. Physiol. Rev. 38:337-356.

Hicks, S. P., and C. J. D'Amato. 1963. Low dose radiation of the developing brain. Science 141:903-905.

Hicks, S. P., and C. J. D'Amato. 1966. Effects of ionizing radiations on mammalian development, pp. 195-250. In D. H. M. Woollam, Ed. Advances in Teratology, Vol. 1. London: Logos Press.

Hicks, S. P., C. J. D'Amato, M. A. Coy, E. D. O'Brien, J. M. Thurston, and D. L. Joftes. 1961. Migrating cells in the developing nervous system studied by their radiosensitivity and tritiated thymidine uptake, pp. 246-261. In Fundamental Aspects of Radiosensitivity. Brookhaven Symposia in Biology, No. 14. Upton, N.Y.: Brookhaven National Laboratory.

Hicks, S. P., C. J. D'Amato and J. L. Falk. 1962. Some effects of radiation on structural and behavioral development. Int. J. Neurol. 3:535-548.

Hilf, G., and W. E. Merz. 1985. Influence of cyclic nucleotides on receptor binding, immunological activity and microheterogeneity of human choriogonadotropin synthesized in placantal tissue culture. Mol. Cell Endocrinol. 39:151-159.

Hill, R. M., W. M. Verniaud, M. G. Horning, L. B. McCulley, and N. F. Morgan. 1974. Infants exposed in utero to antiepileptic drugs. Prospective study. Am. J. Disab. Child 127:645-653.

Hinton, B. T. 1980. The epididymal microenvironment: A site of attack for a male contraceptive? Invest. Urol. 18:1-10.

Hinton, B. T. 1985. The blood-epididymis barrier, pp. 371-382. In T. J. Lobl and E. S. E. Hafez, Eds. Male Fertility and Its Regulation. Boston: MTP Press.

Hinton, B. T., and S. S. Howards. 1982. Micropuncture and microperfusion techniques for the study of testicular physiology. Ann. N.Y. Acad. Sci. 383:29-43.

Hinton, B. T., and D. A. Keefer. 1985. Binding of [^{3}H] aldosterone to a single population of cells within the rat epididymis. J. Steroid Biochem. 23:231-233.

Hinton, B. T., and B. P. Setchell. 1980a. Concentrations of glycerophosphocholine, phosphocholine and free inorganic phosphate in the luminal fluid of the rat testis and epididymis. J. Reprod. Fertil. 58:401-406.

Hinton, B. T., and B. P. Setchell. 1980b. Concentration and uptake of carnitine in the rat epididymis: A micropuncture study, pp. 237-249. In R. A. Frankel and J. D. McGarry, Eds. Carnitine Biosynthesis, Metabolism, and Functions. New York: Academic Press.

Hinton, B. T., R. W. White, and B. P. Setchell. 1980. Concentrations of myo-inositol in the luminal fluid of the mammalian testis and epididymis. J. Reprod. Fertil. 58:395-399.

Hinton, B. T., D. E. Brooks, H. M. Dott, and B. P. Setchell. 1981. Effects of carnitine and some related compounds on the motility of rat spermatozoa from the caput epididymis. J. Reprod. Fertil. 61:59-64.

Hirsch, P. J., I. L. C. Fergusson, and R. J. B. King. 1977. Protein composition of human endometrium and its secretions at different stages of the menstrual cycle. Ann. N.Y. Acad. Sci. 286:233-248.

Hoar, R. M., and I. W. Monie. 1981. Comparative development of specific organ systems, pp. 13-33. In C. A. Kimmel and J. Buelke-Sam, Eds. Developmental Toxicology. New York: Raven Press.

Hobbins, J. C., P. A. Grannum, R. Romero, E. A. Reece, and M. J. Mahoney. 1985. Percutaneous umbilical blood sampling. Am. J. Obstet. Gynecol. 152:1-6.

Hodgson, B. J., and C. J. Pauerstein. 1976. Comparison of oviductal transport of fertilized and unfertilized ova after hCG or coitus-induced ovulation in rabbits. Biol. Reprod. 14:377-380.

Hoffer, A. P., and D. W. Hamilton. 1974. Phagocytosis of sperm by the epithelial cells in ductuli efferentes of experimental rats. Anat. Rec. 178:376-377.

Hoffer, A. P., and B. T. Hinton. 1984. Morphological evidence for a blood-epididymis barrier and the effects of gossypol on its integrity. Biol. Reprod. 30:991-1004.

Hoffer, A. P., D. W. Hamilton, and D. W. Fawcett. 1973. The ultrastructural pathology of the rat epididymis after administration of alpha-chlorhydrin (U-5897). I. Effects of a single high dose. Anat. Rec. 175:203-229.

Hoffer, A. P., D. W. Hamilton, and D. W. Fawcett. 1975. Phagocytosis of spermatozoa by the epithelial cells of the ductuli efferentes after epididymal obstruction in the rat. J. Reprod. Fertil. 44:1-9.

Hoffer, B., L. Olson, H. Björklund, A. Henschen, and M. Palmer. 1984. Some toxic effects of lead and other metals in the nervous system: In oculo experimental models, pp. 141-152. In T. Narahashi, Ed. Cellular and Molecular Neurotoxicology. New York: Raven Press.

Hoffer, B. J., L. Olson, and M. R. Palmer. 1987. Toxic effects of lead in the developing nervous system. In oculo experimental models. Environ. Health Perspect. 74:169-175.

Hoffman, H. S. 1984. Methodological factors in the behavioral analysis of startle: the use of reflex modification procedures and the assessment of threshold, pp. 267-285. In R. C. Eaton, Ed. Neural Mechanisms of Startle Behavior. New York: Plenum Press.

Hoffman, R. E., P. A. Stehr-Green, K. B. Webb, R. G. Evans, A. P. Knutsen, W. F. Schramm, J. L. Staake, B. B. Gibson, and K. K. Steinberg. 1986. Health effects of long-term exposure to 2,3,7,8-tetrachlorodibenzo-p-dioxin. J. Am. Med. Assoc. 255:2031-2038.

Holinka, C. F., Y. C. Tseng, and C. E. Finch. 1979. Reproductive aging in C57BL/6J mice: Plasma progesterone, viable embryos and resorption frequency during pregnancy. Biol. Reprod. 20:1201-1211.

Holson, R. R., J. Adams, J. Buelke-Sam, B. Gough, and C. A. Kimmel. 1985. d-Amphetamine as a behavioral teratogen: Effects depend on dose, sex, age, and task. Neurobehav. Toxicol. Teratol. 7:753-758.

Holstein, A. F. 1978. Spermatophagy in the seminiferous tubules and excurrent ducts of the testis in Rhesus monkey and in man. Andrologia 10:331-352.

Homonnai, Z. T., G. Paz, J. N. Weiss, and M. P. David. 1980a. Quality of semen obtained from 627 fertile men. Int. J. Androl. 3:217-228.

Homonnai, Z. T., G. F. Paz, J. N. Weiss, and M. P. David. 1980b. Relation between semen quality and fate of pregnancy: Retrospective study on 534 pregnancies. Int. J. Androl. 3:574-584.

Hook, E. B. 1981. Prevalence of chromosome abnormalities during human gestation and implications for studies of environmental mutagens. Lancet 2:169-172.

Hopper, B. R., and S. S. C. Yen. 1975. Circulating concentrations of dehydroepiandrosterone and dehydroepiandrosterone sulfate during puberty. J. Clin. Endocrinol. Metab. 40:458-461.

Horne, C. H., and A. D. Nisbet. 1979. Pregnancy proteins: A review. Invest. Cell. Pathol. 2:217-231.

Horning, S. J., R. T. Hoppe, H. S. Kaplan, and S. A. Rosenbert. 1981. Female reproductive potential after treatment for Hodgkin's disease. N. Engl. J. Med. 304:1377-1382.

Hoshina, M., R. Hussa, R. Pattillo, H. M. Camel, and I. Boime. 1982. The role of trophoblast differentiation in the control of the hCG and hPL genes. Adv. Exp. Med. Biol. 176:290-312.

Hoshina, M., M. Boothby, R. Hussa, R. Pattillo, H. M. Camel, and I. Boime. 1985. Linkage of human chorionic gonadotrophin and placental lactogen biosynthesis to trophoblast differentiation and tumorigenesis. Placenta 6:163-172.

Hoskins, D. D., and E. R. Casillas. 1975. Function of cyclic nucleotides in mammalian spermatozoa, pp. 453-460. In D. W. Hamilton and R. O. Greep, Eds. Handbook of Physiology, Section 7: Endocrinology. Volume V. Male Reproductive System. Washington, D.C.: American Physiological Society.

Hoskins, D. D., H. Brandt, and S. Acott. 1978. Initiation of sperm motility in the mammalian epididymis. Fed. Proc. 37:2534-2542.

Hotchkiss, R. S., E. K. Brunner, and P. Grenley. 1938. Semen analyses of 200 fertile men. Am. J. Med. Sci. 196:362-384.

Houck, P. R., C. F. Reynolds, 3d., U. Kopp, and I. Hanin. 1988. Red blood cell/plasma choline ratio in elderly depressed and demented patients. Psychiatry Res.(Ireland) 24:109-116.

Hougie, C. 1985. Circulating anticoagulants. Recent Adv. Blood Coag. 4:63.

Howards, S. S. 1983. The epididymis, sperm maturation, and capacitation, pp.121-134. In L. I. Lipshultz and S. S. Howards, Eds. Infertility in the Male. New York: Churchill Livingstone.

Howards, S. S., S. J. Jessee, and A. L. Johnson. 1976. Micropuncture studies of the blood-seminiferous tubule barrier. Biol. Reprod. 14:264-269.

Hsi, B.-L., C.-J. G. Yeh, and W. P. Faulk. 1982. Human amniochorion: Tissue-specific markers, transferrin receptors and histocompatibility antigens. Placenta 3:1-11.

Hsi, B.-L., C.-J. G. Yeh, and W. P. Faulk. 1984a. Class I antigens of the major histocompatibility complex on cytotrophoblast of human chorion laeve. Immunology 52:621-629.

Hsi, B.-L., C.-J. G. Yeh, and W. P. Faulk. 1984b. Characterization of antibodies to antigens of the human amnion. Placenta 5:513-521.

Hsu, A. F., and P. Troen. 1978. An androgen-binding protein in the testicular cytosol of human testis: Comparison with human testosterone estrogen binding globulin. J. Clin. Invest. 61:1611-1619.

Hsu, T. C., F. Elder, and S. Pathak. 1979. Method for improving the yield of spermatogonial and meiotic metaphases in mammalian testicular preparations. Environ. Mutatgen. 1:291-294.

Hsueh, A. J. W., and J. M. Schaeffer. 1985. Gonadotropin-releasing hormone as a paracrine hormone and neurotransmitter in extra-pituitary sites. J. Steroid Biochem. 23:757-764.

Hudec, T., J. Thean, D. Kuehl, and R. C. Dougherty. 1981. Tris(dichloropropyl)phosphate, a mutagenic flame retardant: Frequent cocurrence in human seminal plasma. Science 211(4485):951-952.

Hughes, J. A., and S. B. Sparber. 1979. d-Amphetamine unmasks postnatal consequences of exposure to methylmercury in utero: Methods for studying behavioral teratogenesis. Pharmacol. Biochem. Behav. 8:365-375.

Huhtanemi, I. T., C. C. Korenbrot, and R. B. Jaffe. 1977. HCG binding and stimulation of testosterone biosynthesis in the human fetal testis. J. Clin. Endocrinol. Metab. 44:963-967.

Hulten, M. N. Saadallah, B. M. N. Wallace, and D. J. Cockburn. 1985. Meiotic investigations of aneuploidy in the human, pp. 75-90. In V. L. Dellarco, P. E. Voytek, and A. Hollaender, Eds. Aneuploidy: Etiology and Mechanisms. New York: Plenum.

Hunter, J., M. A. Urbanowicz, W. Yule, and R. Lansdown. 1985. Automated testing of reaction time and its association with lead in children. Int. Arch. Occup. Environ. Health 57:27-34.

Huret, J. L. 1986. Nuclear chromatin decondensation of human sperm: A review. Arch. Androl. 16:97-109.

Hurley, T. W., C. M. Kuhn, S. M. Schanberg, and S. Handwerger. 1980. Differential effects of placental lactogen, growth hormone and prolactin on rat liver ornithine decarboxylase activity in the perinatal period. Life Sci. 27:2269-2275.

Hutchings, D. E. 1983. Behavioral teratology: A new frontier in neurobehavioral research, pp. 207-235. In E. M. Johnson and D. M. Kochar, Eds. Teratogenesis and Reproductive Toxicology. Handbook of Experimental Pharmacology, Vol. 65. Berlin: Springer-Verlag.

IARC (International Agency for Research on Cancer).

1984. Monitoring Human Exposure to Carcinogenic and Mutagenic Agents. IARC Scientific Publications, No. 59. Lyon: International Agency for Research on Cancer. 457 pp.

Ilgren, E. B. 1983. Control of trophoblastic growth. Placenta 4:307-328.

Inano, H., A. Machino, and B.-I. Tamaoki. 1969. In vitro metabolism of steroid hormones by cell-free homogenates of epididymides of adult rats. Endocrinology 84:997-1003.

Inskeep, P. B., and R. H. Hammerstedt. 1982. Changes in metabolism of ram sperm associated with epididymal transit or induced by exogenous carnitine. Biol. Reprod. 27:735-743.

Irwin, M., N. Nicholson, J. T. Haywood, and G. R. Pourier. 1983. Immunoflourescent localization of a murine seminal vesicle proteinase inhibitor. Biol. Reprod. 28:1201-1206.

Isaacs, W. B., and D. S. Coffey. 1984. The predominant protein of canine seminal plasma is an enzyme. J. Biol. Chem. 259:11520-11526.

Isaacs, W. B., and J. H. Shaper. 1983. Isolation and characterization of the major androgen-dependent glycoprotein of canine prostatic fluid. J. Biol. Chem. 258:6610-6615.

Isaacs, W. B., and J. H. Shaper. 1985. Immunological localization and quantitation of the androgen-dependent secretory protease of the canine prostate. Endocrinology 117:1512-1520.

Ismail, A. A. A., P. Astley, W. A. Burr, K. Wakelin, and M. J. Wheeler. 1986. The role of testosterone measurement in the investigation of androgen disorders. Ann. Clin. Biochem. 23:113-134.

Ison, J. R. 1984. Reflex modification as an objective test for sensory processing following toxicant exposure. Neurobehav. Toxicol. Teratol. 6:437-445.

Israel, R., D. R. Mishell, Jr., S. C. Stone, I. H. Thorneycroft, and D. L. Moyer. 1972. Single luteal phase serum progesterone assay as an indicator of ovulation. Am. J. Obstet. Gynecol. 112:1043-1046.

Iversen, P., L. Kjaer, C. Thomsen, and O. Henriksen. 1988. Magnetic resonance imaging of the prostate. Scand. J. Urol. Nephrol. Suppl. 107:14-18.

Iwata, K. 1980. Neuroophthalmologic indices of Minamata Disease in Nigata, pp. 165-185. In W. H. Merigan and B. Weiss, Eds. Neurotoxicity of the Visual System. New York: Raven Press.

Jaakkola, U. M. 1983. Regional variations in transport of the luminal contents of the rat epididymis in vivo. J. Reprod. Fertil. 68:465-470.

Jaakkola, U. M., and A. Talo. 1981. Effects of oxytocin and vasopressin on electrical and mechanical activity of the rat epididymis in vitro. J. Reprod. Fertil. 63:47-51.

Jaakkola, U. M., and A. Talo. 1983. Movements of the luminal contents in two different regions of the caput epididymidis of the rat in vitro. J. Physiol. 336:453-463.

Jackson, L. 1985. Prenatal genetic diagnosis by chorionic villus sampling (CVS). Semin. Perinatol. 9:209-218.

Jacob, M. P., T. Fulop, Jr., G. Foris, and L. Robert. 1987. Effect of elastin peptides on ion fluxes in mononuclear cells, fibroblasts, and smooth muscle cells. Proc. Natl. Acad. Sci. USA 84:995-999.

Jacobs, H. S., U. A. Knuth, M. G. R. Hull, and S. Franks. 1977. Post-"pill" amenorrhea: Cause or coincidence? Br. Med. J. 2:940-942.

Jacobson, J. L., S. W. Jacobson, G. G. Fein, P. M. Schwartz, and J. K. Dowler. 1984. Prenatal exposure to an environmental toxin: A test of the multiple effects model. Develop. Psychol. 20:523-532.

Jacobson, M. A. 1978. Developmental Neurobiology, 2nd ed. New York: Plenum Press.

Jacobson, S. W., G. G. Fein, J. L. Jacobson, P. M. Schwartz, and J. K. Dowler. 1985. The effect of intrauterine PCB exposure on visual recognition memory. Child Devel. 56:853-860.

Jaffe, R. B. 1986. Protein hormones of the placenta, decidua, and fetal membranes, pp. 758-769. In S. S. C. Yen and R. B. Jaffe, Eds. Reproductive Endocrinology: Physiology, Pathophysiology, and Clinical Management, 2nd ed. Philadelphia: W. B. Saunders Co.

Jagiello, G., and S. J. Lin. 1974. Oral contraceptive compounds and mammalian oocyte meiosis. Am. J. Obstet. Gynecol. 120:390-406.

Jagiello, G., M. Ducayen, J-S. Fang, and J. Graffeo. 1975. Cytogenetic observations in mammalian oocytes. Chromosomes Today 5:43-63,

Jaiswal, A. K., F. J. Gonzalez, and D. W. Nebert. 1985a. Human dioxin-inducible cytochrome P_1-450: Complimentary DNA and amino acid sequence. Science 228:80-82.

Jaiswal, A. K., F. J. Gonzalez, and D. W. Nebert. 1985b. Human P-450 gene sequence and correlation of mRNA with genetic differences in benzo[a]pyrene metabolism. Nucleic Acid Res. 13:4503-4520.

Jeannet, M., P. Bischof, B. Bourrit, and P. Vuagnat. 1985. Sharing of HLA antigens in fertile, subfertile and infertile couples. Transplant Proc. 17:903.

Jenderny, J., and G. Röhrborn. 1987. Chromosome analysis of human sperm. I. First results with a modified method. Hum. Genet. 76:385-388.

Jenkins, A. D., C. P. Lechene, and S. S. Howards. 1980. Concentrations of seven elements in the intraluminal fluids of the rat seminiferous tubules, rete testis, and epididymis. Biol. Reprod. 23:981-987.

Jenkins, A. D., C. P. Lechene, and S. S. Howards. 1983. The effect of spironolactone on the elemental composition of the intraluminal fluids of the seminiferous tubules, rete testis and epididymis of the rat. J. Urol. 129:851-854.

Jensen, J. C., and W. G. Thilly. 1986. Spontaneous and induced chromosomal aberrations and gene mutations in human lymphoblasts: Mitomycin C, methlynitrosourea, and ethylnitrosourea. Mutat. Res. 160:95-102.

Jensen, R. H., W. L. Bigbee, and R. G. Langlois. 1987. In vivo somatic mutations in the glycophorin A locus of human erythroid cells, pp. 149-159. In M. M. Moore, D. M. Demarini, and K. R. Tindall, Eds. Banbury Report 28: Mammalian Cell Mutagenesis. Cold Spring Harbor, N.Y.: Cold Spring Harbor Laboratory Press.

Jequier, A. M. 1986. Infertility in the Male. Edinburgh: Churchill Livingston. 154 pp.

Jeyendran, R. S., H. H. Van der Ven, M. Perez-Pelaez, B. G. Crabo, and L. J. D. Zaneveld. 1984. Development of an assay to assess the functional integrity of the human sperm membrane and its relationship to other semen characteristics. J. Reprod. Fertil. 70:219-228.

Jick, H., and J. Porter. 1977. Relation between smoking and age of natural menopause. Lancet 1:1354-1355.

Johnson, A., and L. Godmilow. 1988. Genetic amniocentenesis at 14 weeks or less. Clin. Obstet. Gynceol. 31:345-352.

Johnson, A. L., and S. S. Howards. 1976. Intratubular hydrostatic pressure in testis and epididymis before and after long-term vasectomy in the guinea pig. Biol. Reprod. 14:371-376.

Johnson, I. R., E. M. Symonds, D. M. Kean, B. S. Worthington, F. Broughton Pipkin, R. C. Hawkes, and M. Gyngell. 1984. Imaging the pregnant human uterus with nuclear magnetic resonance. Am. J. Obstet. Gynecol. 148:1136-1139.

Johnson, L. 1982. A re-evaluation of daily sperm output of men. Fertil. Steril. 37:811-816.

Johnson, L. 1986. A new approach to quantification of Sertoli cells that avoids problems associated with the irregular nuclear surface. Anat. Rec. 214:231-237.

Johnson, L., C. S. Petty, and W. B. Neaves. 1980a. The relationship of biopsy evaluations and testicular measurements to over-all daily sperm production in human testes. Fertil. Steril. 34:36-40.

Johnson, L., R. P. Amann, and B. W. Pickett. 1980b. Maturation of equine epididymal spermatozoa. Am. J. Vet. Res. 41:1190-1196.

Johnson, L., R. S. Zane, C. S. Petty, and W. B. Neaves. 1984a. Quantification of the human Sertoli cell population: Its distribution, relation to germ cell numbers, and age-related decline. Biol. Reprod. 31:785-795.

Johnson, L., C. S. Petty, and W. B. Neaves. 1984b. Influence of age on sperm production and testicular weights in men. J. Reprod. Fertil. 70:211-218.

Johnson, M. H., and J. Rossant. 1981. Molecular studies on cells of trophectodermal lineage of the postimplantation mouse embryo. J. Embryol. Exp. Morphol. 61:103-116.

Johnson, M. H., H. P. M. Pratt, and A. H. Handyside. 1981. The generation and recognition of positional information in the preimplantation mouse embryo, pp. 55-74. In S. R. Glasser and D. W. Bullock, Eds. Cellular and Molecular Aspects of Implantation. New York: Plenum Press.

Jones, E. C., and P. L. Krohn. 1961. The relationships between age, numbers of oocytes and fertility in virgin and multiparous mice. J. Endocrinol. 21:469-495.

Jones, G. S., S. Aksel, and A. C. Wentz. 1974. Serum progesterone values in the luteal phase defects. Effect of chorionic gonadotropin. Obstet. Gynecol. 44:26-34.

Jones, K. L., and G. F. Chernoff. 1984. Effects of chemical and environmental agents, pp. 189-200. In R. K. Creasy and R. Resnik, Eds. Maternal-Fetal Medicine: Principles and Practice. Philadelphia: W. B. Saunders.

Jones, K. L., D. W. Smith, C. N. Ulleland, and A. P. Streissguth. 1973. Pattern of malformation in offspring of chronic alcoholic mothers. Lancet 1:1267-1271.

Jones, L. S., and W. E. Berndtson. 1986. A quantitative study of Sertoli cell and germ cell populations as related to sexual development and aging in the stallion. Biol. Reprod. 35:138-148.

Jones, R., C. R. Brown, K. I. Von Glos, and M. G. Parker. 1980. Hormonal regulation of protein synthesis in the rat epididymis. Characterization of androgen-dependent and testicular fluid-dependent proteins. Biochem. J. 188:667-676.

Jones, R., C. Pholpramool, B. P. Setchell, and C. R. Brown. 1981. Labelling of membrane glycoproteins on rat spermatozoa collected from different regions of the epididymis. Biochem. J. 200:457-460.

Jones, R., K. I. von Glos, and C. R. Brown. 1983. Changes in the protein composition of rat spermatozoa during maturation in the epididymis. J. Reprod. Fertil. 67:299-306.

Joseph, A. M., J. R. Gosden, and A. C. Chandley. 1984. Estimation of aneuploidy levels in human spermatozoa using chromosome specific probes and in situ hybridisation. Hum. Genet. 66:234-238.

Joshi, S. G. 1983. A progestagen-associated protein of the human endometrium: Basic studies and potential clinical applications. J. Steroid Biochem. 19:751-757.

Josimovich, J. B. 1983. Placental lactogen and pituitary prolactin, pp. 144-160. In F. Fuchs and A. Klopper, Eds. Endocrinology of Pregnancy, 3rd ed. Philadelphia: Harper & Row.

Juchau, M. R. 1980. Drug transformation in the placenta. Pharmacol. Ther. 8:501-524.

Kajino, T., J. A. McIntyre, W. P. Faulk, D. S. Cai, and W. D. Billington. 1988. Antibodies to trophoblast in normal pregnant and secondary aborting women. J. Reprod. Immunol. 14:267-282.

Kaler, L. W., and W. B. Neaves. 1978. Attrition of the human Leydig cell population with advancing age. Anat. Rec. 192:513-518.

Kallen, B. 1965. Degeneration and regeneration in the vertebrate central nervous system during embryogenesis, pp. 77-96. In M. Singer and J. P. Schadé, Eds. Degeneration Patterns in the Nervous System. Progress in Brain Research, Vol. 14. Amsterdam: Elsevier.

Kandala, J.C., M. K. Kistler, R. P. Lawther, and W. S. Kistler. 1983. Characterization of a genomic clone rat seminal vesicle secretory protein IV. Nucleic Acids Res. 11:3169-3186.

Kandel, R. E., and J. H. Schwartz. 1981. Principles of Neural Science. New York: Elsevier North-Holland.

Kaplan, M., L. D. Russell, R. N. Peterson, and J. Martan. 1984. Boar sperm cytoplasmic droplets: Their ultrastructure, their numbers in the epididymis and at ejaculation and their removal during isolation of sperm plasma membranes. Tissue Cell 16:455-468.

Kaplan, S., and M. Grumbach. 1981. Chorionic somatommotropin in primates: Secretion and physiology, pp. 127-139. In M. J. Novy and J. A. Resko, Eds. Fetal Endocrinology. New York: Academic Press.

Kapp, R. W., Jr., and C. B. Jacobson. 1980. Analysis of human spermatozoa for Y chromosomal nondisjunction. Teratogenesis Carcinog. Mutagen. 1:193-211.

Kapp, R. W., Jr., D. J. Picciano, and C. B. Jacobson. 1979. Y-chromosomal nondisjunction in dibromochloropropane-exposed workmen. Mutat. Res. 64:47-51.

Kapp, W. W., Jr. 1979. Detection of aneuploidy in human sperm. Environ. Health Perspect. 31:27-31.

Karkun, T., M. Rajalakshmi, and M. R. N. Prasad. 1974. Maintenance of the epididymis in the castrated golden hamster by testosterone and dihydrotestosterone. Contraception 9:471-485.

Karp, W. B., A. F. Robertson, and H. C. Davis. 1984. Relationship of unbound bilirubin concentration to reserve albumin-binding concentration for bilirubin in human neonatal plasma. Biol. Neonate 46:105-109.

Karp, W. B., S. B. Subramanyam, C. K. Ho, and A. F. Robertson. 1985. Drugs affecting bilirubin uptake by human erythrocyte ghosts. Am. J. Med. Sci. 289:236-239.

Karr, J. P., R. Y. Kirdani, G. P. Murphy, and A. A. Sandberg. 1974. Effects of testosterone and estradiol on ventral prostate and body weights of castrated rats. Life Sci. 15:501-513.

Kashimoto, T., H. Miyata, S. Kunita, T.-C. Tung, S.-T. Hsu, K.-J. Chang, G. Ohi, J. Nakagawa, and S.-I. Yamamoto. 1981. Role of polychlorinated dibenzofuran in Yusho (PCB poisoning). Arch. Environ. Health 36:321-326.

Kasson, B. G., E. Y. Adashi, and A. J. W. Hsueh. 1986. Arginine vasopressin in the testis: An intra-gonadal peptide control system. Endocrine Rev. 7:156-168.

Katayama, K. P., and M. R. Roesler. 1986. 500 cases of amniocentensis without bloody tap. Obstet. Gynecol. 68:70-73.

Katz, D. F., and J. W. Overstreet. 1981. Sperm motility assessment by videomicrography. Fertil. Steril. 35:188-193.

Katz, D. F., J. W. Overstreet, and F. W. Hanson. 1980. A new quantitative test for sperm penetration into cervical mucus. Fertil. Steril. 33:179-186.

Katz, D. F., J. W. Overstreet, and F. W. Hanson. 1981. Variations within and amongst normal men of movement chracteristics of seminal spermatozoa. J. Reprod. Fertil. 62:221-228.

Katz, D. F., L. Diel, and J. W. Overstreet. 1982. Differences in the movement of morphologically normal and abnormal human seminal spermatozoa. Biol. Reprod. 26:566-570.

Kaufman, D. W., D. Slone, L. Rosenberg, O. S. Mieittinen, and S. Shapiro. 1980. Cigarette smoking and age at natural menopause. Am. J. Public Health 70:420-422.

Kaufmann, S. L. 1966. An autoradiographic study of the generation cycle in the ten-day mouse neural tube. Exp. Cell Res. 42:67-73.

Kaul, B., G. Slavin, and B. Davidow. 1983. Free erythrocyte protoporphyrin and zinc protoporphyrin measurements compared as primary screening methods for detection of lead poisoning. Clin. Chem. 29:1467-1470.

Kavlock, R. J., and C. T. Grabowski, Eds. 1983. Abnormal Functional Development of the Heart, Lungs, and Kidneys: Approaches to Functional Teratology. New York: Liss.

Kawamoto J. C., S. R. Overmann, D. E. Woolley, and V. K. Vijayjan. 1984. Morphometric effects of preweaning lead exposure on the hippocampal formation of adult rats. NeuroToxicology 5:125-148.

Kawata, M., J. R. Parnes, and L. A. Herzenberg. 1984. Transcriptional control of HLA-A,B,C, antigens in human placental cytotrophoblast isolated using trophoblast- and HLA-specific monoclonal antibodies and the fluorescence-activated cell sorter. J. Exp. Med. 160:633-651.

Kay, H. H., and D. R. Mattison. 1986. Nuclear magnetic resonance spectroscopy and imaging in perinatal medicine, pp. 269-323. In P. W. Nathanielsz, Ed. Animal Models in Fetal Medicine (V). Parturition. Monographs in Fetal Physiology, Vol. 6. Ithaca: Perinatology Press.

Keegan, K. A., Jr., and R. H. Paul. 1980. Antepartum fetal heart rate testing. IV. The nonstress test as a primary approach. Am. J. Obstet. Gynecol. 136:75-80.

Kellermann, G., M. Luyten-Kellermann, and C. R. Shaw. 1973a. Genetic variation of aryl hydrocarbon hydroxylase in human lymphocytes. Am. J. Hum. Genet. 25:327-331.

Kellermann, G., C. R. Shaw, and M. Luyten-Kellermann. 1973b. Aryl hydrocarbon hydroxylase inducibility and bronchogenic carcinoma. N. Engl. J. Med. 289:934-937.

Kelly, P. A., R. P. C. Shiu, M. C. Robertson, and H. G. Friesen. 1975. Characterization of rat chorionic mammotropin. Endocrinology 96:1187-1195.

Kenagy, G. J. 1979. Rapid surgical technique for measurement of testis size in small mammals. J. Mammal. 60:636-638.

Kennedy, A. R., A. H. M. Heagerty, J. P. Ortonne, B. L. Hsi, C. J. G. Yeh, and R. A. J. Eady. 1985. Abnormal binding of an anti-amnion antibody to epidermal basement membrane provides a novel diagnostic probe for junctional epidermolysis bullosa. Br. J. Dermatol. 113:651-659.

Kennedy, G. L., Jr. 1986. Biological effects of acetamide,

formamide, and their monomethyl and dimethyl derivatives. CRC Crit. Rev. Toxicol. 17:129-182.

Kephart, G. M., G. J. Gleich, D. A. Conner, D. W. Gibson, and S. Ackerman. 1984. Deposition of eosinophil granule major basic protein onto microfilariae of onchocerca volvulus in the skin of patients treated with diethylcarbamazine. Lab. Invest. 50:51-61.

Kessler, M. J., M. S. Reddy, R. H. Shah, and O. P. Bahl. 1979. Structures of N-glycosidic carbohydrate units of human chorionic gonadotropin. J. Biol. Chem. 254:7901-7908.

Kidroni, G., R. Har-Nir, J. Menezel, I. W. Frutkoff, Z. Palti, and M. Ron. 1983. Vitamin D3 metabolites in rat epididymis: High 24, 25-dihydroxy vitamin D3 levels in the cauda region. Biochem. Biophys. Res. Commun. 113:982-989.

Kiger, N., G. Chaouat, J. P. Kolb, T. G. Wegmann, and J. L. Guenet. 1985. Immunogenetic studies of spontaneous abortion in mice. Preimmunization of females with allogeneic cells. J. Immunol. 134:2966-2970.

Kimbrough, R. D. 1987. Human health effects of polychlorinated biphenyls (PCBs) and polybrominated biphenyls (PBBs). Annu. Rev. Pharmacol. Toxicol. 27:87-111.

Kimmel, C. A., and J. Buelke-Sam, Eds. 1981. Developmental Toxicology. New York: Raven Press.

Kimmel, C. A., L. D. Grant, G. L. West, C. M. Martinez-Vargas, E. E. McConnell, B. A. Fowler, J. S. Woods, and J. L. Howard. 1978. Multidisciplinary approach to the assessment of development toxicity associated with chronic lead exposure, pp. 396-409. In D. D. Mahlum, M. R. Sikov, P. L. Hackett, and F. D. Andrew, Eds. Developmental Toxicology of Energy-Related Pollutants: Proceedings of the 17th Annual Hanford Biological Symposium at Richland, Wash., Oct. 17-19, 1977. DOE Symposium Series 47. Washington, D.C.: U.S. Department of Energy, Technical Information Center.

King, M. T, and D. Wild. 1979. Transplacental mutagenesis: The micronucleus test on fetal mouse blood. Hum. Genet. 51:183-194.

King, M. T., D. Wild, E. Gocke, and K. Eckhardt. 1982. 5-Bromodexyuridine tablets with improved depot effect for analysis in vivo of sister-chromatid exchanges in bone-marrow and spermatogonial cells. Mutat. Res. 97:117-129.

Kirton, K. T., C. Desjardins, and H. D. Hafs. 1967. Distribution of sperm in male rabbits after various ejaculation frequencies. Anat. Rec. 158:287-292.

Klawans, H. L., Jr., G. W. Paulson, S. P. Ringel, and A. Barbeau. 1972. Use of L-dopa in the detection of presymptomatic Huntington's chorea. N. Engl. J. Med. 286:1332-1334.

Kleene, K. C., R. J. Distel, and N. B. Hecht. 1983. cDNA clones encoding cytoplasmic poly(A) + RNAs which first appear at detectable levels in haploid phases of spermatogenesis in the mouse. Dev. Biol. 98:455-464.

Kleene, K. C., R. J. Distel, and N. B. Hecht. 1984. Translational regulation and coordinate deadenylation of a protamine mRNA during spermiogenesis in the mouse. Dev. Biol. 105:71-79.

Kleene, K. C., R. J. Distel, and N. B. Hecht. 1985. The nucleotide sequence of a cDNA clone encoding mouse protamine 1. Biochemistry 24:719-722.

Klein, N. W., M. A. Vogler, C. L. Chatot, and L. J. Pierro. 1980. The use of cultured rat embryos to evaluate the teratogenic activity of serum: Cadmium and cyclophosphamide. Teratology 21:199-208.

Klein, N. W., J. D. Plenefisch, S. W. Carey, W. T. Fredrickson, G. P. Sackett, T. M. Burbacher, and R. M. Parker. 1982. Serum from monkeys with histories of fetal wastage causes abnormalities in cultured rat embryos. Science 215:66-69.

Kleinberg, D. L. 1980. Lactation and galactorrhea, pp. 70-89. In J. J. Gold and J. B. Josimovich, Eds. Gynecologic Endocrinology, 3rd ed. Hagerstown, Md.: Harper & Row.

Knobil, E. 1980. The neuroendocrine control of the menstrual cycle. Recent Prog. Horm. Res. 36:53-88.

Knochel, J. Q., T. G. Lee, M. G. Melendez, and S. C. Henderson. 1983. Fetal anomalies involving the thorax and abdomen, pp. 61-80. In P. W. Callen, Ed. Ultrasonography in Obstetrics and Gynecology. Philadelphia: W. B. Saunders.

Koering, M. J., D. L. Healy, and G. D. Hodgen. 1986. Morphologic response of endometrium to a progesterone receptor antagonist, RU486, in monkeys. Fertil. Steril. 45:280-287.

Kohama, S., P. C. May, and C. E. Finch. 1986. Oral administation of estradiol induces age-like reproductive acyclicity in C57BL/6J mice. Soc. Neurosci. Abstr. 12:1466.

Koizumi, A., L. Hasegawa, R. Walford, and T. Imamura. 1986. H-2, Ah, and aging: The immune response and the inducibility of P-450 mediated monooxygenase activities, xanthine oxidase, and lipid peroxidation in H-2 congenic mice on C57BL/10, C3H, and A strain backgrounds. Mech. Ageing Dev. 37:119-136.

Komlos, L., R. Zamir, H. Joshua, and I. Halbrecht. 1977. Common HLA antigens in couples with repeated abortions. Clin. Immunol. Immunopathol. 7:330-335.

Koninckx, P. R., W. Heyns, G. Verhoeven, H. Van Baelen, W. D. Lissens, P. De Moor, and I. A. Brosens. 1986. Biochemical characterization of peritoneal fluid in women during the menstrual cycle. J. Clin. Endocrinol. Metab. 51:1239-1244.

Konoeda, Y., P. Terasaki, A. Wakisaka, M. S. Park, and M. R. Mickey. 1986. Public determinants of HLA indicated by pregnancy antibodies. Transplantation 41:253-259.

Koos, B. J. 1985. Central stimulation of breathing movements in fetal lambs by prostaglandin synthetase inhibitors. J. Physiol. 362:455-466.

Koos, B. J., and L. D. Longo. 1976. Mercury toxicity

in the pregnant woman, fetus, and newborn infant. Am. J. Obstet. Gynecol. 126:390-409.

Kopin, I. J., and S. P. Markey. 1988. MPTP toxicity: Implications for research in Parkinson's disease. Annu. Rev. Neurosci. 11:91-96.

Kopp, S. J., A. A. Daar, R. C. Prentice, J. P. Tow, and J. M. Feliksik. 1986. 31P NMR studies of the intact perfused rat heart: A novel analytical approach for determining functional-metabolic correlates, temporal relationships, and intracellular actions of cardiotoxic chemicals nondestructively in an intact organ model. Toxicol. Appl. Pharmacol. 82:200-210.

Koskimies, A. I., and M. Kormano. 1975. Proteins in fluids from different segments of the rat epididymis. J. Reprod. Fertil. 43:345-348.

Koulischer, L., and R. Schoysman. 1974. Chromosomes and human infertility. I. Mitotic and meiotic chromosome studies in 202 consecutive male patients. Clin. Genet. 5:116-126.

Kouri, R. E., C. E. McKinney, A. S. Levine, B. K. Edwards, E. S. Vesell, D. W. Nebert, and T. L. McLemore. 1984. Variations in aryl hydrocarbon hydroxylase activities in mitogen-activated human and nonhuman primate lymphocytes. Toxicol. Pathol. 12:44-48.

Kovacs, L., M. Sas, B. A. Resch, G. Ugocsai, M. L. Swahn, M. Bygdeman, and P. J. Rowe. 1984. Termination of very early pregnancy by RU 486: An antiprogestational compound. Contraception 29:399-410.

Kram, D., G. D. Bynum, G. C. Senula, and E. L. Schneider. 1979. In utero sister chromatid exchange analysis for detection of transplacental mutagens. Nature 279:531.

Kram, D., G. D. Bynum, G. C. Senula, C. K. Bickings, and E. L. Schneider. 1980. In utero analysis of sister chromatid exchange: Alterations in susceptibility to mutagenic damage as a function of fetal cell type and gestational age. Proc. Natl. Acad. Sci. USA 77:4784-4787.

Kramer, J. M., and R. P. Erickson. 1981. Developmental program of PGK-1 and PGK-2 isozymes in spermatogenic cells of the mouse: Specific activities and rates of synthesis. Dev. Biol. 15:37-45.

Krauer, B., F. Krauer, F. E. Hytten, and E. del Pozo, Eds. 1984. Drugs and Pregnancy: Maternal Drug Handling-Fetal Drug Exposure. London: Academic Press. 281 pp.

Kreiss, K. 1985. Studies on populations exposed to polychlorinated biphenyls. Environ. Health. Perspect. 60:193-199.

Kremer, J. 1965. A simple sperm penetration test. Int. J. Fertil. 10:209-215.

Krichevsky, A., E. G. Armstrong, J. Schlatterer, S. Birken, S. Silverberg, J. W. Lustbader, and R. E. Canfield. 1988. Preparation and characterization of antibodies to the urinary fragment of the human chorionic gonadotropin beta-subunit. Endocrinology 123:584-593.

Kronenberg, M. S., and J. H. Clark. 1985. Changes in keratin expression during estrogen-mediated dif-

ferentiation of rat vaginal epithelium. Endocrinology 117:1480-1489.

Kuchera, L. K. 1974. Postcoital contraception with diethyl-stilbesterol—updated. Contraception 10:47-54.

Kuemmerle, H. P., and K. Brendel, Eds. 1984. Clinical Pharmacology in Pregnancy. Fundamentals and Rational Pharmacotherapy. New York: Thieme-Stratton. 386 pp.

Kuida, C. A., G. D. Braunstein, P. Shintaku, and J. W. Said. 1988. Human chorionic gonadotropin expression in lung, breast, and renal carcinomas. Arch. Path. Lab. Med. 112:282-285.

Laehdetie, J., and M. Parvinen. 1981. Meiotic micronuclei induced by X-rays in early spermatids of the rat. Mutat. Res. 81:103-115.

Lambert, G. H., D. A. Schoeller, A. N. Kotake, C. Flores, and D. Hay. 1986. The effect of age, gender, and sexual maturation on the caffeine breath test. Devel. Pharmacol. Ther. 9:375-388.

Lancranjan, I., H. I. Popescu, O. Gavanescu, I. Klepsch, and M. Serbanescu. 1975. Reproductive ability of workmen occupationally exposed to lead. Arch. Environ. Health 30:396-401.

Landrigan, P., E. L. Baker, Jr., R. G. Feldman, D. H. Cox, K. V. Eden, W. A. Orenstein, J. A. Mather, A. J. Yankel, and I. H. Von Lindern. 1976. Increased lead absorption with anemia and slowed nerve conduction in children near a lead smelter. J. Pediatr. 89:904-910.

Langlois, R. G., W. L. Bigbee, and R. H. Jensen. 1986. Measurements of the frequency of human erythrocytes with gene expression loss phenotypes at the glycophorin A locus. Hum. Genet. 74:353-362.

Langlois, R. G., W. L. Bigbee, S. Kyoizumi, N. Nakamura, M. A. Bean, M. Akiyama, and R. H. Jensen. 1987. Evidence for increased somatic cell mutations at the glycophorin A locus in atomic bomb survivors. Science 236:445-448.

Langlois, R. G., W. L. Bigbee, R. H. Jensen, and J. German. 1989. Evidence for increased in vivo mutation and somatic recombination in Bloom's syndrome. Proc. Natl. Acad. Sci. USA 86:670-674.

Langman, J., and C. C. Haden. 1970. Formation and migration of neuroblasts in the spinal cord of the chick embryo. J. Comp. Neurol. 138:419-432.

Langman, J., and G. W. Welch. 1967. Excess vitamin A and development of the cerebral cortex. J. Comp. Neurol. 131:15-26.

Langman, J., R. L. Guerrant, and B. G. Freeman. 1966. Behavior of neuro-epithelial cells during closure of the neural tube. J. Comp. Neurol. 127:399-411.

Langston, J. W., P. Ballard, J. Tertud, and I. Irwin. 1983. Chronic Parkinsonism in humans due to a product of meperidine-analog synthesis. Science 219:979-980.

Lasley, B. L., J. K. Hodges, and N. M. Czekala. 1980. Monitoring the female reproductive cycle of great apes and other primate species by oestrogen and LH in small volumes of urine. J. Reprod. Fertil. 28(Suppl.):121-129.

Latt, S. A., J. Allen, S. E. Bloom, A. Carrano, E. Falke, D. Kram, E. Schneider, R. Schreck, R. Tice, B. Whitfield, and S. Wolff. 1981. Sister-chromatid exchanges: A report of the GENE-TOX program. Mutat. Res. 87:17-62.

Lauritsen, J. G., J. Jorgensen, and F. Kissmeyer-Nielsen. 1976. Significance of HLA and blood-group incompatibility in spontaneous abortion. Clin. Genet. 9:575-582.

Lauwerys, R. R. 1983. Industrial Chemical Exposure: Guidelines for Biological Monitoring. Davis, Calif.: Biomedical Publications. 150 pp.

Lavery, J. P. 1982. Nonstress fetal heart rate testing. Clin. Obstet. Gynecol. 25:689-705.

Lea, O. A., and F. S. French. 1981. Characterization of an acidic glycoprotein secreted by principal cells of the rat epididymis. Biochim. Biophys. Acta 668:370-376.

Lea, O. A., P. Petrusz, and F. S. French. 1979. Prostatein. A major secretory protein of the rat ventral prostate. J. Biol. Chem. 254:6196-6202.

Leblond, C. P., and Y. Clermont. 1952. Definition of the stages of the cycle of the seminiferous epithelium in the rat. Ann. N.Y. Acad. Sci. 55:548-573.

Leenders, K. L., A. J. Palmer, N. Quinn, J. C. Clark, G. Firnau, E. S. Garnett, C. Nahmins, T. Jones, and C. D. Marsden. 1986. Brain dopamine metabolism in patients with Parkinson's disease measured with positron emission tomography. J. Neurol. Neurosurg. Psychiatry 49:853-860.

Legault, Y., M. Bouthillier, G. Bleau, A. Chapdelaine, and K. D. Roberts. 1979. The sterol and sterol sulfate content of the male hamster reproductive tract. Biol. Reprod. 20:1213-1219.

Lejeune, B., and F. Leroy. 1980. Role of the uterine epithelium in inducing the decidual cell reaction, pp. 92-101. In F. Leroy, C. A. Finn, and A. Psychoyos, Eds. Blastocyst-Endometrium Relationships. Progress in Reproductive Biology and Medicine, Vol. 7. Basel: S Karger AG.

LeMaire, W. J., N. S. T. Yang, H. H. Behrman, and J. M. Marsh. 1973. Preovulatory changes in the concentration of prostaglandins in rabbit Graafian follicles. Prostaglandins 3:367-376.

Lemasters, G. K., A. Hagen, and S. J. Samuels. 1985. Reproductive outcomes in women exposed to solvents in 36 reinforced plastics companies. I. Menstrual dysfunction. J. Occup. Med. 27:490-494.

Leridon, H. 1977. Human Fertility: The Basic Components. Chicago: University of Chicago Press. 202 pp.

Lerman, L. S. 1985. A Proposal for the Detection of Mutations in a Large-Scale Sampling of the Human Genome Using Two-Dimensional Denaturing Gradient Separations. Typescript, contract report prepared for the Office of Technology Assessment, U. S. Congress, Washington, D.C.

Levine, M. S., N. L. Fox, B. Thompson, W. Taylor, A. C. Darlington, J. Van der Hoeden, E. A. Emmett, and W. Rutten. 1986. Inhibition of esterase activity and an undercounting of circulating monocytes in a population of production workers. J. Occup. Med. 28:207-211.

Levine, N., and D. J. Marsh. 1971. Micropuncture studies of the electrochemical aspects of fluid and electrolyte transport in individual seminiferous tubules, the epididymis and the vas deferens in rats. J. Physiol. 213:557-570.

Levine, R. J. 1983. Methods for detecting occupational causes of male infertility. Reproductive history versus semen analysis. Scand. J. Work Environ. Health 9:371-376.

Levine, R. J., M. J. Symons, S. A. Balogh, D. M. Arndt, N. R. Kaswandik, and J. W. Gentile. 1980. A method for monitoring the fertility of workers. I. Method and pilot studies. J. Occup. Med. 22:781-791.

Levine, R. J., P. B. Blunden, R. D. DalCorso, T. B. Starr, and C. E. Ross. 1983. Superiority of reproductive histories to sperm counts in detecting infertility at a dibromochloropropane manufacturing plant. J. Occup. Med. 25:591-597.

Lewin, L. M., R. Weissenberg, J. S. Sobel, Z. Marcus, and L. Nebel. 1979. Differences in concanavalin A-FITC binding to rat spermatozoa during epididymal maturation and capacitation. Arch. Androl. 2:279-281.

Lewis, E. L., M. O. Rasor, and J. W. Overstreet. 1985. Measurement of human testicular consistency by tonometry. Fertil. Steril. 43:911-916.

Lewis, S. E., and F. M. Johnson. 1986. The nature of spontaneous and induced electrophoretically detected mutations in the mouse, pp. 359-365. In C. Ramel, B. Lambert, and J. Magnusson, Eds. Genetic Toxicology of Environmental Chemicals, Part B: Genetic Effects and Applied Mutagenesis. Proceedings of the Fourth International Conference on Environmental Mutagens held in Stockholm, Sweden, June 24-25, 1985. New York: Liss.

Liber, H. L., S. L. Danheiser, and W. G. Thilly. 1985. Mutation in single-cell systems induced by low-level mutagen exposure. Basic Life Sci. 33:169-204.

Lightfoot, R. J., K. P. Croker, and H. G. Neil. 1967. Failure of sperm transport in relation to ewe infertility following prolonged grazing on oestrogenic pastures. Aust. J. Agric. 18:755-765.

Lilja, H., J. Oldbring, G. Rannevik, and C. B. Laurell. 1987. Seminal vesicle-secreted proteins and their reactions during gelation and liquefaction of human semen. J. Clin. Invest. 80:281-285.

Limic, N., and V. Valkovic. 1986. Environmental influence on trace element levels in human hair. Bull. Environ. Contam. Toxicol. 37:925-930.

Lindquist, O., and C. Bengtsson. 1979. The effect of smoking on menopausal age. Maturitas 1:171-173.

Livingston, G. K. 1984. An overview of chromosomal, micronucleus, heritable effects, dominent lethal, heritable translocation, and specific locus tests. Prog. Clin. Biol. Res. 160:417-427.

Lloyd, C. W., J. Lobotsky, J. Weisz, D. T. Baird, J. A. McCracken, M. Pupkin, J. Zanartu, and J. Puga. 1971. Concentration of unconjugated estrogens, androgens and gestagens in ovarian and peripheral venous plasma of women: The normal menstrual cycle. J. Clin. Endocrinol. Metab. 32:155-166.

Lodge, A. 1976. Developmental findings with infants born to mothers on methadone maintenance: A preliminary report, pp. 79-85. In G. Beschner and R. Brotman, Eds. Symposium on Comprehensive Health Care for Addicted Families and Their Children. Rockville, Md.: National Institute on Drug Abuse.

Longo, L. D. 1972. Placental transfer mechanisms—an overview. Obstet. Gynceol. Annu. 1:103-138.

Longo, L. D. 1982. Some health consequences of maternal smoking: Issues without answers, pp. 13-31. In W. L. Nyhan and K. L. Jones, Eds. Prenatal Diagnosis and Mechanisms of Teratogenesis. March of Dimes Birth Defects Foundation, Original Article Series 18(3a). New York: Liss.

Longo, L. D. 1983. Maternal blood volume and cardiac output during pregnancy: A hypothesis of endocrinolic control. Am. J. Physiol. 245(5 Pt. 1):R720-R729.

Loumaye, E., J. Donnez, and K. Thomas. 1985. Ovulation instantaneously modifies women's peritoneal fluid characteristics: A demonstration from an in vitro fertilization program. Fertil. Steril. 44:827-829.

Lower, G. M., Jr., and M. S. Kanarek. 1982. Risk, susceptibility and the epidemiology of proliferative neoplastic disease: Descriptive vs. mechanistic approaches. Med. Hypotheses 9:33-49.

Lower, G. M., L. E. Stevens, J. S. Najarian, and K. Reemtsma. 1971. Problems from immunosupressives during pregnancy. Am. J. Obstet. Gynecol. 111:1120-1121.

Lu, L.-J. W., R. M. Disher, M. V. Reddy, and K. Randerath. 1986. ^{32}P-postlabeling assay in mice of transplacental DNA damage induced by the environmental carcinogens safrole, 4-aminobiphenyl, and benzo(a)pyrene. Cancer Res. 46:3046-3054.

Lu, Y.-C., and P.-N. Wong. 1984. Dermatologic, medical, and laboratory findings of patients in Taiwan and their treatments. Am. J. Ind. Med. 5:81-115.

Lubet, R. A., M. J. Brunda, D. Taramelli, D. Dansie, D. W. Nebert, and R. E. Kouri. 1984. Induction of immunotoxicity of polycyclic hydrocarbons: Role of the Ah locus. Arch. Toxicol. 56:18-24.

Luciano, A. A., K. S. Hauser, F. K. Chapler, and B. M. Sherman. 1981. Danazol: Endocrine consequences in healthy women. Am. J. Obstet. Gynecol. 141:723-727.

Lutwak-Mann, C. 1964. Observations on progeny of thalidomide-treated male rabbits. Br. Med. J. 1:1090-1091.

Maathuis, J. B., and R. J. Aitken. 1978. Cycle variation in the concentrations of protein and hexose in human uterine flushings collected by an improved technique. J. Reprod. Fertil. 52:289-295.

Maathuis, J. B., P. F. A. Van Look, and E. A. Michie. 1978. Changes in volume, total protein and ovarian steroid concentrations of peritoneal fluid throughout the human menstrual cycle. J. Endocrinol. 76:123-133.

Macdonald, R. R. 1969. Cyclic changes in cervical mucus. 1. Cyclic changes in cervical mucus as an indication of ovarian function. J. Obstet. Gynaecol. Br. Commonw. 76:1090-1094.

MacLeod, J. 1964. Human seminal cytology as a sensitive indicator of the germinal epithelium. Int. J. Fertil. 9:281-295.

MacLeod, J., and R. Z. Gold. 1951. The male factor in fertility and infertility: Spermatozoon counts in 1000 men of known fertility and in 1000 cases of infertile marriage. J. Urol. 66:436-449.

MacLeod, J., and Y. Wang. 1979. Male fertility potential in terms of semen quality: A review of the past, a study of the present. Fertil. Steril. 31:103-116.

MacLeod, S. M., and I. C. Radde. 1985. Textbook of Pediatric Clinical Pharmacology. Littleton, Mass.: PSG Publishing. 467 pp.

MacMahon, B., and T. F. Pugh. 1970. Epidemiology: Principles and Methods. Boston: Little, Brown and Co. 376 pp.

MacMahon, B., and J. Worcester. 1966. Age at menopause: United States, 1960-1962. Vital Health Stat. 11(19):1-20.

MacQueen, J. M., and F. P. Sanfilippo. 1984. The effect of parental HLA compatibility on the expression of paternal haplotypes in offspring. Hum. Immunol. 11:155-161.

Maddox, D. E., J. H. Butterfield, S. J. Ackerman, C. B. Coulam, and G. J. Gleich. 1983. Elevated serum levels in human pregnancy of a molecule immunochemically similar to eosinophil granule major basic protein. J. Exp. Med. 158:1211-1226.

Maddox, D. E., G. M. Kephart, C. B. Coulam, J. H. Butterfield, K. Benirschke, and G. J. Gleich. 1984. Localization of a molecule immunochemically similar to eosinophil major basic protein in human placenta. J. Exp. Med. 160:29-41.

Magrini, G., G. Chiodoni, F. Rey, and J. P. Felber. 1986. Further evidence for the usefulness of the salivary testosterone radioimmunoassay in the assessment of androgenicity in man in basal and stimulated conditions. Horm. Res. 23:65-73.

Mahesh, V. B., R. B. Greenblatt, and R. F. Coniff. 1968. Adrenal hyperplasia—a case report of delayed onset of the congenital form or an acquired form. J. Clin. Endocrinol. Metab. 28:619-623.

Mahony, M. C., N. J. Alexander, and R. J. Swanson. 1988. Evaluation of semen parameters by means of automated sperm motion analyzers. Fertil. Steril. 49:876-880.

Maier, D., and S. Kuslis. 1987. Uterine Universal Fluid (ULF) Volume and Prolactin (UPRL) in Normal Menstrual Cycles. Presented at the 34th Annual meeting of the Society for Gynecological Investigation held March 12-21, 1987 in Atlanta, Georgia. Abstract No. 251, p. 155.

Mailhes, J. B., R. J. Preston, and K. S. Lavappa. 1986. Mammalian in vivo assays for aneuploidy in female germ cells. Mutat. Res. 167:139-148.

Makler, A., I. Zaidise, E. Paldi, and J. M. Brandes. 1979a. Factors affecting sperm motility. I. In vitro change in motility with time after ejaculation. Fertil. Steril. 31:147-154.

Makler, A., J. Itskovitz, J. M. Brandes, and E. Paldi. 1979b. Sperm velocity and percentage of motility in 100 normospermic specimens analyzed by the multiple exposure photography (MEP) method. Fertil. Steril. 31:155-161.

Malmborg, A. S. 1978. Antimicrobial drugs in human seminal plasma. J. Antimicrob. Chemother. 4:483-485.

Malone, J. D., M. Richards, and A. J. Kahn. 1986. Human peripheral monocytes express putative receptors for neuroexcitatory amino acids. Proc. Natl. Acad. Sci. USA 83:3307-3310.

Manchester, D., and E. Jacoby. 1984. Decreased placental monoxygenase activities associated with birth defects. Teratology 30:31-37.

Manchester, D. K., N. B. Parker, and C. M. Bowman. 1984. Maternal smoking increases xenobiotic metabolism in placenta but not umbilical vein endothelium. Pediatr. Res. 18:1071-1075.

Manchester, D. K., S. K. Gordon, C. L. Golas, E. A. Roberts, and A. B. Okey. 1987. Ah receptor in human placenta: Stabilization by molybdate and characterization of binding of 2,3,7,8-tetrachlorodibenzo-p-dioxin, 3-methylcholanthrene, and benzo(a)pyrene. Cancer Res. 47:4861-4868.

Maneely, R. B. 1959. Epididymal structure and function. A historical and critical review. Acta Zool. 40:1-21.

Mann, T., and C. Lutwak-Mann. 1981. Male Reproductive Function and Semen: Themes and Trends in Physiology, Biochemistry and Investigative Andrology. New York: Springer-Verlag. 495 pp.

Marcal, J. M., N. J. Chew, D. S. Salomon, and M. I. Sherman. 1975. Δ^5,3-ß-hydroxysteroid dehydrogenase activities in rat trophoblast and ovary during pregnancy. Endocrinology 96:1270-1279.

Markkula-Viitanen, M., V. Nikkanen, and A. Talo. 1979. Electrical activity and intraluminal pressure of the cauda epididymidis of the rat. J. Reprod. Fertil. 57:431-435.

Markoff, E., P. Zeitler, S. Peleg, and S. Handwerger. 1983. Characterization of the synthesis and release of prolactin by an enriched fraction of human decidual cells. J. Clin. Endocrinol. Metab. 56:962-968.

Marshall, E. 1986. Immune system theories on trial. Science 334:1490-1492.

Marshall, J. R., C. B. Hammond, G. T. Ross, A. Jacobsen, P. Rayford, and W. D. Odell. 1968. Plasma and urinary chorionic gonadotropin during early human pregnancy. Obstet. Gynecol. 32:760-764.

Marshall, J. F., M. C. Drew, and K. A. Neve. 1983. Recovery of function after mesotelencephalic dopaminergic injury in senescence. Brain Res. 259:249-260.

Marshall, W. A., and J. M. Tanner. 1969. Variations in pattern of pubertal changes in girls. Arch. Dis. Child. 44:291-303.

Martal, J., J. Djiane, and M. P. Dubois. 1977. Immunofluorescent localization of ovine placental lactogen. Cell Tissue Res. 184:424-433.

Martin, R. H. 1983. Sperm chromosomal analysis after radiotherapy. Hum. Genet. 79:392-393.

Martin, R. H., W. Balkan, K. Burns, A. W. Rademaker, C. C. Lin, and N. L. Rudd. 1983. The chromosome constitution of 1000 human spermatozoa. Hum. Genet. 63:305-309.

Martin, R. H., A. W. Rademaker, K. Hildebrand, L. Long-Simpson, D. Peterson, and J. Yamamoto. 1987. Variation in the frequency and type of sperm chromosomal abnormalities among normal men. Hum. Genet. 77:108-114.

Martin-Deleon, P. A., E. L. Shaver, and E. B. Gammal. 1973. Chromosome abnormalities in rabbit blastocysts resulting from spermatozoa aged in the male tract. Fertil. Steril. 24:212-219.

Martz, F., C. Failinger, III, and D. A. Blake. 1977. Phenytoin teratogenesis: Correlation between embryopathic effect and covalent binding of putative arene oxide metabolite in gestational tissue. J. Pharmacol. Expt. Therap. 203:231-239.

Maruo, T., S. J. Segal, and S. S. Koide. 1979. Studies on the apparent human chorionic gonadotropin-like factor in the crab *Ovalipes ocellatus*. Endocrinology 104:932-939.

Maslar, I. A., and D. H. Riddick. 1979. Prolactin production by human endometrium during the normal menstrual cycle. Am. J. Obstet. Gynecol. 135:751-754.

Maslar, I. A., P. Powers-Craddock, and R. Ansbacher. 1986. Decidual prolactin production by organ cultures of human endometrium: Effects of continuous and intermittment progesterone treatment. Biol. Reprod. 34:741-750.

Mason, A. J., J. S. Hayflick, N. Ling, F. Esch, N. Ueno, S.-Y. Ying, R. Guillemin, H. Niall, and P. H. Seeburg. 1985. Complementary DNA sequences of ovarian follicular fluid inhibin show precursor structure and homology with transforming growth factor-beta. Nature 318:659-663.

Masuda, Y., T. Yamaryo, K. Haraguchi, M. Kuratsune, and S.-T. Hsu. 1982. Comparison of causal agents in Taiwan and Fukuoka PCB poisonings. Chemosphere 11:199-206.

Mathur, S., E. R. Baker, H. O. Williamson, F. C. Derrick, K. J. Teague, and H. H. Fudenberg. 1981. Clinical significance of sperm antibodies in infertility. Fertil. Steril. 36:486-495.

Matsumoto, A. M. 1988. The testis, pp. 1404-1421. J. B. Wyngaarden and L. H. Smith, Jr., Eds. In Cecil Textbook of Medicine, 18th ed. Philadelphia: W. B. Saunders Company.

Matter, L., and Faulk, W. P. 1980. Fibrinogen degradation products and Factor VIII consumption in normal pregnancy and pre-eclampsia: Role of the placenta, pp. 357-369. In J. Bonnar, I. MacGillivray, and M. Symonds, Eds. Pregnancy Hypertension. Proceedings of the First Congress of the International Society for the Study of Hypertension in Pregnancy, held at University College, Dublin, on 27-29 September 1978. Lancaster, Eng.: MTP Press.

Mattison, D. R. 1982. The effects of smoking on fertility from gametogenesis to implantation. Environ. Res. 28:410-433.

Mattison, D. R. 1985. Clinical manifestations of ovarian toxicity, pp. 109-130. In R. L. Dixon, Ed. Reproductive Toxicology. Target Organ Toxicology Series. New York: Raven Press.

Mattison, D. R. 1986. Physiologic variations in pharmacokinetics during pregnancy, pp. 37-102. In S. Fabro and A. R. Scialli, Eds. Drug and Chemical Action in Pregnancy: Pharmacologic and Toxicologic Principles. Reproductive Medicine, Vol. 8. New York: Marcel Dekker.

Mattison, D. R., and T. Angtuaco. 1988. Magnetic resonance imaging in prenatal diagnosis. Clin. Obstet. Gynecol. 31:353-389.

Mattison, D. R., and S. S. Thorgeirsson. 1977. Genetic differences in mouse ovarian metabolism of benzo-(a)pyrene and oocyte toxicity. Biochem. Pharmacol. 26:909-912.

Mattison, D. R., M. S. Nightingale, and K. Shiromizu. 1983. Effects of toxic substances on female reproduction. Environ. Health Perspect. 48:43-52.

Mattison, D. R., H. H. Kay, R. K. Miller, and T. Angtuaco. 1988. Magnetic resonance imaging: A noninvasive tool for fetal and maternal physiology. Biol. Reprod. 38:39-49.

Maurissen, J. P. 1985. Psychophysical testing in human populations exposed to neurotoxicants. Neurobehav. Toxicol. Teratol. 7:309-317.

Maurissen, J. P., and B. Weiss. 1980. Vibration sensitivity as an index of somatosensory function. In P. H. Spencer and H. H. Schaumberg, Eds. Experimental and Clinical Neurotoxicology. Baltimore: Williams & Wilkins.

Maxam, A. M., and W. Gilbert. 1977. A new method for sequencing DNA. Proc. Natl. Acad. Sci. USA 74:560-564.

May, P. C., and S. G. Kohama. 1986. Systemic treatment of adult female C57BL/6J mice with the neuronal excitotoxin N-methyl-aspartic acid mimics the age-related lengthening of the estrous cycle, p. 233. In Endocrine Society 68th Annual Meeting. Anaheim, California. Program and Abstracts.

Mayorga, L. S., and F. Bertini. 1982. Effect of androgens on the activity of acid hydrolases in rat epididymis. Int. J. Androl. 5:345-352.

Mazurkiewicz, J. E., J. F. Bank, and S. G. Joshi. 1981. Immunocytochemical localization of a progestagen-associated endometrial protein in the human decidua.

J. Clin. Endocrinol. Metab. 52:1006-1008.

McCarthy, S., and F. Haseltine. 1987. Magnetic Resonance of the Reproductive System. Thorofare, N.J.: Slack. 230 pp.

McConnachie, P. R., and J. A. McIntyre. 1984. Maternal antipaternal immunity in couples predisposed to repeated pregnancy losses. Am. J. Reprod. Immunol. 5:145-150.

McCormack, S. A., and S. R. Glasser. 1978. Ontogeny and regulation of a rat placental estrogen receptor. Endocrinology 102:273-280.

McCormack, S. A., and S. R. Glasser. 1980. Differential response of individual uterine cell types from immature rats treated with estradiol. Endocrinology 106:1634-1649.

McCullagh, D. R. 1932. Dual endocrine activity of the testes. Science 76:19-20.

McDonough, P. G. 1985. Applications of molecular biology to perinatal medicine. Sem. Perinatol. 9:250-256.

McIlree, M. E., W. H. Price, W. M. Brown, W. S. Tulloch, J. E. Newsam, and N. Maclean. 1966. Chromosome studies on testicular cells from 50 subfertile men. Lancet 2(454):69-71.

McIntire, M. S., and C. R. Angle. 1972. Air lead: Relation to lead in blood of black school children deficient in glucose-6-phosphate dehydrogenase. Science 177:520-521.

McIntyre, J. A., and W. P. Faulk. 1979a. Trophoblast modulation of maternal allogeneic recognition. Proc. Natl. Acad. Sci. USA 76:4029-4032.

McIntyre, J. A., and W. P. Faulk. 1979b. Antigens of human trophoblast: Effects of heterologous anti-trophoblast sera on lymphocyte responses in vitro. J. Exp. Med. 149:824-836.

McIntyre, J. A., and W. P. Faulk. 1979c. Maternal blocking factors in human pregnancy are found in plasma not serum. Lancet 2(8147):821-823.

McIntyre, J. A., and W. P. Faulk. 1982. Allotypic trophoblast-lymphocyte cross-reactive (TLX) cell surface antigens. Hum. Immunol. 4:27-35.

McIntyre, J. A., and W. P. Faulk. 1983. Recurrent spontaneous abortion in human pregnancy: Results of immunogenetical, cellular, and humoral studies. Am. J. Reprod. Immunol. 4:165-170.

McIntyre, J. A., and W. P. Faulk. 1985. Laboratory and clinical aspects of research in chronic spontaneous abortion. Diagn. Immunol. 3:163-170.

McIntyre, J. A., P. R. McConnachie, C. G. Taylor, and W. P. Faulk. 1984a. Clinical, immunologic, and genetic definitions of primary and secondary recurrent spontaneous abortions. Fertil. Steril. 42:849-855.

McIntyre, J. A., W. P. Faulk, S. J. Verhulst, and J. A. Colliver. 1984b. Human trophoblast-lymphocyte cross-reactive (TLX) antigens define a new alloantigen system. Science 222:1135-1137.

McIntyre, J. A., W. P. Faulk, V. R. Nichols-Johnson, and C. G. Taylor. 1986. Immunologic testing and immunotherapy in recurrent spontaneous abortion.

Obstet. Gynecol. 67:169-175.

McKay, D. J., B. S. Renaux, and G. H. Dixon. 1985. The amino acid sequence of human sperm protamine P1. Biosci. Rep. 5:383-391.

McKay, D. J., B. S. Renaux, and G. H. Dixon. 1986. Human sperm protamines. Amino acid sequences of two forms of protamine P2. Eur. J. Biochem. 156:5-8.

McKinlay, S. M., N. L. Bifano, and J. B. McKinlay. 1985. Smoking and age at menopause in women. Ann. Inter. Med. 103:350-356.

McKusick, V. A. 1975. Mendelian Inheritance in Man, 4th ed. Baltimore, Md.: Johns Hopkins University Press. 837 pp.

McNatty, K. P., and D. T. Baird. 1978. Relationship between follicle-stimulating hormone, androstenedione and oestradiol in human follicular fluid. J. Endocrinol. 76:527-531.

McNatty, K. P., D. T. Baird, A. Bolton, P. Chambers, C. S. Corker, and H. McLean. 1976. Concentration of oestrogens and androgens in human ovarian venous plasma and follicular fluid throughout the menstrual cycle. J. Endocrinol. 71:77-85.

McWilliams, D., and I. Boime. 1980. Cytological localization of placental lactogen messenger ribonucleic acid in syncytiotrophoblast layers of human placenta. Endocrinology 107:761-765.

Meistrich, M. L. 1977. Separation of spermatogenic cells and nuclei from rodent testes, pp. 15-24. In D. M. Prescott, Ed. Methods in Cell Biology, Vol. XV. New York: Academic Press.

Meistrich, M. L. 1982. Quantitative correlation between testicular stem cell survival, sperm production, and fertility in the mouse after treatment with different cytotoxic agents. J. Androl. 3:58-68.

Meistrich, M. L. 1986. Critical components of testicular function and sensitivity to disruption. Biol. Reprod. 34:17-28.

Meistrich, M. L. 1987. Comparative male gonadal toxicity from cytotoxic cancer therapies. In J. J. Mulvihill and R. H. Sherins, Eds. Reproduction and Cancer. New York: Raven.

Meistrich, M. L., and C. C. Brown. 1983. Estimation of the increased risk of human infertility from alterations in semen characteristics. Fertil. Steril. 2:220-230.

Meistrich, M. L., T. H. Hughes, and W. R. Bruce. 1975. Alteration of epididymal sperm transport and maturation in mice by oestrogen and testosterone. Nature 258:145-147.

Meistrich, M. L., P. K. Trostle, M. Frapart, and R. P. Erickson. 1977. Biosynthesis and localization of lactate dehydrogenase X in pachytene spermatocytes and spermatids of mouse testes. Dev. Biol. 60:428-441.

Meistrich, M. L., L. S. Goldstein, and A. J. Wyrobek. 1985. Long-term infertility and dominant lethal mutations in male mice treated with adriamycin. Mutat. Res. 152:53-65.

Memo, M., C. Missale, M. O. Carruba, and P. F. Spano.
1986. Pharmacology and biochemistry of dopamine receptors in the central nervous system and peripheral tissue. J. Neural Transm. 22(Suppl.):19-32.

Menken, J., J. Trussell, and U. Larsen. 1986. Age and infertility. Science 233:1389-1394.

Mes, J., D. J. Davies, and D. Turton. 1982. Polychlorinated biphenyl and other chlorinated hydrocarbon residues in adipose tissue of Canadians. Bull. Environ. Contam. Toxicol. 28:97-104.

Metcalf, M. G., R. A. Donald, and J. H. Livesey. 1981. Pituitary-ovarian function in normal women during the menopausal transition. Clin. Endocrinol. (Oxf.) 14:245-255.

Meyer-Bahlburg, H. F., A. A. Ehrhardt, J. F. Feldman, L. R. Rosen, N. P. Verdiano, and I. Zimmerman. 1985. Sexual activity level and sexual functioning in women prenatally exposed to diethylstilbestrol. Psychosom. Med. 47:497-511.

Michelson, A. M., G. A. P. Bruns, C. C. Morton, and S. H. Orkin. 1985. The human phosphoglycerate kinase multigene family. HLA-associated sequences and an X-linked locus containing a processed pseudogene and its functional counterpart. J. Biol. Chem. 260:6982-6992.

Michnovicz, J. J., R. J. Hershcopf, H. Naganuma, H. L. Bradlow, and J. Fishman. 1986. Increased 2-hydroxylation of estradiol as a possible mechanism for the anti-estrogenic effect of cigarette smoking. N. Eng. J. Med. 315:1305-1309.

Mikhail, G. 1970. Hormone secretion by the human ovaries. Gynecol. Invest. 1:5-20.

Millan, J. L., C. E. Driscoll, K. M. Levan, and E. Goldberg. 1987. Epitopes of human testis-specific lactate dehydrogenase deduced from a cDNA sequence. Proc. Natl. Acad. Sci. USA 84:5311-5315.

Miller, J. F., E. Williamson, J. Glue, Y. B. Gordon, J. G. Grudzinskas, and A. Sykes. 1980. Fetal loss after implantation: A prospective study. Lancet 2:554-556.

Miller, J. R. 1983. International Commission for Protection against Environmental Mutagens and Carcinogens. ICPEMC Working paper 5/4. Perspectives in mutation epidemiology. 4: General principles and considerations. Mutat. Res. 114:425-447.

Miller, M. W. 1986. Effects of alcohol on the generation and migration of cerebral cortical neurons. Science 233:1308-1311.

Miller, R. K. 1983. Perinatal toxicology: Its recognition and fundamentals. Am. J. Ind. Med. 4:205-244.

Miller, R. K., and C. K. Kellogg. 1985. The pharmacodynamics of prenatal chemical exposure. Natl. Inst. Drug Abuse Res. Mongr. Ser. 60:39-57.

Miller, R. K., and Z. Shaikh. 1983. Perinatal metabolism: Metals and metallothionein, pp. 151-204. In T. Clarkson, G. Nordberg, and P. Sager, Eds. Developmental and Reproductive Toxicity of Metals. New York: Plenum Press.

Miller, R. K., T. R. Koszalka, and R. L Brent. 1976. The transport of molecules across placental membranes, pp. 145-223. In G. Poste and G. L. Nicolson, Eds. The Cell Surface in Animal Embryogenesis and Development. Cell Surface Reviews, Vol. 1. Amsterdam: North-Holland.

Miller, R. K., C. K. Kellogg, and R. A. Saltzman. 1987a. Reproductive and perinatal toxicology, pp. 195-309. In T. J. Haley, and W. O. Berndt, Eds. Handbook of Toxicology. Washington, D.C.: Hemisphere.

Miller, R. K., D. R. Mattison, M. Panigel, T. Ceckler, R. Bryant, and P. Thomford. 1987b. Kinetic assessment of manganese using magnetic resonance imaging in the dually perfused human placenta *in vitro.* Environ. Health Perspect. 74:81-91.

Miller, R. K., D. Mattison, and D. Plowchalk. 1988. Biological monitoring of the human placenta, pp. 567-602. In T. W. Clarkson, L. Freiberg, G. Nordberg, and P. Saeger, Eds. Biological Monitoring of Toxic Metals. New York: Plenum Press.

Mills, J. L., and D. Alexander. 1986. Teratogens and "litogens." N. Engl. J. Med. 315:1234-1236.

Mirmiran, M., E. Brenner, J. van der Gugten, and D. F. Swaab. 1985. Neurochemical and electrophysiological disturbances mediate developmental behavioral alterations produced by medicines. Neurobehav. Toxicol. Teratol. 7:677-683.

Mishell, D. R., Jr. 1979. Oral steroids, pp. 487-523. In D. R. Mishell, Jr. and V. Davajan, Eds. Reproductive Endocrinology, Infertility, and Contraception. Philadelphia: F. A. Davis.

Mizejewski, G. J., and I. H. Porter, Eds. 1985. Alpha-Fetoprotein and Congenital Disorders. Orlando, Fla.: Academic Press. 363 pp.

Mjones, H. 1949. Paralysis agitans: Clinical and genetic study. Acta Psychiatr. Neurol. Scand. (Suppl. 54):1-195.

Mobbs, C. V., K. Flurkey, D. M. Gee, K. Yamamoto, Y. N. Sinha, and C. E. Finch. 1984. Estradiol-induced adult anovulatory syndrome in female C57BL/6J mice: Age-like neuroendocrine, but not ovarian, impairments. Biol. Reprod. 30:556-563.

Mobbs, C. V., L. S. Kannegieter, and C. E. Finch. 1985. Delayed anovulatory syndrome induced by estradiol in female C57BL/6J mice: Age-like neuroendocrine, but not ovarian, impairments. Biol. Repro. 32:1010-1017.

Mocarelli, P., A. Marocchi, P. Brambilla, P. Gerthoux, D. S. Young, and N. Mantel. 1986. Clinical laboratory manifestations of exposure to dioxin in children. A six-year study of the effects of an environmental disaster near Seveso, Italy. J. Am. Med. Assoc. 256:2687-2695.

Moghissi, K. S. 1966. Cyclic changes of cervical mucus in normal and progestin-treated women. Fertil. Steril. 17:663-675.

Moghissi, K. S. 1973. Composition and function of cervical secretion, pp. 25-48. In R. O. Greep, Ed. Handbook of Physiology, Section 7: Endocrinology. Volume II. Female Reproductive System, Part 2. Washington, D.C.: American Physiological Society.

Moghissi, K. S. 1976. Postcoital test: Physiologic basis, technique and interpretation. Fertil. Steril. 27:117-129.

Moghissi, K. S., F. N. Syner, and T. N. Evans. 1972. A composite picture of the menstrual cycle. Am. J. Obstet. Gynecol. 114:405-418.

Molinatti, G. M., F. Massara, M. Messina, and G. Anselmo. 1964. Postpubertal adrenal virilism due to a defect in 21-hydrosylation. Folia Endocrinol. 17:630-636.

Moore, H. D. M. 1981. Effects of castration on specific glycoprotein secretions of the epididymis in the rabbit and hamster. J. Reprod. Fertil. 61:347-354.

Morgan, F. J., S. Birken, and R. E. Canfield. 1973. Human chorionic gonadotropin: A proposal for the amino acid sequence. Mol. Cell. Biochem. 2:79-99.

Morgan, F. J., S. Birken, and R. E. Canfield. 1975. The amino acid sequence of human chorionic gonadotropin (The α subunit and β subunit). J. Biol. Chem. 250:5247-5258.

Morgan, W. F., and S. Wolff. 1984. Effect of 5-bromodeoxyuridine substitution on sister chromatid exchange induction by chemicals. Chromosoma 89:285-289.

Mori, H., and A. K. Christensen. 1980. Morphometric analysis of Leydig cells in the normal rat testis. J. Cell Biol. 84:340-354.

Mori, H., N. Hiromoto, M. Nakahara, and T. Shiraishi. 1982. Stereological analysis of Leydig cell ultrastructure in aged humans. J. Clin. Endocrinol. Metab. 55:634-641.

Morley, A. A., K. J. Trainor, R. Seshadri, and R. G. Ryall. 1983. Measurement of in vivo mutations in human lymphocytes. Nature 302:155-156.

Morris, N. M., J. R. Udry, F. Khan-Dawood, and M. Y. Dawood. 1987. Marital sex frequency and midcycle female testosterone. Arch. Sex Behav. 16:27-37.

Mortimer, D., A. A. Templeton, E. A. Linton, and R. A. Coleman. 1982. Influence of abstinence and ejaculation-to-analysis delay on semen analysis parameters of suspected infertile men. Arch. Androl. 8:251-256.

Morton, H., V. Hegh, and G. J. Clunie. 1976. Studies of the rosette inhibition test in pregnant mice: Evidence for immunosuppression? Proc. R. Soc. Lond. [Biol.] 193(1113):413-419.

Morton, H., B. Rolfe, and G. J. A. Clunie. 1977. An early pregnancy factor detected in human serum by the rosette inhibition test. Lancet 1:394-397.

Morton, H., C. D. Nancarrow, R. J. Scaramuzzi, B. M. Evison, and G. J. A. Clunie. 1979. Detection of early pregnancy in sheep by the rosette inhibition test. J. Reprod. Fertil. 56:75-80.

Morton, H., B. E. Rolfe, and A. L. Cavanaugh. 1982. Early pregnancy factor: Biology and clinical significance, pp. 391-405. In G. Grudzinskas, B. Teisner, and M. Seppala, Eds. Pregnancy Proteins. Sydney: Academic Press.

Moruzzi, J. F., A. J. Wyrobek, B. H. Mayall, and B. L.

Gledhill. 1988. Quantification and classification of human sperm morphology by computer-assisted image analysis. Fertil. Steril. 50:142-152.

Moses, M., and P. G. Prioleau. 1985. Cutaneous histologic findings in chemical workers with and without chloracne with past exposure to 2,3,7,8-tetrachlorodibenzo-para-dioxin. J. Am. Acad. Dermatol. 12:497-506.

Mosher, W. D. 1980. Reproductive impairments among currently married couples: United States, 1976, pp. 1-11. In Advance Data from Vital and Health Statistics of the National Center for Health Statistics. Rockville, Md.: U.S. Department of Health, Education, and Welfare.

Mowbray, J. F. 1987. Genetic and immunological factors in human recurrent abortion. Am. J. Reprod. Immunol. Microbiol. 15:138-140.

Mowbray, J. F., and J. L. Underwood. 1985. Immunology of abortion. Clin. Exp. Immunol. 60:1-7.

Mowbray, J. F., C. R. Gibbing, A. S. Sidgwick, M. Ruszkiewicz, and R. W. Beard. 1983. Effects of transfusion in women with recurrent spontaneous abortion. Transplant Proc. 15:896-899.

Moyle, W. R., C. Lin, R. L. Corson, and P. H. Ehrlich. 1983a. Quantitative explanation for increased affinity shown by mixtures of monoclonal antibodies: Importance of a circular complex. Mol. Immunol. 20:439-452.

Moyle, W. R., D. M. Anderson, and P. H. Ehrlich. 1983b. A circular antibody-antigen complex is responsible for increased affinity shown by mixtures of monoclonal antibodies to human chorionic gonadotropin. J. Immunol. 131:1900-1905.

Müller, H. J. 1927. Artificial transmutation of the gene. Science 66:84-87.

Mulvihill, J. J., and J. Byrne. 1985. Offspring of long-time survivors of childhood cancer. Late Effects in Sucessfully Treated Children with Cancer, Clinics in Oncology 4(2):333-343.

Mulvihill, J. J., and A. Czeizel. 1983. Perspectives in mutations epidemiology 6: A 1983 view of sentinel phenotypes. Mutat. Res. 123:345-361.

Mulvihill, J. J., and A. M. Yeager. 1976. Fetal Alcohol Syndrome. Teratology 13:345-348.

Murakami, M., A. Sugita, and M. Hamasaki. 1982a. Scanning electron microscopic observations of the vas deferens in man and monkey with special reference to spermiophagy in its ampullary region. Scan. Electron Microsc. 3:1333-1339.

Murakami, M., A. Sugita, and M. Hamasaki. 1982b. The vas deferens in man and monkey: Spermiophagy in its ampulla, pp. 187-195. In E. S. E. Hafez and P. Kenemans, Eds. Atlas of Human Reproduction By Scanning Electron Microscopy. Lancaster, U.K.: MTP Press.

Murakami, M., T. Nishida, S. Iwanaga, and M. Shiromoto. 1984. Scanning and transmission electron microscopic evidence of phagocytosis of spermatozoa in the terminal region of the vas deferens of the cat. Experientia 40:958-960.

Murphy, D. L. 1976. Clinical, genetic, hormonal and drug influences on the activity of human platelet monoamine oxidase, pp. 341-353. In Monoamine Oxidase and Its Inhibition. CIBA Foundation Symposium 39 (New Series). Amsterdam: Elsevier.

Murphy, J. B., R. C. Emmott, L. L. Hicks, and P. C. Walsh. 1980. Estrogen receptors in the human prostate, seminal vesicle, epididymis, testis, and genital skin: A marker for estrogen-responsive tissues? J. Clin. Endocrinol. Metab. 50:938-948.

Murthy, Y. S., G. H. Arronet, and M. C. Parekh. 1970. Luteal phase inadequacy. Obstet. Gynecol. 36:758-761.

Musto, N. A., F. Larrea, S.-L. Cheng, N. Kotite, G. Gunsalus, and C. W. Bardin. 1982. Extracellular androgen-binding proteins: Species comparison and structure-function relationships. Ann. N.Y. Acad. Sci. 383:343-359.

Myers, R. M., Z. Larin, and T. Maniatis. 1985. Detection of single base substitutions by ribonuclease cleavage at mismatches in RNA:DNA duplexes. Science 230:1242-1246.

Myles, D. G., and P. Primakoff. 1984. Localized surface antigens of guinea pig sperm migrate to new regions prior to fertilization. J. Cell Biol. 99:1634-1641.

Myles, D. G., P. Primakoff, and A. R. Bellve. 1981. Surface domains of the guinea pig sperm defined with monoclonal antibodies. Cell 23:433-439.

Myrianthopoulos, N. C., F. N. Waldrop, and B. L. Vincent. 1969. A repeat study of hereditary predisposition in drug-induced parkisonism, pp. 486-491. In A. Barbeau and J. R. Brunette, Eds. Progress in Neurogenetics. International Congress Series, No. 175. Amsterdam: Excerpta Medica Foundation.

Naeye, R. L., W. Blanc, W. Leblanc, and M. Khatamee. 1973. Fetal complications of maternal heroin addiction: Abnormal growth, infections, and episodes of stress. J. Pediatr. 83:1055-1061.

Nagao, T. 1987. Frequency of congenital defects and dominant lethals in the offspring of male mice treated with methylnitrosourea. Mutat. Res. 177:171-178.

Nagl, W. 1978. Endopolyploidy and Polyteny in Differentiation and Evolution. Towards an Understanding of Quantitative and Qualitative Variation of Nuclear DNA in Ontogeny and Phylogeny. Amsterdam: North-Holland. 283 pp.

Nahoul, K., L. V. Rao, and R. Scholler. 1986. Saliva testosterone time-course response to hCG in adult normal men. Comparison with plasma levels. J. Steroid Biochem. 24:1011-1015.

Nancarrow, C. D., B. M. Evison, R. J. Scaramuzzi, and K. E. Turnbull. 1979. Detection of induced death of embryos in sheep by the rosette inhibition test. J. Reprod. Fertil. 57:385-389.

Narod, S. A., G. R. Douglas, E. R. Nestmann, and D. H. Blakey. 1988. Human mutagens: Evidence from paternal exposure? Environ. Mol. Mutagen. 11:401-415.

NCCLS (National Committee for Clinical Laboratory Standards). 1981. List of Standards. Villanova, Penn.: National Committee for Clinical Laboratory Standards.

NCCLS (National Committee for Clinical Laboratory Standards). 1985. List of Standards. Villanova, Penn.: National Committee for Clinical Laboratory Standards.

Neaves, W. H. 1975. Biological aspects of vasectomy, pp. 383-404. In D. W. Hamilton and R. O. Greep, Eds. Handbook of Physiology, Section 7: Endocrinology. Volume V. Male Reproductive System. Washington, D.C.: American Physiological Society.

Nebert, D. W., and N. M. Jensen. 1979. The Ah locus: Genetic regulation of the metabolism of carcinogens, drugs, and other environmental chemicals by cytochrome P-450- mediated monooxygenases. CRC Crit. Rev. Biochem. 6:401-437.

Needleman, H. L. 1987. Introduction: Biomarkers in neurodevelopmental toxicology. Environ. Health Perspect. 74:149-152.

Needleman, H. L. 1988. The persistant threat of lead: Medical and sociological issues. Curr. Probl. Pediatr. 18:697-744.

Needleman, H. L., O. C. Tuncay, and I. M. Shapiro. 1972. Lead levels in decidous teeth of urban and suburban American children. Nature 235:111.

Needleman, H. L., I. Davidson, E. M. Sewell, and I. M. Shapiro. 1974. Subclinical lead exposure in Philadelphia schoolchildren: Identification by dentine lead analysis. N. Engl. J. Med. 290:245-248.

Needleman, H. L., C. Gunnoe, A. Leviton, R. Reed, H. Peresie, C. Maher, and P. Barrett. 1979. Deficits in psychologic and classroom performance of children with elevated dentine lead levels. N. Engl. J. Med. 300:689-695.

Needleman, H. L., M. Rabinowitz, A. Leviton, S. Linn, and S. Schoenbaum. 1984. The relationship between prenatal exposure to lead and congenital anomalies. J. Am. Med. Assoc. 251:2956-2959.

Neel, J. V., H. W. Mohrenweiser, and M. H. Meisler. 1980. Rate of spontaneous mutation at human loci encoding protein structure. Proc. Natl. Acad. Sci. USA 77:6037-6041.

Neel, J. V., H. Mohrenweiser, S. Hanash, B. Rosenblum, S. Sternberg, K.-H. Wurzinger, E. Rothman, C. Satoh, K. Goriki, T. Krasteff, M. Long, M. Skolnick, and R. Krzesicki. 1983. Biochemical approaches to monitoring human populations for germinal mutation rates. I: Electrophoresis, pp. 71-93. In F. J. de Serres and W. Sheridan, Eds. Utilization of Mammalian Specific Locus Studies in Hazard Evaluation and Estimation of Genetic Risk. New York: Plenum Press.

Neel, J. V., B. B. Rosenblum, C. F. Sing, M. M. Skolnick, S. M. Hanash, and S. Sternberg. 1984. Adapting two-dimensional gel electrophoresis to the study of human germ-line mutation rates, pp. 259-306. In J. E. Celis and R. Bravo, Eds. Two-Dimensional Gel Electrophoresis of Proteins. Orlando: Academic Press.

Neel, J. V., C. Satoh, K. Goriki, M. Fujita, N. Takahashi, J. Asakawa, and R. Hazama. 1986. The rate with which spontaneous mutation alters the electrophoretic mobility of polypeptides. Proc. Natl. Acad. Sci. USA 83:389-393.

Nelson, J. F., and L. S. Felicio. 1987. Reproductive aging in the female: An etiological perspective updated. Rev. Biol. Res. Aging 3:359-381.

Nelson, J. F., L. S. Felicio, P. K. Randall, C. Sims, and C. E. Finch. 1982. A longitudinal study of estrous cyclicity in aging C57BL/6J mice. I. Cycle frequency, length and vaginal cytology. Biol. Reprod. 27:327-329.

Nestor, A., and M. A. Handel. 1984. The transport of morphologically abnormal sperm in the female reproductive tract of mice. Gamete Res. 10:119-125.

New, M. I. 1985. Premature thelarch and estrogen intoxication, pp. 349-357. In J. A. McLachlan, Ed. Estrogens in the Environment II: Influences on Development. New York: Elsevier.

Newbold, R. R., and J. A. McLachlan. 1982. Vaginal adenosis and adenocarcinoma in mice exposed prenatally or neonatally to diethylstilbestrol. Cancer Res. 42:2003-2011.

Newman, C. G. 1985. Teratogen update: Clinical aspects of thalidomide embryopathy—a continuing preoccupation. Teratology 32:133-144.

Nichols, N. R., J. N. Masters, P. C. May, S. L. Millar, and C. E. Finch. 1986. Altered RNA abundance in rat hippocampus in response to corticosterone, p. 265. In Endocrine Society 68th Annual Meeting. Anaheim, California. Program and Abstracts.

Nicoletti, O., F. Ambrosi, C. Giammartino, L. Fedeli, C. Mannarelli, and P. Filipponi. 1986. Catecholamines and pituitary function. V. Effect of low-dose dopamine infusion on basal and gonadotropin-releasing hormone stimulated gonadotropin release in normal cycling women and patients with hyperprolactinemic amenorrhea. Horm. Metab. Res. 18:479-484.

Nicolson, G. L., and R. Yanagimachi. 1979. Cell surface changes associated with the epididymal maturation of mammalian spermatozoa, pp.187-194. In D. W. Fawcett and J. M. Bedford, Eds. The Spermatozoon. Maturation, Motility, Surface Properties and Comparative Aspects. Baltimore: Urban & Schwarzenberg.

NIEHS (National Institute for Environmental Health Sciences), Task Force 3. 1985. Biochemical and Cellular Markers of Chemical Exposure and Preclinical Indicators of Disease. Washington, D.C.: U.S. Department of Health and Human Services.

Niendorf, von F. 1964. Polyspermie und spontanabort. Gynaecologia 158:35-41.

Nikolopoulou, M., D. A. Soucek, and J. C. Vary. 1985. Changes in the lipid content of boar sperm plasma membranes during epididymal maturation. Biochim. Biophys. Acta 815:486-498.

Nilsson, O. 1970. Some ultrastructrual aspects of ovo-implantation, pp. 52-72. In P. O. Hubinont, F. Leroy, C. Robyn, and P. Leleux, Eds. Ovo-Implantation,

Human Gonadotropins and Prolactin. Basel: S. Karger.

Nishikawa, Y., and Y. Waida. 1952. Studies on the maturation of spermatozoa. I. Natl. Inst. Agr. Sci. B. Ser. G:Anim. Husb. 3:69-81.

Nistal Martin de Serrano, M., and R. Paniagua Gomez-Alvarez. 1984. Testicular and Epididymal Pathology. New York: Thieme-Stratton. 358 pp.

Norppa, H., M. Sorsa, H. Vainio, P. Gröhn, E. Heinonen, L. Holsti, and E. Nordman. 1980. Increased sister chromatid exchange frequencies in lymphocytes of nurses handling cytostatic drugs. Scand. J. Work Environ. Health 6:299-301.

Noyes, R. W., A. W. Hertig, and J. A. Rock. 1950. Dating the endometrial biopsy. Fertil. Steril. 1:3-25.

Noyes, R. W., Z. Dickmann, L. L. Doyle, and A. H. Gates. 1063. Ovum transfers, synchronous and asynchronous, in the study of implantation, pp. 197-212. In A. C. Enders, Ed. Delayed Implantation. Published for William Marsh Rice University by the University of Chicago Press.

NRC (National Research Council). 1982. Identifying and Estimating the Genetic Impact of Chemical Mutagens. Washington, D. C.: National Academy Press. 316 pp.

NRC (National Research Council). 1983. Risk Assessment in the Federal Government: Managing the Process. Washington, D.C.: National Academy Press. 191 pp.

NRC (National Research Council). 1985. Epidemiology and Air Pollution. Washington, D.C.: National Academy Press. 224 pp.

NRC (National Research Council). 1986. Environmental Tobacco Smoke: Measuring Exposures and Assessing Health Effects. Washington, D.C.: National Academy Press. 337 pp.

NRC (National Research Council). In press. Drinking Water and Health, Volume 9. Washington, D.C.: National Academy Press.

O'Brien, W. F. 1984. Midtrimester genetic amniocentesis: A review of the fetal risks. J. Reprod. Med. 29:59-63.

O'Callaghan, J. P., and D. B. Miller. 1983. Nervous-system specific proteins as biochemical indicators of neurotoxicity. Trends Phar. 4:338-390.

O'Callaghan, J. P., and D. B. Miller. 1984. Neuron-specific phosphoproteins as biochemical indicators of neurotoxicity: Effects of acute administration of trimethyltin to the adult rat. J. Pharmacol. Exp. Ther. 231:736-743.

O'Donnell, J. P., and M. Van Tuinan. 1979. Behavior problems of preschool children: Dimensions and congenital correlates. J. Abnorm. Child Psychol. 7:61-75.

Oakberg, E. F. 1975. Radiation response of spermatogonial stem-cells in mouse. Radiat. Res. 59:43-44. (Abstract.)

Oakberg, E. F. 1978. Differential spermatogonial stem-cell survival and mutation frequency. Mutat. Res. 50:327-340.

Ojeda, S. R., W. W. Andrews, J. P. Advis, and S. S. White. 1980. Recent advances in the endocrinology of puberty. Endocr. Rev. 1:228-257.

Okey, A. B., E. A. Roberts, P. A. Harper, and M. S. Denison. 1986. Induction of drug-metabolizing enzymes: Mechanisms and consequences. Clin. Biochem. 19:132-141.

Olson, G. E. 1982. The human spermatozoon, pp. 72-111. In D. Hamilton and F. Naftolin, Eds. Basic Reproductive Medicine. Volume 2: Reproductive Function in Men. Cambridge, Mass.: MIT Press.

Olson, G. E., and M.-C. Orgebin-Crist. 1982. Sperm surface changes during epididymal maturation. Ann. N.Y. Acad. Sci. 383:372-392.

Omenn, G. S. 1986. Susceptibility to occupational and environmental exposure to chemicals. Prog. Clin. Biol. Res. 214:527-545.

Orentreich, N., J. L. Brind, R. L. Rizer, and J. H. Vogelman. 1984. Age changes and sex differences in serum dehydroepiandrosterone sulfate concentrations throughout adulthood. J. Clin. Endocrinol. Metab. 59:551-555.

Orgebin-Crist, M.-C. 1967a. Sperm maturation in rabbit epididymis. Nature 216:816-818.

Orgebin-Crist, M.-C. 1967b. Maturation of spermatozoa in the rabbit epididymis: Fertilizing ability and embryonic mortality in does inseminated with epididymal spermatozoa. Ann. Biol. Anim. Biochim. Biophys. 7:373-389.

Orgebin-Crist, M.-C. 1968. Maturation of spermatozoa in the rabbit epididymis: Delayed fertilization in does inseminated with epididymal spermatozoa. J. Reprod. Fertil. 16:29-33.

Orgebin-Crist, M.-C. 1969. Studies on the function of the epididymis. Biol. Reprod. (Suppl. 1):155-175.

Orgebin-Crist, M.-C. 1981. Epididymal physiology and sperm maturation, pp. 80-89. In C. Bollack and A. Clavert, Eds. Epididymis and Fertility: Biology and Pathology. Progress in Reproductive Biology, Vol. 8. Switzerland: S. Karger.

Orgebin-Crist, M.-C. 1984. Physiologie de l'épididyme, pp. 51-61. In G. Schaison, P. Bouchard, J. Mahoudeau, and F. Labrie, Eds. Médecine de la Reproduction Masculine. Paris: Flammarion.

Orgebin-Crist, M.-C., and J. Djiane. 1979. Properties of a prolactin receptor from the rabbit epididymis. Biol. Reprod. 21:135-139.

Orgebin-Crist, M.-C., and S. Fournier-Delpech. 1982. Sperm-egg interaction: Evidence for maturational changes during epididymal transit. J. Androl. 3:429-433.

Orgebin-Crist, M.-C., and G. E. Olson. 1984. Epididymal sperm maturation, pp. 80-102. In M. Courot, Ed. The Male in Farm Animal Reproduction. Amsterdam: Martinus Nijhoof.

Orgebin-Crist, M.-C., B. J. Danzo, and J. Davies. 1975. Endocrine control of the development and maintenance of sperm fertilizing ability in the epididymis, pp. 319-338. In D. W. Hamilton and R. O. Greep, Eds. Handbook of Physiology, Section 7: Endocrinology. Volume V. Male Reproductive System. Washington, D.C.: American Physiological Society.

Orgebin-Crist, M.-C., B. C. Eller, and B. J. Danzo. 1983. The effects of estradiol, tamoxifen, and testosterone on the weights and histology of the epididymis and accessory sex organs of sexually immature rabbits. Endocrinology 113:1703-1715.

Orr, W. C., and H. J. Hoffman. 1974. A 90-min cardiac biorhythm: Methodology and data analysis using modified periodograms and complex demodulation. IEEE Trans. Biomed. Eng. 21:130-143.

Orsini, L. F., S. Salardi, G. Pilu, L. Bovicelli, and E. Cacciari. 1984. Pelvic organs in premenarcheal girls: Real-time ultrasonography. Radiology 153:113-116.

Orth, J. M., and G. L. Gunsalus. 1987. The Sertoli cell depleted rat: Relationship among Sertoli cell (SC) population size, spermatogenesis and ABP production, pp. 61. In M.-C. Orgebin-Crist and B. J. Danzo, Eds. Cell Biology of the Testis and Epididymis. Annals of the New York Academy of Sciences, Vol. 513. New York: New York Academy of Sciences.

Osterman-Golkar, S., and L. Ehrenberg. 1983. Dosimetry of electrophilic compounds by means of hemoglobin alkylation. Annu. Rev. Public Health 4:397-402.

Ostrowski, M. C., M. K. Kistler, and W. S. Kistler. 1979. Purification and cell-free synthesis of a major protein from rat seminal vesicle secretion. J. Biol. Chem. 254:383-390.

Ostrowski, M. C., M. K. Kistler, and W. S. Kistler. 1982. Effect of castration on the synthesis of seminal vesicle secretory protein IV in the rat. Biochemistry 21:3525-3529.

OTA (U.S. Congress, Office of Technology Assessment). 1986. Technologies for Detecting Heritable Mutations in Human Beings. OTA-H-298. Washington, D. C.: U.S. Government Printing Office. 144 pp.

Otto, D., and L. Reiter. 1984. Developmental changes in slow cortical potentials of young children with elevated body lead burden. Neurophysiological considerations. Ann. N.Y. Acad. Sci. 425:377-383.

Otto, D., V. A. Benignus, K. E. Muller, and C. N. Barton. 1981. Effects of age and body lead burden on CNS function in young children. I. Slow cortical potentials. Electroencephalogr. Clin. Neurophysiol. 52:229-239.

Otto, D., V. Benignus, K. Muller, C. Barton, K. Seiple, J. Prah, and S. Schroeder. 1982. Effects of low to moderate lead exposure on slow cortical potentials in young children: Two-year follow-up study. Neurobehav. Toxicol. Teratol. 4:733-737.

Otto, D., G. Robinson, S. Baumann, S. Schroeder, P. Mushak, D. Kleinbaum, and L. Boone. 1985. 5-year follow-up study of children with low-to-moderate lead absorption: Electrophysiological evaluation. Environ. Res. 38:168-186.

Overstreet, J. W., R. Yanagimachi, D. F. Katz, K. Hayashi, and F. W. Hanson. 1980. Penetration of human spermatozoa into the human zona pellucida and the zona-free hamster egg: A study of fertile donors and infertile patients. Fertil. Steril. 33:534-542.

Overstreet, J. W., M. J. Price, W. F. Blazak, E. L. Lewis, and D. F. Katz. 1981. Simultaneous assessment of human sperm motility and morphology by videomicrography. J. Urol. 126:357-360.

Owen, J. R., S. F. Irani, and A. W. Blair. 1972. Effect of diazepam administered to mothers during labour on temperature regulation of the neonate. Arch. Dis. Child. 47:107-110.

Page, R. D., D. Kirkpatrick-Keller, and R. L. Butcher. 1983. Role of age and length of oestrous cycle in alteration of the oocyte and intrauterine environment in the rat. J. Reprod. Fertil. 69:23-28.

Palm, J. 1974. Proceedings: Maternal-fetal histoincompatibility in rats: An escape from adversity. Cancer Res. 34:2061-2065.

Pan, Y. C., F. S. Sharief, M. Okabe, S. Huang, and S. S. L. Li. 1983. Amino acid sequence studies on lactate dehydrogenase C4 isozymes from mouse and rat testes. J. Biol. Chem. 258:7005-7016.

Panigel, M., G. Wolfe, and A. Zeleznick. 1988. Magnetic resonance imaging of the placenta in rhesus monkeys, *Macaca mulatta*. J. Med. Primatol. 17:3-18.

Papanicolaou, G. N. 1933. The sexual cycle in the human female as revealed by vaginal smears. Am. J. Anat. [Suppl. 52]:519-637.

Paris, F. X., J. Henry-Suchet, L. Tesquier, T. Loysel, V. Loffredo, and J. P. Pez. 1984. Le traitement médical des grossesses extra-utétrines par le RU486. Un moyen d'éviter la chirugie. Presse Med. 13(19):1219.

Paris, F. X., J. Henry-Suchet, L. Tesquier, T. Loysel, J. P. Pez, V. Loffredo, M. Roger, and J. De Brux. 1986. Interet d'un steroide action antiprogesterone dans le trautenebt de la grissesse extra-uterine. [The value of an antiprogesterone steroid in the treatment of extra-uterine pregnancy. Preliminary results.] Rev. Fr. Gynecol. Obstet. 81:33-35.

Parks, D. R., V. M. Bryan, V. T. Oi, and L. A. Herzenberg. 1979. Antigen-specific identification and cloning of hybridomas with a fluorescence-activated cell sorter. Proc. Natl. Acad. Sci. USA 76:1962-1966.

Parks, J. S. 1988. Short stature and delayed puberty, pp. 531-534. In J. W. Hurst, Ed. Medicine for the Practicing Physician, 2nd ed. Boston: Butterworths.

Parks, J. S., P. V. Nielsen, L. A. Sexton, and E. H. Jorgensen. 1985. A effect of gene dosage on production of human chorionic somatomammotropin. J. Clin Endocrinol. Metab. 60:994-997.

Pedersen, R. A., and F. Mangia. 1978. Ultraviolet-light-induced unscheduled DNA synthesis by resting and growing mouse oocytes. Mutat. Res. 49:425-429.

Pellegrino, L. J., and J. Altman. 1979. Effects of differential interference with postnatal cerebellar neurogenesis on motor performance, activity level, and maze learning of rats: A developmental study. J. Comp. Physiol.

Psychol. 93:1-33.

Perera, F. 1986. New approaches in risk assessment for carcinogens. Risk Anal. 6:195-201.

Perera, F. P., and I. B. Weinstein. 1982. Molecular epidemiology and carcinogen-DNA adduct detection: New approaches to studies of human cancer causation. J. Chronic Dis. 35:581-600.

Perera, F., R. Santella, and M. Poirier. 1986. Biomonitoring of workers exposed to carcinogens: Immunoassays to benzo(a)pyrene-DNA adducts as a prototype. J. Occup. Med. 28:1117-1123.

Peress, M. R., C. C. Tsai, R. S. Mathur, and H. O. Williamson. 1982. Hirsutism and menstrual patterns in women exposed to diethylstilbestrol in utero. Am. J. Obstet. Gynecol. 144:135-140.

Pert, C. B., M. R. Ruff, R. J. Weber, and M. Herkenham. 1985. Neuropeptides and their receptors: A psychosomatic network. J. Immunol. 135(2 Suppl.):820s-825s.

Pert, C. B., J. Hill, M. R. Ruff, R. M. Berman, W. B. Robby, L. O. Arthur, F. W. Ruscetti, and W. L. Farrar. 1986. Octapeptides deduced from the neuropathic receptor-like pattern of antigen T-4 in brain potentially inhibit human immunodeficiency virus receptor binding and human T cell infectivity. Proc. Natl. Acad. Sci. USA 83:9254-9258.

Peters, P. J., and R. L. Brunner. 1976. Increased running wheel activity and dyadic behavior of rats with hippocampal granule cell deficits. Behav. Biol. 16:91-97.

Peters, J., S. T. Ball, and S. J. Andrews. 1986. The detection of gene mutations by electrophoresis, and their analysis, pp. 367-374. In C. Ramel, B. Lambert, and J. Magnusson, Eds. Genetic Toxicology of Environmental Chemicals, Part B: Genetic Effects and Applied Mutagenesis. Proceedings of the Fourth International Conference on Environmental Mutagens held in Stockholm, Sweden, June 24-28, 1985. New York: Liss.

Peters, M. S., M. Rodriguez, and G. J. Gleich. 1986. Localization of human eosinophil granule major basic protein, eosinophil cationic protein, and eosinophil-derived neurotoxin by immunoelectron microscopy. Lab. Invest. 54:656-666.

Petit, L., D. P. Alfano, and J. C. LeBoutillier. 1983. Early lead exposure and the hippocampus: A review and recent advances. NeuroToxicology 4:79-94.

Peto, R., F. J. Roe, P. N. Levy, and J. Clack. 1975. Cancer and ageing in mice and men. Br. J. Cancer 32:411-426.

Peyser, M. R., D. Ayalon, A. Harell, R. Toaff, and T. Cardova. Stress induced delay of ovulation. Obstet. Gynecol. 42:667-671.

Phadke, A. M. 1975. Spermiophage cells in man. Fertil. Steril. 26:760-774.

Phillips, D. M. 1972. Comparative analysis of mammalian sperm motility. J. Cell Biol. 53:561-573.

Phillips, D. M. 1975. Mammalian sperm structure, pp. 405-419. In D. W. Hamilton and R. O. Greep, Eds. Handbook of Physiology, Section 7: Endocrinology.

Volume V. Male Reproductive System. Washington, D.C.: American Physiological Society.

Pholpramool, C., and N. Triphrom. 1984. Effects of cholinergic and adrenergic drugs on intraluminal pressures and contractility of the rat testis and epididymis in vivo. J. Reprod. Fertil. 71:181-188.

Pholpramool, C., R. W. White, and B. P. Setchell. 1982. Influence of androgens on inositol secretion and sperm transport in the epididymis of rats. J. Reprod. Fertil. 66:547-553.

Pholpramool, C., O. A. Lea, P. V. Burrow, H. M. Dott, and B. P. Setchell. 1983. The effects of acidic epididymal glycoprotein (AEG) and some other proteins on the motility of rat epididymal spermatozoa. Int. J. Androl. 6:240-248.

Pholpramool, C., N. Triphrom, and A. Din-Udom. 1984. Intraluminal pressures in the seminiferous tubules and in different regions of the epididymis in the rat. J. Reprod. Fertil. 71:173-179.

Picard, J.-Y., C. Goulut, R. Bourrillon, and N. Josso. 1986. Biochemical analysis of bovine testicular anti-Mullerian hormone. FEBS Lett. 195:73-76.

Pierce, J. G., and T. F. Parsons. 1981. Glycoprotein hormones: Structure and function. Annu. Rev. Biochem. 50:465-495.

Pierce, J. G., T. Liao, S. M. Howard, B. Shome, and J. S. Cornell. 1971. Studies on the structure of thyrotropin: Its relationship to luteinizing hormone. Recent Prog. Horm. Res. 27:165-212.

Pijnenborg, R., W. B. Robertson, and I. Brosens. 1985. Morphological aspects of placental ontogeny and phylogeny. Placenta 6:155-162.

Piomelli, S., B. Davidow, V. F. Guinee, P. Young, and G. Gay. 1973. The FEP (free erythrocyte porphryins) test: A screening micromethod for lead poisoning. Pediatrics 51:254-259.

Platt, L. D., C. A. Walla, R. H. Paul, M. E. Trujillo, C. V. Loesser, N. D. Jacobs, and P. M. Broussard. 1985. A prospective trial of the fetal biophysical profile versus the nonstress test in the management of high-risk pregnancies. Am. J. Obstet. Gynecol. 153:624-633.

Pletscher, A. 1968. Metabolism, transfer, and storage of 5-hydroxytryptamine in blood platelets. Br. J. Pharmacol. 32:1-16.

Pohl, C. R., D. W. Richardson, J. S. Hutchison, J. A. Germak, and E. Knobil. 1983. Hypophysiotropic signal frequency and the functioning of the pituitary-ovarian system in the rhesus monkey. Endocrinology 112:2076-2080.

Poland, A., and J. C. Knutson. 1982. 2,3,7,8-tetrachlorodibenzo-p-dioxin and related halogenated aromatic hydrocarbons: Examination of the mechanism of toxicity. Annu. Rev. Pharmacol. Toxicol. 22:517-554.

Pommerrenke, W. T. 1946. Cyclic changes in the physical and chemical properties of cervical mucus. Am. J. Obstet. Gynecol. 52:1023-1031.

Ponzetto, C., and D. J. Wolgemuth. 1985. Haploid expression of a unique c-abl transcript in the mouse male germ line. Mol. Cell Biol. 5:1791-1794.

Porter, I. H. 1986. Genetic aspects of preventive medicine, pp. 1427-1422. In J. M. Last, Ed. Maxcy-Rosenau Public Health and Preventive Medicine. Norwalk, Conn.: Appleton-Century-Crofts.

Porter, S. B., D. E. Ong, F. Chytil, and M. -C. Orgebin-Crist. 1985. Localization of cellular retinol-binding protein and cellular retinoic acid-binding protein in the rat testis and epididymis. J. Androl. 6:197-212.

Poskanzer, D. C., and R. S. Schwab. 1961. Studies in the epidemiology of Parkinson's disease predicting its disappearance as a major clinical entity by 1980. Trans. Am. Neurol. Assoc. 86:234-235.

Pratt, H. P. M. 1977. Uterine proteins and the activation of embryos from mice during delayed implantation. J. Reprod. Fertil. 50:1.

Pratt, R., R. I. Grove, and W. D. Willis. 1982. Prescreening for environmental teratogens using cultured mesenchymal cells from the human embryonic palate. Teratogenesis Carcinog. Mutagen. 2:313-318.

Pretsch, W. 1986. Protein-charge mutations in mice, pp. 383-388. In C. Ramel, B. Lambert, and J. Magnusson, Eds. Genetic Toxicology of Environmental Chemicals, Part B: Genetic Effects and Applied Mutagenesis. Proceedings of the Fourth International Conference on Environmental Mutagens held in Stockholm, Sweden, June 24-28, 1985. New York: Liss.

Psychoyos, A. 1973. Endocrine control of egg implantation, pp. 187-215. In R. O. Greep, Ed. Handbook of Physiology, Section 7: Endocrinology. Volume II. Female Reproductive System, Part 2. Washington, D.C.: American Physiological Society.

Pulsinelli, W. A. 1985. Selective neuronal vulnerability: Morphological and molecular characteristics. Prog. Brain. Res. 63:29-37.

Purvis, K., L. Cusan, H. Attramadal, A. Ege, and V. Hansson. 1982. Rat sperm enzymes during epididymal transit. J. Reprod. Fertil. 65:381-387.

Quinn, P. O., and J. L. Rapoport. 1974. Minor physical anomalies and neurologic status in hyperactive boys. Pediatrics 53:742-747.

Quinn, P. O., M. Renfield, C. Burg, and J. L. Rapoport. 1977. Minor physical anomalies: A newborn screening and 1-year follow-up. J. Am. Acad. Child Psychiatry 16:662-669.

Rabinowitz, M. B., G. W. Wetherill, and J. D. Kopple. 1976. Kinetic analysis of lead metabolism in healthy humans. J. Clin. Invest. 58:260-270.

Rabkin, J. G., and E. L. Struening. 1976. Life events, stress, and illness. Science 194:1013-1020.

Radwanska, E., H. H. G. McGarrigle, and G. I. M. Swyer. 1976. Plasma progesterone and oestradiol estimations in the diagnosis and treatment of luteal insufficiency in menstruating infertile women. Acta Eur. Fertil. 7:39-47.

Radwanska, E., J. Hammond, and P. Smith. 1981. Single midluteal progesterone assay in the management of ovulatory infertility. J. Reprod. Med. 26:85-89.

Rakoff, A. E. 1961. Hormonal cytology in gynecology.

Clin. Obstet. Gynecol. 4:1045-1061.

Ramaley, J. A. 1978. The adrenal rhythm and puberty onset in the female rat. Life Sci. 23:2079-2087.

Randerath, K., M. V. Reddy, and R. C. Gupta. 1981. ^{32}P-labeling test for DNA damage. Proc. Natl. Acad. Sci. USA 78:6126-6129.

Randerath, K., E. Randerath, H. P. Agrawal, R. C. Gupta, M. E. Schurdak, and M. V. Reddy. 1985. Postlabeling methods for carcinogen-DNA adduct analysis. Environ. Health Perspect. 62:57-65.

Randerath, K., M. V. Reddy, and R. M. Disher. 1986. Age- and tissue-related DNA modifications in untreated rats: Detection by ^{32}P-post labeling assay and possible significance for spontaneous tumor induction and aging. Carcinogenesis 7:1615-1617.

Rannevik, G. 1979. Hormonal, metabolic and clinical effects of danazol in the treatment of endometriosis. Postgrad. Med. J. 55(Suppl. 5):14-20.

Rapoport, J. L., P. O. Quinn, and F. Lamprecht. 1974. Minor physical anomalies and plasma dopamine-beta-hydroxylase activity in hyperactive boys. Am. J. Psychiatry 131:386-390.

Rappold, G. A., T. Cremer, H. D. Hager, K. E. Davies, C. R. Mueller, and T. Yang. 1984. Sex chromosome positions in human interphase nuclei as studied by in situ hybridization with chromosome specific DNA probes. Hum. Genet. 67:317-325.

Rastogi, R. K., M. Milone, M. Di Meglio, M. F. Caliendo, and G. Chieffi. 1979. Effects of castration, 5alpha-dihydrotestosterone and cyproterone acetate on enzyme activity in the mouse epididymis. J. Reprod. Fertil. 57:73-77.

Ratcliffe, J. M., S. M. Schrader, K. Steenland, D. E. Clapp, T. Turner, and R. W. Hornung. 1987. Semen quality in papaya workers with long term exposure to ethylene dibromide. Br. J. Ind. Med. 44:317-326.

Rebar, R. N. 1986. Practical evaluation of hormonal status, pp. 683-733. In S. C. C. Yen and R. B. Jaffe, Eds. Reproductive Endocrinology: Physiology, Pathophysiology and Clinical Management. Philadelphia: W. B. Saunders Company.

Reddy, M. V., and K. Randerath. 1987. ^{32}P-postlabeling assay for carcinogen-DNA adducts: Nuclease P_1-mediated enhancement of its sensitivity and applications. Environ. Health Perspect. 76:41-47.

Reddy, M. V., and K. Randerath. 1988. DNA adduct formation in white blood cells in relation to internal organs of mice given benzo(a)pyrene, dibenzo(c,g)carbazole, safrole or cigarette smoking condensate. Proc. Annu. Meet. Am. Assoc. Cancer Res. 29:A357. (Abstract.)

Reed, E., V. Bonagura, P. Kung, D. W. King, and N. Suciu-Foca. 1983. Anti-idiotypic antibodies to HLA-DR4 and DR2. J. Immunol. 131:2890-2894.

Reed, E., M. Hardy, C. Lattes, J. Brensilver, R. McCabe, K. Reemtsma, and N. Suciu-Foca. 1985. Anti idiotypic antibodies and their relevance to transplantation. Transplant Proc. 17:735-738.

Reeves, D. S., R. C. G. Rowe, M. E. Snell, and A. B.

W. Thomas. 1973. Further studies on the secretion of antibiotics in the prostatic fluid of the dog, pp. 197-205. In W. Brumfitt, and A. W. Asscher, Eds. Urinary Tract Infection. London: Oxford University Press.

Reff, M. E., and E. L. Schneider. 1982. Biologic Markers of Aging: Proceedings of Conference on Nonlethal Biological Markers of Physiological Aging, June 19-20, 1981. NIH Publication No. 82-2221. Washington, D.C.: National Institutes of Health. 252 pp.

Reich, P. 1987. Medical aspects of sexuality, pp. 220-223. In Harrison's Principles of Internal Medicine, 11th ed. New York: McGraw Hill.

Reiter, E. O., and M. M. Grumbach. 1982. Neuroendocrine control mechanisms and the onset of puberty. Annu. Rev. Physiol. 44:595-613.

Reiter, L. W. 1982. Developmental neurotoxicity of lead: Experimental studies, pp. 43-54. In J. J. Chisolm, Jr. and D. M. O'Hara, Eds. Lead Absorption in Children. Management, Clinical, and Environmental Aspects. Baltimore: Urban & Schwarzenberg.

Remuzzi, G., D. Marchesi, G. Mecca, R. Misiani, E. Rossi, M. B. Donati, and G. de Gaetano. 1980. Reduction of fetal vascular prostacyclin activity in pre-eclampsia (letter). Lancet 2:310.

Reuhl, K. R., and L. W. Chang. 1979. Effects of methylmercury on the development of the nervous system: A review. NeuroToxicology 1:21-55.

Rhoads, G. G., L. G. Jackson, S. E. Schlesselman, F. F. de la Cruz, R. J. Desnick, M. S. Golbus, D. H. Ledbetter, H. A. Lubs, M. J. Mahoney, E. Pergament, J. L. Simpson, R. J. Carpenter, S. Elias, N. A. Ginsberg, J. D. Goldberg, J. C. Hobbins, L. Lynch, P. H. Shiono, R. J. Wapner, and J. M. Zachary. 1989. The safety and efficacy of chorionic villus sampling for early prenatal diagnosis of cytogenetic abnormalities. N. Eng. J. Med. 320:609-617.

Riad-Fahmy, D., G. F. Read, R. F. Walker, and K. Griffiths. 1982. Steroids in saliva for assessing endocrine function. Endocr. Rev. 3:367-395.

Rich, K. A., and D. M. De Kretser. 1977. Effect of differing degrees of destruction of the rat seminiferous epithelium on levels of serum follicle stimulating hormone and androgen binding protein. Endocrinology 101:959-968.

Richards, J. S., J. J. Ireland, M. C. Rao, G. A. Bernath, A. R. Midgley, Jr., and E. Reichert, Jr. 1976. Ovarian follicular development in the rat: Hormone receptor regulation by estradiol, follicle stimulating hormone and luteinizing hormone. Endocrinology 99:1562-1570.

Richardson, S. J., V. Senikas, and J. F. Nelson. 1987. Follicular depletion during the menopausal transition: Evidence for accelerated loss and ultimate exhaustion. J. Clin. Endocrinol. Metabol. 65:1231-1237.

Riddick, D. H., and C. B. Hammond. 1975. Adrenal virilism due to 21-hydroxylase deficiency in the postmenarchial female. Obstet. Gynecol. 45:21-24.'

Riley, E. P., and C. Vorhees, Eds. 1986. Handbook of Behavioral Teratology. New York: Plenum Press. 522 pp.

Riley, G. M., E. Dontas, and B. Gill. 1955. Use of serial vaginal smears in detecting time of ovulation. Fertil. Steril. 6:86-102.

Ritzen, E. M., S. N. Nayfeh, F. S. French, and M. C. Dobbins. 1971. Demonstration of androgen-binding components in rat epididymis cytosol and comparison with binding components in prostate and other tissues. Endocrinology 89:143-151.

Riva, M., S. Memet, J. Y. Micouin, J. Huet, I. Treich, J. Dassa, R. Young, J. M. Buhler, A. Sentenac, and P. Fromageot. 1986. Isolation of structural genes for yeast RNA polymerases by immunological screening. Proc. Natl. Acad. Sci. USA 83:1554-1558.

Robaire, B., and B. F. Hales. 1982. Regulation of epididymal glutathione S-transferases: Effects of orchidectomy and androgen replacement. Biol. Reprod. 24:559-565.

Robaire, B., and L. Hermo. 1987. Efferent ducts, epididymis and vas deferens: Structure, functions and their regulation. In E. Knobil and J. Neill, Eds. Physiology of Reproduction. New York: Raven Press.

Robaire, B., and B. R. Zirkin. 1981. Hypophysectomy and simultaneous testosterone replacement: Effects on male rat reproductive tract and epididymal $\Delta 4$-5α-reductase and 3α-hydroxysteroid dehydrogenase. Endocrinology 109:1225-1233.

Robaire, B., L. L. Ewing, B. R. Zirkin, and D. C. Irby. 1977. Steroid $\Delta 4$-5α-reductase and 3α-hydroxysteroid dehydrogenase in the rat epididymis. Endocrinology 101:1379-1390.

Robaire, B., L. L. Ewing, D. C. Irby, and C. Desjardins. 1979. Interactions of testosterone and estradiol-17beta on the reproductive tract of the male rat. Biol. Reprod. 21:455-463.

Robaire, B., H. Scheer, and C. Hachey. 1981. Regulation of epididymal steroid metabolizing enzymes, pp.487-498. In G. Jagiello and H. J. Vogel, Eds. Bioregulators of Reproduction. New York: Academic Press.

Roberts, E. A., N. H. Shear, A. B. Okey, and D. K. Manchester. 1985. The Ah receptor and dioxin toxicity. Chemosphere 14:661-674.

Roberts, G. P., J. M. Parker, and S. R. Henderson. 1976. Proteins in human uterine fluid. J. Reprod. Fertil. 48:153-157.

Roberts, M. L., W. H. Scouten, and S. E. Nyquist. 1976. Isolation and characterization of the cytoplasmic droplet in the rat. Biol. Reprod. 14:421-424.

Roberts, N. S., L. K. Dunn, S. Weiner, L. Godmilow, and R. Miller. 1983. Mid-trimester amniocentesis: Indications, technique, risks and potential for prenatal diagnosis. J. Reprod. Med. 28:167-188.

Robertson, D. M., H. Suginami, H. H. Montes, C. P. Puri, S. K. Choi, and E. Diczfalusy. 1978. Studies on a human chorionic gonadotropin-like material present in non-pregnant subjects. Acta Endocrinol. 89:492-505.

Robertson, M. C., B. Gillespie, and H. G. Friesen. 1982.

Characterization of the two forms of rat placental lactogen (rPL): rPL-I and rPL-II. Endocrinology 111:1862-1866.

Robertson, M. C., R. E. Owens, J. Klindt, and H. G. Friesen. 1984. Ovariectomy leads to a rapid increase in rat placental lactogen secretion. Endocrinology 114:1805-1811.

Robison, A. K., W. A. Schmidt, and G. M. Stancel. 1985. Estrogenic activity of DDT: Estrogen-receptor profiles and the responses of individual uterine cell types following o,p'-DDT administration. J. Toxicol. Environ. Health 16:493-508.

Rocklin, R. E., J. L. Kitzmiller, and M. R. Garvoy. 1982. Maternal-fetal relation. II. Further characterization of an immunologic blocking factor that develops during pregnancy. Clin. Immunol. Immunopathol. 22:305-315.

Rodeck, C. H., and K. H. Nicolaides. 1983a. Fetoscopy and fetal tissue sampling. Br. Med. Bull. 39:332-337.

Rodeck, C. H., and K. H. Nicolaides. 1983b. Ultrasound guided invasive procedures in obstetrics. Clin. Obstet. Gynaecol. 10:515-539.

Rogan, W. J. 1982. PCBs and cola-colored babies: Japan, 1968, and Taiwan, 1979. Teratology 26:259-261.

Rogan, W. J., J. R. Reigart, and B. C. Gladen. 1986. Association of amino levulinate dehydratase and ferrochelatase inhibition in childhood lead exposure. J. Pediatr. 109:60-64.

Rogers, B. J., H. Van Campen, M. Ueno, H. Lambert, R. Bronson, and R. Hale. 1979. Analysis of human spermatozoal fertilizing ability using zona-free ova. Fertil. Steril. 32:664-670.

Rolfe, B. E. 1982. Detection of fetal wastage. Fertil. Steril. 37:655-660.

Rolfe, B., H. Morton, A. Cavanagh, and R. A. Gardiner. 1983. Detection of an early pregnancy factor-like substance in sera of patients with testicular germ cell tumors. Am. J. Reprod. Immunol. 3:97-100.

Rolfe, B., A. Cavanagh, C. Forde, F. Bastin, C. Chen, and H. Morton. 1984. Modified rosette inhibition test with mouse lymphocytes for detection of early pregnancy factor in human pregnancy serum. J. Immunol. Methods 70:1-11.

Roman-Franco, A. A., M. Turiello, B. Albini, E. Ossi, F. Milgrom, and G. A. Andres. 1978. Anti-basement antibodies and antigen-antibody complexes in rabbits injected with mercuric chloride. Clin. Immunol. Immunopathol. 9:464-481.

Romrell, L. J., A. R. Bellve, and D. W. Fawcett. 1976. Separation of mouse spermatogenic cells by sedimentation velocity. A morphological characterization. Dev. Biol. 49:119-131.

Rondell, P. 1974. Role of steroid synthesis in the process of ovulation. Biol. Reprod. 10:199-215.

Roosen-Runge, E. C. 1969. Comparative aspects of spermatogenesis. Biol. Reprod. 1(Suppl.):24-31.

Rosen, S. W. 1986. New placental proteins: Chemistry,

physiology and clinical use. Placenta 7:575-594.

Rosenberg, J. B., and G. M. Weller. 1973. Minor physical anomalies and academic performance in young school children. Dev. Med. Child Neurol. 15:131-135.

Rosenberg, M. J., A. J. Wyrobek, J. Ratcliffe, L. A. Gordon, G. Watchmaker, S. Fox, D. Moore II, and R. W. Hornung. 1985. Sperm as an indicator of reproductive risk among petroleum refinery workers. Br. J. Ind. Med. 42:123-127.

Rosenfeld, D. L., S. Chudow, and R. A. Bronson. 1980. Diagnosis of luteal phase inadequacy. Obstet. Gynecol. 56:193-196.

Rosenfeld, R. S., B. J. Rosenberg, D. K. Fukushima, and L. Hellman. 1975. 24-Hour secretory pattern of dehydroisoandrosterone and dehydroisoandrosterone sulfate. J. Clin. Endocrinol. Metab. 40:850-855.

Rossant, J., and L. Ofer. 1977. Properties of extra-embryonic ectoderm isolated from postimplantation mouse embryos. J. Embryol. Exp. Morphol. 39:183-194.

Rossant, J., and W. Tamura-Lis. 1981. Effect of culture conditions on diploid to giant-cell transformation in postimplantation mouse trophoblast. J. Embryol. Exp. Morphol. 62:217-227.

Rosvold, H. E., A. F. Mirsky, I. Sarason, E. D. Bransome, and L. H. Beck. 1956. A continuous performance test of brain damage. J. Consult. Psychol. 20:343-350.

Rothstein, P., and J. B. Gould. 1974. Born with a habit: Infants of drug-addicted mothers. Pediatr. Clin. North Am. 21:307-321.

Rowley, M. J., F. Teshima, and C. G. Heller. 1970. Duration of transit of spermatozoa through the human male ductular system. Fertil. Steril. 21:390-396.

Rowley, M. J., D. R. Leach, G. A. Warner, and C. G. Heller. 1974. Effect of graded doses of ionizing radiation on the human testis. Radiat. Res. 59:665-678.

Rudak, E., P. A. Jacobs, and R. Yanagimachi. 1978. Direct analysis of the chromosome constitution of human spermatozoa. Nature 274:911-913.

Rugh, R. 1953. Vertebrate radiobiology: Embryology. Ann. Rev. Nuclear Sci. 3:271-302.

Rugh, R. 1959. Vertebrate radiobiology (embryology). Ann. Rev. Nuclear Sci. 9:493-522.

Rui, H., L. Morkas, and K. Purvis. 1986. Time- and temperature-related alterations in seminal plasma constituents after ejaculation. Int. J. Androl. 9:195-200.

Rumack, C. M., M. A. Guggenheim, B. H. Rumack, R. G. Peterson, M. L. Johnson, and W. R. Braithwaite. 1981. Neonatal intracranial hemorrhage and maternal use of aspirin. Obstet. Gynecol. 58:52S-56S.

Rumke, P. 1968. Sperm agglutinating autoantibodies in relation to male infertility. Proc. R. Soc. Med. 61:275-278.

Rumke, P., and G. Hellinga. 1959. Autoantibodies against spermatozoa in sterile men. Am. J. Clin. Pathol. 32:357-363.

Rummo, J. H., D. K. Routh, N. J. Rummo, and J. F. Brown. 1979. Behavioral and neurological effects of symptomatic and asymptomatic lead exposure in children. Arch. Environ. Health 34:120-124.

Russell, L. B. 1985. Experimental approaches for the detection of chromosomal malsegregation occuring in the germline of mammals, pp. 377-396. In V. L. Dellarco, P. E. Voytik, and A. Hollaender, Eds. Aneuploidy: Etiology and Mechanisms. New York: Plenum Press.

Russell, L. B., and W. L. Russell. 1954. An analysis of the changing radiation response of the developing mouse embryo. J. Cell. Comp. Physiol. 43(Suppl. 1):103-149.

Russell, L. B., and M. D. Shelby. 1985. Tests for heritable genetic damage and for evidence of gonadal exposure in mammals. Mutat. Res. 154:69-84.

Russell, L. D., J. P. Malone, and S. L. Karpas. 1981. Morphological pattern elicited by agents affecting spermatogenesis by disruption of its hormonal stimulation. Tissue Cell. 13:369-380.

Russell, M. J., G. M. Switz, and K. Thompson. 1980. Olfactory influences on the human menstrual cycle. Pharmacol. Biochem. Behav. 13:737-738.

Rutter, M. 1980. Raised lead levels and impaired cognitive/behavioural functioning: A review of evidence. Dev. Med. Child Neurol. Suppl. 42:1-36.

Sadovsky, E., and W. Z. Polishuk. 1977. Fetal movements in utero: Nature, assessment, prognostic value, timing of delivery. Obstet. Gynecol. 50:49-55.

Safe, S. H. 1986. Comparative toxicology and mechanism of action of polychlorinated dibenzo-p-dioxins and dibenzofurans. Annu. Rev. Pharmacol. Toxicol. 26:371-399.

Safe, S., S. Bandiera, T. Sawyer, L. Robertson, L. Safe, A. Parkinson, P. E. Thomas, D. E. Ryan, L. M. Reik, W. Levin, M. A. Denomme, and T. Fujita. 1985. PCBs: Structure-function relationships and mechanism of action. Environ. Health Perspect. 60:47-56.

Saling, P. M. 1982. Development of the ability to bind zonae pellucidae during epididymal maturation: Reversible immobilization of mouse spermatozoa by lanthanum. Biol. Reprod. 26:429-436.

Saling, P. M. 1986. Mouse sperm antigens that participate in fertilization. IV. A monoclonal antibody prevents zona penetration by inhibition of the acrosome reaction. Dev. Biol. 117:511-519.

Saling, P. M., G. Irons, and R. Waibel. 1985. Mouse sperm antigens that participate in fertilization. I. Inhibition of sperm fusion with the egg plasma membrane using monoclonal antibodies. Biol. Reprod. 33:515-526.

Salmon, S. E. 1982. Drugs & the immune system, pp. 712-728. In B. G. Katzung, Ed. Basic & Clinical Pharmacology, 2nd edition. Los Altos, Calif.: Lange Medical Publications.

Samuels, S. W. 1986. Medical surveillance: Biological, social, and ethical parameters. J. Occup. Med. 28:572-577.

Sanger, F., S. Nicklen, and A. R. Coulson. 1977. DNA sequencing with chain-terminating inhibitors. Proc. Natl. Acad. Sci. USA 74:5463-5467.

Sankaranarayanan, K. 1982. Genetic Effects of Ionizing Radiation in Multicellular Eukaryotes and the Assessment of Genetic Radiation Hazards in Man. Amsterdam: Elsevier Biomedical Press. 385 pp.

Santella, R. M. 1987. In vitro testing for carcinogens and mutagens. State Art Rev. Occup. Med. 2(1):39-46.

Sapolsky, R. M. 1985. A mechanism for glucocorticoid toxicity in the hippocampus: Increased neuronal vulnerability to metabolic insults. J. Neurosci. 5:1228-1232.

Sapolsky, R. M. 1986. Glucocorticoid toxicity in the hippocampus: Reversal by supplementation with brain fuels. J. Neurosci. 6:2240-2244.

Satoh, C., A. A. Awa, J. V. Neel, W. J. Schull, H. Kato, H. B. Hamilton, M. Otake, and K. Goriki. 1982. Genetic effects of atomic bombs. Prog. Clin. Biol. Res. 103(Pt. A.):267-276.

Saxena, B. B., S. H. Hasan, F. Haour, and M. Schmidt-Gollwitzer. 1974. Radioreceptor assay of human chorionic gonadotropin: Detection of early pregnancy. Science 184:793-795.

Schacter, B., A. Muir, M. Gyves, and M. Tasin. 1979. HLA-A,B compatibility in parents of offspring with neural-tube defects or couples expressing involuntary fetal wastage. Lancet 1:796-799.

Scheer, H., and B. Robaire. 1980. Steroid delta 4-5 alpha-reductase and 3 alpha-hydroxysteroid dehydrogenase in the rat epididymis during development. Endocrinology 107:948-953.

Schipper, H., J. R. Brawer, J. F. Nelson, L. S. Felicio, and C. E. Finch. 1981. The role of gonads in the histologic aging of the hypothalamic arcuate nucleus. Biol. Repro. 25:413-419.

Schlafke, S., and A. C. Enders. 1975. Cellular basis of interaction between trophoblast and uterus at implantation. Biol. Reprod. 12:41-65.

Schleicher, G., U. Drews, W. E. Stumpf, and M. Sar. 1984. Differential distribution of dihydrotestosterone and estradiol binding sites in the epididymis of the mouse. An autoradiographic study. Histochemistry 81:139-147.

Schmassmann, A., G. Mikuz, G. Bartsch, and H. Rohr. 1982. Spermiometrics: Objective and reproducible methods for evaluating sperm morphology. Eur. Urol 8:274-279.

Schomberg, D. W. 1979. Steroidal modulation of steroid secretion in vitro: An experimental approach to intrafollicular regulatory mechanisms. Adv. Exp. Med. Biol. 112:155-165.

Schou, M., M. D. Goldfield, M. R. Weinstein, and A. Villeneuve. 1973. Lithium and pregnancy. I. Report from the Register of Lithium Babies. Br. Med. J. 2:135-156.

Schrag, S. D., and R. L. Dixon. 1985. Occupational exposures associated with male reproductive dysfunction. Annu. Rev. Pharmacol. Toxicol. 25:567-592.

Schull, W. J., M. Otake, and J. V. Neel. 1981a. Genetic effects of the atomic bombs: A reappraisal. Science 213:1220-1227.

Schull, W. J., M. Otake, and J. V. Neel. 1981b. Hiroshima and Nagasaki: A reassessment of the mutagenic effect of exposure to ionizing radiation. In E. B. Hook and I. H. Porter, Eds. Population and Biological Aspects of Human Mutation. New York: Academic Press.

Schwartz, D., A. Laplanche, P. Jouannet, and G. David. 1979. Within-subject variability of human semen in regard to sperm count, volume, total number of spermatozoa and length of abstinence. J. Reprod. Fertil. 57:391-395.

Schwartz, J., P. J. Landrigan, R. G. Feldman, E. K. Silbergeld, E. L. Baker, Jr., and I. H. von Lindern. 1988. Threshold effect in lead-induced neuropathy. J. Pediatr. 112:12-17.

Schwartz, M., and R. Jewelewicz. 1981. The use of gonadotropins for induction of ovulation. Fertil. Steril. 35:3-12.

Searle, A. G., and C. V. Beechey. 1974. Sperm-count, egg-fertilization and dominant lethality after X-irradiation. Mutat. Res. 22:63-72.

Searle, A. G., and C. V. Beechey. 1985. Noncomplementation phenomena and their bearing on nondisjunctional effects, pp. 363-376. In V. L. Dellarco, P. E. Voytek, and A. Hollaender, Eds. Aneuploidy: Etiology and Mechanisms. New York: Plenum Press.

Sega, G. A., and J. G. Owens. 1983. Methylation of DNA and protamine by methyl methanesulfonate in the germ cells of male mice. Mutat. Res. 111:227-244.

Sega, G. A., and J. G. Owens. 1987. Binding of ethylene oxide in spermiogenic germ cell stages of the mouse after low-level inhalation exposure. Environ. Mol. Mutagen. 10:119-127.

Sega, G. A., A. E. Sluder, L. S. McCoy, J. G. Owens, and E. E. Generoso. 1986. The use of alkaline elution procedures to measure DNA damage in spermiogenic stages of mice exposed to methyl methanesulfonate. Mutat. Res. 159:55-63.

Sengelaub, D. R., R. P. Dolan and R. L. Finlay. 1986. Cell generation, death, and retinal growth in the development of the hamster retinal ganglion cell layer. J. Comp. Neurol. 246:527-543.

Seppalainen, A. 1978. Diagnostic utility of neuroelectric measures in environmental and occupational medicine, pp. 448-452. In D. A. Otto, Ed. Multidisciplinary Perspectives in Event Related Brain Potential Research. EPA 600/9-77-043. Washington, D.C.: U.S. Environmental Protection Agency.

Seppalainen, A., and S. Hernberg. 1980. Subclinical lead neuropathy. Am. J. Ind. Med. 1:413-420.

Seppalainen, A., C. Raitta, and M. S. Huuskonen. 1979. n-Hexane-induced changes in visual evoked potentials and electroretinograms of industrial workers. Electroencephalogr. Clin. Neurophysiol. 47:492-498.

Seppalainen, A., K. Savolainen, and T. Kovala. 1981. Changes induced by xylene and alcohol in human evoked potentials. Electroencephalogr. Clin. Neurophysiol. 51:148-155.

Seshadri, R., E. Baker, and G. R. Sutherland. 1982. Sister-chromatid exchange (SCE) analysis in mothers exposed to DNA-damaging agents and their newborn infants. Mutat. Res. 97:139-146.

Setchell, B. P., and G. M. H. Waites. 1975. The blood-testis barrier, pp. 143-172. In D. W. Hamilton and R. O. Greep, Eds. Handbook of Physiology, Section 7: Endocrinology. Volume V. Male Reproductive System. Washington, D.C.: American Physiological Society.

Sever, J. L., and R. L. Brent. 1986. Teratogen Update. Environmentally Induced Birth Defect Risks. New York: Liss. 248 pp.

Sever, L. W., and N. A. Hessol. 1985. Toxic effects of occupational and environmental chemicals on the testes, pp. 211-248. In J. A. Thomas, K. S. Korach and J. A. McLachlan, Eds. Endocrine Toxicology. Target Organ Toxicology Series. New York: Raven Press.

Shackleton, C. H. L. 1986. Profiling steroid hormones and urinary steroids. J. Chromatogr. 379:91-156.

Shaha, C., A. S. Liotta, D. T. Krieger, and C. W. Bardin. 1984. The ontogeny of immunoreactive ß-endorphin in fetal, neonatal, and pubertal testes from mouse and hamster. Endocrinology 114:1584-1591.

Shalgi, R., J. Dor, E. Rudak, A. Lusky, B. Goldman, S. Mashiach, and L. Nebel. 1985. Penetration of sperm from teratospermic men into zona-free hamster eggs. Int. J. Androl. 8:285-294.

Shamsuddin, A. K. M., N. T. Sinopoli, K. Hemminki, R. R. Boesch, and C. C. Harris. 1985. Detection of benzo[a]pyrene: DNA adducts in human white blood cells. Cancer Res. 45:66-68.

Shangold, M. M., M. L. Gatz, and B. Thysen. 1981. Acute effects of exercise on plasma concentrations of prolactin and testosterone in recreational women runners. Fertil. Steril. 35:699-702.

Shapiro, G., and S. Evron. 1980. A novel use of spironolactone: Treatment of hirsutism. J. Clin. Endocrinol. Metab. 51:429-432.

Shapiro, S. S., and P. Thiagarajan. 1982. Lupus anticoagulants. Prog. Hemost. Thromb. 6:263-285.

Sheehan, H. L. 1968. Neurohypophysis and hypothalamus, pp. 12-74. In J. M. B. Bloodworth, Jr., Ed. Endocrine Pathology. Baltimore, Md.: Williams & Wilkins.

Shepard, T. H. 1986. Human teratogenicity. Adv. Pediatr. 33:225-268.

Sherins, R. J., and S. S. Howards. 1986. Male infertility, pp. 640-697. In Campbell's Urology, 5th ed. Philadelphia: W. B. Saunders Company.

Sherins, R. J., D. Brightwell, and P. M. Sternthal. 1977. Longitudinal analysis of semen of fertile and infertile men, pp. 473-488. In P. Troen and H. R. Nankin, Eds. The Testis in Normal and Infertile Men. New

York: Raven Press.

Sherins, R. J., A. P. Patterson, D. Brightwell, R. Udelsman, and J. Sartor. 1982. Alterations in the plasma testosterone:estradiol ratio: An alternative to the inhibin hypothesis. Ann. N.Y. Acad. Sci. 383:295-306.

Sherman, B. M., and S. G. Korenman. 1975. Hormonal characteristics of the human menstrual cycle throughout reproductive life. J. Clin. Invest. 55:699-706.

Sherman, B. M., J. H. West, and S. G. Korenman. 1976. The menopausal transition: Analysis of LH, FSH, estradiol, and progesterone concentrations during menstrual cycles of older women. Endrocinol. Metab. 42:629-636.

Sherman, M. I., and L. R. Wudl. 1976. The implanting mouse blastocyst, pp. 81-125. In G. Poste and G. L. Nicolson, Eds. The Cell Surface in Animal Embryogenesis and Development. Amsterdam: Elsevier-North Holland.

Sherman, M. I., A. McLaren, and P. M. B. Walker. 1972. Mechanism of accumulation of DNA in giant cells of mouse trophoblast. Nature 238:175-176.

Shiu, R. P. C., and H. G. Friesen. 1980. Mechanism of action of prolactin in the control of mammary gland function. Annu. Rev. Physiol. 42:83-96.

Short, R. V. 1962. Steroids in the follicular fluid and the corpus luteum of the mare. A "two-cell type" theory of ovarian steroid synthesis. J. Endocrinol. 24:59-63.

Short, R. V. 1979. When conception fails to become a pregnancy, pp. 377-387. In Maternal Recognition of Pregnancy. CIBA Foundation Symposium 64 (New Series). Amsterdam: Excerpta Medica.

Shum, S., N. M. Jensen, and D. W. Nebert. 1979. The murine Ah locus: In utero toxicity and teratogenesis associated with genetic differences in benzo[a]pyrene metabolism. Teratology 20:365-376.

Sidman, R. L. 1961. Histogenesis of mouse retina studied with thymidine-H^3, pp. 487-506. In G. K. Smelser, Ed. Structure of the Eye. New York: Academic Press.

Sidman, R. L., I. L. Miale and N. Feder. 1959. Cell proliferation and migration in the primitive ependymal zone: An autoradiographic study of histogenesis in the nervous system. Exp. Neurol. 1:322-333.

Sigler, A. T., B. H. Cohen, A. M. Lilienfeld, J. E. Westlake, and W. H. Hetznecker. 1967. Reproductive and marital experience of parents of children with Down's Syndrome (mongolism). J. Pediatr. 70:608-614.

Silber, S. J., and L. J. Rodriguez-Rigau. 1981. Quantitative analysis of testicle biopsy: Determination of partial obstruction and prediction of sperm count after surgery for obstruction. Fertil. Steril. 36:480-485.

Silbergeld, E. K. 1983. Experimental studies of lead neurotoxicity: Implications for mechanisms, dose-response and reversibility, pp. 191-216. In M. Rutter and R. Russell Jones, Eds. Lead Versus Health: Sources and Effects of Low Level Lead. New York:

John Wiley & Sons.

Silbergeld, E. K. 1985. The relevance of animal models for neurotoxic disease states. Int. J. Mental Health 14:26-43.

Silbergeld, E. K., and J. J. Chisolm, Jr. 1976. Lead poisoning: Altered urinary catecholamine metabolites as indicators of intoxication in mice and children. Science 192(4235):153-155.

Silbergeld, E. K., and R. E. Hruska. 1980. Neurochemical investigations of low lead exposure, pp. 135-152. In H. L. Needleman, Ed. Low Level Lead Exposure: The Clinical Implications of Current Research. New York: Raven Press.

Silbergeld, E. K., J. Schwartz, and K. Mahaffey. 1988. Lead and Osteoporosis: Mobilization of lead from bone in postmenopausal women. Environ. Res.

Silkworth, J. B., L. Antrim, and L. S. Kaminsky. 1984. Correlations between polychlorinated biphenyl immunotoxicity, the aromatic hydrocarbon locus, and liver microsomal enzyme induction in C57BL/6 and DBA/2 mice. Toxicol. Appl. Pharmacol. 75:156-165.

Simonds, J. F., and L. Aston. 1981. Relationship between minor physical anomalies, perinatal complications, and psychiatric diagnoses in children. Psychiatry Res. 4:181-188.

Simpson, J. L., M. S. Golbus, A. O. Martin, and G. E. Sarto. 1982. Genetics in Obstetrics and Gynecology. New York: Grune & Stratton. 322 pp.

Sims, D. G., and G. A. Neligan. 1975. Factors affecting the increasing incidence of severe non-hemolytic neonatal jaundice. Br. J. Obstet. Gyencol. 82:863-867.

Singhal, R. L., and J. A. Thomas, Eds. 1980. Lead Toxicity. Baltimore: Urban & Schwarzenberg. 514 pp.

Sinosich, M. J., J. G. Grudzinskas, and D. M. Saunders. 1985. Placental proteins in the diagnosis and evaluation of the "elusive" early pregnancy. Obstet. Gynecol. Surv. 40:273-282.

Skare, J. A., and K. R. Schrotel. 1984. Alkaline elution of rat testicular DNA: Detection of DNA strand breaks after in vivo treatment with chemical mutagens. Mutat. Res. 130:283-294.

Skinner, M. K., and M. D. Griswold. 1982. Secretion of testicular transferrin by cultured Sertoli cells is regulated by hormones and retinoids. Biol. Reprod. 27:211-221.

Skinner, M. K., L. Cosand, and M. D. Griswold. 1984. Purification and characterization of testicular transferrin secreted by rat Sertoli cells. Biochem. J. 218:313-320.

Smalley, S. L., R. F. Asarnow, and M. A. Spence. 1988. Autism and genetics. A decade of research. Arch. Gen. Psychiatry 45:953-961.

Smith, D. M., and A. E. Smith. 1971. Uptake and incorporation of amino acids by cultured mouse embryos: Estrogen stimulation. Biol. Reprod. 4:66-73.

Smith, F. W., A. H. Adam, and W. D. P. Phillips. 1983.

NMR imaging in pregnancy (letter). Lancet 1:61-62.

Smith, K. D., L. J. Rodriquez-Rigau, and E. Steinberger. 1977. Relation between indices of semen analysis and pregnancy rate in infertile couples. Fertil. Steril. 28:1314-1319.

Smith, N. J. 1980. Excessive weight loss and food aversion in athletes simulating anorexia nervosa. Pediatrics 66:139-142.

Smith, P. J. 1985. Behavioral toxicology: Evaluating cognitive function. Neurobehav. Toxicol. Teratol. 7:345-350.

Snyder, S. H. 1984. Neurosciences: An integrated discipline. Science 225:1255-1257.

Soares, M. J., and F. Talamantes. 1985. Placental lactogen secretion in the mouse: In vitro responses and hormonal influences. J. Exp. Zool. 234:97-104.

Soares, M. J., P. Colosi, and F. Talamantes. 1982. The development and characterization of a homologous radioimmunoassay for mouse placental lactogen. Endocrinology 110:668-670.

Soares, M. J., J. A. Julian, and S. R. Glasser. 1985. Trophoblast giant cell release of placental lactogens: temporal and regional characteristics. Dev. Biol. 107:520-526.

Soberon, J., J. J. Calderon, and J. W. Goldzieher. 1966. Relation of parity to age at menopause. Am. J. Obstet. Gynecol. 96:96-100.

Sommer, B. 1978. Stress and menstrual distress. J. Hum. Stress 4(3):5-10,41-47.

Sommer, G., T. Brosnan, Q. Cao, D. Nishimura, A. Macovski, and J. McNeal. 1988. Noise-reduced prostatic MR imaging: Work in progress. Radiology 169:347-350.

Song, B. J., H. V. Gelboin, S. S. Park, G. C. Tsokos, and F. K. Friedman. 1985. Monoclonal antibody-directed radioimmunoassay detects cytochrome P-450 in human placenta and lymphocytes. Science 228(4698):490-492.

Soules, M. R., R. H. Wiebe, S. Aksel, and C. B. Hammond. 1977. The diagnosis and therapy of luteal phase deficiency. Fertil. Steril. 28:1033-1037.

Soules, M. R., G. P. Sutton, C. B. Hammond, and A. F. Haney. 1980. Endocrine changes at operation under general anesthesia: Reproductive hormone fluctuations in young women. Fertil. Steril. 33:364-371.

Soyka, L. F., and J. M. Joffe. 1980. Male mediated drug effects on offspring. Prog. Clin. Biol. Res. 36:49-66.

Speight, A. N. 1977. Floppy-infant syndrome and maternal diazepam and/or nitrazepam [letter]. Lancet 2:878.

Speroff, C., R. H. Glass, and N. G. Kase. 1983. Clinical Gynecologic Endocrinology and Infertility. 3rd ed. Baltimore: Williams & Wilkins.

Spyker, J. M., S. B Sparber, and A. M. Goldberg. 1972. Subtle consequences of methylmercury exposure: Behavioral deviations in offspring of treated mothers.

Science 177:621-623.

Stamey, T. A., S. R. M. Bushby, and J. Bragonje. 1973. The concentration of trimethoprim in prostatic fluid: Nonionic diffusion or active transport? J. Infect. Dis. 128(Suppl.):686-692.

Stanker, L. H., A. Wyrobek, and R. Balhorn. 1987a. Monoclonal antibodies to human protamines. Hybridoma 6:293-303.

Stanker, L. H., B. Watkins, N. Rogers, and M. Vanderlaan. 1987b. Monoclonal antibodies for dioxin: Antibody characterization and assay development. Toxicology 45:229-243.

Stave, U. 1978. Perinatal Physiology. New York: Plenum Medical Book. 851 pp.

Steenhout, A., and M. Pourtois. 1981. Lead accumulation in teeth as a function of age with different exposures. Br. J. Ind. Med. 38:297-303.

Steg, J., and J. L. Rapoport. 1975. Minor phyiscal anomalies in normal, neurotic, learning disabled, and severely disturbed children. J. Autism Child Schizophr. 5:299-303.

Steinberger, A., and E. Steinberger. 1976. Secretion of an FSH-inhibiting factor by cultured Sertoli cells. Endocrinology 99:918-921.

Stewart, T. A., A. R. Bellve, and P. Leder. 1984. Transcription and promoter usage of the myc gene in normal somatic and spermatogenic cells. Science 226:707-710.

Stock, R. J., J. B. Josimovich, B. Kosor, A. Klopper, and G. R. Wilson. 1971. The effect of chorionic gonadotropin and chorionic somatomammotropin on steroidogenesis in the corpus luteum. J. Obstet. Gynaecol. Br. Commonw. 78:549-553.

Stoetzer, H., A. Niggenschulze, W. Froelke, and D. Elich. 1987. The action of methylthiouracil on the thyroid of rats of the F1 and F2 generation, pp. 109-120. In T. Fujii and P. M. Adams, Eds. Functional Teratogenesis: Functional Effects on the Offspring after Parental Drug Exposure. Tokyo: Teikyo University Press.

Stone, M. L., L. J. Salerno, M. Green, and C. Zelson. 1971. Narcotic addiction in pregnancy. Am. J. Obstet. Gynecol. 109:716-723.

Streissguth, A. P., H. M. Barr, and D. C. Martin. 1984. Alcohol exposure in utero and functional deficits in children during the first four years of life, pp. 176-196. In Mechanisms of Alcohol Damage in utero. CIBA Foundation Symposium 105. London: Pitman.

Strobino, B. R., J. Kline, and Z. Stein. 1978. Chemical and physical exposure of parents: Effects on human reproduction. Early Hum. Dev. 1:371-399.

Stumpf, W. E., M. Sar, and L. D. Grant. 1980. Autoradiographic localization of ^{210}Pb amid its decay products in rat forebrain. Neurotoxicology 1:593-606.

Suciu-Foca, N., E. Reed, C. Rohowsky, P. Kung, and D. W. King. 1983. Anti-idiotypic antibodies to anti-HLA receptors induced by pregancy. Proc. Natl. Acad. Sci. USA 80:830-834.

Surani, M. A. H. 1975. Hormonal regulation of proteins in the uterine secretion of ovariectmized rats and the implications for implantation and embryonic diapause. J. Reprod. Fertil. 43:411-417.

Suskind, R. R. 1985. Chloracne, "the hallmark of dioxin intoxication". Scand. J. Work Environ. Health 11(3 Spec. No.):165-171.

Sutherland, J. M. 1959. Fatal cardiovascular collapse of infants receiving large amounts of chloramphenicol. A.M.A. J. Dis. Child. 97:761-767.

Suzuki, F., and T. Nagano. 1978. Regional differentiation of cell junctions in the excurrent duct epithelium of the rat testis as revealed by freeze-fracture. Anat. Rec. 191:503-519.

Sved, A. F. 1983. Precursor control of the function of monoaminergic neurons, pp. 223-275. In R. J. Wurtman and J. J. Wurtman, Eds. Nutrition and the Brain, Vol. 6. Physiological and Behavioral Effects of Food Constituents. New York: Raven Press.

Swaab, D. F., and E. Fliers. 1985. A sexually dimorphic nucleus in the human brain. Science 228:1112-1115.

Swerdloff, R. S., and P. C. Walsh. 1973. Testosterone and oestradiol suppression of LH and FSH in adult male rats: Duration of castration, duration of treatment and combined treatment. Acta Endocrinol. 73:11-21.

Swierstra, E. E. 1971. Sperm production of boars as measured from epididymal sperm reserves and quantitative testicular histology. J. Reprod. Fertil. 27:91-99.

Sylvan, P. E., D. T. MacLaughlin, G. S. Richardson, R. E. Scully, and N. Nikrui. 1981. Human uterine fluid proteins associated with secretory phase endometrium: Progesterone-induced products? Biol. Reprod. 24:423-429.

Sylvester, S. R., M. K. Skinner, and M. D. Griswold. 1984. A sulfated glycoprotein synthesized by Sertoli cells and by epididymal cells is a component of the sperm membrane. Biol. Reprod. 31:1087-1101.

Talamantes, F., L. Ogren, E. Markoff, S. Woodward, and J. Madrid. 1980. Phylogenetic distribution, regulation of secretion, and prolactin-like effects of placental lactogens. Fed. Proc. 39:2582-2587.

Tanner, J. M. A. 1981. A History of the Study of Human Growth. Cambridge: Cambridge University Press.

Tate, S. S., and J. Orlando. 1979. Conversion of glutathione to glutathione disulfide, a catalytic function of gamma-glutamyl transpeptidase. J. Biol. Chem. 254:5573-5575.

Tates, A. D., and P. deBoer. 1984. Further evaluation of a micronucleus method for detection of meiotic micronuclei in male germ cells of mammals. Mutat. Res. 140:187-191.

Taylor, C., and W. P. Faulk. 1981. Prevention of recurrent abortion with leucocyte transfusions. Lancet 2:68-70.

Taylor, C., W. P. Faulk, and J. A. McIntyre. 1985. Prevention of recurrent spontaneous abortions by leucocyte transfusion. J. R. Soc. Med. 78:623-627.

Templado, C., F. Vidal, J. Navarro, and J. Egozcue. 1986. Improved technique for the study of meiosis in ejaculate: Results of the first 50 consecutive cases. Hum. Genet. 72:275-277.

Terasawa, E. 1985. Developmental changes in the positive feedback effect of estrogen on luteinizing hormone release in ovariectomized female rhesus monkeys. Endocrinology 117:2490-2497.

Teratology Society. 1987. Recomendations for Vitamin A use in pregnancy. Teratology Society Position Paper. Teratology 35:269-275.

Thau, R. B., Y. Yamamoto, M. Goldstein, P. H. Ehrlich, S. S. Witkin, G. N. Burrow, R. E. Canfield, and C. W. Bardin. 1983. Characterization of a human anti-hCG antiserum: A proposed standard for laboratories involved with the development of hCG vaccines. Contraception 27:627-637.

Thijssen, J. H., J. Poortman, F. Schwarz, and F. de Waard. 1975. Post-menopausal estrogen production, with special reference to patients with mammary carcinoma. Front. Horm. Res. 3:45-62.

Thomas, M. L., J. H. Harger, D. K. Wagener, B. S. Rabin, and T. J. Gill, 3d. 1985. HLA sharing and spontaneous abortion in humans. Am. J. Obstet. Gynecol. 151:1053-1058.

Tice, R. R., B. Lambert, K. Morimoto, and A. Hollaender. 1984. A review of the International Symposium on Sister Chromatid Exchanges: Twenty-five years of experimental research. Environ. Mutagen. 6:737-752.

Tietz, N. W., O. Sheremeta, and E. Bukay. 1971. Gas chromatographic separation and determination of 17-ketosteroids, prenanediol and pregnanetriol, p. 525-537. In F. W. Sunderman, and F. W. Sunderman, Jr., Eds. Laboratory Diagnosis of Endocrine Disease. St. Louis, Mo.: Warren H. Green.

Tilson, H. A., and D. C. Wright. 1985. Interpretation of behavioral teratology data. Neurobehav. Toxicol. Teratol. 7:667-668.

Tilson, H. A., G. J. Davis, J. A. McLachlan, and G. W. Lucier. 1979. The effects of polychlorinated biphenyls given prenatally on the neurobehavioral development of mice. Environ. Res. 18:466-474.

Timourian, H., J. B. Bishop, and A. J. Wyrobek. 1983. Sperm shape abnormalities in the offspring of male mice treated with triethylenemelamine (TEM). Environ. Mutagen. 5:399.

Tindall, D. J., V. Hansson, W. S. McLean, E. M. Ritzen, S. N. Nayfeh, and F. S. French. 1975. Androgen-binding proteins in rat epididymis: Properties of a cytoplasmic receptor for androgen similar to the androgen receptor in ventral prostate and different from androgen-binding protein (ABP). Mol. Cell. Endocrinol. 3:83-101.

Tinneberg, H. R., R. P. Staves, V. Hanf, W. Scholz, K. Semm, and L. Mettler. 1985. Use of a modified test system to determine early pregnancy factor (EPF) levels in patients with normal first trimester pregnancy and after therapeutic abortion. Ann. N.Y. Acad. Sci.

442:551-557.

Tolis, G., D. Ruggere, D. R. Popkin, J. Chow, M. E. Boyd, A. De Leon, A. B. Lalonde, A. Asswad, M. Hendelman, V. Scali, R. Koby, G. Arronet, B. Yufe, F. J. Tweedie, P. R. Fournier, and F. Naftolin. 1979. Prolonged amenorrhea and oral contraceptives. Fertil. Steril. 32:265-268.

Toman, Z., C. Dambly-Chaudiere, L. Tenenbaum, and M. Radman. 1985. A system for detection of genetic and epigenetic alterations in Escherichia coli induced by DNA-damaging agents. J. Mol. Biol. 186:97-105.

Toowicharanont, P., and M. Chulavatnatol. 1983. Measurement of anionic sites of rat epididymal spermatozoa using tritiated polycationized ferritin. J. Reprod. Fertil. 69:303-306.

Torry, D. S., J. A. McIntyre, W. P. Faulk, and P. R. McConnachie. 1986. Inhibitors of complement-mediated cytotoxicity in normal and secondary aborter sera. Am. J. Reprod. Immunol. Microbiol. 10:53-57.

Trapp, M., V. Baukloh, H. G. Bohnet, and W. Heeschen. 1984. Pollutants in human follicular fluid. Fertil. Steril. 42:146-148.

Trasler, J. M., B. F. Hales, and B. Robaire. 1985. Paternal cyclophosphamide treatment of rats causes fetal loss and malformations without affecting male fertility. Nature 316:144-146.

Trasler, J. M., B. F. Hales, and B. Robaire. 1986. Chronic low-dose cyclophosphamide treatment of adult male rats: Effects on fertility, pregnancy outcome and progeny. Biol. Reprod. 34:275-283.

Trasler, J. M., B. F. Hales, and B. Robaire. 1987. A time-course study of chronic paternal cyclophosphamide treatment in rats: Effects on pregnancy outcome and the male reproductive and hematologic systems. Biol. Reprod. 37:317-326.

Trasler, J. M., L. Hermo, and B. Robaire. 1988. Morphological changes in the testis and epididymis of rats treated with cyclophosphamide: A quantitative approach. Biol. Reprod. 38:463-479.

Treloar, A. E., R. E. Boynton, B. G. Behn, and B. W. Brown. 1970. Variation of the human menstrual cycle throughout reproductive life. Int. J. Fertil. 12:77-126.

Trillingsgaard, A., O. N. Hansen, and I. Besse. 1985. The Bender-Gestalt Test as a neurobehavioral measure of preclinical visual-motor integration deficits in children with low-level lead exposure, pp. 189-193. In Neurobehavioral Methods in Occupational and Environmental Health. WHO Environmental Health Document 3. Copenhagen: World Health Organization.

Trimble, G. X., F. A. Finnerty, N. S. Assali, and S. L. Leiken. 1964. Thiazide and neonatal thrombocytopenia (letter). N. Engl. J. Med. 271:160.

Trounson, A. O., M. Mahadevan, J. Wood, and J. F. Leeton. 1980. Studies on the deep-freezing and artificial insemination of human semen, pp. 173-185. In D. W. Richardson, D. Joyce, and E. M. Symonds, Eds. Frozen Human Semen. The Hague: Martinus Nijhoff.

Tsai, S. P., and C. P. Wen. 1986. A review of methodological issues of the standardized mortality ratio (SMR) in occupational cohort studies. Int. J. Epidemiol. 15:8-21.

Tsong, S. D., D. Phillips, N. Halmi, A. S. Liotta, A. Margioris, C. W. Bardin, and D. T. Krieger. 1982a. ACTH and ß-endorphin-related peptides are present in multiple sites in the reproductive tract of the male rat. Endocrinology 110:2204-2206.

Tsong, S. D., D. M. Phillips, N. Halmi, D. Krieger, and C. W. Bardin. 1982b. ß-endorphin is present in the male reproductive tract of five species. Biol. Reprod. 27:755-764.

Turk, J. L. 1975. Delayed Hypersensitivity, 2nd ed. Frontiers of Biology, Vol. 4. Amsterdam: North Holland. 319 pp.

Turner, T. T. 1979. On the epididymis and its function. Invest. Urol. 16:311-321.

Turner, T. T., and D. M. Cesarini. 1983. The ability of the rat epididymis to concentrate spermatozoa: Responsiveness to aldosterone. J. Androl. 4:197-202.

Turner, T. T., and R. D. Giles. 1982. Sperm motility-inhibiting factor in rat epididymis. Am. J. Physiol. 242:R199-R203.

Turner, T. T., and S. S. Howards. 1985. The tenacity of the blood-testis and blood-epididymal barriers, pp. 383-393. In T. J. Lobl and E. S. E. Hafez. Male Fertility and Its Regulation. New York: MTP Press.

Turner, T. T., and A. D. Johnson. 1973. The metabolic activity of the bovine epididymis. I. Utilization of glucose and fructose. J. Reprod. Fertil. 34:201-213.

Turner, T. T., R. D. Giles, and S. S. Howards. 1981. Effect of oestradiol valerate on the rat blood—testis and blood—epididymal barriers to [^{3}H] inulin. J. Reprod. Fertil. 63:355-358.

Turner, T. T., C. E. Jones, S. S. Howards, L. L. Ewing, B. Zegeye, and G. L. Gunsalus. 1984. On the androgen microenvironment of maturing spermatozoa. Endocrinology 115:1925-1932.

Tyler, J. P. P., N. G. Crockett, and L. Driscoll. 1982. Studies of human seminal parameters with frequent ejaculation. I. Clinical characteristics. Clin. Reprod. Fertil. 1:273-285.

Tzartos, S. J., and M. S. H. Surani. 1979. Affinity of uterine luminal proteins for rat blastocysts. J. Reprod. Fertil. 56:579-586.

Ullsa-Aguirre, A., and S. C. Chappel. 1982. Multiple species of follicle-stimulating hormone exist within the anterior pituitary gland of male golden hamsters. J. Endocrinol. 95:257-266.

Ulstein, M. 1972. Sperm penetration of cervical mucus as a criterion of male fertility. Acta Obstet. Gynecol. Scand. 51:335-340.

Unander, A. M., and L. B. Olding. 1983. Habitual abortion: Parental sharing of HLA antigens, absence of maternal blocking antibody, and suppression of maternal lymphocytes. Am. J. Reprod. Immunol. 4:171-178.

UNSCEAR (United Nations Scientific Committee on the Effects of Atomic Radiation). 1982. Report, Sources and Effects of Ionizing Radiation. New York: United Nations.

U.S. EPA (United States Environmental Protection Agency). 1986. Air Quality Criteria for Lead. EPA-600/8-83/028aF. Research Triangle Park, N.C.: U.S. Environmental Protection Agency. 4 vols.

U.S. EPA (United States Environmental Protection Agency). 1987. Health effects testing: Reproduction and fertility effects. Code of Federal Regulations, Title 40, Chap. 1, Pt. 798, section 4700. Revised July 1, 1987. Washington, D.C.: U.S. Government Printing Office.

U.S. EPA (United States Environmental Protection Agency). 1989. Ambient Air Quality Criteria Document. 1989 Addendum. Washington, D.C.: U.S. Environmental Protection Agency.

U.S. FDA (Food and Drug Administration). 1988. Good laboratory practice for nonclinical laboratory studies. Code of Federal Regulations, Title 21, Chap. 1, Pt. 58. Washington, D.C.: U.S. Government Printing Office.

Usselman, M. C., and R. A. Cone. 1983. Rat sperm are mechanically immobilized in the caudal epididymis by "immobilin," a high molecular weight glycoprotein. Biol. Reprod. 29:1241-1253.

Uzman, L. L. 1960. The histogenesis of the mouse cerebellum as studied by its tritiated thymidine uptake. J. Comp. Neurol. 114:137-160.

Vaitukaitis, J. L., G. D. Braunstein, and G. T. Ross. 1972. A radioimmunoassay which specifically measures human chorionic gonadotropin in the presence of human luteinizing hormone. Am. J. Obstet. Gynecol. 113:751-758.

Valenca, M. M., and A. Negro-Vilar. 1986. Pro-opiomelanocortin-derived peptides in testicular interstitial fluid: Characterization and changes in secretion after human chorionic gonadotropin or luteinizing hormone-releasing hormone analog treatment. Endocrinology 118:32-37.

Van Beurden-Lamers, W. M. O., A. O. Brinkmann, E. Mulder, and H. J. Van der Molen. 1974. High-affinity binding of oestradiol-17β by cytosols from testis interstitial tissue, pituitary, adrenal, liver and accessory sex glands of the male rat. Biochem. J. 140:495-502.

Vanderlaan, M., J. Van Emon, B. Watkins, and L. Stanker. 1987. Monoclonal antibodies for the detection of trace chemicals, pp. 597-602. In R. Greenhalgh and T. R. Roberts, Eds. Pesticide Science and Biotechnology. Blackwell Scientific Publications.

van der Spuy, Z. M., D. L. Jones, C. S. Wright, B. Piura, D. B. Paintin, V. H. James, and H. S. Jacobs. 1985. Inhibition of 3-beta-hydroxy steroid dehydrogenase activity in first trimester human pregnancy with trilostane and Win 32729. Clin. Endocrinol. (Oxf.) 19:521-531.

Van Dilla, M. A., and M. L. Mendelsohn. 1979. Introduction and resumé of flow cytometry and sorting, pp. 11-37. In M. R. Melamed, P. F. Mullaney, and M. L. Mendelsohn, Eds. Flow Cytometry and Sorting. New York: John Wiley & Sons.

van Dop, P. A., R. J. Scholtmeijer, P. H. J. Kurver, J. P. A. Baak, J. Oort, and L. A. M. Stolte. 1980a. A quantitative structural model of the testis of fertile moles with normal sperm counts. Int. J. Androl. 3:153-169.

van Dop, P. A., P. H. J. Kurver, R. J. Scholtmeijer, J. P. A. Baak, J. Oort, and L. A. M. Stolte. 1980b. Correlations between the quantitative morphology of the human testis and sperm production. Int. J. Androl. 3:170-176

Vanha-Perttula, T., J. P. Mather, C. W. Bardin, S. B. Moss, and A. R. Bellve. 1985. Localization of the angiotensin-converting enzyme activity in testis and epididymis. Biol. Reprod. 33:870-877.

Van Keep, P. A., P. C. Brand, and P. H. Lehert. 1979. Factors affecting the age at menopause. J. Biosoc. Sci. 6(Suppl.):37-55.

Van Thiel, D. H., R. Lester, and R. J. Sherins. 1974. Hypogonadism in alcoholic liver disease: Evidence for a double defect. Gastroenterology 67:1188-1199.

van Zeeland, A. A. 1986. Molecular dosimetry of alkylating agents: Quantitative comparison of genetic effects on the basis of DNA-adduct formation. Unpublished manuscript. Department of Radiation Genetics & Chemical Mutagenesis, Sylvius Laboratories, State University of Leiden. The Netherlands. (December)

Venetianer A., D. L. Schiller, T. Magin, and W. W. Franke. 1983. Cessation of cytokeratin expression in a rat hepatoma cell line lacking differentiated functions. Nature 305:730-733.

Verjans, H. L., F. H. de Jong, B. A. Cooke, H. J. Van der Molen, and K. B. Eik-Nes. 1974. Effect of oestradiol benzoate on pituitary and testis function in the normal and adult male rat. Acta Endocrinol. 77:636-642.

Vermeulen, A., L. Verdonch, M. Van der Straeten, and N. Orie. 1969. Capacity of the testosterone-binding globulin in human plasma and influence of specific binding of testosterone on its metabolic clearance rate. J. Clin. Endocrinol. Metab. 29:1470.

Vermeulen, A., T. Stoica, and L. Verdonck. 1971. The apparent free testosterone concentration, an index of androgenicity. J. Clin. Endocrinol. Metab. 33:759-767.

Vernon, R. B., C. H. Muller, J. C. Herr, F. A. Feuchter, and E. M. Eddy. 1982. Epididymal secretion of a mouse sperm surface component recognized by a monoclonal antibody. Biol. Reprod. 26:523-535.

Vernon, R. B., M. S. Hamilton, and E. M. Eddy. 1985. Effects of in vivo and in vitro fertilization environments on the expression of a surface antigen of the mouse sperm tail. Biol. Reprod. 32:669-680.

Vesell, E. S., G. T. Passananti, F. E. Greene, and J. G. Page. 1971. Genetic control of drug levels and of the induction of drug-metabolizing enzymes in man: Individual variability in the extent of allupurinol and nortriptyline inhibition of drug metabolism. Ann. N.Y. Acad. Sci. 79:752-773.

Vessey, M., R. Doll, R. Peto, B. Johnson, and P. Wiggins. 1976. A long-term follow-up study of women using different methods of contraception: An interim report. J. Biosoc. Sci. 8:373-427.

Vestel, R. E., and G. W. Dawson. 1985. Pharmacology and aging, pp. 744-819. In C. E. Finch and E. L. Schneider, Eds. Handbook of the Biology of Aging, 2nd ed. New York: Van Nostrand Reinhold.

Vigersky, R. A., A. E. Andersen, R. H. Thompson, and D. L. Loriaux. 1977. Hypothalamic dysfunction in secondary amenorrhea associated with simple weight loss. N. Engl. J. Med. 297:1141-1145.

Virji, N. 1985. LDH-C4 in human seminal plasma and its relationship to testicular function. I. Methodogical aspects. Int. J. Androl. 8:193-200.

Vogel, F., and R. Rathenberg. 1975. Spontaneous mutation in man. Adv. Hum. Genet. 5:223-318.

Voglmayr, J. K. 1975. Metabolic changes in spermatozoa during epididymal transit, pp. 437-451. In D. W. Hamilton and R. O. Greep, Eds. Handbook of Physiology, Section 7: Endocrinology. Volume V. Male Reproductive System. Washington, D.C.: American Physiological Society.

Voglmayr, J. K., and I. G. White. 1979. Effects of rete testis and epididymal fluid on the metabolism and motility of testicular and post-testicular spermatozoa of the ram. Biol. Reprod. 20:288-293.

Voisin, G., R. G. Kinsky, and H. T. Duc. 1972. Immune status of mice tolerant of living cells. II. Continous presence and nature of facilitation-enhancing antibodies in tolerant animals. J. Exp. Med. 135:1185-1203.

Vom Saal, F. S., and C. L. Moyer. 1985. Prenatal effects on reproductive capacity during aging in female mice. Biol. Reprod. 32:1116-1126.

Vorhees, C. V. 1983. Fetal anticonvulsant syndrome in rats: Dose- and period-response relationships of prenatal diphenylhydantoin, trimethadione, and phenobarbital exposure on the structural and functional development of the offspring. J. Pharmacol. Exp. Ther. 227:274-287.

Vrbova, G., T. Gordon, and R. Jones. 1978. Nerve-Muscle Interaction. London: Chapman and Hall. 233 pp.

Wald, N. J., and H. S. Cuckle. 1980. Alpha fetoprotein in the antenatal diagnosis of open neural tube defects. Br. J. Hosp. Med. 23:473-480.

Waldrop, M., and C. Halverson. 1972. Minor physical anomalies: Their incidence and relation to behavior in a normal and deviant sample, pp. 146-156. In R. C. Smart and M. S. Smart, Eds. Readings in Child Development and Relationships. New York: Macmillan.

Waldrop, M. F., F. A. Pederson, and R. Q. Bell. 1968. Minor physical anomalies and behavior in preschool children. Child Dev. 39:391-400.

Waldrop, M. F., R. Q. Bell, and J. D. Goering. 1976. Minor physical anomalies and inhibited behavior in elementary school girls. J. Child Psychol. Psychiatry 17:113-122.

Walker, J. S., H. Winet, and M. Freund. 1982. A comparison of subjective and objective sperm motility evaluation. J. Androl. 3:184-192.

Walsh, P. C., and J. D. Wilson. 1976. The induction of prostatic hypertrophy in the dog with androstanediol. J. Clin. Invest. 57:1093-1097.

Walsh, P. C., and J. D. Wilson. 1987. Impotence and infertility in men, pp. 217-220. In Harrison's Principles of Internal Medicine, 11th ed. New York: McGraw Hill.

Walters, M. R., W. Hunziker, and A. W. Norman. 1982. 1, 25-dihydroxyvitamin D_3 receptors: Exchange assay and presence in reproductive tissues, pp. 91-93. In A. W. Norman, K. Schaefer, D. V. Herrath, H. G. Grigoleit, and W. de Gruyter, Eds. Vitamin D, Chemical, Biochemical and Clinical Endocrinology of Calcium Metabolism. Berlin: W. de Gruyter.

Ward, R. H. T., B. Modell, M. Petrou, F. Karagozlu, and E. Douratsos. 1983. Method of sampling chorionic villi in first trimester of pregnancy under guidance of real time ultrasound. Br. Med. J. 286:1542-1544.

Warkany, J. 1971. Congenital Malformations. Chicago: Year Book Medical Publishers. 1309 pp.

Warkany, J. 1976. Warfarin embryopathy. Teratology 14:205-209.

Warren, M. P. 1980. The effects of exercise on pubertal progression and reproductive function in girls. J. Clin. Endocrinol. Metab. 51:1150-1157.

Warren, M. P. 1982. The effects of altered nutritional states, stress, and systemic illness on reproduction in women, pp. 177-206. In J. L. Vaitukaitis, Ed. Clinical Reproductive Neuroendocrinology. New York: Elsevier Biomedical.

Warren, M. P. 1983. Effects of undernutrition on reproductive function in the human. Endocr. Rev. 4:363-377.

Warren, M. P., R. Jeweiewicz, I. Dyrenfurth, R. Ans, S. Khalaf, and R. L. Vande Wiele. 1975. The significance of weight loss in the evaluation of pituitary response of LH-RH in women with secondary amenorrhea. J. Clin. Endocrinol. Metab. 40:601-611.

Wasmoen, T. L., C. B. Coulam, K. M Leiferman, and G. J. Gleich. 1987. Increases of plasma eosinophil major basic protein levels late in pregnancy predict onset of labor. Proc. Natl. Acad. Sci. USA 84:3029-3032.

Waters, S. H., R. J. Distel, and N. B. Hecht. 1985. Mouse testes contain two size classes of actin mRNA that are differentially expressed during spermatogenesis. Mol. Cell Biol. 5:1649-1654.

Watkins, W. B. 1978. Use of immunocytochemical techniques for the localization of human placental lactogen.

J. Histochem. Cytochem. 26:288-292.

Watkins, W. B., and S. Reddy. 1980. Ovine placental lactogen in the cotyledonary and intercotyledonary placenta of the ewe. J. Reprod. Fertil. 58:411-414.

Weaver, D. D. 1988. Survey of prenatally diagnosed disorders. Clin. Obstet. Gynecol. 31:253-269.

Webb, P. D., and S. R. Glasser. 1984. Implantation, pp. 341-364. In D. P. Wolf and M. M. Quigley, Eds. Human In Vitro Fertilization and Embryo Transfer. New York: Plenum Press.

Webster, M. A., S. L. Phipps, and M. D. Gillmer. 1985. Interruption of first trimester human pregnancy following Epostane therapy. Effect of prostaglandin in E_2 pessaries. Br. J. Obstet. Gynaecol. 92:963-968.

Wehmann, R. E., S. M. Harman, S. Birken, R. E. Canfield, and B. C. Nisula. 1981. Convenient radioimmunoassay for urinary human choriogonadotropin without interference by urinary human lutropin. Clin. Chem. 27:1997-2001.

Weiss, B., and R. A. Doherty. 1975. Methylmercury poisoning. Teratology 12:311-313.

Weissenberg, R., A. Eshkol, E. Rudak, and B. Lunenfeld. 1983. Inability of round acrosomeless human spermatozoa to penetrate zona-free hamster ova. Arch. Androl. 11:167-169.

Weitzman, E. D., R. M. Boyar, S. Kapen, and L. Hellman. 1975. The relationship of sleep and sleep stages to neuroendocrine secretion and biological rhythms in man. Rec. Prog. Horm. Res. 31:399-446.

Welch, L. S., S. M. Schrader, T. W. Turner, and M. R. Cullen. 1988. Effects of exposure to ethylene glycol ethers on shipyard painters: II. Male reproduction. Am. J. Ind. Med. 14:509-526.

Welch, R. M, Y. E. Harrison, B. W. Gommi, P. J. Poppers, M. Finster, and A. H. Conney. 1969. Stimulatory effect of cigarette smoking on the hydroxylation of 3,4-benzpyrene and the N-demethylation of 3-methyl-4-monomethylaminoazobenzene by enzymes in human placenta. Clin. Pharmacol. Ther. 10:100-109.

Wells, M., B. L. Hsi, and W. P. Faulk. 1984a. Class I antigens of the major histocompatibility complex on cytotrophoblast of the human placental basal plate. Am. J. Reprod. Immunol. 6:167-174.

Wells, M., B. L. Hsi, C. J. G. Yeh, and W. P. Faulk. 1984b. Spiral (uteroplacental) arteries of the human placental bed show the presence of amniotic basement membrane antigens. Am. J. Obstet. Gynecol. 150:973-977.

Werb, Z., M. J. Banda, and P. A. Jones. 1980. Degradation of connective tissue matrices by macrophages. I. Proteolysis of elastin, glycoproteins and collagen by proteases isolated from macrophages. J. Exp. Med. 152:1340-1357.

Werler, M. M., B. R. Pober, and L. B. Holmes. 1985. Smoking and pregnancy. Teratology 32:473-481.

Wewer, U. M., M. Faber, L. A. Liotta, and R. Albrechtson. 1985. Immunochemical and ultrastructural assessment of the nature of the pericellular basement membrane of human decidual cells. Lab. Invest.

53:624-633.

Wheat, T. E., J. A. Shelton, V. Gonzales-Prevatt, and E. Goldberg. 1985. The antigenicity of synthetic peptide fragments of lactate dehydrogenase C4. Mol. Immunol. 22:1195-1199.

White, T., R. A. Saltzman, R. K. Miller, R. Sutherland, P. A. diSant'Agnese, and P. Keng. 1988a. Human choriocarcinoma (JAr) cells grown as multicellular spheriods: A system for studying the trophoblast. Placenta (in press).

White, T., P.A. di Sant'Agnese, and R. K. Miller. 1988b. Coculture of three dimensional trophoblast cultures with human endometrial cells. Trophoblast Research (in press).

Whitfield, C. R. 1976. Rhesus haemolytic disease. J. Clin. Pathol. 10:54-62.

Whitlock, J. P. 1986. The regulation of cytochrome P-450 gene expression. Annu. Rev. Pharmacol. Toxicol. 26:333-369.

Whittaker, P. G., A. Taylor, and T. Lind. 1983. Unsuspected pregnancy loss in healthy women. Lancet 1(8334):1126-1127.

WHO (World Health Organization). 1980a. Laboratory Manual for the Examination of Human Semen and Semen-Cervical Mucus Interaction. Singapore: Press Concern.

WHO (World Health Organization). 1980b. A Retrospective Analysis of Steroid Glucuronide Excretion in Ovulatory Women. Special Programme of Research Development and Research Training in Human Reproduction. Project No. 77930.

WHO (World Health Organization). 1986. Effects of Occupational Health Hazards on Reproductive Functions. Geneva: Office of Occupational Health, World Health Organization.

WHO (World Health Organization). 1987. Laboratory Manual for the Examination of Human Semen and Semen-Cervical Mucus Interaction. Cambridge: Cambridge University Press.

Whorton, D., R. M. Krauss, S. Marshall, and T. H. Milby. 1977. Infertility in male pesticide workers. Lancet 2(8051):1259-1261.

Whorton, D., T. H. Milby, R. M Krauss, and H. A. Stubbs. 1979. Testicular function in DBCP exposed pesticide workers. J. Occup. Med. 21:161-166.

Whorton, M. D., and C. R. Meyer. 1984. Sperm count results from 861 American chemical/agricultural workers from 14 separate studies. Fertil. Steril. 42:82-86.

Wide, L. 1962. An immunological method for the assay of human chorionic gonadotropin. Acta. Endocrinol. 4l(Suppl. 70):1-111.

Wide, L. 1981. Male and female forms of human follicle-stimulating hormones in serum. J. Clin. Endocrinol. Metab. 55:682-687.

Wide, L., and C. A. Gemzell. 1960. An immunological pregnancy test. Acta. Endocrinol. 35:261-267.

Wieben, E. D. 1981. Regulation of the synthesis of lactate dehydrogenase-X during spermatogenesis in the mouse. J. Cell. Biol. 88:492-498.

Wilcox, A. J., C. R. Weinberg, R. E. Wehmann, E. G. Armstrong, R. E. Canfield, and B. C. Nisula. 1985. Measuring early pregnancy loss: Laboratory and field methods. Fertil. Steril. 44:366-374.

Wilcox, A. J., C. R. Weinberg, E. G. Armstrong, and R. E. Canfield. 1987a. Urinary human chorionic gonadotropin among intrauterine device users: Detection with a highly specific and sensitive assay. Fertil. Steril. 47:265-269.

Wilcox, A. J., D. D. Baird, C. R. Weinberg, E. G. Armstrong, P. I. Musey, R. E. Wehmann, and R. E. Canfield. 1987b. The use of biochemical assays in epidemiologic studies of reproduction. Environ. Health Perspect. 75:29-35.

Wildt, K., R. Eliasson, and M. Berlin. 1983. Effects of occupational exposure to lead on sperm and semen, pp. 279-300. In T. W. Clarkson, G. F. Norberg, and P. R. Sager, Eds. Reproductive and Developmental Toxicity of Metals. New York: Plenum.

Wilkinson, C. F. 1987. Being more realistic about chemical carcinogenesis. Environ. Sci. Technol. 21:843-847.

Williams, G. M. 1977. Detection of chemical carcinogens by unscheduled DNA synthesis in rat liver primary cell cultures. Cancer Res. 37:1845-1851.

Wilson, E. M., and F. S. French. 1980. Biochemical homology between rat dorsal prostate and coagulating gland. Purification of a major androgen-induced protein. J. Biol. Chem. 2551:10946-10953.

Wilson, E. M., D. H. Viskochil, R. J. Bartlett, O. A. Lea, C. M. Noyes, P. Petrusz, D. W. Stafford, and F. S. French. 1981. Model systems for studies on androgen-dependent gene expression in the rat prostate, pp. 351-380. In G. P. Murphy, A. A. Sandberg, and J. P. Karr, Eds. The Prostatic Cell: Structure and Function Part A. Morphologic, Secretory, and Biochemical Aspects. New York: Liss.

Wilson, G. S., R. McCreary, J. Kean, and J. C. Baxter. 1979. The development of preschool children of heroin-addicted mothers: A controlled study. Pediatrics 63:135-141.

Wilson, J. G. 1954. Differentiation and the reaction of rat embryos to radiation. J. Cell. Comp. Physiol. 43(Suppl. 1):11-38.

Wilson, J. G. 1973. Environment and Birth Defects. New York: Academic Press. 305 pp.

Wilson, R. S., W. Smead, and F. Char. 1978. Diphenylhydantoin teratogenicity: Ocular manifestations and related deformities. J. Pediatr. Ophthal. Stabismus. 15:137-140.

Winder, C., L. L. Garten, and P. D. Lewis. 1983. The morphological effects of lead on the developing central nervous system. Neuropathol. Appl. Neurobiol. 9:87-108.

Wing, T.-Y., and A. K. Christensen. 1982. Morphometric studies on rat seminiferous tubules. Am. J. Anat. 165:13-25.

Winneke, G., G. Foder, and H. Schlipkoter. 1978. Carbon monoxide, trichloroethylene, and alcohol: Reliability and validity of neurobehavioral effects, pp. 461-469. In D. A. Otto, Ed. Multidisciplinary Perspectives in Event Related Brain Potential Research. EPA 600/9-77-043. Washington, D.C.: U.S. Environmental Protection Agency.

Witkin, S. S., J. M. Richards, and J. M. Bedford. 1983. Influence of epididymal maturation on the capacity of hamster and rabbit spermatozoa for complement activation. J. Reprod. Fertil. 69:517-521.

Wogan, G. N. 1988. Detection of DNA damage in studies on cancer etiology and prevention, pp. 32-54. In H. Bartsch, K. Hemminki, and I. K. O'Neill, Eds. Methods for Detecting DNA Damaging Agents in Humans: Applications in Cancer Epidemiology and Prevention. IARC Scientific Publications, No. 89. Lyon: International Agency for Research on Cancer.

Wogan, G. N., and N. J. Gorelick. 1985. Chemical and biochemical dosimetry of exposure to genotoxic chemicals. Environ. Health Perspect. 62:5-18.

Wolf, D. P., and S. Mastroianni, Jr. 1975. Protein composition of human uterine fluid. Fertil. Steril. 26:240-247.

Wong, D. F., H. N. Wagner, Jr., L. E. Tune, R. F. Dannals, G. D. Pearlson, J. M. Links, C. A. Tamminga, E. P. Broussolle, H. T. Ravert, and A. A. Wilson. 1986. Positron emission tomography reveals elevated D2 dopamine receptors in drug-naive schizophrenics. Science 234:1558-1563.

Wong, O., M. D. Utidjian, and V. S. Karten. 1979. Retrospective evaluation of reproductive performance of workers exposed to ethylene dibromide (EDB). J. Occup. Med. 21:98-102.

Wong, P. Y. D., and W. M. Lee. 1982. Effects of spironolactone (aldosterone antagonist) on electrolytes and water content of the cauda epididymidis and fertility of male rats. Biol. Reprod. 27:771-777.

Wong, P. Y. D., and C. H. Yeung. 1977. Hormonal regulation of fluid reabsorption in isolated rat cauda epididymis. Endocrinology 101:1391-1397.

Wong, T. K., R. B. Everson, and S-T. Hsu. 1985a. Potent induction of human placental mono-oxygenase activity by previous dietary exposure to polychlorinated biphenyls and their thermal degradation production. Lancet 1:721-724.

Wong, T. K., T. E. Blanton, C. K. Hunnicutt, and R. B. Everson. 1985b. Quantification of aryl hydrocarbon hydroxylase and 7-ethoxycoumarin O-deethylase activity in human placentae: Development of a protocol suitable for studying effects of environmental exposures on human metabolism. Placenta 6:297-310.

Wooding, F. B. P. 1981. Localization of ovine placental lactogen in sheep placentomes by electron microscope immunocytochemistry. J. Reprod. Fertil. 62:15-19.

Wooding, F. B. P. 1982. The role of the binucleate cell in ruminant placental structure. J. Reprod. Fertil. 31(Suppl.):32-39.

Working, P. K., and M. E. Hurtt. 1987. Computerized

videomicrographic analysis of rat sperm motility. J. Androl. 8:330-337.

Wramsby, H., K. Fregda, and P. Liedholm. 1987. Chromosome analysis of human oocytes recovered from preovulatory follicles in stimulated cycles. N. Engl. J. Med. 316:121-124.

Wright, W. W., N. A. Musto, J. P. Mather, and C. W. Bardin. 1981. Sertoli cells secrete both testis-specific and serum proteins. Proc. Natl. Acad. Sci. USA 78:7565-7569

Wyrobek, A. J., and D. Pinkel. 1986. Detecting chromosomes in human sperm by fluorescent hybridization. Am. J. Hum. Genet. 39:A102. (Abstract.)

Wyrobek, A. J., G. Watchmaker, L. Gordon, K. Wong, D. H. Moore II, and D. Whorton. 1981. Sperm shape abnormalities in carbaryl-exposed employees. Environ. Health Perspect. 40:255-265.

Wyrobek, A. J., L. A. Gordon, G. Watchmaker, and D. H. Moore II. 1982. Human sperm morphology testing: Description of a reliable method and its statistical power, pp. 527-541. In B. A. Bridges, B. E. Butterworth, and I. B. Weinstein, Eds. Banbury Report 13: Indicators of Genotoxic Exposure. [Cold Spring Harbor, N.Y.]: Cold Spring Harbor Laboratory.

Wyrobek, A. J., L. A. Gordon, J. G. Burkhart, M. W. Francis, R. W. Kapp, G. Letz, H. V. Malling, J. C. Topham, and M. D. Whorton. 1983a. An evaluation of human sperm as indicators of chemically induced alterations of spermatogenic function. A report of the U. S. Environmental Protection Agency Gene-Tox Program. Mutat. Res. 115:73-148.

Wyrobek, A. J., L. A. Gordon, J. G. Burkhart, M. W. Francis, R. W. Kapp, Jr., G. Letz, H. V. Malling, J. C. Topham, and M. D. Whorton. 1983b. An evaluation of the mouse sperm morphology test and other sperm tests in non-human mammals. A report of the U. S. Environmental Protection Agency Gene-Tox Program. Mutat. Res. 115:1-72.

Wyrobek, A. J., G. Watchmaker, and L. Gordon. 1984. An evaluation of sperm tests as indicators of germ-cell damage in men exposed to chemical or physical agents, pp. 385-405. In Reproduction: The New Frontier in Occupational and Environmental Health Research. New York: Liss.

Wyrobek, A. J., P. Buffler, D. Houston, and T. Connor. In press. Antifertility effects, mutagenicity, and adverse reproductive outcomes in animals and human beings exposed to dibromochloropropane. Am. J. Indust. Med.

Yager, J. W., C. J. Hines, and R. C. Spear. 1983. Exposure to ethylene oxide at work increases sister chromatid exchanges in human peripheral lymphocytes. Science 29:1221-1223.

Yanagimachi, R., H. Yanagimachi, and B. J. Rogers. 1976. The use of zona-free animal ova as a test-system for the assessment of the fertilizing capacity of human spermatozoa. Biol. Reprod. 15:471-476.

Yanagimachi, R. 1984. Zona-free hamster eggs: Their use in assessing fertilizing capacity and examining chromosomes of human spermatozoa. Gamete Res. 10:187-232.

Yeh, C. J. G., B. L. Hsi, and W. P. Faulk. 1984. Histocompatibility antigens, transferrin receptors and extra-embryonic markers of human amniotic epithelial cells in vitro. Placenta 4:361-368.

Yelick, P. C., R. Balhorn, P. A. Johnson, M. Corzett, C. H. Mazrimas, K. C. Kleene, and N. B. Hecht. 1987. Mouse protamine 2 is synthesized as a precursor whereas mouse protamine 1 is not. Mol. Cell. Biol. 7:2173-2179.

Yen, S. S. C. 1977. The biology of menopause. J. Reprod. Med. 18:287-296.

Yodaiken, R. E. 1986. Surveillance, monitoring, and regulatory concerns. J. Occup. Med. 28:569-571.

Yoneda, T., and R. M. Pratt. 1981. Mensenchymal cells from the human embryonic palate are highly responsive to epidermal growth factors. Science 213:563-565.

Yoshimi, T., C. A. Strott, J. R. Marshall, and M. B. Lipsett. 1969. Corpus luteum function in early pregnancy. J. Clin. Endocrinol. 29:225.

Younes, M. A., and C. G. Pierrepoint. 1981. Androgen steroid-receptor binding in the canine epididymis. Prostate 2:133-142.

Younes, M., B. A. J. Evans, N. Chaisiri, Y. Valotaire, and C. G. Pierrepoint. 1979. Steroids receptors in the canine epididymis. J. Reprod. Fertil. 56:45-52.

Young, M. D., and W. P. Blackmore. 1977. The use of danazol in the management of endometriosis. J. Int. Med. Res. 5(Suppl. 3):86-91.

Young, R. W. 1985. Cell proliferation during postnatal development of the retina in the mouse. Brain Res. 353:229-239.

Younglai, E. V., and R. V. Short. 1970. Pathways of steroid biosynthesis in the intact Graafian follicle of mares in oestrus. J. Endocrinol. 47:321-331.

Yule, Q., R. Lansdown, I. B. Millar, and M. A. Urbanowicz. 1981. The relationship between blood lead concentrations, intelligence and attainment in a school population: A pilot study. Dev. Med. Child Neurol. 23:567-576.

Yussman, M. A., and M. L. Taymor. 1970. Serum levels of follicle stimulating hormone and luteinizing hormone and of plasma progesterone related to ovulation by corpus luteum biopsy. J. Clin. Endocrinol. Metab. 30:396-399.

Zamenhof, S., E. Van Marthens, and L. Grauel. 1972. DNA (cell number) and protein in rat brain. Second generation (F2) alteration by maternal (F0) dietary protein restriction. Nutr. Metab. 14:262-270.

Zaneveld, L. J. D. 1978. The biology of human spermatozoa. Obstet. Gynecol. Annu. 7:15-40.

Zappoli, R., G. Giuliano, L. Rossi, M. Papini, O. Ronchi, A. Ragazzoni, and A. Amantini. 1978. CNV and SEP in shoe industry workers with neuropathy resulting from toxic effect of adhesive solvents, pp. 476-480. In D. A. Otto, Ed. Multidisciplinary Perspectives in Event Related Brain Potential Research. EPA 600/9-77-043. Washington, D.C.: U.S. Environmental

Protection Agency.

Zec, R. F., and D. R. Weinberger. 1986. Relationship between CT scan findings and neuropsychological performance in chronic schizophrenia. Psychiatr. Clin. North Am. 9:49-61.

Zeisler, R., S. H. Harrison, and S. A. Wise, Eds. 1983. The Pilot National Environmental Specimen Bank: Analysis of Human Liver. National Bureau of Standards Special Publication 656. Washington, D.C.: U.S. Department of Commerce, National Bureau of Standards. 128 pp.

Zelson, C., E. Rubio, and E. Wasserman. 1971. Neonatal narcotic addiction: 10 year observation. Pediatrics 48:178-189.

Zenick, H., and E. D. Cleeg. 1989. Assessment of male reproductive toxicity: A risk assessment approach, pp. 275-310. In A. W. Hayes, Ed. Principles and Methods in Toxicology, 2nd ed. New York: Raven Press.

Zenz, C. 1988. Occupational Medicine. Principles and Practical Applications, 2nd ed. Chicago: Year Book Medical Publishers.

Zhu, L.P., D. Chen, S. Z. Zhang, and S. L. Liu. 1985. Study on the effect of calf thymic peptides preparation (TP) on human B lymphocytes. Immunol. Invest. 14:115-129.

Zirkin, B. R., and J. D. Strandberg. 1984. Quantitative changes in the morphology of the aging canine prostate. Anat. Rec. 208:207-214.

Zirkin, B. R., L. L. Ewing, N. Kromann, and R. C. Cochran. 1980. Testosterone secretion by rat, rabbit, guinea pig, dog, and hamster testes perfused in vitro: correlation with Leydig cell ultrastructure. Endocrinology 107:1867-1874.

Zirkin, B. R., R. Gross, and L. L. Ewing. 1985a. Effects of lead acetate on male rat reproduction, pp. 138-145. In F. Homburger and A. M. Goldberg, Eds. In vitro Embrytoxicity and Teratogenicity Tests. Concepts in Toxicology, Vol. 3. Basel: S. Karger.

Zirkin, B. R., D. A. Soucek, T. S. K. Chang, and S. D. Perreault. 1985b. In vitro and in vivo studies of mammalian sperm nuclear decondensation. Res. Gamete 11:349-365.

Zorn, J. R., M. Roger, M. Savale, and J. Grenier. 1982. Steroid hormone levels in peritoneal fluid during the periovulatory period. Fertil. Steril. 38:162-165.

Zukerman, Z., L. J. Rodriguez-Rigau, K. D. Smith, and E. Steinberger. 1977. Frequency distribution of sperm counts in fertile and infertile males. Fertil. Steril. 28:1310-1313.

Biographies

DONALD R. MATTISON, Chairman, Subcommittee on Reproductive and Neurodevelopmental Toxicology, is professor of obstetrics and gynecology at the University of Arkansas for Medical Sciences and medical officer at the National Center for Toxicological Research. His research interests, as reflected in his numerous publications, include reproductive pharmacology and toxicology, nuclear magnetic resonance imaging and spectroscopy, and risk assessment. He has served on committees for the National Research Council, the National Institutes of Health, the U.S. Environmental Protection Agency, and the U.S. Food and Drug Administration. Dr. Mattison also serves on the editorial board of *Reproductive Toxicology, Developmental Pharmacology and Toxicology*, and *Risk Analysis*.

JUDITH L. BUELKE-SAM is an associate within the Department of Genetic and Developmental toxicology at Lilly Research Laboratories, Eli Lilly and Company, where she is involved in the design, conduct, and interpretation of preclinical toxicity studies for Japanese submission. She is the current president of the Behavioral Teratology Society, a member of the Behavioral Toxicology Task Force of the Drug Safety Subsection, Pharmaceutical Manufacturers' Association, and on the editorial advisory board of *Neurotoxicology and Teratology*. Ms. Buelke-Sam was an organizer and principal investigator for the collaborative behavioral teratology study, coordinated by the National Center for Toxicological Research, for which she received the FDA Commissioner's Special Citation. Her current research addresses the role of behavioral evaluation in characterizing the pharmacology and toxicology of new pharmaceutical agents.

ROBERT E. CANFIELD is professor of medicine and director of the Irving Center for Clinical Research at the Columbia-Presbyterian Medical Center in New York. Dr. Canfield's primary research interests include the chemistry and immunology of gonadotropin hormones and fibrinogen, as well as research related to metabolic bone disease. He is an author or coauthor of many research publications in all three fields.

J. DAVID ERICKSON has been at the Centers for Disease Control since 1974, serving first as an epidemiologist and later, Chief in the Birth Defects Branch, Chief of the Etiologic Studies Section, Director of the Agent Orange Projects, Chief of the

Cancer Branch, and most recently, as Chief of the Birth Defects and Genetic Diseases Branch. He has also served on numerous professional and government advisory committees, including the National Institute of Occupational Safety and Health Committee on Reproductive Effects of the Workplace, the Surgeon General's Advisory Committee on the Health Consequences of Using Smokeless Tobacco, and the Veterans Administration Advisory Committee on Health-Related Effects of Herbicides. He has also served on the Panel on Reproductive and Developmental Toxicology of the National Research Council and was a recent editor of the epidemiology section in the publication *Teratology*. He is currently chairman of the International Clearinghouse for Birth Defects Monitoring Programs.

Dr. Erickson's research interests include the causes and prevention of human birth defects and human genetics. Board certified by the American Board of Dental Public Health, he belongs to numerous professional societies, including the American Epidemiological Society, the American College of Epidemiology, the Society for Epidemiologic Research, the American Society of Human Genetics, and the Teratology Society. He holds a D.D.S. degree from the University of Alberta, an M.P.H. degree from the University of Minnesota, and a Ph.D. degree from the University of Washington.

LARRY L. EWING, chairman of the male reproductive toxicology panel, is a professor of Population Dynamics at the Johns Hopkins University School of Hygiene and Public Health. Dr. Ewing has served as editor-in-chief of *Biology of Reproduction*, associate editor of *The Physiology of Reproduction*, and editorial board member of the professional journals, *American Journal of Physiology*, *Endocrinology*, *Biology of Reproduction*, and the *Journal of Andrology*.

Dr. Ewing has served as president of the Society of the Study of Reproduction and the American Society of Andrology and was a member of the Reproductive Biology Section for the National Institutes of Health. He is presently chairman of Subcommittee 3 of the Clinical Sciences Study Section. He has also served on the National Research Council of the National Academy of Sciences and with numerous federal advisory groups that include the U. S. Environmental Protection Agency, the Department of Agriculture, the Veterans Administration, the Office of Technology Assessment, and the Food and Drug Administration. His current research is in male reproductive biology.

W. PAGE FAULK is director of experimental pathology at the Center for Reproduction and Transplantation Immunology at Methodist Hospital, Indianapolis, Indiana. He was formerly professor of immunology at the University of Nice and director of research at the Royal College of Surgeons' Blond-McIndoe Transplantation Institute at East Grinstead, Sussex, England. His career in pregnancy immunology research has included positions with the World Health Organization and the British Medical Research Council and professorships at the Medical University of South Carolina, the University of Texas, and Brisbane University. He is cofounder, and for many years, coeditor of *Placenta*. His research is primarily concerned with interlocking aspects of immunological responses in pregnancy, organ transplantation, and cancer.

CALEB E. FINCH is ARCO/William F. Kieschnick Professor in the Department of Neurobiology of Aging at the Andrus Gerontology Center of the University of Southern California. He is a neurobiologist with major interests in the neuroendocrinology of aging and in Alzheimer's disease. Dr. Finch is also on the National Advisory Council to the National Institute on Aging.

WALDERICO M. GENEROSO is a senior scientist in the biology division of the Oak Ridge National Laboratory in Tennessee. His research includes the mechanisms for induction of heritable mutations in mice, genetic risk evaluation, and understanding the etiology of mutagen-induced developmental anomalies. He was leader of the U.S. Environmental

Protection Agency Gene-Tox Working Group on Heritable Translocation Tests in Mice and of the International Group on Commission for Protection Against Environmental Mutagens and Carcinogens Working Group on Dominant-Lethal Tests in Rodents. He served as a member of the EPA Gene-Tox Committee on Risk Assessment, the National Academy of Sciences/National Research Council Panel on Anticholinesterase Chemicals and their Panel on Cholinesterase Reactivators. Dr. Generoso has also served as councilor of the Environmental Mutagen Society. He is a member of the editorial boards of *Mutation Research* and *Teratogenesis, Carcinogenesis, and Mutagenesis*, and was senior editor of the book, *DNA Repair and Mutagenesis in Eukaryotes*. He also served as coeditor of the books, *Molecular and Cellular Mechanisms of Mutagenesis* and *Cellular Mutation, Cancer, and Malformation*.

JAMES E. GIBSON is director of toxicology affairs, the Dow Chemical Company, Midland, Michigan and has been, until recently, vice-president and director of research for the Chemical Industry Institute of Toxicology, Research Triangle Park, North Carolina. He also serves as an adjunct professor in toxicology at the University of North Carolina and North Carolina State University, and in pharmacology at the Duke University Medical Center. Dr. Gibson holds an undergraduate degree from Drake University and M.S. and Ph.D degrees in pharmacology from the University of Iowa. He is a Diplomate, Academy of Toxicological Sciences, and holds leadership posts in many professional societies. He also serves in editorial and advisory capacities for many professional publications. In addition, Dr. Gibson serves on numerous professional, industry, and government advisory boards, the most recent of which include the Scientific Review Panel of the National Library of Medicine, the Science Advisory Board of the Environmental Protection Agency, the Environmental Health Sciences Review Committee of the National Institute of Environmental Health Sciences, the Scientific Board of Scientific Directors, ILSI Risk Science Institute, and the Advisory Committee for the Joint Industry Research Project on Benzene. Dr. Gibson was also a panelist on the National Academy of Sciences/National Research Council's Subcommittee on Toxicology, Safe Drinking Water Committee.

STANLEY R. GLASSER is a faculty member in the Department of Cell Biology and Center for Population Research and Studies in Reproductive Biology at the Baylor College of Medicine. His laboratory is charged with the analysis of the cellular and molecular biological mechanisms that regulate the reciprocal relationships between the early mammalian embryo and individual cell types of the uterine endometrium that allow their temporally circumscribed interaction.

Dr. Glasser received a B.A. degree in zoology and a B.S. degree in animal husbandry from Cornell University, and a Ph.D. degree from Rutgers University. He was a member of the radiation biology faculty at the University of Rochester Medical School and in the obstetrics and gynecology faculty at Vanderbilt Medical School. He has also been a fellow of the Weizmann Institute and has served on advisory panels of Task Force III of the National Institute of Environmental Health Sciences and the National Science Foundation. Dr. Glasser has organized a number of international symposia focusing on early mammalian development and has served on the editorial boards of his scientific societies. His publications include two books devoted to the regulatory biology of early pregnancy.

BERNARD D. GOLDSTEIN is chairman of the Department of Environmental and Community Medicine at the University of Medicine and Dentistry of New Jersey-Robert Wood Johnson Medical School. He is also director of the Environmental and Occupational Health Sciences Institute, director of the graduate program in public health, and director of an NIEHS Center of Excellence, all joint programs of Rutgers University and UMDNJ.

Dr. Goldstein, a physician board-certified in internal medicine and hematology, was a faculty member in the departments of Environmental Medicine and Medicine at New York University Medical Center until 1980, when he joined the staff of the UMDNJ. From 1983 to 1985, Dr. Goldstein was assistant administrator for research and development of the U.S. Environmental Protection Agency. He has been a member and chairman of the National Institutes of Health Toxicology Study Section and of EPA's Clean Air Scientific Advisory Committee. He was also chairman of the National Academy of Sciences Institute of Medicine Committee on the Role of the Physician in Occupational and Environmental Medicine, and currently chairs the NRC Committee on the Biomarkers in Environmental Health Research.

ARTHUR F. HANEY is an associate professor and the director of the Division of Reproductive Endocrinology and Infertility in the Department of Obstetrics and Gynecology with a joint appointment in Radiology at the Duke University Medical Center, Durham, North Carolina.

He has served as a consultant for the National Institutes of Health, the Environmental Protection Agency, and the National Toxicology Program. He serves as an examiner for the American Board of Obstetrics and Gynecology and represented the United States in a scientific exchange with the People Republic of China. He is an active clinician, primarily involved in the assisted reproductive technologies. His research interests include the evaluation of reproductive toxicity with *in vitro* systems, the characterization of genital tract teratogenicity, the reproductive consequences of prenatal diethylstilbestrol exposure, and treatments for infertility.

MAUREEN C. HATCH is assistant professor of epidemiology at the Columbia University School of Public Health. She is also affiliated with the Gertrude H. Sergievsky Center at Columbia and the Columbia Comprehensive Cancer Center. Dr. Hatch has also served on expert and research review groups for the National Institute for Occupational Safety and Health, the U. S. Environmental Protection Agency, the National Institute for Child health and Human Development, and the task force on environmental cancer and heart and lung disease. Her current research is on stress in pregnancy and low-level radiation and cancer. She is co-editor, with Dr. Zena Stein, of *Reproductive Problems in the Workplace*.

ROGENE F. HENDERSON is a senior scientist and supervisor of the Chemistry and Biochemical Toxicology Group at the Lovelace Inhalation Toxicology Research Institute in Albuquerque, New Mexico. She has done extensive research on the analysis of bronchoalveolar lavage fluid to evaluate lung injury in animal toxicology studies. She has also headed up studies on the disposition and metabolic fate of inhaled vapors to aid in planning and interpretation of long-term carcinogenicity studies in rodents. Dr. Henderson has been a member of the Committee on Toxicology of the Board on Environmental Studies and Toxicology of the National Research Council since 1985.

JOHN E. HOBBIE is director of the Ecosystems Center of the Marine Biological Laboratory, Woods Hole, Massachusetts. He is past president of the American Society for Limnology and Oceanography and of the Association of Ecosystem Research Centers. He currently serves on the Board of Trustees of the Marine Biological Laboratory. During 1988-89 he holds the Tage Erlander Visiting Professorship of the National Research Council of Sweden. His current research is on models of ecosystem response to global change, on the processes in arctic ecosystems, and on the use of stable isotopes in ecological research.

PHILIP J. LANDRIGAN is professor of Community Medicine and director of the Division of Environmental and Occupational Medicine at Mt. Sinai School of Medicine where he is responsible for directing a research program in environmental and occupational medicine, for the training of residents, and for the teaching of medical students. Dr. Landrigan also holds a professorship in pediatrics at Mt. Sinai. He obtained his medical degree from the Harvard Medical School, interned at Cleveland Metropolitan General Hospital, and completed a residency in pediatrics at the Children's Hospital Medical Center in Boston.

From 1970-1985, Dr. Landrigan, as a commissioned officer in the United States Public Health Service, served as a medical epidemiologist with the Centers for Disease Control. In this capacity, he established and directed the Environmental Hazards Branch of the Cancer and Birth Defects Division of the Bureau of Epidemiology. From 1979-1985, as director of the Division of Surveillance, Hazard Evaluations and Field Studies of the National Institute for Occupational Safety and Health, he directed the national program in occupational epidemiology.

Dr. Landrigan has a long-standing research interest in the clinical and epidemiologic evaluation of human diseases caused by toxic environmental and occupational exposures. His research has included studies of heavy metal poisoning, pesticide poisoning, solvent neuropathy, chronic lung disease, chemically induced renal disease, and occupational carcinogenesis.

LAWRENCE D. LONGO is professor of physiology, professor of obstetrics and gynecology, and head of the Division of Perinatal Biology, Loma Linda University, Loma Linda, California. Under Dr. Longo's leadership, this division has grown into a prolific and well-regarded national research center for fetal and newborn biology. His primary areas of research have involved the dynamics and regulation of fetal oxygenation and the neuroendocrine regulation of the fetus. As a result of his investigation of maternal-fetal carbon monoxide exchange, he was asked to write the Surgeon General's report enumerating the negative effect of maternal cigarette smoking on the fetus. Dr. Longo has published numerous papers, review articles, and book chapters as a result of his research.

RICHARD K. MILLER (Chairman, Panel on Pregnancy) is professor of obstetrics and gynecology and of toxicology at the University of Rochester, School of Medicine and Dentistry. He is also director of the Division of Research at the same institution. He has had appointments as an NIH Fogarty Senior International Fellow and a Fulbright Distinguished Professor. He is also a member of the Board of Scientific Counselors for the National Toxicology Program and serves on numerous committees for the U.S. Environmental Protection Agency, the U.S. National Institutes of Health, and the National Library of Medicine. Dr. Miller is currently editor-in-chief of the professional journal, *Trophoblast Research*, and has served on the editorial boards of *Placenta*, *Reproductive Toxicology*, and *Teratology*. His numerous books and research publications are concerned with reproduction, placental function, teratology, and reproductive pharmacology and toxicology.

HERBERT L. NEEDLEMAN is professor of psychiatry and associate professor of pediatrics at the University of Pittsburgh School of Medicine. He is medical director of the Allegheny County Lead Screening Program, and director of lead research at the university. He has served as consultant to the World Health Organization in designing studies of low level lead toxicity and on many advisory boards in environmental health.

C. ALVIN PAULSEN, MD, is professor of medicine and the director of the Population Center for Research in Reproduction at the University of Washington School of Medicine, Seattle, Washington. He has served as chief of research at the U.S. Public Health Service Pacific Medical Center, where he currently holds the position of chief of endocrinology. He received his BA in 1947 from the University of Washington and his MD in 1952 from the University of Oregon. Following service in the U.S. Navy, he became director of laboratories at the Pacific Northwest Research Foundation.

Dr. Paulsen's research has concentrated on studies of the male reproductive system with major research contributions that have included the recognition of sex chromosomal mosaicism and its concomitant clinical and pathologic manifestations in patients with Klinefelter's Syndrome, the delineation of the clinical characteristics and treatment modalities of patients with hypogonadotropic enunochoidism, and documentation of the relationship between varicocele and abnormal spermatogenesis with infertility. In these studies, Dr. Paulsen and his colleagues demonstrated the details of testicular function in those men with varicocele who remained fertile.

Dr. Paulsen has contributed numerous articles to professional and scientific journals and has served on advisory boards for the National Institute of Health and on the Male Task Force for the World Health Organization. He is past president of the Pacific Coast Fertility Society, the American Society of Andrology, and the American Fertility Society.

FREDERICA PERERA is associate professor in the Division of Environmental Sciences at the Columbia University School of Public Health. She is project director of several molecular epidemiology studies of chemical carcinogenesis in humans and has published extensively in the areas of molecular epidemiology, biomonitoring, risk assessment, and public health policy. Dr. Perera has served on numerous committees including those of the National Research Council, the National Toxicology Program, and the U.S. Environmental Protection Agency.

EMIL PFITZER is assistant vice-president and group director of the Department of Toxicology and Pathology at Hoffmann-LaRoche Inc., where he is responsible for the design, conduct, and interpretation of toxicologic chemicals. He holds appointments as adjunct professor at Rutgers University and the New York University Institute of Environmental Medicine. He was president of the Society of Toxicology in 1985–1986 and is a member of a number of national scientific organizations. He was certified in general toxicology by the American Board of Toxicology, Inc., in 1980. In addition to his work at Hoffmann-LaRoche Inc., he has served on the National Institute of Environmental Health Sciences Training Grant Review Committee, on advisory boards for the national Center for Toxicological Research and the Brookhaven National Laboratory Medical Department, and on several National Research Council committees. Dr. Pfitzer's publications include several book chapters on the principles of dose-effect and dose-response relationships.

BERNARD ROBAIRE is professor of pharmacology and therapeutics at McGill University, Montreal, Quebec, Canada. He is also appointed in the McGill Department of Obstetrics and Gynecology. From 1982–1987 he served as director of the McGill Centre for the Study of Reproduction. Dr. Robaire has held offices in several learned societies and in 1988 he was president of the Canadian Fertility and Andrology Society. His scientific publications have focused on the regulation of epididymal functions, steroidal regulation of the hypothalamic pituitary testicular axis and male mediated teratogenesis.

NEENA B. SCHWARTZ is William Deering Professor of Biological Sciences and director of the Center for Reproductive Science at Northwestern University, Evanston, Illinois. She has served as president of the Society for the Study of Reproduction and of the Endocrine Society. In 1985, she received the Williams Distinguished Service Award of the Endocrine Society, and is a fellow of the American Association for the Advancement of Science. She has served on the NICHD population research and training committee and is presently a member of the National Institutes of Health endocrinology study section. Her research focuses on the control of gonadotropic control hormone secretion by the anterior pituitary as it is regulated by brain neuropeptides and gonadal steroid and peptide hormone feedback.

RICHARD J. SHERINS, M.D., a specialist in male reproduction and andrology, is director of the Division of Andrology at the Genetics & IVF Institute in Fairfax, Virginia. He was formerly the Chief of the Section on Reproductive Endocrinology in the Developmental Endocrinology Branch of the National Institute of Child Health and Human Development, the National Institutes of Health. He holds an undergraduate degree from UCLA and a medical degree from University of California, San Francisco Medical School. He received his clinical training in internal medicine at UCLA and served a fellowship in endocrinology at the University of Washington.

Dr. Sherins has served on the editorial boards of numerous professional and scientific publications and has been adviser to industry and government organizations. He holds wide membership in professional societies and organizations and has served as president of the American Society of Andrology and chairman of the male reproduction/urology committee of the American Fertility Society. His current research interests are in the hormonal regulation of human spermatogenesis, male infertility, and sperm biology.

ELLEN K. SILBERGELD is chief scientist of the Toxicological Branch for the Environmental Defense Fund in Washington, DC and serves as a neuropharmacologist with the National Institutes of Health Division of Communicable Disorders and Stroke. She also serves as an adjunct professor at the University of Maryland in the toxicology program and is on the faculty of Johns Hopkins University in the Department of Health Policy Management at the School of Hygiene and Public Health, the Johns Hopkins Medical Institute. Her professional appointments include the Science Advisory Board for the Environmental Protection Agency, the Scientific Advisory Panel for the Michigan Agent Orange Commission, and the Scientific Advisory Panel for the International Joint Commission on the Great Lakes. Among the many professional societies and organizations to which Dr. Silbergeld belongs are the American Public Health Association, the Association of Women in Science, the Society for Occupational and Environmental Health, the Society for Neuroscience and the Society of Toxicology. She has published extensively in scientific journals and is currently preparing a book for the World Health Organization on the health problems of lead in the environment. Dr. Silbergeld holds a Ph.D degree from the Johns Hopkins University in environmental engineering sciences.

RICHARD G. SKALKO is professor and chairman of the Department of Anatomy at the Quillen-Dishner College of Medicine, East Tennessee State University. He has served as director of the embryology laboratory at the Birth Defects Institute, New York State Department of Health, and was professor of anatomy and toxicology at the Albany Medical College. He has also taught anatomy and toxicology at the Louisiana State and Cornell University Medical Centers.

Dr. Skalko received undergraduate degrees from Providence College and St. John's University, and his Ph.D. from the University of Florida. His professional activities include membership in the Science Advisory Board for the National Center for Toxicological Research, in the Reproductive and Developmental Toxicology Review Committee for the National Toxicology Program, and in the Toxicology Study Section of the National Institutes of Health. He also serves as a member of the editorial boards of professional journals and as president of the Reproductive and Developmental Toxicology Specialty Section of the Society of Toxicology.

STEPHEN P. SPIELBERG is associate professor of pediatrics and pharmacology at the University of Toronto and director of the Division of Clinical Pharmacology and

Toxicology at the Hospital for Sick Children. He also serves director of the University of Toronto Centre for Drug Safety Research. He has served on pharmacology and toxicology grant review panels of both the National Institutes of Health and the Medical Research Council of Canada, as well as on the Committee on Drugs of the American Academy of Pediatrics. Dr. Spielberg also holds a Scholarship from the Medical Research Council of Canada.

His research interests include the mechanisms of adverse effects of drugs and environmental chemicals in the human population, pharmacogenetics, and the development of biological markers of chemical exposure and toxicity. Current research activities include investigation of the mechanisms of toxicity and pharmacogenetic susceptibility to adverse reactions from sulfonamides and aromatic anticonvulsants as well as from environmental chemicals such as the polychlorinated biphenyls and dioxins.

ANDREW WYROBEK is senior staff biophysicist in the Molecular Biology Section of the Biomedical Sciences Division at the Lawrence Livermore National Laboratory in Livermore, California. He received his B.S. in physics in 1970 from the University of Notre Dame and his Ph.D. in 1975 in medical biophysics from the University of Toronto, Canada. He was a commissioned officer in the U.S. Air Force and served at the Medical Research Laboratories at Wright-Patterson Air Force Base, in Dayton, Ohio. Dr. Wyrobek has been a member of numerous national science committees, advisory groups, and editorial boards. His research has included investigations of the effects of toxic chemicals and ionizing radiation on sperm production, and he developed quantitative morphometric approaches for human semen analysis. At Lawrence Livermore National Laboratory, he directs three research projects for developing new molecular methods to detect male reproductive toxicity as well as human somatic and germinal mutations.

Index

A

Abortion, spontaneous, 1, 8, 10, 15, 30, 123, 127, 171, 173, 201—202, 208, 219, 221—222
  amniocentesis and, 315, 316
  DBCP and, 109
  fetoscopy and, 245
  genetic damage and, 123, 126, 127, 139, 159
  immunology and, 218—219, 222
  trophoblast antigens and, 219—220
Abstinence period, sperm and, 90—92, 94
Accessory sex organs, male, 64, 70, 72, 77—81, *fig.* 78, 143
Acetylcholine, 67, 277, 290
Acid phosphatase, 80, 103
Acromegaly, 305
Acrosin, 103
Acrosome reaction, 101—103
Acrylamide, 107, 124, 287,
ACTH, *see* Adrenocorticotropin
Actins, testicular, 61
Activin, 182
Adenine, 132
Adenohypophysis, 52
Adenosine triphosphate (ATP), 103, 307
Adipose tissue, 181, 183, 251, 266, 267, 270
Adluminal compartment, seminiferous epithelium, 52, 53
Adrenal glands, 154—155, 207, 208, 211—212, 270
Adrenarche, 154—155, 170
Adrenocorticotropin (ACTH), fetal, 208, 260
Adriamycin, 98
AFP, *see* Alpha-fetoprotein
Agarose gel electrophoresis, 132
Agency for Toxic Substances and Disease Registry, 2, 16, 202
Agglutination, sperm, 97

Aging, female, markers of, 8, 159—161, 169—177, 199—200, 202
AHH, *see* Arylhydrocarbon hydroxylase
AIDS virus, 306
Albumin, 68, 70, 267
Alcohol, 15, 16, 43, 51, 283
  and male reproductive function, 43, 51
  neurodevelopmental effects, 293, 294
  and pregnancy, 213, 242
Aldosterone, 74
Alkaline elution, sperm-DNA, 137—138, 145
Alkaline phosphatase, 257
Allergic reactions, sperm and, 94
Allogeneic recognition reactions, 216, 220
Alpha-fetoprotein (AFP), 5, 10, 17, 26, 241, 244, 254, 255, 257, 260, 311—313, 315—316
Alveolar macrophages, 305
Alzheimer's disease, 174, 176, 177, 307
Amino acids, 53, 128, 235, 273, 278—279, 305
Aminolevulinic acid dehydrase (ALAD), 21, 300, 301
Amniocentesis, 24, 206, 258, 260
  in detecting genetic damage, 211, 244, 262
  and fetal death, 315, 316
Amniochorion, 215, 220
Amnion, 256
Amniotic antigen-3 (AA3), 220, 261
Amniotic cavity, 244, 274
Amniotic fluid, 244, 248, 250, 251, 262
  AFP in, 17, 241, 311, 312, 315, 316
  hPRL in, 230
Ampullae, vas deferens, 77, 79
AMSA therapy, 48, *fig.* 49, 86
Androgen-binding protein (ABP), 53—54, 65, 71, 73, 74, 143
Androgens
  and epididymal function, 64, 70, 72—74, 75, 76
  in female reproductive system, 154, 175, 181, 186